Lecture Notes in Computer Science 7927

Commenced Publication in 1973
Founding and Former Series Editors:
Gerhard Goos, Juris Hartmanis, and Jan van Leeuwen

Editorial Board

José-Manuel Colom   Jörg Desel (Eds.)

Application and Theory
of Petri Nets
and Concurrency

34th International Conference, PETRI NETS 2013
Milan, Italy, June 24-28, 2013
Proceedings

 Springer

Volume Editors

José-Manuel Colom
Universidad de Zaragoza
Departamento de Informática e Ingeniería de Sistemas
María de Luna, 1, 50018 Zaragoza, Spain
E-mail: jm@unizar.es

Jörg Desel
Fernuniversität Hagen, IZ
Universitätsstraße 1, 58097 Hagen, Germany
E-mail: joerg.desel@fernuni-hagen.de

ISSN 0302-9743   e-ISSN 1611-3349
ISBN 978-3-642-38696-1   e-ISBN 978-3-642-38697-8
DOI 10.1007/978-3-642-38697-8
Springer Heidelberg Dordrecht London New York

Library of Congress Control Number: 2013939457

CR Subject Classification (1998): D.2.2, D.2.4, F.1, F.3, G.3, J.1, E.1

LNCS Sublibrary: SL 1 – Theoretical Computer Science and General Issues

Typesetting: Camera-ready by author, data conversion by Scientific Publishing Services, Chennai, India

Printed on acid-free paper

Springer is part of Springer Science+Business Media (www.springer.com)

Preface

This volume constitutes the proceedings of the 34th International Conference on Application and Theory of Petri Nets and Concurrency (PETRI NETS 2013). The Petri Net conferences serve as annual meeting places to discuss the progress in the field of Petri nets and related models of concurrency. They provide a forum for researchers to present and discuss both applications and theoretical developments in this area. Novel tools and substantial enhancements to existing tools can also be presented. The satellite program of the conference comprised three workshops, a Petri net course including basic and advanced tutorials and an additional tutorial on the work of Carl Adam Petri and Anatol W. Holt.

The PETRI NETS 2013 conference was organized by the Università degli Studi di Milano - Bicocca. It took place in Milan, Italy, during June 24–28, 2013. We would like to express our deepest thanks to the Organizing Committee chaired by Lucia Pomello for the time and effort invested in the local organization of the conference. This year the number of submitted papers amounted to 56, which included 52 full papers and 4 tool papers. The authors of the papers represented 26 different countries. We thank all the authors who submitted papers. Each paper was reviewed by at least four referees. The Program Committee (PC) meeting took place electronically, using the EasyChair conference system for the paper selection process. The PC selected 20 papers (18 regular papers and 2 tool papers) for presentation. After the conference, some authors were invited to publish an extended version of their contribution in the *Fundamenta Informaticae* journal.

We thank the PC members and other reviewers for their careful and timely evaluation of the submissions before the meeting, and the fruitful discussions during the electronic meeting. The Springer LNCS Team and the EasyChair system provided high-quality support in the preparation of this volume. We are also grateful to the invited speakers for their contribution: Kees van Hee, Kurt Jensen, Moshe Vardi, Stéphane Lafortune, and Catuscia Palamidessi. Manuscripts of the keynotes from Kees van Hee and from Stéphane Lafortune are included in this volume.

Finally, we would like to pay tribute to Philippe Darondeau, member of the Program Committee of this conference and distinguished colleague. We were deeply shocked and saddened to learn of Philippe's death on March 18th, just a few days after the completion of the review process. He was an extremely friendly and highly-respected colleague, who remained modest and was always ready to help. Most members of our community knew and liked him and many of us collaborated with him. His absence will be felt for many years to come.

June 2013

José-Manuel Colom
Jörg Desel

Organization

Steering Committee

W. van der Aalst, The Netherlands
J. Billington, Australia
G. Ciardo, USA
J. Desel, Germany
S. Donatelli, Italy
S. Haddad, France
K. Hiraishi, Japan
K. Jensen, Denmark
J. Kleijn, The Netherlands
F. Kordon, France

M. Koutny, UK (chair)
C. Lin, China
W. Penczek, Poland
L. Pomello, Italy
W. Reisig, Germany
G. Rozenberg, The Netherlands
M. Silva, Spain
A. Valmari, Finland
A. Yakovlev, UK

Program Committee

Hassane Alla	Université de Grenoble, France
Marco Beccuti	Università degli Studi di Torino, Italy
Jonathan Billington	University of South Australia, Australia
Josep Carmona	Universitat Politècnica Catalunya, Spain
Gianfranco Ciardo	University of California at Riverside, USA
José-Manuel Colom[1] (Co-chair)	Universidad de Zaragoza, Spain
Philippe Darondeau †	INRIA Rennes-Bretagne Atlantique, France
Jörg Desel (Co-chair)	Fernuniversität in Hagen, Germany
Raymond Devillers	Université Libre de Bruxelles, Belgium
Zhenhua Duan	Xidian University, China
Javier Esparza	Technische Universität München, Germany
Maria Pia Fanti	Politecnico di Bari, Italy
Luís Gomes	Universidade Nova de Lisboa, Portugal
Serge Haddad	École Normale Supérieure de Cachan, France
Henri Hansen	Tampere University of Technology, Finland
Kunihiko Hiraishi	Japan Advanced Institute of Science and Technology, Japan
Victor Khomenko	Newcastle University, UK
Ekkart Kindler	Technical University of Denmark, Denmark
Hanna Klaudel	Université d'Evry-Val d'Essonne, France

---

[1] Sponsored by Spanish Ministry of Science and Innovation (MICINN) [TIN2011-27479-C04-01].

Jetty Kleijn   Universiteit Leiden, The Netherlands
Ranko Lazic   University of Warwick, UK
Chuang Lin   Tsingnhua University, China
Niels Lohmann   Universität Rostock, Germany
Irina Lomazova   Higher School of Economics, Moscow, Russia
Andrew Miner   Iowa State University, USA
Lucia Pomello   Università di Milano-Bicocca, Italy
Wolfgang Reisig   Humboldt-Universität zu Berlin, Germany
Carla Seatzu   Università di Cagliari, Italy
Christian Stahl   Technische Universiteit Eindhoven,
                                     The Netherlands
Satoshi Taoka   Hiroshima University, Japan
Antti Valmari   Tampere University of Technology, Finland
Michael Westergaard   Technische Universiteit Eindhoven,
                                     The Netherlands

Organizing Committee

Lucia Pomello (Chair), Italy

Workshops and Tutorials Chairs

Serge Haddad, France   Alex Yakovlev, UK

Tools Exhibition Chair

Luca Bernardinello, Italy

Publicity Chairs

Lucia Pomello, Italy   Luca Bernardinello, Italy

Additional Reviewers

Alekseyev, Arseniy   Bernardinello, Luca   Fabre, Eric
André, Étienne   Bollig, Benedikt   Fahland, Dirk
Badouel, Eric   Bonsangue, Marcello   Ferigato, Carlo
Barros, Joao Paulo   Caillaud, Benoit   Fernandes, João M.
Bashkin, Vladimir   Carvalho, Rafael   Finkel, Alain
Basile, Francesco   Chatain, Thomas   Fornari, Chiara
Basu, Samik   Clarisó, Robert   Franceschinis, Giuliana
Bérard, Béatrice   Costa, Anikó   Fronc, Łukasz

Gabarró, Joaquim
Geeraerts, Gilles
Genest, Blaise
Gierds, Christian
Haddad, Axel
Hoogeboom, Hendrik J.
Huiquan, Zhu
Iacobellis, Giorgio
Jiménez, Emilio
Jin, Xiaoqing
Júlvez, Jorge
Kikuchi, Shinji
Kilinç, Görkem
Kobayashi, Koichi
Kohkichi, Tsuji
Kuzmin, Egor
Le Cornec, Yves-Stan
Liu, Guan Jun
Liu, Yan
Mahulea, Cristian
Mangini, Agostino M.

Mangioni, Elisabetta
Markowitch, Olivier
Melliti, Tarek
Miyamoto, Toshiyuki
Moelle, Andre
Mokhov, Andrey
Müller, Richard
Nakamura, Morikazu
Ohta, Atsushi
Pawłowski, Wiesław
Peschanski, Frédéric
Pinna, G. Michele
Poliakov, Ivan
Pommereau, Franck
Praveen, M.
Prüfer, Robert
Reynier, Pierre-Alain
Ribeiro, Oscar
Rosa-Velardo, Fernando
Sangnier, Arnaud
Schmitz, Sylvain

Schwoon, Stefan
Sené, Sylvain
Siirtola, Antti
Siri, Silvia
Sokolov, Danil
Song, Songzheng
Sproston, Jeremy
Steggles, Jason
Sürmeli, Jan
Takahashi, Koji
Ter Beek, Maurice H.
Tsuji, Kohkichi
Verbeek, Fons
Vázquez, Carlos Renato
Wan, Min
Wimmel, Harro
Yamaguchi, Shingo
Yamane, Satoshi
Zhao, Yang
Zheng, Manchun

Table of Contents

Tool Papers

The Right Timing: Reflections
on the Modeling and Analysis of Time

Kees van Hee and Natalia Sidorova

Department of Mathematics and Computer Science
Technische Universiteit Eindhoven
P.O. Box 513, 5600 MB Eindhoven, The Netherlands
{k.m.v.hee,n.sidorova}@tue.nl

Abstract. In this paper we discuss several approaches to time in Petri nets. If time is considered for performance analysis, probability distributions for choices should be included into the model and thus we need Petri nets with time and stochastics. In literature, most attention is paid to models where the time is expressed by delaying transitions and for the stochastic case to continuous time models with exponential enabling distributions, known by its software tools as GSPN. Here we focus on discrete models where the time is expressed by delaying tokens and the probability distributions are discrete, because this model class has some advantages. We show how model checking methods can be applied for the non-stochastic case. For the stochastic case we show how Markov techniques can be used. We also consider structural analysis techniques, which do not need the state space.

1 Introduction

Over the last 25 years, the modeling and analysis of time has been studied extensively in the context of Petri nets as well as of other models of concurrency. Most of the Petri net models with time are extensions of classical Petri nets, so if we discard the time aspects we obtain a classical Petri net. There are many different approaches to model time such as: delaying of instantaneous transitions, duration of transitions, aging or delays in tokens, timers and clocks. The earliest papers seem to be [27] which introduces delaying of transitions and [30] introducing duration of transitions. These model classes are often referred to as Merlin time (or Time Petri Nets) and Timed Petri Nets respectively. Many authors have contributed to these models with results on *expressiveness*, e.g. [13,15,20,36] and on *model checking* [10,9,34,3,22]. Next to the verification questions, *performance* analysis (c.f. [35,35,31,16]) is an important reason for extending models with time features. While verification is concerned with the extremities of behavior (like "will every service request be processed within 5 time units?"), performance analysis is concerned with "average" behavior or "normal" behavior, i.e. behavior within bounds with a specified probability (like "will 90% of service requests be processed within 5 time units?"). To make performance analysis possible, non-deterministic choices in models should be endowed with a probability.

J.-M. Colom and J. Desel (Eds.): PETRI NETS 2013, LNCS 7927, pp. 1–20, 2013.

There is an overwhelming literature on timed automata and their relationship to Petri nets with time,e.g. [5,24,14,12]. Most of these papers refer to timed automata with clocks as incorporated in the UPPAAL tool [40,6]. There is also a very extensive literature on stochastic Petri nets where the execution time of transitions is exponentially distributed which leads to continuous time Markov processes, see e.g. [4,26,18]. For the class of Generalized Stochastic Petri Nets, having also instantaneous transitions, there is a famous software tool for modeling and analysis GSPN [39]. This class is in fact the stochastic extension of Merlin time, so the approach with delaying transitions. Different approaches proposed for modeling time dimension in Petri nets all have their own merits and weaknesses. Some of the approaches are questionable from the modeling point of view but are handy for analysis purposes, while others are good for modeling but bad for analysis.

In this paper we focus on the approach with time in tokens since we have the feeling that this class did not obtain enough attention although it is a powerful class both for modeling and for analysis. Extending tokens with time information is studied in a number of works, see [19,2,21,8,23,11]. There are multiple examples of industrial applications of this model of time in Petri nets, often called Interval Timed Petri Nets. We restrict our focus to the discrete time setting and call the corresponding class DTPN (Discrete Time Petri Nets). The software tools CPN Tools [37] and ExSpect [38] are using a simplification of it. Next to that, we will consider a stochastic variant of DTPN, called DSPN that can be seen as a discrete alternative to the GSPN model. This class has *discrete* time, i.e. finite or countable delay sets and discrete probability distributions over the choices and it encompasses several well-known subclasses.

We start with preliminaries in Section 2. In Section 3, we compare different options for introducing timed elements in Petri nets, without claiming to be complete. We do not consider continuous Petri [32] nets there because the underlying untimed net is not a classical Petri net any more. In Section 4, we define DTPN and show how the subclasses of DTPN are related. We show by modeling inhibitor arcs that several subclasses of DTPN are Turing complete. In Section 5, we consider a stochastic version of DTPN and we show how classical Markov techniques can be used to analyse them. In particular, we consider three questions: the probability of reaching a set of states, the expected time to reach a set of states and the equilibrium distribution. For these methods we need the whole state space. For workflow nets, there are structural techniques that can be combined with the others.

2   Preliminaries

We denote the set of reals, rationals, integers and naturals by $\mathbb{R}$, $\mathbb{Q}$, $\mathbb{Z}$ and $\mathbb{N}$ (with $0 \in \mathbb{N}$), respectively. We use superscripts $+$ for the corresponding subsets containing all the non-negative values, e.g. $\mathbb{Q}^+$. A set with one element is called a *singleton*. Let $\inf(A)$, $\sup(A)$, $\min(A)$ and $\max(A)$ of the set A have the usual meaning and $\mathcal{P}(A)$ is the power set of A. We define $\max(\emptyset) = \infty$ and $\min(\emptyset) =$

$-\infty$ and we call a set $\{x \in \mathbb{Q} | a \leq x \leq b\}$ with $a, b \in \mathbb{Q}^+$ a *closed rational interval*. The set B is a *refinement* of set A denoted by $A \lhd B$ if and only if $A \subseteq \mathbb{Q}^+ \wedge A \subseteq B \wedge \sup(A) = \sup(B) \wedge \inf(A) = \inf(B)$.

A *labeled transition system* is a tuple $(S, A, \rightarrow, s_0)$ where S is the set of states, A is a finite set of *action names*, $\rightarrow \subseteq S \times A \times S$ is a transition relation and $s_0 \in S$ is an initial state. We write $(s, a, s') \in \rightarrow$ when $s \xrightarrow{a} s'$. An action $a \in A$ is called *enabled* in a state $s \in S$, denoted by $s \xrightarrow{a}$, if there is a state s' such that $s \xrightarrow{a} s'$. If $s \xrightarrow{a} s'$, we say that state s' is reachable from s by an action labeled a.

We lift the notion of reachability to sequences of actions. We say that a non-empty finite sequence $\sigma \in A^*$ of length $n \in \mathbb{N}$ is a *firing sequence*, denoted by $s_0 \xrightarrow{\sigma} s_n$, if there exist states $s_i, s_{i+1} \in S$ such that $s_i \xrightarrow{\sigma(i)} s_{i+1}$ for all $0 \leq i \leq n - 1$. We write $s \xrightarrow{*} s'$ if there exists a sequence $\sigma \in A^*$ such that $s \xrightarrow{\sigma} s'$ and say that s' is reachable from s.

Given two transition systems $N_1 = (S_1, A_1, \rightarrow, s_0)$ and $N_2 = (S_2, A_2, \rightarrow, s_0')$, a binary relation $R \subseteq S_1 \times S_2$ is a *simulation* if and only if for all $s_1 \in S_1, s_2 \in S_2, a \in A_1, (s_1, s_2) \in R$ and $s_1 \xrightarrow{a} s_1'$ implies that there exist $s_2' \in S_2$ such that $s_2 \xrightarrow{a} s_2'$ and $(s_1', s_2') \in R$. We write $N_2 \preceq N_1$ if a simulation relation R exists. If R and R^{-1} are both simulations, relation R is called a *bisimulation* denoted by $\simeq$. If a simulation relation R' exists such that $N_1 \preceq N_2$ then N_1 and N_2 are *simulation equivalent*. We use the notion of *branching bisimulation* as defined in [17].

A *Petri net* is a tuple $N = (P, T, F)$, where P is the set of *places*, T is the set of *transitions*, with $P \cap T = \emptyset$, and $F \subseteq (P \times T) \cup (T \times P)$ is a *flow relation*. An *inhibitor net* is a tuple (P, T, F, ι), where (P, T, F) is a Petri net and $\iota : T \rightarrow \mathcal{P}(P)$ specifies the inhibitor arcs. We refer to elements of $P \cup T$ as *nodes* and elements from F as *arcs*. We define the preset of a node n as ${}^\bullet n = \{m | (m, n) \in F\}$ and the postset as $n^\bullet = \{m | (n, m) \in F\}$.

The state of a Petri net $N = (P, T, F)$ is determined by its marking which represents the distribution of tokens over the places of the net. A marking m of a Petri net N is a bag over its places P. A transition $t \in T$ is enabled in m if and only if ${}^\bullet t \leq m$. If N is an inhibitor net then for the enabling of transition t in a marking m, we additionally require that $m(p) = 0$ for all places $p \in \iota(t)$. An enabled transition may fire, which results in a new marking $m' = m - {}^\bullet t + t^\bullet$, denoted by $m \xrightarrow{t} m'$.

3　Overview of Petri Nets with Time

In this section we describe several options for extending Petri nets with time. To make the syntax uniform, we use the same definition for different classes of timed Petri nets; in this definition we add delay sets to the arcs of a classical Petri net. The semantics of these delay sets differ significantly in different model classes, and additional constraints on delays will be imposed in certain cases.

Definition 1 (Timed Petri net). *A* timed Petri net *(TPN) is a tuple* (P, T, F, δ)*, where* (P, T, F) *is a Petri net,* $\delta : F \rightarrow \mathcal{P}(\mathbb{R})$ *is a function assigning delay sets of non-negative delays to arcs.*

We consider $\delta(p,t)$, $(p,t) \in F$, as an *input delay* for transition t and $\delta(t,p)$, $(t,p) \in \cap F$, as an *output delay* for transition t. We distinguish multiple subclasses of TPNs using different combinations of the following restrictions of the delay functions:

In-zero: Input arcs are non-delaying, i.e. for any $(p,t) \in F$, $\delta(p,t) = 0$.
Out-zero: Output arcs are non-delaying, i.e. for any $(t,p) \in F$, $\delta(t,p) = 0$.
In-single: Input delays are fixed values, i.e. for any $(p,t) \in F$, $|\delta(p,t)| = 1$.
Out-single: Output delays are fixed values, i.e. for any $(t,p) \in F$, $|\delta(t,p)| = 1$.
In-fint: Input delays are finite rational sets.
Out-fint: Output delays are finite rational sets.
In-rint: Input delays are closed rational intervals.
Out-rint: Output delays are closed rational intervals.
Tr-equal: For every transition, the delays on its input arcs are equal, i.e. for any $t \in T, (p_1,t), (p_2,t) \in F$, $\delta(p_1,t) = \delta(p_2,t)$
Pl-equal: For every place, the delays on its input arcs are equal, i.e. for any $p \in P, (p,t_1), (p,t_2) \in F$, $\delta(p,t_1) = \delta(p,t_2)$.

There are models of time for Petri nets placing timed elements on places or transitions instead of arcs. *Tr-equal* allows to define a delay interval for a transition and *Pl-equal* – a delay interval for a place. We can combine the restrictions (e.g. *In-zero, Out-single*).

There are several *dimensions* on which different models of time for Petri nets differ semantically, such as:

– *Timed tokens vs untimed tokens*: Some models extend tokens, and thus also markings, with the time dimension, e.g. with timestamps to indicate at which moment of time these tokens become consumable or/and till which time the tokens are still consumable. A transition may fire if it has consumable tokens on its input places. Other models keep tokens untimed, meaning in fact that tokens are always consumable. The time semantics is then captured by time features of places/transitions only.
– *Instantaneous firing vs prolonged firing:* In some models, the firing of a transition takes time, i.e. the tokens are removed from the input places of a transition when the firing starts and they are produced in the output places of the transition when the firing is finished. In an alternative semantics, a potential execution delay is selected from a delay interval of a transition, but the transition firing is instantaneous.
– *Eager/urgent/lazy firing semantics*: In the eager semantics, the transition that can fire at the earliest moment is the transition chosen to fire; with the urgent semantics the transition does not have to fire at the earliest moment possible, but the firing may become urgent, i.e. there is a time boundary for the firing; in the lazy semantics the transitions do not have to fire even if they loose their ability to fire as a consequence (because e.g. the tokens on the input places are getting too old and thus not consumable any more).
– *Preemption vs non-preemption*: Preemption assumes that if a transition t gets enabled and is waiting for its firing for a period of time defined by its

delay and another transition consumes one of the input tokens of t, then the clock or timer resumes when t is enabled again, and thus its firing will be delayed only by the waiting period left from the previous enabling. The alternative is called non-preemption.

– *Non-deterministic versus stochastic choices* in the delays or order of firing.

We introduce the notion of a *timed marking* for all timed Petri nets. We assume the existence of a *global clock* that runs "in the background". The "current time" is a sort of cursor on a time line that indicates where the system is on its journey in time. We give the tokens a unique *identity* in a marking and a *timestamp*. The timestamps have the meaning that a token cannot be consumed by a transition before this time.[1]

Definition 2 (Marking of a TPN). *Let I denote a countably infinite set of identifiers. A* marking *of a TPN is a partial function $m : I \nrightarrow P \times \mathbb{Q}$ with a finite domain. For $i \in dom(m)$ with $m(i) = (p, q)$ we say that (i, p, q) is a token on place p with timestamp q. We denote the set of all markings of a TPN by $\mathbb{M}$ and we define the projection functions π, τ as $\pi((p, q)) = p$ and $\tau((p, q)) = q$.*

The semantics of the different model classes are given by different transition relations. However, they have a commonality: if we abstract from the time information, the transition relation is a subset of the transition relation of the classical Petri net. This requirement to the semantics is expressed in the following definition.

Definition 3 (Common properties of TPNs). *A transition relation $\rightarrow \subseteq \mathbb{M} \times T \times \mathbb{M}$ of a TPN satisfies the following property: if $m \xrightarrow{t} m'$ then:*

– *$dom(m' \backslash m) \cap dom(m \backslash m') = \emptyset$ (identities of new tokens differ from consumed tokens),*
– *$\{\pi(m(i)) | i \in dom(m \backslash m')\} = {}^\bullet t$ and $|m \backslash m'| = |{}^\bullet t|$ (consumption from the pre-places of t), and*
– *$\{\pi(m'(i)) | i \in dom(m' \backslash m)\} = t^\bullet$ and $|m' \backslash m| = |t^\bullet|$ (production to the post-places of t).*

Given a marking $m_0 \in \mathbb{M}$, a sequence of transitions $\langle t_1, \ldots, t_n \rangle$ is called a firing sequence *if $\exists m_1, \ldots, m_n \in \mathbb{M} : m_0 \xrightarrow{t_1} m_1 \ldots \xrightarrow{t_n} m_n$. We denote the finite set of all firing sequences of a TPN N from a marking $m \in \mathbb{M}$ by $FS(N, m)$.*

There are three main classes of TPN, based on the time passing scheme chosen:

– *Duration of firing* ($M1$)
   Time is connected to transitions (Tr-equal). As soon as transitions are enabled one of them is selected and for that transition one value is chosen from

---

[1] By extending the time domain to *Time* $\times$ *Time* with *Time* $\subseteq \mathbb{R}$ we could talk about "usability" of tokens, with tokens having the earliest and the latest consumption times.

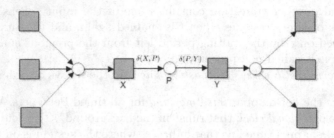

Fig. 1. Modeling task execution times

the input delay interval. Then the tokens are consumed immediately and new tokens for the output places are produced after the delay. The global clock moves till either the end of the delay or to an earlier moment where another transition is enabled. This is the Timed Petri Net model.

– *Delaying of firing (M2)*
Time is connected to transitions (Tr-equal). As soon as transitions are enabled, for each of them a delay is selected from the input delay interval. One of the transitions having the minimal of those delays is chosen to fire after the delay has passed. This is often called Merlin time and race semantics.

– *Delaying of tokens (M3)*
Time is connected to tokens (Out-rint, In-single). This is the DTPN model where a transition can fire at the time of the maximal timestamp of its input tokens. One of the earliest enabled transitions will fire. The input delay is added to the timestamp in order to determine if it is consumable or not.

For modeling systems in practice, it is often possible to model synchronization by instantaneous transitions with zero input and zero output delays. Activities that take time are modeled by a *begin* transition X and an *end* transition Y, as in Fig. 1. In this case, all model classes are equal since if X fires at time t then Y will fire at $t + a + b$, where $a \in \delta(X, P)$ and $b \in \delta(P, Y)$. Furthermore, input delays are handy for modeling time-outs.

Timed automata are another important class of timed models (see [6]). For model classes $M1$ and $M2$, it is possible to translate them to timed automata [25,14]. Some models of model class $M3$ (DTPN) can also be translated to timed automata, but because of eagerness the translations is only possible for the subclass (*Tr-equal, In-single, Out-rint*) (see [7]).

Generalized Stochastic Petri Nets (GSPN) is the famous class of stochastic Petri nets. It belongs to the class of $M2$ (race semantics with non-preemption). In general, non-preemption is a strange property from a practical point of view, because independent parts of a system are "working" while waiting for the firing of their transitions, and in case the input tokens of some transition t, being in preparation to its firing, are taken by another transition, the preparation work done by t is lost. However, preemption and non-preemption coincide in case of exponentially distributed delays: if a transition with a delayed firing is interrupted, the rest-distribution in case of preemption is the same as the original waiting time. Therefore these models can be transformed into a continuous time Markov process and the analysis techniques for Markov processes can be applied. However, exponential distributions are very restrictive for modeling in practice. It is possible to approximate an arbitrary continuous distribution by a *phase type* distributions, which are combinations of exponential distributions, but this is not a nice solution to model arbitrary transition delay distributions, because it blows up the state space and it disturbs the preemption property. Therefore, in Section 5, we will introduce a stochastic version of the class of $M3$, which we call *DSPN*.

4   Discrete Timed Petri Nets

In this section we first define the semantics of model class DTPN and we consider two subclasses of DTPN: sDTPN and fDTPN. The subclass sDTPN satisfies (*Out-single, In-single*) and an fDTPN satisfies (*In-single, Out-fint*). We show that DTPN, sDTPN and fDTPN are equivalent and we will transform sDTPN, by reducing the time component, into a strongly bisimilar labeled transition system called rDTPN. For rDTPN we show that it has a finite reachability graph if the underlying Petri net is bounded. So rDTPN can be used for model checking and since it is equivalent to the general DTPN we are able to model check them as well. For the proofs of the formal theorems we refer to [7].

4.1   Semantics of DTPN

In order to define the firing rule of a DTPN we introduce an *activator*. For a transition t in a marking m the activator is a minimal subset of the marking that enables this transition. Like in classical Petri nets, an activator has exactly one token on every input place of t and no other tokens.

Definition 4 (Activator). *Consider a TPN with the set of all markings* $\mathbb{M}$. *A marking* $a \in \mathbb{M}$ *is called an* activator *of transition* $t \in T$ *in a marking* $m \in \mathbb{M}$ *if (1)* $a \subseteq m$ *(a coincides with m on* $dom(a)$*), (2)* $\{\pi(m(i)) \mid i \in dom(a)\} = {}^{\bullet}t$ *and* $|a| = |{}^{\bullet}t|$ *(t is classically enabled), and (3) for any* $i \in dom(a) : \tau(m(i)) = \min\{\tau(m(j))|j \in dom(m) \wedge \pi(a(j)) = \pi(a(i))\}$ *(i is the oldest token on that place). We denote the set of all activators of a transition* t *in a marking* m *by* $A(m,t)$.

The *enabling time* of a transition with an activator a is the earliest possible time this transition enabled by an activator can fire. The *firing time* of a marking is the earliest possible time one of the transitions, enabled by this marking, can fire. We are assuming an *eager* system.

The time information of a DTPN is contained in its marking. This makes it possible to consider the system only at the moments when a transition fires. For this reason, we do not represent time progression as a state transition.

Definition 5 (Enabling time, firing time). *Let a be an activator of a transition $t \in T$ in a marking m of a DTPN. The* enabling time *of transition t by activator a is defined as $et(a,t) = \max\{\tau(a(i)) + \delta(\pi(a(i)),t) \mid i \in dom(a)\}$. The* firing time *of a marking m is defined as $ft(m) = \min\{et(a,t) \mid t \in T, a \in A(m,t)\}$.*

The firing time for a marking is thus completely determined by the marking itself.

The firing of a transition and its effect on a marking are described by the *transition relation*. Produced tokens are "fresh", i.e. they have new identities.

Definition 6 (Transition relation). *The* transition relation *of a DTPN satisfies Def. 3 and moreover, for any $m \xrightarrow{t} m'$:*

- *there is an $a \in A(m,t)$ such that $a = m \setminus m' \wedge ft(m) = et(a,t) < \infty$, and*
- *$\forall i \in dom(m' \setminus m) : \tau(m'(i)) = ft(m) + x$ with $x \in \delta(t, \pi(m(i)))$.*

Since the delays on arcs are nonnegative, a system cannot go back in time, i.e. $\forall m, m' \in \mathbb{M}, t \in T : m \xrightarrow{t} m' \Rightarrow ft(m) \leq ft(m')$.

In order to reduce infinite delay intervals to finite ones, we introduce the *refinement* of a DTPN, which is in fact a refinement of output delay intervals. Input delays are singletons and each singleton is a refinement of itself.

Definition 7 (Refinement of a DTPN). *Let $N_1 = (P,T,F,\delta)$ and $N_2 = (P,T,F,\delta')$ be DTPN. Then N_2 is a refinement of N_1 denoted by $N_1 \lhd N_2$ iff $\forall (x,y) \in F : \delta(x,y) \lhd \delta'(x,y)$.*

The refinement of a DTPN might introduce new firing sequences in the system. Due to this property, we are able to show that a refined DTPN can simulate its original DTPN with identical markings but not vice versa.

Theorem 8 (Simulation by refinement). *Consider two DTPN's N_1 and N_2 such that $N_1 \lhd N_2$. Then $\forall m \in \mathbb{M} : (N_2, m) \preceq (N_1, m)$ w.r.t. identity relation. So $\forall m \in \mathbb{M} : FS(N_1, m) \subseteq FS(N_2, m)$.*

A similar result holds for the untimed version $\overline{N}$ of a DTPN N: $\forall m \in \mathbb{M} : (\overline{N}, \overline{m}) \curvearrowleft (N, m) \wedge FS(N, m) \subseteq FS(\overline{N}, \overline{m})$ where $\overline{m}$ is the untimed version of m.

4.2   Relationship between sDTPN, fDTPN and DTPN

For a DTPN N_1 and a fDTPN N_2 such that $N_2 \lhd N_1$, we have $N_1 \preceq N_2$. The question we will address here is: *Is there an fDTPN such that $N_2 \preceq N_1$?* The answer is positive. In order to show this, we introduce a function φ assigning to a DTPN a fDTPN, called its *proxy*, which has a finite grid on each delay interval. This is only interesting for output delays, but we define it for all delay intervals. The *grid distance* is the smallest value such that all the bounds of delay sets and timestamps in the initial marking are multiples of it.

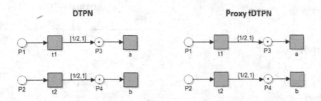

Fig. 2. Counter Example: DTPN simulates fDTPN

Definition 9 (Proxy fDTPN). *A DTPN $N = (P, T, F, \delta)$ with an initial marking m_0 has a proxy fDTPN $\varphi(N)$ with the same initial marking such that $\varphi(N) = (P, T, F, \delta')$ where:*

- $A = \{\tau(m(i)) \mid i \in dom(m_0)\} \cup \{\min(\delta(x, y), \max(\delta(x, y) \mid (x, y) \in F\}$,
- $\forall(x, y) \in F : \delta'(x, y) = \{i/d \mid i \in \mathbb{Z} \wedge min(\delta(x, y))/d \leq i/d \leq max(\delta(x, y)/d)\}$, *where* $d = min\{k \in \mathbb{N} \mid \{k.a \mid a \in A\} \subseteq \mathbb{N}\}$ *and* $1/d$ *is the* grid distance.

The grid distance is the least common multiple of the denominators of the non-zero elements of A expressed as non-reducible elements of $\mathbb{Q}$. We introduce the *round-off* of a marking, which is obtained if the timestamps are round-off to a grid.

Definition 10 (Round-off Relation). *For the set of all markings $\mathbb{M}$ of a DTPN: $\forall m, \bar{m} \in \mathbb{M} : m \sim \bar{m}$ iff $dom(m) = dom(\bar{m})$ and $\forall i \in dom(m) : \pi(m(i)) = \pi(\bar{m}(i))$ and $\forall i \in dom(m) : \exists k \in \mathbb{Z}, d \in \mathbb{N} :*

$$(\tau(m(i)) \in [k/d, k/d + 1/2d] \wedge \tau(\bar{m}(i)) = k/d) \vee$$
$$(\tau(m(i)) \in (k/d + 1/2d, (k+1)/d] \wedge \tau(\bar{m}(i)) = (k+1)/d)$$

The round-off relation preserves the order of timestamps.

Corollary 11. *Let N be an DTPN with the set of all markings $\mathbb{M}$. Let $\bar{N} = \varphi(N)$ be its proxy fDTPN with the set of all markings $\bar{\mathbb{M}}$. If two markings $m \in \mathbb{M}, \bar{m} \in \bar{\mathbb{M}} : m \sim \bar{m}$ then $\forall i, j \in dom(m) :*

$$\tau(m(i)) = \tau(m(j)) \Rightarrow \tau(\bar{m}(i)) = \tau(\bar{m}(j))$$
$$\tau(m(i)) < \tau(m(j)) \Rightarrow \tau(\bar{m}(i)) \leq \tau(\bar{m}(j))$$
$$\tau(m(i)) > \tau(m(j)) \Rightarrow \tau(\bar{m}(i)) \geq \tau(\bar{m}(j))$$

As a consequence of the preservation of the order of timestamps order we have:

Theorem 12 (Simulation by proxy). *Let N be a DTPN and $\bar{N} = \varphi(N)$ be its proxy fDTPN with the set of all common timed markings $\mathbb{M}$. Then $\forall m \in \mathbb{M} : (\bar{N}, m) \preceq (N, m)$ w.r.t. the round-off relation.*

The opposite is not true, i.e. a DTPN is not simulating its proxy with respect to the round-off relation as illustrated by the example in Fig. 2. The grid distance

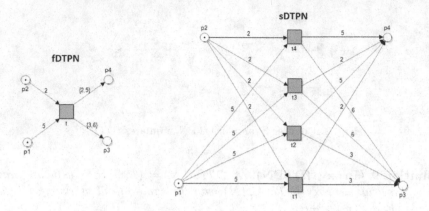

Fig. 3. Construction of a DTPN from an fDTPN

$d = 1/2$. From the initial marking with tokens in places $p1$ and $p2$ with zero timestamps, a marking m with tokens in places $p3$ and $p4$ with timestamps $11/16$ and $10/16$, respectively is reachable. In the proxy, a marking $\bar{m}$ is reachable such that $m \sim \bar{m}$, i.e. with tokens in places $p3$ and $p4$ with timestamps equal to one. From marking m only transition b is enabled but from marking $\bar{m}$ both transitions a and b are enabled. However, by Theorem 8 we know that a DTPN simulates its proxy w.r.t. the identity relation.

Corollary 13 (Simulation Equivalence). *A DTPN N and its proxy fDTPN $\bar{N} = \varphi(N)$ are simulation equivalent for each initial marking m_0 and hence $FS(N, m_0) = FS(\bar{N}, m_0)$.*

The final step in the reduction process is to show that any fDTPN is strongly bisimilar to a sDTPN. This can be done by making a copy of a transition, i.e. a new transition with the same preset and postset as the original one and with singleton delays for each possible combination of output delays from its output intervals. Formally this can be defined using the generalized cartesian product. We give an example in the Fig. 3.

Definition 14 (Reduction of fDTPN to sDTPN). *Let $N = (P, T, F, \delta)$ be a fDTPN. For each $t \in T$, let $A_t = \prod_{p \in t\bullet} \delta(t, p)$ be the generalized cartesian product of all its delay sets. Then, the corresponding sDTPN is the tuple $construct(N) = (P, T', F', \delta')$, where*

$$T' = \{t_x \mid t \in T \land x \in A_t\},$$
$$F' = \{(p, t_x) \mid (p, t) \in F \land x \in A_t\} \cup \{(t_x, p) \mid (t, p) \in F \land x \in A_t\},$$
$$\forall p \in P : \forall t \in T : \forall x \in A_t : \delta'(p, t_x) = \delta(p, t) \land \delta'(t_x, p) = x(p).$$

Theorem 15 (Bisimulation of fDTPN and sDTPN). *Let N be an arbitrary fDTPN. Then $N \simeq construct(N)$.*

As a consequence, for each DTPN there exists an sDTPN that is simulation equivalent.

4.3 Analysis of sDTPN

To analyse the behavior of DTPNs it is sufficient to consider sDTPNs. However, since time is non-decreasing, the reachability graph of an sDTPN is usually infinite. Still, there is a finite *time window* that contains all relevant behavior, because new tokens obtain a timestamp bounded by a maximum in the future, i.e. the maximum of all maxima of output delays and the timestamps of tokens earlier than the current time minus an upperbound of the input delays are irrelevant, i.e. they can be updated to the current time minus this upper bound. So we can reduce the time frame of a DTPN to a finite time window. This is done by defining a reduction function that maps the timestamps into this window. We introduce a labeled transition system for an sDTPN that is strongly bisimilar to it. Therefore we call this labeled transition system rDTPN, although, strictly speaking, it is not a DTPN.

We denote by $\delta_i^\uparrow$ the maximal *incoming* arc delay, i.e. $\delta_i^\uparrow = \max\{\delta(p,t)|(p,t) \in (P \times T) \cap F\}$ and $\delta_o^\uparrow$ the maximal *outgoing* arc delay, i.e. $\delta_o^\uparrow = \max\{\delta(t,p)|(t,p) \in (T \times P) \cap F\}$. A *reduced* marking is obtained by subtracting the firing time of the marking from the timestamp of each token in the marking with a lower bound of $-\delta_i^\uparrow$.

Definition 16 (Reduction function). *Consider an sDTPN with the set of all markings* $\mathbb{M}$. *The reduction function* $\alpha : \mathbb{M} \to \mathbb{M}$ *satisfies* $\forall m \in \mathbb{M}$: $dom(\alpha(m)) = dom(m)$ *and* $\forall i \in dom(m) : \pi(\alpha(m)(i)) = \pi(m(i)) \wedge \tau(\alpha(m)(i)) = \max\{-\delta_i^\uparrow, \tau(m(i)) - ft(m)\}$. *The set of all* reduced *markings of a sDTPN is defined as* $\bar{\mathbb{M}} = \{m \in \mathbb{M} \mid \alpha(m) = m\}$.

The reduction function is either (1) reducing the timestamp of tokens by the firing time, or (b) mapping timestamps less than or equal to $-\delta_i^\uparrow$ to $-\delta_i^\uparrow$.

Corollary 17. *Let N be a sDTPN with the set of all markings* $\mathbb{M}$. *Then* $\forall m \in \mathbb{M}, t \in T : \forall i \in dom(m) : \tau(\alpha(m)(i)) \geq \tau(m(i)) - ft(m) \wedge A(\alpha(m), t) = \{\alpha(e) \mid e \in A(m,t)\}$.

Lemma 18. *Consider a sDTPN with the set of all markings* $\mathbb{M}$. *Then* $\forall m \in \mathbb{M} : ft(\alpha(m)) = 0$, *i.e the firing time of a reduced marking is zero.*

As a consequence of Lemma 18, the reduction function α is idempotent.

Corollary 19. *Let N be a sDTPN. Then* $\forall m \in \mathbb{M} : \alpha(\alpha(m)) = \alpha(m)$.

We will now show that, given a marking with an enabled transition, the same transition is also enabled in its reduced marking and the new marking created by firing this transition from both enabling markings have the same reduced marking. Furthermore, the firing time of a marking reachable from a reduced marking can be used to compute the arrival time in the concrete system.

Lemma 20. *Consider a sDTPN with the set of markings* $\mathbb{M}$. *Let* $m, m' \in \mathbb{M}$: $m \xrightarrow{t} m'$. *Then* $\exists \tilde{m} \in \mathbb{M} : \alpha(m) \xrightarrow{t} \tilde{m}$ *and* $ft(\tilde{m}) = ft(m') - ft(m)$ *and* $\alpha(m') = \alpha(\tilde{m})$.

For an executable firing sequence, the above theorem can be extended to the following corollary.

Corollary 21. *Let* $(N, \mathbb{M}, \rightarrow, m_0)$ *be a labeled transition system. Let* $m_0, m_1 \ldots m_n \in \mathbb{M}$ *and* $m_0 \xrightarrow{t_1} m_1 \xrightarrow{t_2} \ldots \xrightarrow{t_n} m_n$. *Then* $\exists \tilde{m}_0, \tilde{m}_1, \ldots, \tilde{m}_n \in \mathbb{M} : \tilde{m}_0 = m_0 \land \forall j \in \{1, \ldots, n\} : \alpha(m_{j-1}) \xrightarrow{t_j} \tilde{m}_j$ *and* $ft(m_n) = \sum_{j=0}^{n} ft(\tilde{m}_j)$

The *reduced transition relation* defines the relationship between two reduced markings, one reachable from the other.

Definition 22 (Reduced transition relation). *Let* N *be an sDTPN with the set of all markings* $\mathbb{M}$ *and the set of all reduced markings* $\bar{\mathbb{M}}$. *The* reduced transition relation $\rightsquigarrow \subseteq \bar{\mathbb{M}} \times T \times \bar{\mathbb{M}}$ *satisfies* $\forall \bar{m}, \bar{m}' \in \bar{\mathbb{M}} : \bar{m} \xrightarrow{t} \bar{m}' \Leftrightarrow \exists \tilde{m} \in \mathbb{M} : \bar{m} \xrightarrow{t} \tilde{m} \land \alpha(\tilde{m}) = \bar{m}'$.

Definition 23 (rDTPN). *An* rDTPN *is a* labeled transition system $(\mathbb{M}, T, \rightsquigarrow, \bar{m}_0)$, *where* $\bar{m}_0 = \alpha(m_0)$ *is an initial marking.*

For a given sDTPN, its reduced labelled transition system and timed labelled transition system are strongly bisimilar w.r.t. reduction relation. Due to Lemma 20, the time relation in the bisimulation is implicit.

Theorem 24 (Bisimulation sDTPN and rDTPN). *Consider a sDTPN with a timed labeled transition system* $N = (\mathbb{M}, T, \rightarrow, m_0)$ *and its rDTPN* $\bar{N} = (\bar{\mathbb{M}}, T, \rightsquigarrow, \bar{m}_0)$. *Then* $N \simeq \bar{N}$.

The number of different timestamps in the reachability graph of the rDTPN is finite. This is observed in several papers (see [8]). To see this, we consider first only timestamps in $\mathbb{Z}$. Since we have finitely many of them in the initial marking and the only operations we execute on them are: (1) selection of the maximum, (2) adding one of a finite set of delays and (3) subtracting the selected timestamp with a minimum. So the upper bound is the maximal output delay and the lower bound is zero minus the maximal input delay. Hence, we have a finite interval of $\mathbb{Z}$ which means finitely many values for all markings. In case we have delays in $\mathbb{Q}$ we multiply with the lcm of all relevant denominators, like in definition 10 and then we are in the former case.

Theorem 25. *The set of timestamps in the reachability graph of a rDTPN is finite, and so if the underlying Petri net is bounded, the reachability graph of the rDTPN is finite.*

Furthermore, using Corollary 21, given a path in the reachability graph of a rDTPN, we are able to compute the time required to execute this path in the original DTPN.

Finally, we sketch by two constructions how the two basic variants of DTPN, namely DTPN-1 with (In-single, Out-zero) and DTPN-2 with (In-zero, Out-single) can express inhibitor arcs. The models are branching bisimilar with the classical Petri net with inhibitors. In both constructions, we add a special place

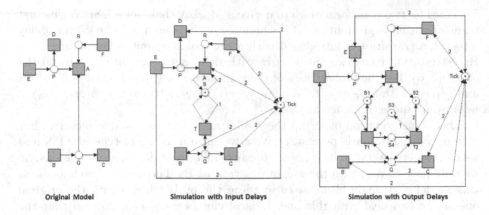

Original Model   Simulation with Input Delays   Simulation with Output Delays

Fig. 4. Simulating Inhibitor Arcs with Input Delays and Output Delays

called *Tick* that is connected to each original transition by an input and an output arc and we replace each inhibitor by a subnet containing a silent transition T. In DTPN-1 and DTPN-2 original transitions fire at even time units while the silent transitions may also fire at odd time units.

5   Discrete Stochastic Petri Nets

In this section we endow the transition firings of a DTPN with probabilities. We do this only for DTPN with (*In-single, Out-fint*) and we assign a probability distribution to these intervals. Additionally, we should have a stochastic mechanism to choose a transition from all enabled transitions. We do this by assigning a non-negative weight to all transitions and draw an enabled transition x with probability $w(x)/\sum_{y:enabled} w(y)$. If there are no priorities for transitions, we may choose all weights to be equal. Note that we could introduce stochastics for the two other main classes, $M1$ and $M2$ in a similar way. For the class $M1$, we first select a classically enabled transition with the weights and then a duration from a finite distribution associated with the transition. For the class $M2$, we select for all classical enabled transitions a delay from a finite distribution associated with the transition and then we select with the weights one of them having the minimal delay.

Definition 26 (DSPN)
A DSPN is 6-tuple $(P, T, F, \delta, w, \phi)$, where

- (P, T, F, δ) *is a DTPN with* (*In-single, Out-fint*).
- $w : T \to \mathbb{R}^+$ *a weight function, used to choose one of the simultaneously enabled transitions,*
- ϕ *is a function with domain $F \cap T \times P$ and for $\forall (t, p) \in F$:*
  $\phi(t, p) : \delta(t, p) \to [0, 1]$ *such that $\sum_{x \in \delta(t,p)} \phi(t, p) = 1$, so $\phi(t, p)$ assigns probabilities to $\delta(t, p)$.*

We consider two transformations to derive a Markov chain for a DSPN. The first transformation is given in Def. 14, where for each value of a finite output delay interval, a transition is introduced with a one point output delay, as in Fig. 3. Here transition t has two output arcs with delay sets, one with $\{2,5\}$ and the other $\{3,6\}$. Let the probabilities of these intervals be (p_1, p_2), with $p_1 + p_2 = 1$ and (q_1, q_2) with $q_1 + q_2 = 1$. So $w(t_1) = w(t).p_1.q_1, w(t_2) = w(t).p_1.q_2, w(t_3) = w(t).p_2.q_1$. and $w(t_4) = w(t).p_2.q_2$.

This transformation blows up the model and gives unreadable pictures, but it is only for automatic processing. Now we have a model of type sDTPN and we can forget the probabilities $\phi(.,.)$ because all output delays are singletons. So we only have to deal with the weight function w. By Theorem 15, we know that this model is strongly bisimilar (discarding the probabilities) with the original one so we can deal with this one. It is obvious by the construction that the probabilities over the delays of produced tokens are the same as well. So after these transformations we can consider a DSPN as a 5-tuple (P, T, F, δ, w).

The next transformation concerns this sDTPN model into the reduced labeled transition system (rDTPN) as in Theorem 24 which is strongly bisimilar with the sDTPN model. The weights can be transferred to this rDTPN model because the underlying Petri net has not changed. We call this new model class *rDSPN*. Remember that if the underlying Petri net is bounded, then rDSPN has a finite reachability graph.

We will now add two values to an arc in the reachability graph of the rDSPN, representing a transition $m, m' \in \mathbb{M} : m \xrightarrow{t} m'$: (1) *probability* of choosing this arc and (2) the *sojourn time* in a marking m' if coming from m. Remember that for each marking the firing time is uniquely determined, but the sojourn time depends on the former marking. The sojourn time can be computed during the reduction process as expressed by Lemma 20.

Definition 27 (Transition probability and sojourn time)
The transition probability $Q : \mathbb{M} \times \mathbb{M} \to [0,1]$ *satisfies:*

$$Q_{m,m'} = \sum_{x:m\xrightarrow{x}m'} w(x) \Big/ \sum_{y:\exists m'':m\xrightarrow{y}m''} w(y).$$

For $m, m' \in \mathbb{M} : r(m, m') = ft(m') - ft(m)$ *is the* sojourn time *in marking* m' *if coming from* m.

The transition probability contains all information of the reachability graph. Finally we are able to define the *Markov chain* that is determined by the reachability graph of the rDSPN endowed with the transition probabilities.

Definition 28 (Markov chain)
Let a rDSPN (P, T, F, δ, w) *be given and let the* Q *be the transition probability over the state space. Then the* Markov chain *of the rDSPN is a sequence of random variables* $\{X_n | n = 0, 1, ...\}$, *where* $X_0 = m_0$ *the initial marking and* X_n *is marking after* n *steps, such that:*

$$\mathbb{P}[X_{n+1} = m'|X_n = m, X_{n-1} = m_{n-1}, ..., X_0 = m_0] = Q(m, m')$$

for arbitrary $m_0, ..., m_{n-1} \in \mathbb{M}$.

The *Markov property* is implied by the fact that only the last marking before the transition firing is taken into account.

Since a marking and an enabled transition determine uniquely the next state, we can also consider another stochastic process $\{Y_n|n \in \mathbb{N}\}$, where $Y_n \in T$, which is a *stochastic firing sequence*. For a firing sequence $\sigma = (t_1, ..., t_n)$ with $m_0 \xrightarrow{t_1} m_1, ..., \xrightarrow{t_n} m_n$ we have

$$\mathbb{P}[Y_1 = t_1, ..., Y_n = t_n|X_0 = m_0] = \mathbb{P}[X_1 = m_1, ..., X_n = m_n|X_0 = m_0].$$

So we can compute the probability for each finite firing sequence.

Markov chains are often endowed with a *cost structure* which is a function assigning to a pair of successive markings a real value, called *cost function*. Then we can express the *total expected cost* when starting in marking m as:

$$\mathbb{E}[\sum_{n=0}^{N} c(X_n, X_{n+1})|X_0 = m].$$

Here, $N \in \mathbb{N}$ or $N = \infty$. In particular we will use the sojourn times as "cost". In fact we may associate a *semi-Markov process* to rDSPN, because the sojourn times themselves are random variables, but in our case they are completely determined if we know the former state. The Markov chain $\{X_n|n = 0, 1, ...\}$ is then the embedded Markov chain of the semi-Markov process [33]. We will use the function $v : \mathbb{M} \to \mathbb{R}$, which is usually called the *value function* for Markov processes (see [29]): $v(m) = \mathbb{E}[\sum_{n=0}^{N} c(X_n, X_{n+1})|X_0 = m]$ for cost functions c.

We will use the Markov chain to answer three types of important questions. We will use the cost function to express the questions and we use the Markov property to translate our questions into *Bellman equations* (see [33] and [29]).

- Probability of reaching a subset,
- Expected time to leave a subset,
- Expected sojourn times in equilibrium.

Note that all these questions concern sequences of markings or equivalently firing sequences. So they belong to LTL (see [28]).

Probability of Reaching a Subset

Let $A, B \subseteq \mathbb{M}$ be a subsets of the state space, $A \cap B = \emptyset$. We are interested in the probability of reaching A from B. Here we choose $c(m, m') = 1$ if $m \in B \wedge m' \in A$ and $c(m, m') = 0$ otherwise. Further we stop as soon as we reach A. Then

$$\forall m \in B : v(m) = \sum_{m' \in A} Q_{m,m'} + \sum_{m' \in B} Q_{m,m'} \cdot v(m').$$

If B is finite, this can be computed, even if the underlying Petri net is unbounded. For example if B is the set of k-bounded markings (i.e. markings with at most k tokens per place), this computation is possible.

Expected Time to Leave a Subset. A of the state space.
Here we use a cost function $c(m, m') = r(m, m')$, the sojourn time in m', if $m, m' \in A$ and $c(m, m') = 0$ elsewhere. If $m \in A$ then

$$v(m) = \sum_{m' \in A} Q_{m,m'}.(r(m, m') + v(m')).$$

This can be computed even if the underlying Petri net is unbounded but A is finite.

Expected Sojourn Times in Equilibrium
We restrict us to the case where the underlying Petri net is bounded. $\mathbb{P}[X_n = m'|X_0 = m] = Q^n(m, m')$ where Q^n is the Q to power n. The *limit of averages* exists: $\pi_m(m') := \lim_{N \to \infty} \sum_{n=0}^{N} Q_{m,m'}^n$ and it satisfies

$$\pi_{m_0}(m') = \sum_{m \in M} \pi_{m_0}(m).Q_{m,m'}.$$

We now assume the system is a strongly connected component (i.e. the Markov chain is irreducible), which implies that the limit distribution π is independent of the initial state m_0. Further, we know that the expected time spent in a marking m depends on the former marking, so the expected time of being in marking m' is:

$$\sum_{m \in M} r(m, m').\mathbb{P}[X_{n-1} = m|X_n = m'],$$

which can be rewritten using Bayes rule to:

$$\sum_{m \in M} r(m, m').\mathbb{P}[X_n = m'|X_{n-1} = m].\mathbb{P}[X_{n-1} = m]/\mathbb{P}[X_n = m'].$$

Thus, the expected sojourn time in some *arbitrary* marking is obtained by multiplying with $\mathbb{P}[X_n = m']$

$$\sum_{m \in M} r(m, m').Q_{m,m'}.\mathbb{P}[X_{n-1} = m].$$

This formula could also be derived immediately as the expected sojourn time in the next marking. For $n \to \infty$, this converges either by the normal limit or limit of averages to:

$$\sum_{m \in M} r(m, m').Q_{m,m'}.\pi(m).$$

If we want to solve these equations using matrix calculations, we need to compute the transition matrix of the reachability graph. However, we can also use the method of *successive approximations* to approximate these values in an iterative way using only two functions (vectors) over the state space. As an example, the probability of reaching a set A from a set B we set: $\forall m \in B : v_0(m) = 0$ and

$$\forall m \in B : v_{n+1}(m) = \sum_{m' \in A} Q_{m,m'} + \sum_{m' \in B} Q_{m,m'}.v_n(m').$$

According to [1] we can derive for specially structured *workflow nets* the distribution of the *throughput time* of a case (i.e. the time a token needs to go from the initial to the final place) analytically in case of DTPN with (*In-zero, Out-fint*). Models of this class can be built by *transition refinement*, using the patterns displayed in Fig. 5. Pattern 1 is a *sequence* construction. Pattern 2 is an *iteration* where we have arc weights q and $1 - q$ for the probability of continuing or ending the loop. Pattern 3 is the *parallel* construction. Pattern 4 is the *choice*, which has also arc weights q and $1 - q$ representing the probabilities for the choices. In Fig. 5 the intervals $[a, b], [c, d]$ indicate the finite probability distributions. In order to be a model of this class, it must be possible to construct it as follows. We start with an initial net and we may replace all transitions t with $|{}^\bullet t| = |t^\bullet| = 1$ using one of the four rules. There should be a proper parse tree for a net of this class. We associate to all transitions with output delay sets a random variable; for the initial net the random variable U with distribution on $[a, b]$ and similarly random variable Y and Z for the patterns.

If we have such a net, we can apply the rules in the reversed order. If we have at some stage a subnet satisfying to one of the four patterns, with the finite distributions as indicated, we can replace it by an initial subnet with a "suitable" distribution on the output delay interval. For the initial subnet we have a random output variable U. For the sequential construction (rule 1) we have two independent random variables Y and Z with discrete distributions on $[a, b]$ and $[c, d]$ respectively. So $U = Y + Z$ and the distribution of U is the *convolution* of the distributions of Y and Z. For the parallel construction (rule 3) we have $U = \max(Y, Z)$ which is the product distribution, i.e. $\mathbb{P}[U \leq x] = \mathbb{P}[Y \leq x].\mathbb{P}[Z \leq x]$. For the choice (rule 4) it is a *mixture* of two distributions, $\mathbb{P}[U \leq x] = \mathbb{P}[Y \leq x].q + \mathbb{P}[Z \leq x].(1 - q)$. The most difficult one is the iteration (rule 2), since here we have the distribution of $U := \sum_{n=0}^{N} (Y_n + Z_n)$ where N is a geometrically distributed random variable with distribution $\mathbb{P}[N = n] = q^{n-1}.(1 - q)]$ indicating the number of iterations and Y_n and Z_n are random variables from the distributions on $[a, b]$ and $[c, d]$ respectively. All these random variables are independent. The distribution of U can be derived using the Fourier transform (see [1]). This is an approximation, since the domain of U is infinite, even if Y_n and Z_n have finite domains. However, we can cut the infinite domain with a controllable error.

Thus, we are able to reduce a complex DSPN. This method is only applicable if the original net is safe, otherwise different cases can influence each other and so the independency assumptions are violated.

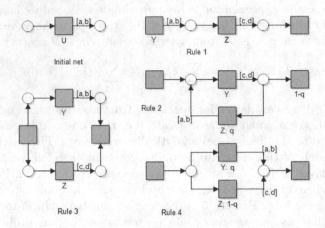

Fig. 5. Refinement rules

6   Conclusions

In this paper we considered Petri nets with time and stochastics, and in particular, one class of them that did not get much attention in literature: the class
with the timestamps for tokens. We call this class DTPN and show how we can
analyse the behaviour of the nets from this class in case the underlying Petri
net is bounded. We did not consider complexity but only computability, since
history has shown that methods can become feasible due to increase in computing power and smart heuristics. We considered several subclasses of DTPN and
we showed that they all are Turing complete, because they can express inhibitor
arcs, but that some have better *modeling comfort*, i.e. they are easier for modeling. The DTPN class can easily be extended to deal with stochastics, as we have
shown. Here, we have the advantage above the GSPN model that we can use
arbitrary finite distributions, while only exponential distributions can be used
in GSPN. The analysis of stochastic behavior is based on Markov chains and so
it is similar to the approach in GSPN. We also showed how analytical methods
for stochastic analysis can be incorporated in model checking.

References

1. van der Aalst, W.M.P., van Hee, K.M., Reijers, H.A.: Analysis of discrete-time
   stochastic Petri nets. Statistica Neerlandica 54(2), 237–255 (2000)
2. van der Aalst, W.M.P.: Interval Timed Coloured Petri Nets and their Analysis.
   PhD thesis, Eindhoven University of Technology (1993)
3. Abdulla, P.A., Nylén, A.: Timed Petri nets and BQOs. In: Colom, J.-M., Koutny,
   M. (eds.) ICATPN 2001. LNCS, vol. 2075, pp. 53–70. Springer, Heidelberg (2001)
4. Marsan, M.A., Conte, G., Balbo, G.: A class of generalized stochastic Petri nets
   for the performance evaluation of multiprocessor systems. ACM Trans. Comput.
   Syst. 2(2), 93–122 (1984)

5. Alur, R., Dill, D.L.: A theory of timed automata. Theor. Comput. Sci. 126(2), 183–235 (1994)
6. Bengtsson, J., Larsen, K.G., Larsson, F., Pettersson, P., Yi, W.: Uppaal - a tool suite for automatic verification of real-time systems. In: Alur, R., Sontag, E.D., Henzinger, T.A. (eds.) HS 1995. LNCS, vol. 1066, pp. 232–243. Springer, Heidelberg (1996)
7. Bera, D., van Hee, K.M., Sidorova, N.: Discrete timed Petri nets. Computer Science Report 13-03, Technische Universiteit Eindhoven, P.O. Box 513, 5600 MB Eindhoven, The Netherlands (April 2013)
8. Berthelot, G., Boucheneb, H.: Occurrence graphs for interval timed coloured nets. In: Valette, R. (ed.) ICATPN 1994. LNCS, vol. 815, pp. 79–98. Springer, Heidelberg (1994)
9. Berthomieu, B., Diaz, M.: Modeling and verification of time dependent systems using time Petri nets. IEEE Trans. Softw. Eng. 17(3), 259–273 (1991)
10. Berthomieu, B., Menasche, M.: An enumerative approach for analyzing time Petri nets. In: Proceedings IFIP, pp. 41–46. Elsevier Science Publishers (1983)
11. Boucheneb, H.: Interval timed coloured Petri net: efficient construction of its state class space preserving linear properties. Form. Asp. Comput. 20(2), 225–238 (2008)
12. Bouyer, P., Haddad, S., Reynier, P.-A.: Undecidability results for timed automata with silent transitions. Fundam. Inf. 92(1-2), 1–25 (2009)
13. Boyer, M., Roux, O.H.: On the compared expressiveness of arc, place and transition time Petri nets. Fundam. Inf. 88(3), 225–249 (2008)
14. Cassez, F., Roux, O.-H.: Structural translation from time Petri nets to timed automata. Electron. Notes Theor. Comput. Sci. 128(6), 145–160 (2005)
15. Cerone, A., Maggiolio-Schettini, A.: Time-based expressivity of timed Petri nets for system specification. Theor. Comput. Sci. 216(1-2), 1–53 (1999)
16. Ghezzi, C., Mandrioli, D., Morasca, S., Pezze, M.: A unified high-level Petri net formalism for time-critical systems. IEEE Trans. Softw. Eng. 17(2), 160–172 (1991)
17. van Glabbeek, R.: The linear time-branching time spectrum. In: Baeten, J.C.M., Klop, J.W. (eds.) CONCUR 1990. LNCS, vol. 458, pp. 278–297. Springer, Heidelberg (1990)
18. Haddad, S., Moreaux, P.: Sub-stochastic matrix analysis for bounds computation - theoretical results. European Journal of Operational Research 176(2), 999–1015 (2007)
19. van Hee, K.M., Somers, L.J., Voorhoeve, M.: Executable specifications for distributed information systems. In: Proceedings of the IFIP TC 8/WG 8.1, pp. 139–156. Elsevier (1989)
20. Jantzen, M.: Language theory of Petri nets. In: Brauer, W., Reisig, W., Rozenberg, G. (eds.) APN 1986. LNCS, vol. 254, pp. 397–412. Springer, Heidelberg (1987)
21. Jensen, K.: An introduction to the theoretical aspects of coloured Petri nets. In: de Bakker, J.W., de Roever, W.-P., Rozenberg, G. (eds.) REX 1993. LNCS, vol. 803, pp. 230–272. Springer, Heidelberg (1994)
22. Knapik, M., Penczek, W., Szreter, M., Polrola, A.: Bounded parametric verification for distributed time Petri nets with discrete-time semantics. Fundam. Inf. 101(1-2), 9–27 (2010)
23. Christensen, S., Kristensen, L.M., Mailund, T.: Condensed state spaces for timed petri nets. In: Colom, J.-M., Koutny, M. (eds.) ICATPN 2001. LNCS, vol. 2075, pp. 101–120. Springer, Heidelberg (2001)
24. Lime, D., Roux, O.H.: Model checking of time Petri nets using the state class timed automaton. Discrete Event Dynamic Systems 16(2), 179–205 (2006)

25. Lime, D., Roux, O.H.: Model checking of time Petri nets using the state class timed automaton. Discrete Event Dynamic Systems 16(2), 179–205 (2006)
26. Ajmone Marsan, M., Balbo, G., Conte, G., Donatelli, S., Franceschinis, G.: Modelling with generalized stochastic Petri nets. SIGMETRICS Perform. Eval. Rev. 26(2) (August 1998)
27. Merlin, P.M., Farber, D.J.: Recoverability of communication protocols: Implications of a theoretical study. IEEE Trans. Comm. (September 1976)
28. Pnueli, A.: The temporal logic of programs. In: Proceedings of the 18th Annual Symposium on Foundations of Computer Science, SFCS 1977, pp. 46–57. IEEE Computer Society, Washington, DC (1977)
29. Puterman, M.L.: Markov Decision Processes: Discrete Stochastic Dynamic Programming, 1st edn. John Wiley and Sons, Inc., New York (1994)
30. Ramachandani, C.: Analysis of asynchronous concurrent systems by timed Petri nets. PhD thesis, Massachusetts Institute of Tenchnology, Cambridge, MA (1974)
31. Ramamoorthy, C.V., Ho, G.S.: Performance evaluation of asynchronous concurrent systems using Petri nets. IEEE Transactions on Software Engineering 6(5), 440–449 (1980)
32. Recalde, L., Haddad, S., Silva, M.: Continuous Petri nets: Expressive power and decidability issues. Int. J. Found. Comput. Sci. 21(2), 235–256 (2010)
33. Ross, S.M.: Introduction to Probability Models, 9th edn. Academic Press, Inc., Orlando (2006)
34. Valero Ruiz, V., de Frutos Escrig, D., Cuartero Gomez, F.: On non-decidability of reachability for timed-arc Petri nets. In: Proc. 8th. International Workshop on Petri Nets and Performance Models, pp. 188–196 (1999)
35. Sifakis, J.: Use of Petri nets for performance evaluation. In: Proceedings of the Third International Symposium on Measuring, Modelling and Evaluating Computer Systems, Bonn - Bad Godesberg, Germany, October 3-5, pp. 75–93. North-Holland (1977)
36. Starke, P.H.: Some properties of timed nets under the earliest firing rule. In: Rozenberg, G. (ed.) APN 1989. LNCS, vol. 424, pp. 418–432. Springer, Heidelberg (1990)
37. CPN website, http://www.cpntools.org/
38. ExSpecT website, http://www.exspect.com/
39. Great SPN website, http://www.di.unito.it/~greatspn/index.html
40. UPPAAL website, http://www.uppaal.org/

Eliminating Concurrency Bugs in Multithreaded Software: An Approach Based on Control of Petri Nets*

Stéphane Lafortune[1], Yin Wang[2], and Spyros Reveliotis[3]

[1] Department of EECS, University of Michigan, USA
[2] HP Laboratories, USA
[3] School of ISyE, Georgia Tech, USA

Abstract. We describe the Gadara project, a research effort whose goal is to eliminate certain classes of concurrency bugs in multithreaded software by controlling the execution of programs at run-time. The Gadara process involves three stages: modeling of the source code at compile time in the form of a Petri net, feedback control synthesis, and control logic implementation into the source code. The feedback control logic is synthesized using techniques from supervisory control of discrete event systems, where the specification captures the avoidance of certain types of concurrency bugs, such as deadlocks. We focus on the case of circular-wait deadlocks in multithreaded programs employing mutual exclusion locks for shared data. The application of the Gadara methodology to other classes of concurrency bugs is briefly discussed.

1 Introduction

The concepts and techniques of control engineering find numerous applications in computer and software engineering. For instance, classical control theory, for time-driven systems with continuous state spaces, has been applied to computer systems problems involving quantitative properties, such as throughput stabilization; see, e.g., [1]. However, many important problems in computer and software engineering involve qualitative specifications, such as deadlock avoidance, and their solution requires control-theoretic approaches for event-driven systems with discrete state spaces, i.e., Discrete Event Systems (DES). In the last few years, there has been increased interest in solving discrete-event control problems that arise in software and embedded systems; see, e.g., [2–10]. In particular, the paradigm of *controlling software execution to avoid software defects at run-time* is receiving increased attention in the control engineering, programming languages, and operating systems communities.

* Research partially supported by the U.S. National Science Foundation, under grants CCF-0819882, CMMI-0928231, and CCF-1138860 (Expeditions in Computing project ExCAPE: Expeditions in Computer Augmented Program Engineering), and by HP Laboratories, under an Open Innovations Award.

J.-M. Colom and J. Desel (Eds.): PETRI NETS 2013, LNCS 7927, pp. 21–28, 2013.

We have been investigating how to control software execution to avoid certain classes of concurrency bugs under the so-called "Gadara Project" [11], a multidisciplinary effort centered at the University of Michigan and pursued in collaboration with HP Laboratories and the Georgia Institute of Technology in the U.S. In this effort, control techniques from the field of DES, such as supervisory control [12] and supervision based on place invariants [13], are employed to synthesize control logic that is instrumented into the source code and enforces the desired safety properties at run-time [14]. Since it is model-based and relies on theoretical results in DES, this approach provably guarantees the desired safety properties, subject to model accuracy. The principal safety property of interest in the work to-date is deadlock avoidance in multithreaded programs that use locking primitives to control access to shared data [15, 16]. Recent results address certain types of atomicity violations in multithreaded programs [17, 18].

This paper describes and discusses the key features of the Gadara methodology, with relevant references. It is based on, and complements, the keynote lecture of the first author at the 34th International Conference on Application and Theory of Petri Nets and Concurrency (June 2013).

2   Gadara Methodology

There is a large amount of literature in computer science on the study of deadlock using a variety of modeling and analysis techniques. Petri net models have been used for deadlock analysis in several application domains, including computer and manufacturing systems. In particular, several special classes of Petri nets have been characterized and analyzed for deadlock problems that involve a set of "processes" sharing a set of common "resources" in the context of automated manufacturing applications; see [19, 20]. Such systems are often referred to as Resource Allocation Systems, or RAS. RAS also occur in the context of software systems, where processes may be *threads* and shared resources may be *data objects*. Modeling thread creation/termination and lock/unlock operations on shared data is in fact a classical application of Petri nets [21], and Petri nets have been employed to model multithreaded synchronization primitives in the popular Pthread library for C/C++ programs [22]. Petri nets were also used to analyze deadlocks in Ada programs [23]. A review of the application of Petri nets to computer programming is presented in [24].

The methodology developed in the Gadara project for avoidance of certain classes of deadlocks in multithreaded software is also based on Petri net models. The methodology relies on the extraction of a suitable model of the program source code at compile time in the form of an enhanced Control Flow Graph (CFG) that captures the control flow and the locking behavior of all the program threads. This step generally requires the use of static analysis techniques (see, e.g., [25, 26]) to ensure a more accurate model. This model is then translated into a Petri net in a straightforward process: places in the net represent basic blocks (i.e., branch-free sets of consecutive instructions) in the CFG or locks that will be acquired by the threads; transitions in the net represent transitions in the CFG or lock acquisition and release operations; finally, tokens represent the states of the threads and of the locks. The special class of RAS Petri nets that arises in this context is called *Gadara nets*. The reader is referred to [27] for the formal definition of Gadara nets and for a treatment of their analytical properties in the context of multithreaded programs that use mutually-exclusive locks to control access to shared data. The deadlocks caused by the use of mutually-exclusive locks are circular-mutex-wait deadlocks, where threads in a set are waiting for one another and none can proceed. Avoidance of circular-mutex-wait deadlocks in multithreaded programs is mapped to the problem of *liveness* in Gadara nets in [27]. Liveness here refers to the property that every transition is eventually firable from any reachable state of the net. Due to the structure of Gadara nets, liveness is equivalent to reversibility, i.e., the initial state must be reachable from every reachable state. Algorithms based on solving Mixed Integer Linear Programs (MILP) are presented in [27] for determining if liveness holds or not. The algorithms exploit the structural properties of Gadara nets, in terms of certain classes of siphons. In this regard, we note that many works have considered similar structural analyses for related classes of Petri nets; see, e.g., [20, 28].

The central aspect of the Gadara methodology is its focus on synthesizing a control strategy for the Petri net model so that the controlled system is provably live, with respect to the model. This control strategy, referred to as the *control logic* hereafter, must satisfy four key requirements in addition to liveness. The first two requirements, denoted by (R1) and (R2), pertain to its synthesis and the last two requirements, denoted by (R3) and (R4), pertain to its implementation. (R1): The control logic should not alter the behavior of the program; it should only act by delaying lock acquisition or release operations performed by the threads. (R2): The control logic should only intervene when absolutely necessary; this is referred to as *maximal permissiveness*. A correct strategy could be to force the threads to always execute serially; deadlock would be avoided, but no concurrency would be allowed. (R3): The control logic must be readily translatable to the original source code that is modeled by the Gadara net, thereby allowing code instrumentation as an implementation mechanism. (R4): The runtime overhead of the control logic must be minimized, so that the instrumented program runs almost as fast as the original program.

The supervisory control theory for DES initiated in [12] and widely studied since then is well suited for handling (R1) and (R2). The notion of *uncontrollable* transitions (or events) captures (R1), while maximal permissiveness is handled by the concept of the *supremal controllable sublanguage* of the legal language with respect to the uncontrolled system language. Here, the legal language is the live sublanguage of the uncontrolled system, obtained by deleting states that deadlock and those that are in a livelock, when the initial state is the only marked state. However, using the standard algorithms of this supervisory control theory, as described in [29] for instance, requires building the reachability graph of the Gadara net model of the program. Moreover, the form of the control strategy, which is now a global function over the entire reachability graph, will perform poorly in terms of (R3) and (R4), unless it can be encoded in a different form. These two considerations have motivated the control logic synthesis research performed in the Gadara project, which is overviewed in the next section.

3   Control Logic Synthesis

Requirements (R3) and (R4) of the previous section suggest to use *control places* (also called monitor places) as the control mechanism for the Gadara net model of the program. Control places are connected to the transitions of the net, which in turn can be mapped back to specific lines of code in the program. Instrumented code can then be inserted at the appropriate location to implement the constraint imposed by the control place, which is treated as a global variable. This control mechanism only affects program execution when it reaches a point where the corresponding transition in the Gadara net is connected to a control place. The control synthesis task is therefore to determine a set of control places, their initial marking, and their connectivity to the net, such that the control logic enforced by these control places keeps the Gadara net live in a maximally-permissive manner. Moreover, the control places should never lead to the disablement of an uncontrollable transition in the net. In other words, the control logic enforced by these control places should correspond exactly to the supremal controllable sublanguage of the Gadara net subject to the live sublanguage specification mentioned earlier.

In our efforts so far, we have used the control technique called Supervision Based on Place Invariants (SBPI) to synthesize the desired control places. SBPI is a control logic synthesis framework that uses control places to enforce a set of linear inequality constraints on the reachable states of a given arbitrary Petri net [13]. Each linear inequality corresponds to a weighted sum of the number of tokens in each place of the net, and it will be *exactly* enforced by one control place, if enforceable at all; that is, the control is correct and maximally permissive with respect to the linear inequality. We have pursued two approaches that leverage the SBPI technique.

Assume that we can enumerate the set of reachable states of the Gadara net and calculate the supremal controllable sublanguage solution. This solution corresponds to a partition of the set of reachable states of the Gadara net into *legal* and *illegal* states. It is shown in [27] that this partition can be done using a set of *linear* inequalities on the states; in other words, the set of legal states is linearly separable. In [30], the problem of finding the *minimum* number of linear inequalities for effecting the desired separation of the state space is solved using concepts and techniques borrowed from classification theory. SBPI can then be invoked to obtain control logic that necessitates the minimum number of control places, which is highly correlated to the achievement of requirement (R4). This methodology is referred to as MSCL, for Marking Separation using CLassifiers, in subsequent works.

To avoid the explicit enumeration of the state space of the Gadara net that must be performed to calculate the supremal controllable sublanguage solution, a control logic synthesis technique based on structural analysis of the Gadara net was developed. The general framework of this methodology, called ICOG for Iterative Control Of Gadara nets, is presented in [31], while its customization to the case of programs modeled by Gadara nets is presented in [32] and referred to as ICOG-O therein, since the nets involved remain ordinary throughout the iterations. This approach does not guarantee at the outset that the number of control places will be minimized. However, it leverages the structural properties associated with liveness analysis in Gadara nets from [27] in the context of an iterative scheme that eliminates illegal states by eliminating so-called *resource-induced empty siphons* [20, 27]. ICOG employs siphon analysis, coupled with SBPI, as well as a book-keeping mechanism to ensure convergence in a finite number of iterations. At convergence, a set of control places that separates the set of legal states from the set of illegal states is obtained. ICOG explicitly considers the controllability properties of transitions when synthesizing the control logic, so that no control place has an outgoing arc to an uncontrollable transition. In effect, ICOG computes the supremal controllable sublanguage solution by iterating directly on the Gadara net structure, using the notion of resource-induced empty siphon to capture illegal states.

4 Discussion

The principal focus of the Gadara project so far has been the problem of circular-wait-mutex deadlock in multithreaded software. This is an important problem due to the prevalence of multicore computer architectures. There are numerous other software problems where we believe control engineering techniques from the field of DES hold great promise. These include other types of deadlocks, such as reader-writer deadlock, condition wait/signal deadlock, inter-process deadlock, and other concurrency issues such as race, atomicity violation, and priority inversion. Results on the case of reader-writer deadlocks have recently appeared in [33], while certain types of atomicity violations have been addressed in [17, 18].

A key challenge that is posed by the consideration of reader-writer locks stems from the fact that the underlying state space is not necessarily finite; this is because one can perceive this class of RAS as one where in writing mode, the capacity of the resource is "one," while in reading mode, it is "infinite." This obstacle is addressed in [33] by taking advantage of special structure that exists in the set of inadmissible states, which enables a finite representation of this set through its minimal elements.

Detecting atomicity violations is substantially more difficult than deadlock detection. In [18], a class of atomicity violation bugs called "single-variable atomicity violations" is considered. Gadara nets are employed to capture this class of bugs by control specifications expressed as linear inequalities on the net marking; adjustments to the construction of the Gadara net at modeling time are necessary to make this possible. After adding one monitor place to enforce each linear inequality using SBPI, the ICOG methodology is then directly applied on the resulting controlled Gadara net to eliminate potential deadlocks introduced by these control places. If one is interested in obtaining the minimum-size controller, in terms of number of added control places, then MSCL can be employed, albeit the process is more involved.

5 Conclusion

The application of the control engineering paradigm and of DES techniques to software failure avoidance opens up new avenues of research that cover the gamut from theory to implementation. While some of the above-mentioned opportunities can be solved by existing DES control theory, better customized solutions are often desirable. A crucial issue is scalability, which often necessitates the development of customized algorithms that exploit problem structure. Another crucial issue is the requirement on run-time overhead of the control logic in software applications, which is much more stringent than in other application areas such as manufacturing systems or process control, for instance. This leads to numerous opportunities to advance the state-of-the-art of DES control theory. Collaboration with domain experts is essential to construct suitable models and to understand the implementation constraints of the control logic. We wish to encourage students and researchers to consider contributing to this very promising emerging area of research.

References

1. Hellerstein, J.L., Diao, Y., Parekh, S., Tilbury, D.M.: Feedback Control of Computing Systems. Wiley (2004)
2. Wallace, C., Jensen, P., Soparkar, N.: Supervisory control of workflow scheduling. In: Proc. International Workshop on Advanced Transaction Models and Architectures (1996)
3. Phoha, V.V., Nadgar, A.U., Ray, A., Phoha, S.: Supervisory control of software systems. IEEE Transactions on Computers 53(9), 1187–1199 (2004)
4. Liu, C., Kondratyev, A., Watanabe, Y., Desel, J., Sangiovanni-Vincentelli, A.: Schedulability analysis of Petri nets based on structural properties. In: Proc. International Conference on Application of Concurrency to System Design (2006)
5. Wang, Y., Kelly, T., Lafortune, S.: Discrete control for safe execution of IT automation workflows. In: Proc. ACM EuroSys Conference (2007)
6. Dragert, C., Dingel, J., Rudie, K.: Generation of concurrency control code using discrete-event systems theory. In: Proc. ACM International Symposium on Foundations of Software Engineering (2008)
7. Auer, A., Dingel, J., Rudie, K.: Concurrency control generation for dynamic threads using discrete-event systems. In: Proc. Allerton Conference on Communication, Control and Computing (2009)
8. Gamatie, A., Yu, H., Delaval, G., Rutten, E.: A case study on controller synthesis for data-intensive embedded system. In: Proc. International Conference on Embedded Software and Systems (2009)
9. Iordache, M.V., Antsaklis, P.J.: Concurrent program synthesis based on supervisory control. In: Proc. 2010 American Control Conference, pp. 3378–3383 (2010)

10. Delaval, G., Marchand, H., Rutten, E.: Contracts for modular discrete controller synthesis. In: Proc. ACM Conference on Languages, Compilers and Tools for Embedded Systems (2010)
11. Gadara Team: Gadara project, http://gadara.eecs.umich.edu/
12. Ramadge, P.J., Wonham, W.M.: Supervisory control of a class of discrete event processes. SIAM J. Control Optim. 25(1) (1987)
13. Moody, J.O., Antsaklis, P.J.: Supervisory Control of Discrete Event Systems Using Petri Nets. Kluwer Academic Publishers, Boston (1998)
14. Kelly, T., Wang, Y., Lafortune, S., Mahlke, S.: Eliminating concurrency bugs with control engineering. IEEE Computer 42(12), 52–60 (2009)
15. Wang, Y., Kelly, T., Kudlur, M., Lafortune, S., Mahlke, S.A.: Gadara: Dynamic deadlock avoidance for multithreaded programs. In: Proc. the 8th USENIX Symposium on Operating Systems Design and Implementation, 281–294 (2008)
16. Wang, Y., Lafortune, S., Kelly, T., Kudlur, M., Mahlke, S.: The theory of deadlock avoidance via discrete control. In: Proc. the 36th Annual ACM SIGPLAN-SIGACT Symposium on Principles of Programming Languages, 252–263 (2009)
17. Liu, P., Zhang, C.: Axis: Automatically fixing atomicity violations through solving control constraints. In: International Conference on Software Engineering (2012)
18. Wang, Y., Liu, P., Kelly, T., Lafortune, S., Reveliotis, S., Zhang, C.: On atomicity enforcement in concurrent software via discrete event systems theory. In: Proc. the 51st IEEE Conference and Decision and Control (2012)
19. Li, Z., Zhou, M., Wu, N.: A survey and comparison of Petri net-based deadlock prevention policies for flexible manufacturing systems. IEEE Transactions on Systems, Man, and Cybernetics—Part C 38(2), 173–188 (2008)
20. Reveliotis, S.A.: Real-Time Management of Resource Allocation Systems: A Discrete-Event Systems Approach. Springer, New York (2005)
21. Murata, T.: Petri nets: Properties, analysis and applications. Proceedings of the IEEE 77(4), 541–580 (1989)
22. Kavi, K.M., Moshtaghi, A., Chen, D.: Modeling multithreaded applications using Petri nets. International Journal of Parallel Programming 35(5), 353–371 (2002)
23. Murata, T., Shenker, B., Shatz, S.M.: Detection of Ada static deadlocks using Petri net invariants. IEEE Transactions on Software Engineering 15(3), 314–326 (1989)
24. Iordache, M.V., Antsaklis, P.J.: Petri nets and programming: A survey. In: Proc, American Control Conference, 4994–4999 (2009)
25. Engler, D., Ashcraft, K.: RacerX: Effective, static detection of race conditions and deadlocks. In: Proc. 19th ACM Symposium on Operating Systems Principles (2003)
26. Cho, H.K., Wang, Y., Liao, H., Kelly, T., Lafortune, S., Mahlke, S.: Practical lock/unlock pairing for concurrent programs. In: Proc. 2013 International Symposium on Code Generation and Optimization, CGO 2013 (February 2013)
27. Liao, H., Wang, Y., Cho, H.K., Stanley, J., Kelly, T., Lafortune, S., Mahlke, S., Reveliotis, S.: Concurrency bugs in multithreaded software: Modeling and analysis using Petri nets. Discrete Event Dynamic Systems: Theory & Applications 23 (2013) (published online May 2012)
28. Cano, E.E., Rovetto, C.A., Colom, J.M.: An algorithm to compute the minimal siphons in S^4PR nets. In: Proc. International Workshop on Discrete Event Systems, pp. 18–23 (2010)
29. Cassandras, C.G., Lafortune, S.: Introduction to Discrete Event Systems, 2nd edn. Springer (2008)

30. Nazeem, A., Reveliotis, S., Wang, Y., Lafortune, S.: Designing compact and max-
    imally permissive deadlock avoidance policies for complex resource allocation sys-
    tems through classification theory: The linear case. IEEE Transactions on Auto-
    matic Control 56(8), 1818–1833 (2011)
31. Liao, H., Lafortune, S., Reveliotis, S., Wang, Y., Mahlke, S.: Optimal liveness-
    enforcing control of a class of petri nets arising in multithreaded software. IEEE
    Transactions on Automatic Control 58(5) (2013)
32. Liao, H., Wang, Y., Stanley, J., Lafortune, S., Reveliotis, S., Kelly, T., Mahlke, S.:
    Eliminating concurrency bugs in multithreaded software: A new approach based on
    discrete-event control. IEEE Transactions on Control Systems Technology (2013)
    (published online January 2013)
33. Nazeem, A., Reveliotis, S.: Maximally permissive deadlock avoidance for resource
    allocation systems with r/w-locks. In: Proc. the 11th International Workshop on
    Discrete Event Systems (2012)

Contextual Merged Processes*

César Rodríguez[1], Stefan Schwoon[1], and Victor Khomenko[2]

[1] LSV, ENS Cachan & CNRS, INRIA Saclay, France
[2] School of Computing Science, Newcastle University, U.K.

Abstract. We integrate two compact data structures for representing state spaces of Petri nets: merged processes and contextual prefixes. The resulting data structure, called *contextual merged processes (CMP)*, combines the advantages of the original ones and copes with several important sources of state space explosion: concurrency, sequences of choices, and concurrent read accesses to shared resources. In particular, we demonstrate on a number of benchmarks that CMPs are more compact than either of the original data structures. Moreover, we sketch a polynomial (in the CMP size) encoding into SAT of the model-checking problem for reachability properties.

1   Introduction

Model checking of concurrent systems is an important and practical way of ensuring their correctness. However, the main drawback of model checking is that it suffers from the *state-space explosion (SSE)* problem [23]. That is, even a relatively small system specification can (and often does) yield a very large state space. To alleviate SSE, many model-checking techniques use a condensed representation of the full state space of the system. Among them, a prominent technique are McMillan's Petri net unfoldings (see, e.g. [15,6,11]). They rely on the partial-order view of concurrent computation and represent system states implicitly, using an acyclic *unfolding prefix*.

There are several common sources of SSE. One of them is concurrency, and the unfolding techniques were primarily designed for efficient verification of highly concurrent systems. Indeed, complete prefixes are often exponentially smaller than the corresponding reachability graphs because they represent concurrency directly rather than by multidimensional 'diamonds' as it is done in reachability graphs. For example, if the original Petri net consists of 100 transitions that can fire once in parallel, the reachability graph will be a 100-dimensional hypercube with 2^{100} vertices, whereas the complete prefix will be isomorphic to the net itself. However, unfoldings do not cope well with some other important sources of SSE, and in what follows, we consider two such sources.

One important source of SSE are sequences of choices. For example, the smallest complete prefix of the Petri net in Fig. 1 is exponential in its size since no event can be declared a cutoff — intuitively, each reachable marking 'remembers' its past, and so different runs cannot lead to the same marking.

---

* This research was supported by the EPSRC grant EP/K001698/1 (UNCOVER).

J.-M. Colom and J. Desel (Eds.): PETRI NETS 2013, LNCS 7927, pp. 29–48, 2013.

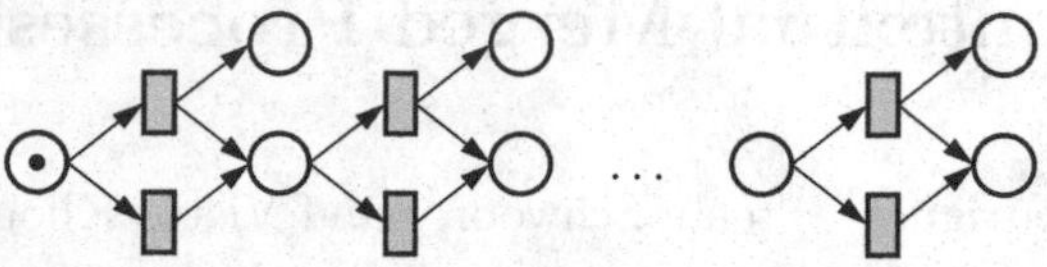

Fig. 1. A Petri net with exponentially large unfolding prefix

Another important source of SSE are concurrent read accesses, that is, multiple actions requiring non-exclusive access to a shared resource. *Contextual nets (c-nets)* are an extension of Petri nets where such read accesses are modelled by a special type of arcs, called *read arcs* and denoted by lines (in contrast to arrows for the traditional consuming and producing arcs). Read arcs allow a transition to check for the presence of a token without consuming it. As concurrent read access to a shared resource is a natural operation in many concurrent systems, c-nets are often the formalism of choice for a wide variety of applications, e.g. to model concurrent database access [18], concurrent constraint programs [16], priorities [9], and asynchronous circuits [24].

The usual way of modelling c-nets using traditional Petri nets is by replacing read arcs by "consume-reproduce loops": a transition consumes a token from a place and immediately puts a token back, see Fig. 2 (a,b). Unfortunately, this makes the unfolding technique inefficient: concurrent transitions of a c-net reading the same place are sequentialised by this encoding, and thus all their interleavings are represented in the unfolding, see Fig. 3 (b). This problem can be mitigated using the place-replication (PR) encoding proposed in [24], which replicates each place that is read by several transitions so that each of them obtains a "private" copy of the place and accesses it using a consume-reproduce loop, see Fig. 2 (a,c). However, the resulting unfolding may still be large, see Fig. 3 (c). Moreover, the PR encoding can significantly increase the sizes of presets of some transitions, considerably slowing down the unfolding algorithm, because (with some reasonable assumptions) the problem of checking if the currently built part of the prefix can be extended by a new instance of a transition t is NP-complete in the prefix size and $|{}^\bullet t|$ [7, Sect. 4.4].

Recently, techniques addressing these sources of SSE emerged. In [14], a new condensed representation of Petri net behaviour called *merged processes (MPs)* was proposed, which copes not only with concurrency, but also with sequences of choices. Moreover, this representation is sufficiently similar to the traditional unfoldings so that a large body of results developed for unfoldings can be re-used. The main idea behind MPs is to fuse some nodes in the unfolding prefix, and use the resulting net as the basis for verification. For example, the unfolding of the net shown in Fig. 1 will collapse back to the original net after the fusion. It turns out that for a safe Petri net, model checking of a reachability-like property (i.e. the existence of a reachable state satisfying a predicate given by a Boolean expression) can be efficiently performed on its MP, and [14] provides a polynomial reduction of this problem to SAT. Furthermore, an efficient *unravelling algorithm* that builds a complete MP of a given safe PN has been proposed in [12].

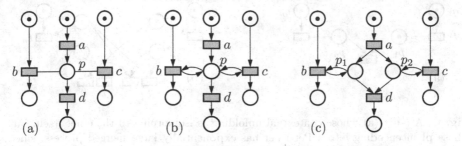

Fig. 2. (a) A c-net; (b) its *plain encoding*; (c) and its *place-replication encoding*

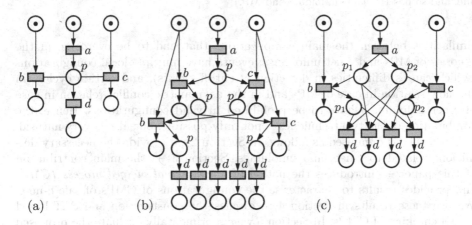

Fig. 3. Unfoldings of (a), (b), and (c) in Fig. 2

The experimental results in [14] indicate that this method is quite practical. Unfortunately, MPs do not cope well with read arcs, as illustrated in Fig. 4.

An extension of the unfolding technique to c-nets was proposed in [3,24], and a practical unfolding algorithm and SAT-based model checking for reachability-like properties have been developed in [1,20]. The idea is to allow read arcs also in the unfolding, which allows for significant compression in some cases — see Fig. 3(a). The experimental results in [1,20] demonstrate that the performance of this method is comparable to the traditional unfoldings when c-nets have no read arcs (i.e. can be directly interpreted as Petri nets), and can be much better (in terms of both the runtime and the size of the generated prefix) than traditional unfolding of plain and PR encodings of c-nets with many read arcs. Unfortunately, this method does not cope with SSE resulting from sequences of choices, e.g. it does not offer any improvement for the Petri net in Fig. 1, as it contains no read arcs.

In this paper we observe that the described techniques for compressing the unfolding prefix are in fact orthogonal, and can be combined into one that copes with all the mentioned sources of SSE, viz. concurrency, sequences of choices and concurrent read accesses to a shared resource. Moreover, there are striking

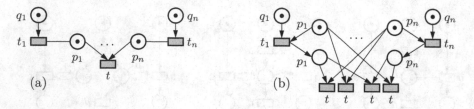

Fig. 4. A c-net (a) whose contextual unfolding is isomorphic to the c-net itself, but whose plain encoding into a Petri net has exponentially large merged process, since no place instances in its unfolding (b) can be merged, and so there are 2^n mp-events corresponding to transition t. (For this c-net the PR encoding coincides with the plain one, and so has the same unfolding and MP.)

similarities between the main complications that had to be overcome in the theories of MPs and c-net unfoldings: events have multiple local configurations (which causes difficulties in detection of cutoff events), and certain cycles (in the flow relation in case of MPs and in the asymmetric conflict relation in case of c-net unfoldings) have to be prohibited in valid configurations. Hence, the combination of the two techniques is not only possible, but also is very natural.

The paper is organised as follows. In Section 2 we provide the necessary definitions related to c-nets and unfoldings. Section 3 — the main contribution of this paper — introduces the notion of a *contextual merged process (CMP)* and provides results to characterise the configurations of CMPs of safe c-nets. We use these results in Section 4 to discuss the construction and SAT-based model checking of CMPs. In Section 5 we experimentally evaluate the proposed approach on a number of benchmark examples. In Section 6 we conclude and outline the directions for future research.

A longer version of this paper, including proofs, is available at [22].

2   Basic Notions

In this section, we set our basic definitions and recall previous results (see [2,21]).

A *multiset* over a set S is a function $M\colon S \to \mathbb{N}$. The *support* of M is the set $\bar{M} := \{x \in S \mid M(x) > 0\}$ of elements in S occurring at least once in M. We write $x \in M$ if x is in the support of M. We say that M is finite iff its support is. Given multisets M and N over S, their sum and difference are $(M + N)(x) := M(x) + N(x)$ and $(M - N)(x) := \max(0, M(x) - N(x))$. We write $M \leq N$ iff $M(x) \leq N(x)$ for all $x \in S$. Any function $f\colon S \to T$ can be lifted to multisets by letting $f(M)(x) := \sum_{y \in f^{-1}(x)} M(y)$; note that this sum is well-defined iff finitely many of its summands are non-zero, which is always the case if, for instance, M has a finite support. Any set can be interpreted as a multiset in the natural way.

A *contextual net* (c-net) is a tuple $N = \langle P, T, F, C, m_0 \rangle$, where P and T are disjoint sets of *places* and *transitions*, $F \subseteq (P \times T) \cup (T \times P)$ is the *flow relation*, $C \subseteq P \times T$ is the *context relation*, and the *initial marking* m_0 is a multiset over

P. A pair $(p, t) \in C$ is called *read arc*. A *Petri net* is a c-net without read arcs. N is called *finite* if P and T are finite sets. Places and transitions together are called *nodes*. Fig. 2 (a) depicts a c-net, where read arcs are drawn as undirected lines, e.g. between p and c.

For $x \in P \cup T$, ${}^\bullet x := \{ y \in P \cup T \mid (y, x) \in F \}$ is the *preset* of x, $x^\bullet := \{ y \in P \cup T \mid (x, y) \in F \}$ is the *postset* of x, and $\underline{x} := \{ y \in P \cup T \mid (y, x) \in C \cup C^{-1} \}$ is the *context* of x. We assume that for each node $x \in P \cup T$ the sets ${}^\bullet x$, $x^\bullet$, and $\underline{x}$ are pairwise disjoint.

A *marking* of N is a multiset m over P. A transition t is *enabled* at m if $m(p) \geq 1$ for all $p \in \underline{t} \cup {}^\bullet t$. Such t can *fire*, leading to the well-defined marking $m' := m - {}^\bullet t + t^\bullet$. The tuple $\langle m, t, m' \rangle$ is called a *step*. A marking m is *reachable* if it can be obtained by a finite sequence of firings starting at m_0. N is k-*bounded* if $m(p) \leq k$ for all reachable m and all $p \in P$, and *safe* if it is 1-bounded. For safe nets, we treat markings as sets of places.

Two distinct transitions t and t' are in *symmetric conflict*, denoted $t \# t'$, if ${}^\bullet t \cap {}^\bullet t' \neq \emptyset$, and in *asymmetric conflict*, written $t \nearrow t'$, if (i) $t^\bullet \cap ({}^\bullet t' \cup \underline{t'}) \neq \emptyset$, or (ii) $\underline{t} \cap {}^\bullet t' \neq \emptyset$, or (iii) $t \# t'$. Intuitively, when $t \nearrow t'$, then if both t, t' fire in a run, t fires before t'. Note that t and t' may not fire together in any run, e.g. if $t \# t'$, where we have $t \nearrow t'$ *and* $t' \nearrow t$ — corresponding to the intuition that t has to fire before t' and vice versa. In Fig. 6 (b) we have $e_3 \nearrow e_5$ due to (i); in Fig. 5 we have $e_1 \nearrow e_2$ due to (ii). For a set of transitions $X \subseteq T$, we write $\nearrow_X$ to denote the relation $\nearrow \cap (X \times X)$.

Let $N' = \langle P', T', F', C', m'_0 \rangle$ be a c-net. A *homomorphism* [24] from N to N' is a function $h \colon P \cup T \to P' \cup T'$ satisfying: $h(P) \subseteq P'$, $h(T) \subseteq T'$, $h(m_0) = m'_0$, and h restricted to ${}^\bullet t$, $t^\bullet$, $\underline{t}$ for all $t \in T$ is a bijection to ${}^\bullet h(t)$, $h(t)^\bullet$ and $\underline{h(t)}$, respectively. Such a homomorphism is a specialisation of Definition 4.20 in [3].

For two nodes x and y we write $x <_i y$ if either $(x, y) \in F$ or $x, y \in T$ and $x^\bullet \cap \underline{y} \neq \emptyset$. We write $<$ for the transitive closure of $<_i$, and $\leq$ for the reflexive closure of $<$. For a node x, we define its set of *causes* as $[x] := \{ t \in T \mid t \leq x \}$. A set $X \subseteq T$ is *causally closed* if $[t] \subseteq X$ for all $t \in X$.

An *occurrence net* is a c-net $O = \langle B, E, G, D, \tilde{m}_0 \rangle$ if (i) O is safe and for any $b \in B$, we have $|{}^\bullet b| \leq 1$; (ii) $<$ is a strict partial order for O; (iii) for all $e \in E$, $[e]$ is finite and $\nearrow_{[e]}$ acyclic; (iv) $\tilde{m}_0 = \{ b \in B \mid {}^\bullet b = \emptyset \}$. As per tradition, we call the elements of B *conditions*, and those of E *events*. A *configuration* of O is a finite, causally closed set of events $\mathcal{C}$ such that $\nearrow_\mathcal{C}$ is acyclic; $Conf(O)$ denotes the set of all configurations. For a configuration $\mathcal{C}$, let $cut(\mathcal{C}) := (\tilde{m}_0 \cup \mathcal{C}^\bullet) \setminus {}^\bullet \mathcal{C}$. A *prefix* of O is a c-net $\mathcal{P} = \langle B', E', G', D', \tilde{m}_0 \rangle$ such that $E' \subseteq E$ is causally closed, $B' = \tilde{m}_0 \cup E'^\bullet$, and G' and D' are the restrictions of G and D to $B' \cup E'$; in such a case we write $\mathcal{P} \sqsubseteq O$.

Fig. 5 shows an occurrence net illustrating why it is necessary to restrict configurations to sets without cycles in $\nearrow$. There are three events, and each pair of them can fire, but not all three. Indeed, $e_1 \nearrow e_2 \nearrow e_3 \nearrow e_1$ is a cycle of asymmetric conflicts.

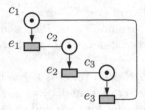

Fig. 5. An occurrence net illustrating circular asymmetric conflict

A *branching process* of N is a pair $\mathcal{P} = \langle O, h \rangle$, where O is an occurrence net and h is a homomorphism from O to N with the property that $h(e) = h(e') \wedge {}^\bullet e = {}^\bullet e' \wedge \underline{e} = \underline{e}'$ implies $e = e'$ for all events $e, e' \in O$. For every N, there is a unique (up to isomorphism) maximal (wrt. $\sqsubseteq$) branching process $\mathcal{U}_N = \langle U, h' \rangle$ that we call the *unfolding* of N [2]. Thus, any branching processes $\langle O, h \rangle$ is characterised by a prefix O of U and the restriction h of h' to the elements of O. For convenience, we shall often equate a branching process with its underlying net and call it an *unfolding prefix*. As usual, for $\mathcal{C} \in Conf(\mathcal{U}_N)$, we define $mark(\mathcal{C}) := h(cut(\mathcal{C}))$.

An unfolding prefix $\mathcal{P}$ is called *marking-complete* if for any marking m reachable in N there exists a marking $\tilde{m}$ reachable in $\mathcal{P}$ with $h(\tilde{m}) = m$. For example, $\mathcal{U}_N$ is marking-complete but in general infinite. For bounded N, it is however possible to compute a finite marking-complete prefix $\mathcal{P}_N$ [2, 21].

The key notion in computing marking-complete prefixes is a *history*. Given a configuration $\mathcal{C} \in Conf(U)$ and some event $e \in \mathcal{C}$, the *history of e in $\mathcal{C}$* is defined as $\mathcal{C}\llbracket e \rrbracket := \{ e' \in \mathcal{C} \mid e' \nearrow_{\mathcal{C}}^* e \}$. For $e \in E$, $Hist(e) := \{ \mathcal{C}\llbracket e \rrbracket \mid \mathcal{C} \in Conf(U) \}$ is the set of *all histories* of e. The construction of a complete prefix discovers events that do not contribute to reaching new markings of N in the prefix: an event e is declared *cutoff* if for every history H of e there exists a configuration $\mathcal{C}$ in $\mathcal{P}$ such that $mark(\mathcal{C}) = mark(H)$ and $\mathcal{C} \prec H$, where $\prec$ is a so-called *adequate order* on configurations.[1] The construction then excludes events that are causal successors of e, thereby ensuring the finiteness of $\mathcal{P}$ while guaranteeing its marking-completeness: for every reachable marking m of N there is a configuration $\mathcal{C}$ of $\mathcal{P}$ such that $mark(\mathcal{C}) = m$ and $\mathcal{C}$ does not include any cutoffs.

3   Contextual Merged Processes

In this section, we introduce the notion of *contextual merged processes* (CMP) and discuss some of their properties. These results generalise those of [14], in particular it turns out that the notions of mp-configuration, defined in [14] for Petri

---

[1] Actually, [2] defines pairs $\langle e, H \rangle$ as cutoffs; above, we chose an equivalent presentation that will be more convenient for defining CMPs. Also, only histories are considered for $\mathcal{C}$ in [2]; we come back to this point in Section 4.

nets, and the notion of a c-net configuration from [2], both of which introduce acyclicity constraints, can be seamlessly integrated into a common framework.

We first show (see [22] for the proofs of all results) that asymmetric conflict, causality, and steps are, among other notions, preserved by homomorphisms.

Lemma 1. *Let N and N' be c-nets, and h be a homomorphism from N to N'. If $\langle m, t, \widehat{m} \rangle$ is a step of N and $h(m)$ is well-defined,[2] then $\langle h(m), h(t), h(\widehat{m}) \rangle$ is a step of N'. Furthermore, for any nodes x, y and transitions t, u of N, $x < y$ implies $h(x) < h(y)$ and $t \nearrow u$ implies $h(t) \nearrow h(u)$.*

As usual, homomorphisms preserve runs and reachable markings: if σ is a run of N that reaches m, then $h(\sigma)$ is a run of N' that reaches $h(m)$, because $h(m_0) = m_0'$ is a well-defined marking and due to Lemma 1.

The first step to define CMPs is the notion of occurrence depth.

Definition 1 (occurrence depth). *Let x be a node of a branching process $\langle O, h \rangle$. The* occurrence depth *of x, denoted $\mathrm{od}(x)$, is the maximum number of $h(x)$-labelled nodes in any path in the directed graph $(\widetilde{m}_0 \cup [x] \cup [x]^{\bullet}, <_i)$ starting at any initial condition and ending in x.*

Recall that the cone $[x]$ is finite and $<_i$ is a partial order, so there is only a finite number of paths to evaluate, and the definition is well-given.

A CMP is obtained from a branching process in two steps. First, all conditions that have the same label and occurrence depth are fused together (their initial markings are totalled); then all events that have the same label and environment (after fusing conditions) are merged. Conditions in the initial marking will have, by definition, occurrence depth 1. If n of them share the same label, they will be fused together, and the resulting condition will be initially marked with n tokens. This is formalised as follows:

Definition 2 (contextual merged process). *Let $N = \langle P, T, F, C, m_0 \rangle$ be a c-net and $\mathcal{P} = \langle \langle B, E, G, D, \widetilde{m}_0 \rangle, h \rangle$ be a branching process of N. Define a net $\mathcal{Q} = \langle \widehat{B}, \widehat{E}, \widehat{G}, \widehat{D}, \widehat{m}_0 \rangle$, where $\widehat{B} \subseteq P \times \mathbb{N}$, $\widehat{E} \subseteq T \times 2^{\widehat{B}} \times 2^{\widehat{B}} \times 2^{\widehat{B}}$, and a homomorphism $\hbar$ from $\mathcal{P}$ to $\mathcal{Q}$ as follows:*

- *for $b \in B$, $\hbar(b) := \langle h(b), \mathrm{od}(b) \rangle$; set $\widehat{B} := \hbar(B)$;*
- *for $e \in E$, $\hbar(e) := \langle h(e), \hbar(^{\bullet}e), \hbar(\underline{e}), \hbar(e^{\bullet}) \rangle$; set $\widehat{E} := \hbar(E)$;*
- *$\widehat{G}, \widehat{D}$ are such that for every $\widehat{e} = \langle t, Pre, Cont, Post \rangle \in \widehat{E}$ we have $^{\bullet}\widehat{e} := Pre$, $\underline{\widehat{e}} := Cont$, $\widehat{e}^{\bullet} := Post$;*
- *$\widehat{m}_0(\langle p, d \rangle) := |\widetilde{m}_0 \cap \{ b \in B \colon h(b) = p,\ \mathrm{od}(b) = d \}|$.*

Moreover, let $\widehat{h}$ be the homomorphism from $\mathcal{Q}$ to N given by projecting the nodes of $\mathcal{Q}$ to their first components. We call $\mathfrak{Merge}(\mathcal{P}) := \langle \mathcal{Q}, \widehat{h} \rangle$ the merged process of $\mathcal{P}$. The merged process $\mathcal{M}_N := \mathfrak{Merge}(\mathcal{U}_N)$ of the unfolding of N is called the unravelling *of N.*

---

[2] That is, $h(m)$ is a well-defined multiset.

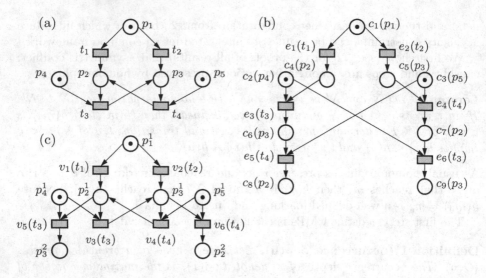

Fig. 6. (a) A net; (b) its unfolding; (c) its unravelling

Fig. 6 shows a 1-safe net (taken from [14]), its unfolding, and its unravelling. For the rest of this section, let $N = \langle P, T, F, C, m_0 \rangle$ be a bounded c-net, $\mathcal{U}_N$ be its unfolding, $\mathcal{P}_N = \langle \langle B, E, G, D, \tilde{m}_0 \rangle, h \rangle$ be a branching process of N, and $\mathcal{Q}_N = \langle \langle \widehat{E}, \widehat{B}, \widehat{G}, \widehat{D}, \tilde{m}_0 \rangle, \widehat{h} \rangle$ be the corresponding merged process, i.e. $\mathfrak{Merge}(\mathcal{P}_N)$. The places of $\mathcal{Q}_N$ are called *mp-conditions* and its transitions *mp-events*. We shall write p^d for an mp-condition $\langle p, d \rangle$. Note that $\widehat{m}_0(p^d)$ equals $m_0(p)$ if $d = 1$ and is 0 otherwise. An mp-event $\widehat{e}$ is an *mp-cutoff* if all events in $\widehat{h}^{-1}(\widehat{e})$ are cutoffs. We denote these mp-events by $\widehat{E}_{\mathrm{cut}}$.

We call a run $t_1 t_2 \ldots$ of a c-net *repetition-free* if no transition occurs more than once in it. Some properties of contextual merged processes follow.

Remark 1. The following properties hold for CMPs or c-net unfoldings:

1. In general, $\mathcal{M}_N$ is not acyclic; see Fig. 6 (c).
2. There can be mp-events consuming conditions in the postset of an mp-cutoff.
3. There is at most one mp-condition p^k resulting from fusing occurrences of place p at depth $k \geq 1$.
4. For two mp-conditions p^k and p^{k+1}, there is a directed path in the $<_i$ relation from the former to the latter.
5. Two different conditions c_1 and c_2 having the same label and occurrence depth are not causally related. Hence, if the original c-net is safe, then $\nearrow_{[c_1] \cup [c_2]}$ contains a cycle.
6. $h = \widehat{h} \circ \hbar$.
7. $\hbar$ and $\widehat{h}$ are homomorphisms.
8. A sequence of transitions σ is a run of N iff there exists a run $\widehat{\sigma}$ of $\mathcal{M}_N$ such that $\sigma = \widehat{h}(\widehat{\sigma})$.

Additionally, if N is safe, we have:

9. $\hbar$ is injective when restricted to the events of a configuration.
10. Property 8 is true if we additionally require $\widehat{\sigma}$ to be repetition-free.

Note that Property 9 is still true when $\hbar$ is restricted to the elements of $\widetilde{m}_0 \cup \mathcal{C} \cup \mathcal{C}^\bullet$. Indeed, $\hbar$ is bijective when restricted to $\widetilde{m}_0$, because $\widehat{m}_0$ is safe, and two conditions $c, c' \in \mathcal{C}^\bullet$ cannot be merged because $\nearrow_{[c] \cup [c']}$ would have cycles and $[c] \cup [c'] \subseteq \mathcal{C}$.

Definition 3 (mp-configuration). *A multiset of mp-events $\widehat{C}$ is an mp-configuration of $\mathcal{Q}_N$ if there exists a configuration $\mathcal{C}$ of $\mathcal{U}_N$ verifying $\hbar(\mathcal{C}) = \widehat{C}$.*

As it is the case for configurations of branching processes, any mp-configuration of a merged process represents a (concurrent) run of its mp-events, i.e. there exists at least one linear ordering of the mp-events of $\widehat{C}$ that is a run of the merged process. This is because the same is true for configurations of the associated branching process and because $\hbar$ is a homomorphism.

Every finite firing sequence of $\mathcal{U}_N$ consists of a set of events that form a configuration $\mathcal{C}$, which, due to Definition 3, corresponds to an mp-configuration $\widehat{C}$ of $\mathcal{M}_N$. However, the inverse statement is not true: a firing sequence of $\mathcal{M}_N$ may consist of a multiset of events X that is not an mp-configuration since no $\mathcal{C} \in Conf(\mathcal{U}_N)$ satisfies $\hbar(\mathcal{C}) = X$. This already holds for nets without read arcs, as the example in Fig. 6 shows: $v_1 v_5$ is a valid firing sequence of $\mathcal{M}_N$ corresponding to events e_1 and e_6 of $\mathcal{U}_N$ (i.e. $\hbar(e_1) = v_1$ and $\hbar(e_6) = v_5$) which do not form a configuration. However, $\widehat{h}$ applied to $v_1 v_5$ still gives a valid firing sequence $t_1 t_3$ of N thanks to Remark 1 (8). Below we formalise these observations.

Definition 4 (marking-complete CMP). *Let X be a finite multiset of mp-events. The* cut *and* marking *of X are respectively defined as the multisets*

$$cut(X) := (\widehat{m}_0 + X^\bullet) - {}^\bullet X \qquad and \qquad mark(X) := \widehat{h}(cut(X)).$$

We call $\mathcal{Q}_N$ marking-complete *if for each reachable marking m of N there exists a cutoff-free mp-configuration $\widehat{C}$ in $\mathcal{Q}_N$ satisfying $mark(\widehat{C}) = m$.*

The intuition behind these definitions is as follows. If X is the multiset of mp-events associated to a finite run (i.e. the multiset M such that $M(\widehat{e}) = n$ if $\widehat{e}$ fires n times) then $cut(X)$ is the marking reached by this run in the CMP, and $mark(X)$ is the $\widehat{h}$-image of $cut(X)$, i.e. the corresponding marking of N.

Observe that in the definition of a marking-complete CMP, one could ask for a finite run (rather than a configuration) that reaches a marking m. The resulting definition would be equivalent, but we preferred the current variant because it (i) mimics the analogous definition for unfoldings and (ii) avoids some unpleasant properties of runs: e.g. finite CMPs can have infinite runs and therefore infinitely many finite runs, which is impossible for configurations.

We now focus on the practically relevant class of safe c-nets. Here, the mapping $\hbar$ lifted to configurations establishes an injective correspondence between the configurations of the unfolding and the mp-configurations of the unravelling. For each mp-configuration $\widehat{C}$ there exists a *unique* configuration C such that $\widehat{C} = \hbar(C)$.

We give, for safe nets, characterisations of sets of mp-events that correspond to reachable markings of N (Proposition 1) and to configurations of $\mathcal{U}_N$ (Proposition 2). They serve to aid CMP-based model-checking, as well as the direct construction of CMPs, see Section 4. We note that the problem of generalising these approaches to bounded, but not safe, nets is still open even for merged processes without read arcs [14].

Proposition 1. *Let $\mathcal{Q}_N$ be a marking-complete CMP of a safe c-net N. Then a marking m is reachable in N iff there exists a cutoff-free set X of mp-events of $\mathcal{Q}_N$ satisfying:*

1. $\forall \widehat{e} \in X : \forall \widehat{c} \in {}^{\bullet}\widehat{e} \cup \underline{\widehat{e}} : (\widehat{c} \in \widehat{m}_0 \vee \exists \widehat{e}' \in {}^{\bullet}\widehat{c} : \widehat{e}' \in X)$, and
2. $\nearrow_X$ is acyclic, and
3. $m = mark(X)$.

Note that the conditions in Proposition 1 do not ensure that X is an mp-configuration; however, they do guarantee that X corresponds to a repetition-free run of $\mathcal{Q}_N$, and thus are sufficient to check reachability (see the comment before Definition 4 for an example). Finally, observe that not every repetition-free run satisfies the first two conditions of Proposition 1: $v_1 v_3 v_4$ is a repetition-free run of Fig. 6 but $\{v_1, v_3, v_4\}$ violates the second condition. This means that Proposition 1 characterizes a strict subset of repetition-free runs that are enough for representing *all* reachable markings of N.

Proposition 2. *If N is safe, a set of mp-events $\widehat{C}$ is an mp-configuration of $\mathcal{Q}_N$ iff it satisfies the following conditions:*

1. $\forall \widehat{e} \in \widehat{C} : \forall \widehat{c} \in {}^{\bullet}\widehat{e} \cup \underline{\widehat{e}} : (\widehat{c} \in \widehat{m}_0 \vee \exists \widehat{e}' \in {}^{\bullet}\widehat{c} : \widehat{e}' \in \widehat{C})$, and
2. $\nearrow_{\widehat{C}}$ is acyclic, and
3. for $k \geq 1$, $p^{k+1} \in \widehat{C}^{\bullet}$ implies $p^k \in \widehat{m}_0 \cup \widehat{C}^{\bullet}$ and there exists a path in the directed graph $(\widehat{m}_0 \cup \widehat{C} \cup \widehat{C}^{\bullet}, <_i)$ between p^k and p^{k+1}.

A key detail in both results is that acyclicity of $\nearrow$ prohibits, at the same time, asymmetric conflicts inherent to c-net unfoldings (Fig. 5) and cycles in the flow relation introduced by merging (Fig. 6 (c)).

4   Computing and Analysing Complete CMPs

In this section, we discuss various algorithmic aspects of CMPs, in particular how to construct a complete CMPs from a given *safe* Petri net N, and how to use the resulting CMP to check properties of N.

4.1   CMP Construction

Recall that a marking-complete CMP is one in which every reachable marking m of N is the image (through $\widehat{h}$) of the cut of some cutoff-free mp-configuration. We wish to construct such a CMP in order to analyse properties of N such as reachability or deadlock.

Indirect Methods. It follows from Section 3 that one can achieve this goal by (i) constructing a marking-complete unfolding prefix $\mathcal{P}$ and (ii) applying the construction from Definition 2 to $\mathcal{P}$. Available options for step (i) are:

1. Directly construct $\mathcal{P}$ from N. This approach is implemented in the tool CUNF [19], which is based on the results from [21].
2. Replace all read arcs by consume-produce loops (cf. Fig. 2 (b)) and unfold the resulting Petri net using, e.g., the tool PUNF [13], obtaining some complete prefix $\mathcal{P}'$. We then apply a "folding" operation to $\mathcal{P}'$ in which we repeatedly carry out the following steps: (i) all conditions that were created due to a consume-produce loop are merged and their flow arcs replaced by a read arc; (ii) all events with the same label and the same preset after (i) are merged, and so are their postsets. The resulting c-net prefix $\mathcal{P}$ has the same reachable markings as $\mathcal{P}'$ and is therefore marking-complete. Indeed, applying this operation to the prefix in Fig. 3 (b), which is the unfolding of Fig. 2 (b), would yield the c-net unfolding from Fig. 3 (a).
3. A similar approach as before, but using the place-replication (PR) encoding and adapting the folding operation accordingly (see Fig. 2 (c) and Fig. 3 (c)).

While the first approach is usually more efficient than the others [21], certain aspects of the currently available tool support make options 2 and 3 interesting for the purposes of comparing the resulting CMP sizes. For instance, as pointed out in Footnote 1, CUNF declares a pair $\langle e, H \rangle$, where H is a history of e, a cutoff if it finds another pair $\langle e', H' \rangle$ with $mark(H') = mark(H)$ and $H' \prec H$; this was motivated by the approach from Petri net unfolding [6], where an event is declared cutoff if its local configuration leads to the same marking as the local configuration of another event. However, PUNF implements an approach for Petri nets in which more general configurations are considered for the role of H' [8], leading to smaller unfolding sizes.

Direct Method. Another option is to construct a CMP directly from the c-net N. A similar approach for nets without read arcs was presented in [12]. No such implementation currently exists for CMPs; in the following we describe some key elements that are required for extending [12] to CMPs.

A procedure for direct CMP construction would start with a CMP containing mp-conditions that represent the initial marking of N and extend it one mp-event at a time. To know whether the current CMP $\mathcal{Q}$ can be extended by an mp-event $\widehat{e}$, one has to identify an mp-configuration $\widehat{\mathcal{C}}$ of $\mathcal{Q}$ and check (i) whether $\widehat{\mathcal{C}} \cup \{\widehat{e}\}$ is an mp-configuration of $\mathcal{M}_N$ and (ii) whether $\widehat{e}$ constitutes a cutoff.

Problem (i) can be formulated as a variant of the model-checking algorithm based on Proposition 2 that can be encoded in SAT, see Section 4.2. For (ii), observe that an mp-configuration $\widehat{H}$ corresponds to some history H of an event e of $\mathcal{U}_N$ with $\hbar(e) = \widehat{e}$ iff $\widehat{e}$ is the maximal element of the relation $\nearrow_{\widehat{H}}$. The problem then corresponds to asking whether for all such $\widehat{H}$ there exists another mp-configuration $\widehat{C}$ such that $mark(\widehat{C}) = mark(\widehat{H})$ and $\widehat{C} \prec \widehat{H}$. For $\widehat{C}, \widehat{H}$ in $\mathcal{Q}$, this problem can be encoded in 2QBF, which is more complicated than SAT but less so than QBF in general, and for which specialised solutions exist [17].

However, as $\mathcal{Q}$ grows, the number of possible candidates for $\widehat{H}$ may increase. In general $\widehat{e}$ cannot be designated a cutoff until the construction has been terminated, instead the possibility of adding $\widehat{e}$ may have to be re-checked periodically.

To summarise, the basic structure of the algorithm from [12] would remain unchanged, however one needs to use the characterisation 2 of mp-configurations rather than the non-contextual one in [12].

4.2   Model Checking CMPs

Let $\mathcal{Q}$ be a CMP. We briefly discuss a possible encoding for runs and mp-configurations of $\mathcal{Q}$ into SAT, using Propositions 1 and 2. Note that [14] discusses the corresponding problems for non-contextual MPs and [20] for contextual unfoldings. Remarkably, both problems require to encode acyclicity for different purposes, which are united into a single acyclicity constraint in our case.

Proposition 1 says that every reachable marking m of N is represented by some $\nearrow$-acyclic run X of $\mathcal{Q}$. Reachability of m reduces, then, to the satisfiability of a SAT formula that has variables c, e, p for mp-conditions, mp-events, and places, respectively, such that e is true iff $\widehat{e} \in X$, c iff $\widehat{c} \in cut(X)$ and p iff $p \in mark(X)$.

Condition 1 of Proposition 1 demands that every event needs a causal predecessor for all non-initial mp-conditions in its preset or context. Condition 3 imposes that the variables for mp-conditions and places be correctly related and that the place variables correspond to m. Both these conditions can easily be encoded in linear size wrt. $|\mathcal{Q}|$. For the acyclicity constraint (Condition 2) there are multiple encodings of polynomial size. We refer the reader to [14, 20], where such encodings are discussed and experimentally evaluated.

Proposition 2, used for constructing CMPs, differs from Proposition 1 in having a more restrictive third condition. This constraint and its encoding is very similar to the "no-gap" constraint from [14] to which we refer the reader for details.

5   Experiments and Case Studies

In this section, we experimentally[3] compare the sizes of CMPs, MPs, and unfoldings for a number of families of c-nets. In Section 5.1, we discuss an artificial

---

[3] All the benchmarks and tools referenced in this section are publicly available from http://www.lsv.ens-cachan.fr/~rodriguez/experiments/pn2013/

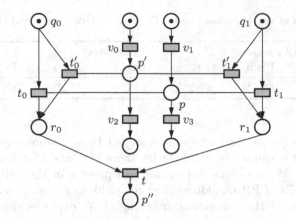

Fig. 7. The c-net 2-GEN

family of examples that allows one to study the effects of read arcs and choice for the various methods in isolation. Section 5.2 presents a case study on Dijkstra's mutual exclusion protocol. Finally, Section 5.3 shows how our methods behave on assorted practical benchmarks.

5.1   Interplay between Read-Arcs and Choice

We study a family of c-net examples called n-GEN, shown for $n = 2$ in Fig. 7. The net represents n processes that concurrently generate resources r_i. Once all r_is are produced, an action t consumes them all. Resource r_i can be produced if one of two conditions is fulfilled, symbolised by transitions t_i or t'_i. Thus, t_i, t'_i share context with transitions t_j and t'_j, respectively, whenever $j \neq i$.

For some $n \geq 1$, let N_c be the c-net n-GEN, N_p its plain encoding, and N_r its PR encoding. The unfoldings of the three nets and the MPs of N_p and N_r blow up due to at least one of the following reasons, which we explain in the sequel: (1) choices between t_i and t'_i or (2) sequentialised read access to p and p'.

For (1), notice that process i can produce r_i in two different ways. At least two occurrences of each r_i are thus present in the unfolding of any of the three nets. Hence there are at least 2^n ways of choosing t's preset, i.e. at least 2^n occurrences of t and p'' in any of the three unfoldings.

Roughly speaking, (2) refers to the same phenomena that were demonstrated in Fig. 2 and Fig. 3. While all t_i are concurrent in N_c, they are sequentialised in N_p: they all consume and produce the same p. This creates conflicts between them, and as a result all their exponentially many interleavings are explicitly present in $\mathcal{U}_{N_p}$. Importantly, any occurrence of t_i that consumes an occurrence of p at depth d, produces an occurrence of p at depth $d + 1$.

In N_r, even if all t_i are still concurrent to each other, their occurrences produce two conditions with occurrence depths 1 and 2, each labelled by their respective private copy of p. For $\mathcal{U}_{N_r}$, this again has the consequence of producing 2^n ways

Table 1. Growth of the unfoldings and MPs of the n-GEN c-nets and their encodings

Merged Processes			Unfoldings		
Ctx	Plain	PR	Ctx	Plain	PR
$\mathcal{O}(n)$	$\mathcal{O}(n^2)$	$\mathcal{O}(2^n)$	$\mathcal{O}(2^n)$	$\mathcal{O}(2^n)$	$\mathcal{O}(2^n)$

of choosing v_3's preset, and 2^n events labelled by v_3. More importantly, the private copies of t_i cannot be merged with those of t_j and they remain in $\mathcal{Q}_{N_r}$. As a result, all 2^n occurrences of v_3 are also present in the MP of N_r. This suggests that MPs of PR unfoldings may not yield, in general, much gain.

While the size of the contextual unfolding of N_c explodes due to (1), it is unaffected by (2). On the other hand, the MP of N_p effectively deals with (1), but only partially with (2). We now see why. Notice that there are $\mathcal{O}(2^n)$ conditions labelled by p in $\mathcal{U}_{N_p}$, all with occurrence depths between 1 and $n+1$. In the MP, they are merged into the $n+1$ mp-conditions $p^1, \ldots, p^{n+1}$. Since all instances of q_i and r_i have occurrence-depth 1, all the exponentially many events labelled by t_i are merged into n mp-events, each consuming some p^j and producing p^{j+1}, for $1 \le j \le n$. This yields an MP of size $\mathcal{O}(n^2)$.

Finally, the CMP of N_c deals effectively with both (1) and (2); it is, in fact, isomorphic to N. Roughly speaking, this is because the unfolding of N_c already deals with (2), as we said, and the '*merging*' solves (1). Thus, the CMP is polynomially more compact than the MP of N_p and exponentially more than the MP of N_r, or the unfoldings of N_c, N_p, or N_r. See Table 1 for a summary.

While this example in itself is artificial, the underlying structures are quite simple and commonly occur in more complex c-nets, which explains some of the experimental results below.

5.2   Dijkstra's Mutual Exclusion Algorithm

In this section we analyse the performance of CMPs on a well-known concurrent algorithm for mutual exclusion due to Dijkstra [5]. What follows is a condensed technical explanation of the algorithm, see [5] for more details.

Dijkstra's algorithm allows n threads to ensure that no two of them are simultaneously in a critical section. Two Boolean arrays b and c of size n, and one integer variable k, satisfying $1 \le k \le n$, are employed. All the entries of both arrays are initialised to *true*, and k's initial value is irrelevant. All threads use the same algorithm, which runs in two phases. During the first, thread i sets $b[i] := false$, and repeatedly checks the value of $b[k]$, setting $k := i$ if $b[k]$ is true, until $k = i$ holds. At this point, thread i starts phase 2, where it sets $c[i] := false$, and enters the critical section if $c[j]$ holds for all $j \ne i$. If the check fails, it sets $c[i] := true$ and restarts in phase 1. After the critical section, $b[i]$ and $c[i]$ are set to *true*. Note that more than one thread could pass phase 1, and phase 2 is thus necessary.

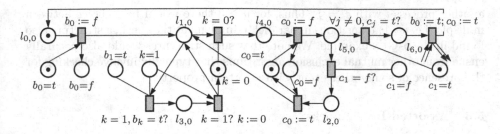

Fig. 8. The fragment of 2-DIJKSTRA that encodes thread 0. Note that arrows from transition $b_0 := t; c_0 := t$ are only partially depicted.

We encoded Dijkstra's algorithm into a c-net as follows. The entries of arrays b, c are represented by two places, e.g. $b_i = t$ and $b_i = f$. Variable k is encoded by n places $k=0, \dots, k=n-1$. Places $l_{0,i}, \dots, l_{6,i}$ encode thread i's instruction pointer. Fig. 8 shows the fragment of 2-DIJKSTRA that encodes thread 0. Roughly, each transition encodes one instruction of the original algorithm [5], updating the instruction pointer and the variables affected by the instruction. Transitions encoding conditional instructions, like $k = 0$?, or $\forall j \neq 0, c_j = t$? employ read arcs to the places coding the variables involved in the predicate.

MPs of n-DIJKSTRA, and in particular CMPs, exhibit a very good growth with respect to n. Table 2 shows the figures, obtained under the same setting as in Section 5.3. While all unfoldings are exponential in n and $|T|$, all the MPs are of polynomial size. The sizes of the plain and PR unfoldings seem to increase by a factor of 5 for each process added. The contextual unfolding reduces this factor down to 3. The plain and PR MPs seems to fit a polynomial curve of degree close to 3. The CMP seems to grow linearly with n^2, i.e. linear with $|T|$, the number of transitions in the net. As it was the case for n-GEN, PR MPs seem to be less efficient than plain MPs on n-DIJKSTRA.

We note that this example exhibits some of the features explained in Section 5.1. For instance, process 0 can transition from $l_{5,0}$ to $l_{2,0}$ if there exists

Table 2. Unfolding and MP sizes of n-DIJKSTRA, its plain, and PR encodings. Last row obtained through regression analysis, see the text.

Net		Merged Processes			Unfoldings		
n	$|T|$	Ctx	Plain	PR	Ctx	Plain	PR
2	18	31	42	40	35	54	54
3	36	64	113	121	131	371	364
4	60	105	220	278	406	2080	1998
5	90	155	375	582	1139	10463	9822
6	126	214	589	1198	3000	49331	44993
		$\mathcal{O}(n^2)$	$\mathcal{O}(n^2)$	$\mathcal{O}(n^3)$	$\mathcal{O}(n^3)$	$\mathcal{O}(3^n)$	$\mathcal{O}(5^n)$	$\mathcal{O}(5^n)$

another process i with $c_i = f$. Thus, for $n \geq 3$ there would be a choice between multiple (i.e. $n-1$) transitions in parallel to implement the check, a structure also found in the n-GEN example. We note that such structures would also naturally ensue from other mutual exclusion algorithms that typically involve checking for the presence of some other event with a certain property.

5.3   Assorted Benchmarks

In this section we present experimental results for a number of benchmark examples circulating in the PN community (collected mostly by Corbett [4]). The following consistent setup was used to produce them:

- The total adequate order proposed in [12] was used.
- All configurations were allowed as cutoff correspondents.
- The cutoff (mp-)events and post-cutoff (mp-)conditions were not counted.

The plain and PR unfolding prefixes were constructed using PUNF [13], and the contextual unfolding prefixes were computed by compressing the PR ones with PRCOMPRESS[4]. The plain and PR MPs have been merged from the corresponding unfolding prefixes with MCI2MP, and the CMPs were merged from the corresponding contextual unfolding prefixes using CMERGE. Note that the direct construction of contextual unfoldings and MPs would yield the same results [12,1].

Recall the following theoretical guarantees:

- The contextual unfolding prefix is never larger than the PR prefix.
- The plain/PR/contextual MP is never larger than the corresponding unfolding prefix.

Table 3 compares the sizes of plain, PR and contextual unfolding prefixes and MPs. The 4th and 5th columns from the left are, respectively, the number of read arcs in the net and place replicas in its PR encoding.[5] The number of conditions and events for the plain and PR unfoldings is normalised wrt. that of the contextual unfolding. Similarly, mp-conditions and mp-events of the plain and PR MPs are normalised wrt. those of the CMPs. The last three columns show the compression gains of CMPs wrt. plain and contextual unfolding prefixes, and the gain of plain MPs wrt. plain unfolding prefixes.

One can see that CMPs are the most compact of all the considered representations.[6] Furthermore, on some benchmarks, notably KEY(4), it has significant advantages over both plain and PR MPs. Interestingly, in this case the PR MP is significantly larger than even the plain MP, which seems to be due to place replication making the subsequent merging much less efficient. As CMPs do not suffer from this problem, they come as a clear winner in such cases.

---

[4] All tools available from the URL indicated in Footnote 3.

[5] More precisely, $\sum_{p \in P, |p| > 1}(|p| - 1)$.

[6] Though the PR MP of RW(1,2) has four mp-events fewer, it has many more mp-conditions.

Table 3. Experimental results, see the text for more information

Benchmark					Unfolding						Merged process						Gains		
Name	Net		Stats.		Plain		PR		Ctx		Plain		PR		Ctx		a/d	b/d	a/c
	$|P|$	$|T|$	$|C|$	Repl.	$|B|$	$|E|^{(a)}$	$|B|$	$|E|$	$|B|$	$|E|^{(b)}$	$|B|$	$|E|^{(c)}$	$|B|$	$|E|$	$|B|$	$|E|^{(d)}$			
Bds	53	59	24	15	4.50	3.79	2.60	2.14	424	252	1.11	1.14	1.36	1.07	70	44	21.73	5.73	19.12
Brujin	86	165	158	142	2.84	1.97	7.00	1.70	286	208	1.33	1.44	3.83	1.31	115	127	3.22	1.64	2.23
Byz	504	409	376	284	2.38	1.80	1.41	1.00	17019	7748	1.23	1.03	1.85	1.22	529	303	46.11	25.57	44.78
Eisenbahn	44	44	6	3	2.19	2.15	2.22	2.04	99	53	1.13	1.30	1.22	1.19	69	43	2.65	1.23	2.04
Ftp	176	529	39	33	1.06	1.04	1.02	1.00	73516	37540	1.02	1.05	1.19	1.00	254	455	85.74	82.51	81.61
Knuth	78	137	114	98	2.62	1.81	5.42	1.62	247	178	1.32	1.31	3.25	1.27	102	112	2.88	1.59	2.20
Mutual	49	41	12	4	1.41	1.23	1.51	1.23	187	121	1.12	1.26	1.23	1.27	92	73	2.04	1.66	1.62
Dme(2)	135	98	132	0	1.61	1.00	1.61	1.00	293	118	1.55	1.04	1.55	1.04	195	90	1.31	1.31	1.26
Dme(4)	269	196	264	0	1.73	1.00	1.73	1.00	1337	636	1.56	1.04	1.56	1.04	389	180	3.53	3.53	3.38
Dme(6)	403	294	396	0	1.79	1.00	1.79	1.00	3517	1794	1.56	1.04	1.56	1.04	583	270	6.64	6.64	6.36
Dme(8)	537	392	528	0	1.83	1.00	1.83	1.00	7217	3832	1.56	1.04	1.56	1.04	777	360	10.64	10.64	10.19
Dme(10)	671	490	660	0	1.86	1.00	1.86	1.00	12821	6990	1.56	1.04	1.56	1.04	971	450	15.53	15.53	14.87
Elev(1)	63	99	18	11	1.02	1.00	1.34	1.00	152	85	1.02	1.00	1.38	1.00	56	46	1.85	1.85	1.85
Elev(2)	146	299	59	47	1.03	1.00	1.85	1.00	858	477	1.01	1.00	2.02	1.00	125	135	3.53	3.53	3.53
Elev(3)	327	783	160	141	1.03	1.00	2.66	1.00	4050	2241	1.00	1.00	2.90	1.00	263	346	6.48	6.48	6.48
Elev(4)	736	1939	405	375	1.04	1.00	4.08	1.00	17360	9567	1.00	1.00	3.98	1.00	556	841	11.38	11.38	11.38
Furn(1)	27	37	9	5	1.12	1.04	1.18	1.00	130	72	1.09	1.03	1.23	1.00	43	33	2.27	2.18	2.21
Furn(2)	40	65	11	6	1.14	1.06	1.14	1.00	582	324	1.07	1.05	1.19	1.02	75	94	3.66	3.45	3.47
Furn(3)	53	99	13	7	1.14	1.08	1.12	1.00	2265	1250	1.07	1.02	1.17	1.01	106	221	6.09	5.66	5.95
Key(2)	94	92	32	31	2.74	2.16	1.72	1.27	302	191	1.22	2.50	1.59	1.17	112	105	3.92	1.82	1.57
Key(3)	129	133	48	47	5.81	4.60	2.81	2.19	1276	806	1.21	2.34	1.64	2.34	151	186	19.93	4.33	4.83
Key(4)	164	174	64	63	11.37	9.08	5.56	4.43	5806	3637	1.21	5.26	1.67	9.92	190	290	113.82	12.54	21.63
Mmgt(1)	50	58	1	0	1.00	1.00	1.00	1.00	79	38	1.00	1.00	1.00	1.00	57	38	1.00	1.00	1.00
Mmgt(2)	86	114	2	0	1.00	1.00	1.00	1.00	502	250	1.00	1.00	1.00	1.00	99	155	1.61	1.61	1.61
Mmgt(3)	122	172	3	0	1.00	1.00	1.00	1.00	2849	1424	1.00	1.00	1.00	1.00	141	355	4.01	4.01	4.01
Mmgt(4)	158	232	4	0	1.00	1.00	1.00	1.00	14900	7450	1.00	1.00	1.00	1.00	183	638	11.68	11.68	11.68
Rw(1,1)	84	208	123	75	1.23	1.00	1.78	1.00	142	94	1.26	1.00	2.03	1.00	77	65	1.45	1.45	1.45
Rw(2,1)	72	88	27	16	1.19	1.01	1.30	1.00	845	554	1.26	1.01	1.45	1.00	113	165	3.39	3.36	3.37
Rw(3,1)	106	270	129	81	1.19	1.04	1.60	1.00	5100	3376	1.24	1.02	1.93	1.00	160	368	9.54	9.17	9.39
Rw(1,2)	209	1482	1132	717	1.21	1.00	3.36	1.00	2836	1838	1.35	1.01	4.76	0.99	159	371	4.95	4.95	4.90
Sentest(25)	104	55	8	5	1.32	1.29	1.35	1.29	188	104	1.06	1.11	1.10	1.11	107	55	2.44	1.89	2.20
Sentest(50)	179	80	8	5	1.23	1.23	1.25	1.23	263	129	1.03	1.08	1.04	1.06	182	80	1.99	1.61	1.85
Sentest(75)	254	105	8	5	1.18	1.19	1.19	1.19	338	154	1.02	1.06	1.04	1.06	257	105	1.75	1.47	1.66
Sentest(100)	329	130	8	5	1.15	1.17	1.16	1.17	413	179	1.02	1.05	1.03	1.05	332	130	1.61	1.38	1.54

6   Conclusions and Future Work

We have developed a new condensed representation of the state space of a contextual Petri net, called contextual merged processes. This representation combines the advantages of merged processes and contextual unfoldings, and copes with several important sources of state space explosion: concurrency, sequences of choices, and concurrent read accesses to shared resources. The experimental results demonstrate that this representation is significantly more compact than either merged processes or contextual unfoldings.

We also proved a number of results which lay the foundation for model checking of reachability-like properties of safe c-nets based on CMPs. In particular, given a CMP, they allow one to reduce (in polynomial time) such a model checking problem to SAT. Furthermore, since the algorithm for direct construction of merged processes of safe Petri nets proposed in [12] is based on model checking, it can be transferred to the contextual case, which would complete the verification flow based on CMPs.

We currently work on implementing the proposed model checking algorithm and on porting the algorithm for direct construction of MPs proposed in [12] to the contextual case. (While the high-level structure of the latter algorithm remains the same, moving from Petri nets to c-nets entails several low-level changes in the nets representation, which pervade the whole code; thus, this porting requires significant implementation effort.)

Another possible direction of future work is to generalise our approach. Normal Petri net unfoldings work very well when systems are entirely concurrent and independent of one another, but many sources of state-space explosion appear when they interact. The approaches that we have combined in this work tackle two such sources; they compress the unfolding and have further commonalities. While Petri net unfoldings are *structurally* acyclic, c-net unfoldings and merged processes have structural cycles but could be said to be *semantically* acyclic: every marking can still be reached by a repetition-free execution and hence one retains the NP-completeness of reachability problem (which is PSPACE-complete for safe Petri nets). This poses the question whether our solutions are a part of a more general phenomenon. The following example suggests that this might be the case. Consider Fig. 9 (a). The token on place p acts as a lock ensuring mutual exclusion between two critical sections represented by places b_1 and b_2.

The two processes are independent of one another, except for the temporal restriction that they cannot possess the lock p at the same time. This imposes a truly semantic sequentialisation constraint (unlike the sequentialisation in Fig. 3 (b), which is merely due to an inadequate semantics-changing encoding). The traditional unfolding techniques cannot take advantage of the fact that the processes are otherwise independent. Indeed, when the example from Fig. 9 is scaled to n processes, a complete unfolding prefix is of size $\mathcal{O}(2^n)$ and a complete MP is still of size $\mathcal{O}(n^2)$ when produced by the tool PUNF.

It is conceivable that this case could be handled by treating locks explicitly and annotating transitions with locking (P) and unlocking (V) actions, like in

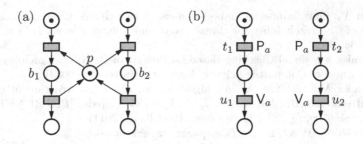

Fig. 9. Two processes competing for lock p: (a) a Petri net (b) a net where lock operations are annotated on transitions

Fig. 9 (b). When multiple locks are involved, their use may introduce circular precedence constraints that can be captured with, e.g. the Lock Causality Graphs of [10]. A suitably defined unfolding for such a case would then unfold both processes independently, only demanding that configurations do not include circular lock constraints. One easily observes that in such a setting, like in ours, an event may have multiple histories that would need to be taken into account to determine cutoffs. In Fig. 9 (b), for instance, t_2 may occur either individually or in a context in which t_1, u_1 *must* have occurred before it. It is therefore quite conceivable that locks could be seamlessly integrated with CMPs as they once again exhibit similar characteristics. To conclude, an interesting perspective for the future research would be to develop a generic framework that handles such effects.

References

1. Baldan, P., Bruni, A., Corradini, A., König, B., Rodríguez, C., Schwoon, S.: Efficient unfolding of contextual Petri nets. Theo. Comp. Sci. 449, 2–22 (2012)
2. Baldan, P., Corradini, A., König, B., Schwoon, S.: McMillan's complete prefix for contextual nets. In: Jensen, K., van der Aalst, W.M.P., Billington, J. (eds.) ToPNoC I, LNCS, vol. 5100, pp. 199–220. Springer, Heidelberg (2008)
3. Baldan, P., Corradini, A., Montanari, U.: Contextual Petri nets, asymmetric event structures, and processes. Inf. Comput. 171(1), 1–49 (2001)
4. Corbett, J.C.: Evaluating deadlock detection methods for concurrent software. IEEE Transactions on Software Engineering 22, 161–180 (1996)
5. Dijkstra, E.W.: Solution of a problem in concurrent programming control. Commun. ACM 8(9), 569 (1965)
6. Esparza, J., Römer, S., Vogler, W.: An improvement of McMillan's unfolding algorithm. Formal Methods in System Design 20, 285–310 (2002)
7. Heljanko, K.: Deadlock and reachability checking with finite complete prefixes. Licentiate's thesis, Helsinki University of Technology (1999)
8. Heljanko, K.: Minimizing finite complete prefixes. In: Proc. CS&P, pp. 83–95 (1999)
9. Janicki, R., Koutny, M.: Invariant semantics of nets with inhibitor arcs. In: Groote, J.F., Baeten, J.C.M. (eds.) CONCUR 1991. LNCS, vol. 527, pp. 317–331. Springer, Heidelberg (1991)

10. Kahlon, V.: Boundedness vs. unboundedness of lock chains: Characterizing decidability of CFL-reachability for threads communicating via locks. In: Proc. LICS, pp. 27–36 (2009)
11. Khomenko, V.: Model Checking Based on Prefixes of Petri Net Unfoldings. Ph.D. thesis, School of Computing Science, Newcastle University (2003)
12. Khomenko, V., Mokhov, A.: An algorithm for direct construction of complete merged processes. In: Kristensen, L.M., Petrucci, L. (eds.) PETRI NETS 2011. LNCS, vol. 6709, pp. 89–108. Springer, Heidelberg (2011)
13. Khomenko, V.: PUNF, http://homepages.cs.ncl.ac.uk/victor.khomenko/tools/punf/
14. Khomenko, V., Kondratyev, A., Koutny, M., Vogler, W.: Merged processes – a new condensed representation of Petri net behaviour. Act. Inf. 43(5), 307–330 (2006)
15. McMillan, K.L.: Using unfoldings to avoid the state explosion problem in the verification of asynchronous circuits. In: Probst, D.K., von Bochmann, G. (eds.) CAV 1992. LNCS, vol. 663, pp. 164–177. Springer, Heidelberg (1993)
16. Montanari, U., Rossi, F.: Contextual occurrence nets and concurrent constraint programming. In: Ehrig, H., Schneider, H.-J. (eds.) Dagstuhl Seminar 1993. LNCS, vol. 776, pp. 280–295. Springer, Heidelberg (1994)
17. Ranjan, D.P., Tang, D., Malik, S.: A comparative study of 2QBF algorithms. In: Proc. SAT (2004)
18. Ristori, G.: Modelling Systems with Shared Resources via Petri Nets. Ph.D. thesis, Department of Computer Science, University of Pisa (1994)
19. Rodríguez, C.: CUNF, http://www.lsv.ens-cachan.fr/~rodriguez/tools/cunf/
20. Rodríguez, C., Schwoon, S.: Verification of Petri Nets with Read Arcs. In: Koutny, M., Ulidowski, I. (eds.) CONCUR 2012. LNCS, vol. 7454, pp. 471–485. Springer, Heidelberg (2012)
21. Rodríguez, C., Schwoon, S., Baldan, P.: Efficient contextual unfolding. In: Katoen, J.-P., König, B. (eds.) CONCUR 2011. LNCS, vol. 6901, pp. 342–357. Springer, Heidelberg (2011)
22. Rodríguez, C., Schwoon, S., Khomenko, V.: Contextual merged processes. Tech. Rep. LSV-13-06, LSV, ENS de Cachan (2013)
23. Valmari, A.: The state explosion problem. In: Reisig, W., Rozenberg, G. (eds.) APN 1998. LNCS, vol. 1491, pp. 429–528. Springer, Heidelberg (1998)
24. Vogler, W., Semenov, A., Yakovlev, A.: Unfolding and finite prefix for nets with read arcs. In: Sangiorgi, D., de Simone, R. (eds.) CONCUR 1998. LNCS, vol. 1466, pp. 501–516. Springer, Heidelberg (1998)

ω-Petri Nets

Gilles Geeraerts[1,*], Alexander Heussner[2], M. Praveen[3], and Jean-François Raskin[1]

[1] Université Libre de Bruxelles (ULB), Belgium
[2] Otto-Friedrich Universität Bamberg, Germany
[3] Laboratoire Spécification et Vérification, ENS Cachan, France

Abstract. We introduce ω-Petri nets (ωPN), an extension of *plain* Petri nets with ω-labeled input and output arcs, that is well-suited to analyse *parametric concurrent systems with dynamic thread creation*. Most techniques (such as the Karp and Miller tree or the Rackoff technique) that have been proposed in the setting of *plain Petri nets* do not apply directly to ωPN because ωPN define transition systems that have *infinite branching*. This motivates a thorough analysis of the computational aspects of ωPN. We show that an ωPN can be turned into a plain Petri net that allows to recover the reachability set of the ωPN, but that does not preserve termination. This yields complexity bounds for the reachability, (place) boundedness and coverability problems on ωPN. We provide a practical algorithm to compute a coverability set of the ωPN and to decide termination by adapting the classical Karp and Miller tree construction. We also adapt the Rackoff technique to ωPN, to obtain the exact complexity of the termination problem. Finally, we consider the extension of ωPN with reset and transfer arcs, and show how this extension impacts the decidability and complexity of the aforementioned problems.

1 Introduction

In this paper, we introduce ω-Petri nets (ωPN), an extension of *plain* Petri nets (PN) that allows input and output arcs to be labeled by the symbol ω, instead of a natural number. An ω-labeled input arc consumes, non-deterministically, any number of tokens in its input place while an ω-labeled output arc produces non-deterministically any number of tokens in its output place. We claim that ωPN are particularly well suited for modeling *parametric concurrent systems* (see for instance our recent work on the Grand Central Dispatch technology [12]), and to perform *parametric verification* [15] on those systems, as we illustrate now by means of the example in Fig 1. The example present a skeleton of a distributed program, in which a `main` function forks P parallel threads (where P is a parameter of the program), each executing the `one_task` function. Many distributed programs follow this abstract skeleton that allows to perform calculations in parallel, and being able to model precisely such concurrent behaviors is an important issue. In particular, we would like that the model captures the fact that P *is a parameter*, so that we can, for instance, check that the execution of the program always terminates (assuming each individual execution of `one_task` does), *for all possible values of P.*

* Partially supported by a 'Crédit aux chercheurs' of the F.R.S/FNRS.

J.-M. Colom and J. Desel (Eds.): PETRI NETS 2013, LNCS 7927, pp. 49–69, 2013.
© Springer-Verlag Berlin Heidelberg 2013

50   G. Geeraerts et al.

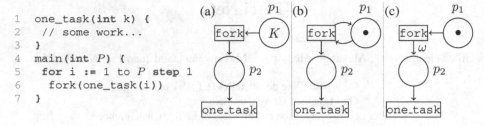

```
1   one_task(int k) {
2     // some work...
3   }
4   main(int P) {
5     for i := 1 to P step 1
6       fork(one_task(i))
7   }
```

Fig. 1. An example of a parametric system with three possible models

Clearly, the Petri net (a) in Fig. 1 does not capture the parametric nature of the example, as place p_1 contains a fixed number K of tokens. The PN (b), on the other hand captures the fact that the program can `fork` an unbounded number of threads, but *does not preserve termination*: $(\texttt{fork})^\omega$ is an infinite execution of PN (b), while the programme terminates (assuming each `one_task` thread terminates) for all values of P, because the `for` loop in line 5 executes exactly P times. Finally, observe that the ωPN (c) has the desired properties: firing transition `fork` creates *non-deterministically* an *unbounded* albeit *finite* number of tokens in p_2 (to model all the possible executions of the for loop in line 5), and all possible executions of this ωPN terminate, because the number of tokens produced in p_2 remains *finite* and no further token creation in p_2 is allowed after the firing of the `fork` transition.

While close to Petri nets, ωPN are sufficiently different that a thorough and careful study of their computational properties is required. This is the main contribution of the paper. A first example of discrepancy is that the semantics of ωPN is an infinite transition system which is *infinitely branching*. This is not the case for plain PN: their transition systems can be infinite but they are finitely branching. As a consequence, some of the classical techniques for the analysis of Petri nets cannot be applied. Consider for example the *finite unfolding of the transition system* [10] that stops the development of a branch of the reachability tree whenever a node with a smaller ancestor is found. This tree is finite (and effectively constructible) for any plain Petri net and any initial marking because the set of markings $\mathbb{N}^k$ is *well-quasi ordered*, and *finite branching* of plain Petri nets allows for the use of König's lemma[1]. However, this technique cannot be applied to ωPN, as they are infinitely branching. Such peculiarities of ωPN motivate our study of three different tools for analysing them. First, we consider, in Section 3, a variant of the Karp and Miller tree [16] that applies to ωPN. In order to cope with the infinite branching of the semantics of ωPN, we need to introduce in the Karp and Miller tree ω's that are not the result of accelerations but the result of ω-output arcs. Our variant of the Karp and Miller construction is *recursive*, this allows us to tame the technicality of the proof, and as a consequence, our proof when applied to *plain* Petri nets, provides a simplification of the original proof by Karp and Miller. Second, in Section 4, we show how to construct, from an ωPN, a plain Petri net that preserves its reachability set. This reduction allows to prove that many bounds on the algorithmic complexity

---

[1] In fact, this construction is applicable to any well-structured transition system which is finitely branching and allows to decide the termination problem for example.

Table 1. Complexity results on ωPN (with the section numbers where the results are proved). ωIPN+R (ωOPN+R) and ωIPN+T (ωOPN+T) denote resp. Petri nets with reset (R) and transfer (T) arcs with ω on input (output) arcs only.

Problem	ωPN	ωPN+T	ωPN+R
Reachability	Decidable and ExPSPACE-hard (4)	Undecidable (6)	Undecidable (6)
Place-boundedness			
Boundedness	ExpSpace-c (4)	Decidable (6)	
Coverability		Decidable and Ackerman-hard (6)	

Problem	ωPN	ωOPN+T, ωOPN+R	ωIPN+T, ωIPN+R
Termination	ExpSpace-c (5)	Undecidable (6)	Decidable and Ackerman-hard (6)

of (plain) PN problems apply to ωPN too. However, it does not preserve *termination*. Thus, we study, in Section 5, as a third contribution, an extension of the self-covering path technique due to Rackoff [20]. This technique allows to provide a direct proof of ExpSpace upper bounds for several classical decision problems, and in particular, this allows to prove ExpSpace completeness of the termination problem.

Finally, in Section 6, as an additional contribution, and to get a complete picture, we consider extensions of ωPN with *reset* and *transfer* arcs [7]. For those extensions, the decidability results for reset and transfer nets (without ω arcs) also apply to our extension with the notable exception of the termination problem that becomes, as we show here, undecidable. The summary of our results are given in Table 1.

Related Works. ωPN are well-structured transition systems [10]. The set saturation technique [1] and so symbolic backward analysis can be applied to them while the finite tree unfolding is not applicable because of the infinite branching property of ωPN. For the same reason, ωPN are *not* well-structured nets [11].

In [3], Brazdil *et al.* extends the Rackoff technique to VASS games with ω output arcs. While this extension of the Rackoff technique is technically close to ours, we cannot directly use their results to solve the termination problem of ωPN.

Several works (see for instance [4,5] rely on Petri nets to model *parametric systems* and perform *parametrised verification*. However, in all these works, the dynamic creation of threads uses the same pattern as in Fig. 1 (b), and does not preserve termination. ωPN allow to model more faithfully the dynamic creation of an unbounded number of threads, and are thus better suited to model new programming paradigms (such as GCD [12]) that have been recently proposed to better support multi-core platforms.

Remark: Missing proofs can be found in the companion technical report [13].

2 ω-Petri Nets

Let us define the syntax and semantics of our Petri net extension, called ω *Petri nets* (ωPN for short). Let ω be a symbol that denotes 'any positive integer value'. We extend the arithmetic and the $\leq$ ordering on $\mathbb{Z}$ to $\mathbb{Z} \cup \{\omega\}$ as follows: $\omega + \omega = \omega - \omega = \omega$; and for all $c \in \mathbb{Z}$: $c + \omega = \omega + c = \omega - c = \omega$; $c - \omega = c$; and $c \leq \omega$. The fact that $c - \omega = c$ might sound surprising but will be justified later when we introduce ωPN. An ω-*multiset* (or simply *multiset*) of elements from S is a function $m : S \mapsto \mathbb{N} \cup \{\omega\}$. We denote multisets m of $S = \{s_1, s_2, \ldots, s_n\}$ by extension using the syntax $\{m(s_1) \otimes s_1, m(s_2) \otimes s_2, \ldots, m(s_n) \otimes s_n\}$ (when $m(s) = 1$, we write s instead of $m(s) \otimes s$, and we omit elements $m(s) \otimes s$ when $m(s) = 0$). Given two multisets m_1 and m_2, and an integer value c we let $m_1 + m_2$ be the multiset s.t. $(m_1 + m_2)(p) = m_1(p) + m_2(p)$; $m_1 - m_2$ be the multiset s.t. $(m_1 - m_2)(p) = m_1(p) - m_2(p)$; and $c \cdot m_1$ be the multiset s.t. $(c \cdot m_1)(p) = c \times m_1(p)$ for all $p \in P$.

Syntax. Syntactically, ωPN extend plain Petri nets [19,21] by allowing (input and output) arcs to be labeled by ω. Intuitively, if a transition t has ω as output (resp. input) effect on place p, the firing of t non-deterministically creates (consumes) a positive number of tokens in p.

Definition 1. *A Petri net with ω-arcs (ωPN) is a tuple $\mathcal{N} = \langle P, T \rangle$ where: P is a finite set of* places; *T a finite set of* transitions. *Each transition is a pair $t = (I, O)$, where: $I : P \rightarrow \mathbb{N} \cup \{\omega\}$ and $O : P \rightarrow \mathbb{N} \cup \{\omega\}$, give respectively the input (output) effect $I(p)$ $(O(p))$ of t on place p.*

By abuse of notation, we denote by $I(t)$ (resp. $O(t)$) the functions s.t. $t = (I(t), O(t))$. When convenient, we sometimes regard $I(t)$ or $O(t)$ as ω-*multisets* of places. Whenever there is p s.t. $O(t)(p) = \omega$ (resp. $I(t)(p) = \omega$), we say that t is an ω-*output-transition* (ω-*input-transition*). A transition t is an ω-transition iff it is an ω-output-transition or an ω-input-transition. Otherwise, t is a *plain* transition. Note that a (plain) Petri net is an ωPN with plain transitions only. Moreover, when an ωPN contains no ω-output-transitions (resp. no ω-input transitions), we say that it is an ω-input-PN (ω-output-PN), or ωIPN (ωOPN) for short. For all transitions t, we denote by *effect*(t) the function $O(t) - I(t)$. Note that *effect*$(t)(p)$ could be ω for some p (in particular when $O(t)(p) = I(t)(p) = \omega$). Intuitively, *effect*$(t)(p) = \omega$ models the fact that firing t can increase the marking of p by an arbitrary number of tokens. Finally, observe that $O(t)(p) = c \neq \omega$ and $I(t)(p) = \omega$ implies *effect*$(t)(p) = c - \omega = c$. This models the fact that firing t can at most increase the marking of p by c tokens. Thus, intuitively, the value *effect*$(t)(p)$ models the *maximal possible effect* of t on p. We extend the definition of *effect* to sequences of transitions $\sigma = t_1 t_2 \cdots t_n$ by letting *effect*$(\sigma) = \sum_{i=1}^{n} effect(t_i)$.

A *marking* is a function $P \mapsto \mathbb{N}$. An ω-*marking* is a function $P \mapsto \mathbb{N} \cup \{\omega\}$, i.e. an ω-multiset on P. Any marking is an ω-marking. For all transitions $t = (I, O)$, I and O are both ω-markings. We denote by $\mathbf{0}$ the marking s.t. $\mathbf{0}(p) = 0$ for all $p \in P$. For all ω-markings m, we let $\omega(m)$ be the set of places $\{p \mid m(p) = \omega\}$, and let nb$\omega\,(m) = |\omega(m)|$. We define *the concretisation* of m as the set of all markings that coincide with m on all places $p \notin \omega(m)$, and take an arbitrary value in any place from $\omega(m)$. Formally: $\gamma(m) = \{m' \mid \forall p \notin \omega(m) : m'(p) = m(p)\}$. We further define a family of orderings on ω-markings as follows. For any $P' \subseteq P$, we let $m_1 \preceq_{P'} m_2$ iff (i) for all $p \in P'$: $m_1(p) \leq m_2(p)$, and (ii) for all $p \in P \setminus P'$: $m_1(p) = m_2(p)$. We abbreviate $\preceq_P$ by $\preceq$ (where P is the set of places of the ωPN). It is well-known

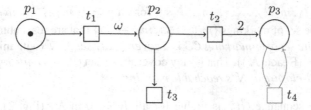

Fig. 2. An example ωPN $\mathcal{N}_1$. The ωPN $\mathcal{N}_1'$ is obtained by removing transition t_4 (gray)

that $\preceq$ is a *well-quasi ordering* (wqo), that is, we can extract, from any infinite sequence $m_1, m_2, \ldots, m_i, \ldots$ of markings, an infinite subsequence $\overline{m}_1, \overline{m}_2, \ldots, \overline{m}_i, \ldots$ s.t. $\overline{m}_i \preceq \overline{m}_{i+1}$ for all $i \geq 1$. For all ω-markings m, we let $\downarrow(m)$ be the *downward-closure* of m, defined as $\downarrow(m) = \{m' \mid m' \text{ is a marking and } m' \preceq m\}$. We extend $\downarrow$ to sets of ω-markings: $\downarrow(S) = \cup_{m \in S} \downarrow(m)$. A set D of markings is *downward-closed* iff $\downarrow(D) = D$. It is well-known that (possibly infinite) downward-closed sets of markings can always be represented by a finite set of ω-markings, because the set of ω-markings forms an *adequate domain of limits* [14]: for all downward-closed sets D of markings, there exists a finite set M of ω-markings s.t. $\downarrow(M) = D$. We associate, to each ωPN, an *intial marking* m_0. From now on, we consider mostly initialised ωPN $\langle P, T, m_0 \rangle$.

Example 2. An example of an ωPN (actually an ωOPN) $\mathcal{N}_1 = \langle P, T, m_0 \rangle$ is shown in Fig. 2. In this example, $P = \{p_1, p_2, p_3\}$, $T = \{t_1, t_2, t_3, t_4\}$, $m_0(p_1) = 1$ and $m_0(p_2) = m_0(p_3) = 0$. t_1 is the only ω-transition, with $O(t_1)(p_2) = \omega$. This ωPN will serve as a running example throughout the section.

Semantics. Let m be an ω-*marking*. A transition $t = (I, O)$ is *firable from* m iff: $m(p) \succeq I(p)$ for all p s.t. $I(p) \neq \omega$. We consider two kinds of possible effects for t. The first is the *concrete semantics* and applies only when m is a *marking*. In this case, firing t yields a new marking m' s.t. for all $p \in P$: $m'(p) = m(p) - i + o$ where: $i = I(t)(p)$ if $I(t)(p) \neq \omega$, $i \in \{0, \ldots, m(p)\}$ if $I(t)(p) = \omega$, $o = O(t)(p)$ if $O(t)(p) \neq \omega$ and $o \geq 0$ if $O(t)(p) = \omega$. This is denoted by $m \xrightarrow{t} m'$. Thus, intuitively, $I(t)(p) = \omega$ (resp. $O(t)(p) = \omega$) means that t consumes (produces) an arbitrary number of tokens in p when fired. Note that, in the concrete semantics, ω-transitions are *non-deterministic*: when t is an ω-transitions that is firable in m, there are *infinitely many* m' s.t. $m \xrightarrow{t} m'$. The latter semantics is the ω-*semantics*. In this case, firing $t = (I, O)$ yields the (unique) ω-marking $m' = m - I + O$ (denoted $m \xrightarrow{t}_\omega m'$). Note that $m \xrightarrow{t} m'$ iff $m \xrightarrow{t}_\omega m'$ when m and m' are markings.

We extend the $\rightarrow$ and $\rightarrow_\omega$ relations to finite or infinite sequences of transitions in the usual way. Also we write $m \xrightarrow{\sigma}$ iff σ is *firable* from m. More precisely, for a finite sequence of transitions $\sigma = t_1 \cdots t_n$, we write $m \xrightarrow{\sigma}$ iff there are $m_1, \ldots, m_n$ s.t. for all $1 \leq i \leq n$: $m_{i-1} \xrightarrow{t_i} m_i$. For an infinite sequence of transitions $\sigma = t_1 \cdots t_j \cdots$, we write $m_0 \xrightarrow{\sigma}$ iff there are $m_1, \ldots, m_j, \ldots$ s.t. for all $i \geq 1$: $m_{i-1} \xrightarrow{t_i} m_i$.

Given an ωPN $\mathcal{N} = \langle P, T, m_0 \rangle$, an *execution* of $\mathcal{N}$ is either a finite sequence of the form $m_0, t_1, m_1, t_2, \ldots, t_n, m_n$ s.t. $m_0 \xrightarrow{t_1} m_1 \xrightarrow{t_2} \cdots \xrightarrow{t_n} m_n$, or an infinite sequence of the form $m_0, t_1, m_1, t_2, \ldots, t_j, m_j, \ldots$ s.t. for all $j \geq 1$: $m_{j-1} \xrightarrow{t_j} m_j$. We denote by Reach($\mathcal{N}$) the set of markings $\{m \mid \exists \sigma \text{ s.t. } m_0 \xrightarrow{\sigma} m\}$ that are reachable from m_0 in $\mathcal{N}$. Finally, a *finite set of ω-markings* $\mathcal{CS}$ is a *coverability set* of $\mathcal{N}$ (with initial marking m_0) iff $\downarrow(\mathcal{CS}) = \downarrow(\text{Reach}(\mathcal{N}))$. That is, any coverability set $\mathcal{CS}$ is a *finite representation of the downward-closure of $\mathcal{N}$'s reachable markings.*

Example 3. The sequence $t_1 t_2^K$ is firable for all $K \geq 0$ in $\mathcal{N}_1$ (Fig. 2). Indeed, for each $K \geq 0$, one possible execution corresponding to $t_1 t_2^K$ is given by $\langle 1, 0, 0 \rangle \xrightarrow{t_1} \langle 0, 3K, 0 \rangle \xrightarrow{t_2} \langle 0, 3K - 1, 2 \rangle \xrightarrow{t_2} \langle 0, 3K - 2, 4 \rangle \xrightarrow{t_2} \cdots \xrightarrow{t_2} \langle 0, 2K, 2K \rangle$. There are other possible executions corresponding to the same sequence of transitions, because the number of tokens created by t_1 in p_2 is chosen non-deterministically. Also, $t_1 t_2 t_4^\omega$ is an infinite firable sequence of transitions. Finally, observe that the set of reachable markings in $\mathcal{N}_1$ is Reach($\mathcal{N}$) $= \{\langle 1, 0, 0 \rangle\} \cup \{\langle 0, i, 2 \times j \rangle \mid i, j \in \mathbb{N}\}$. The set of ω markings $\mathcal{CS} = \{\langle 1, 0, 0 \rangle, \langle 0, \omega, \omega \rangle\}$ is a coverability set of $\mathcal{N}$. Note that $\downarrow(\mathcal{CS}) \supsetneq$ Reach($\mathcal{N}$): for instance, $\langle 0, 1, 1 \rangle \in \downarrow(\mathcal{CS})$, but $\langle 0, 1, 1 \rangle$ is not reachable.

Let us now observe two properties of the semantics of ωPN, that will be useful for the proofs of Section 3. The first says that, when firing a sequence of transitions σ that have non ω-labeled arcs on to and from some place p, the effect of σ on p is as in a plain PN:

Lemma 4. *Let m and m' be two markings and let $\sigma = t_1 \cdots t_n$ be a sequence of transitions of an ωPN s.t. $m \xrightarrow{\sigma} m'$. Let p be a place s.t. for all $1 \leq i \leq n$: $O(t_i)(p) \neq \omega \neq I(t_i)(p)$. Then, $m'(p) = m(p) + \text{effect}(\sigma)(p)$.*

The latter property says that the set of markings that are reachable by a given sequence of transitions σ is upward-closed[2] w.r.t. $\preceq_{P'}$, where P' is the set of places where the effect of σ is ω.

Lemma 5. *Let m_1, m_2 and m_3 be three markings, and let σ be a sequence of transitions s.t. (i) $m_1 \xrightarrow{\sigma} m_2$, (ii) $m_3 \succeq_{P'} m_2$ with $P' = \{p \mid \text{effect}(\sigma)(p) = \omega\}$. Then, $m_1 \xrightarrow{\sigma} m_3$ holds too.*

Problems. We consider the following problems. Let $\mathcal{N} = (P, T, m_0)$ be an ωPN:

1. The *reachability problem* asks, given a marking m, whether $m \in$ Reach(N).
2. The *place boundedness problem* asks, given a place p of $\mathcal{N}$, whether there exists $K \in \mathbb{N}$ s.t. for all $m \in$ Reach($\mathcal{N}$): $m(p) \leq K$. If the answer is positive, we say that p is *bounded* (from m_0).
3. The *boundedness problem* asks whether all places of $\mathcal{N}$ are bounded (from m_0).
4. The *covering problem* asks, given a marking m of $\mathcal{N}$, whether there exists $m' \in$ Reach($\mathcal{N}$) s.t. $m' \succeq m$.
5. The *termination problem* asks whether all executions of $\mathcal{N}$ are finite.

---

[2] A set $U \subseteq S$ is upward-closed wrt to a partial order $\leq$ iff for all $u \in U$ and $s \in S$: $u \leq s$ implies that $s \in U$.

A *coverability set* of the ωPN is sufficient to solve *boundedness, place boundedness* and *covering*, as in the case of Petri nets. If CS is a coverability set of $\mathcal{N}$, then: (i) p is bounded iff $m(p) \neq \omega$ for all $m \in CS$; (ii) $\mathcal{N}$ is bounded iff $m(p) \neq \omega$ for all p and for all $m \in CS$; and (iii), $\mathcal{N}$ can cover m iff there exists $m' \in CS$ s.t. $m \preceq m'$. As in the plain Petri nets case, a sufficient and necessary condition of non-termination is the existence of a *self covering execution*. A *self covering execution* of an ωPN $\mathcal{N} = \langle P, T, m_0 \rangle$ is a *finite* execution of the form $m_0 \xrightarrow{t_1} m_1 \cdots \xrightarrow{t_k} m_k \xrightarrow{t_{k+1}} \cdots \xrightarrow{t_n} m_n$ with $m_n \succeq m_k$:

Lemma 6. *An ωPN terminates iff it admits no self-covering execution.*

Example 7. Consider again the ωPN $\mathcal{N}_1$ in Fig. 2. Recall from Example 3 that, for all $K \geq 0$, $t_1 t_2^K$ is firable and allows to *reach* $\langle 0, 2K, 2K \rangle$. All these markings are thus *reachable*. These sequences of transitions also show that p_2 and p_3 are *unbounded* (hence, $\mathcal{N}_1$ is unbounded too), while p_1 is *bounded*. Marking $\langle 0, 1, 1 \rangle$ is *not reachable* but *coverable*, while $\langle 2, 0, 0 \rangle$ is neither reachable nor coverable. Finally, $\mathcal{N}_1$ does not terminate (because $t_1 t_2 t_4^{\omega}$ is firable), while $\mathcal{N}_1'$ does. In particular, *in $\mathcal{N}_1'$*, t_3 can fire only a *finite* number of times, because t_1 will always create a finite (albeit unbounded) number of tokens in p_2. This an important difference between ωPN and plain PN: no unbounded PNs terminates, while there are unbounded ωPN that terminate, e.g. $\mathcal{N}_1'$.

3   A Karp and Miller Procedure for ωPN

In this section, we present an extension of the classical Karp & Miller procedure [16], adapted to ωPN. We show that the finite tree built by this algorithm (coined the KM tree), allows, as in the case of PNs, to decide *boundedness, place boundedness, coverability* and *termination* on ωPN.

Before describing the algorithm, we discuss intuitively the KM trees of the ωPN $\mathcal{N}_1$ and $\mathcal{N}_1'$ given in Fig. 2. Their respective KM trees (for the initial marking $m_0 = \langle 1, 0, 0 \rangle$) are $\mathcal{T}_1$ and $\mathcal{T}_1'$, respectively the tree in Fig. 3 and its *subtree made of the bold nodes* (i.e., excluding n_7). As can be observed, the nodes and edges of a KM tree are labeled by ω-markings and transitions respectively. The relationship between a KM tree and the executions of the corresponding ωPN can be formalised using the notion of *stuttering path*. Intuitively, a stuttering path is a sequence of nodes $n_1, n_2, \ldots, n_k$ s.t. for all $i \geq 2$: either n_i is a son of n_{i-1}, or n_i is an *ancestor* of n_{i-1} *that has the same label* as n_{i-1}. For instance, $\pi = n_1, n_2, n_4, n_2, n_3, n_6, n_3, n_5, n_3, n_5$ is a stuttering path in $\mathcal{T}_1'$. Then, we claim (i) that *every execution of the ωPN is simulated by a stuttering path* in its KM tree, and that (ii) *every stuttering path in the KM tree corresponds to a family of executions of the ωPN*, where an arbitrary number of tokens can be produced in the places marked by ω in the KM tree. For instance, the execution $m_0, t_1, \langle 0, 42, 0 \rangle, t_3, \langle 0, 41, 0 \rangle, t_2, \langle 0, 40, 2 \rangle, t_3, \langle 0, 39, 2 \rangle, t_2, \langle 0, 38, 4 \rangle, t_2, \langle 0, 37, 6 \rangle$, of $\mathcal{N}_1'$ is witnessed in $\mathcal{T}_1'$ by the stuttering path π given above – observe that the sequence of edge labels in π's equals the sequence of transitions of the execution, and that all markings along the execution are *covered* by the labels of the corresponding nodes in π: $m_0 \in \gamma(n_1)$, $\langle 0, 42, 0 \rangle \in \gamma(n_2)$, and so forth. On the other hand, the

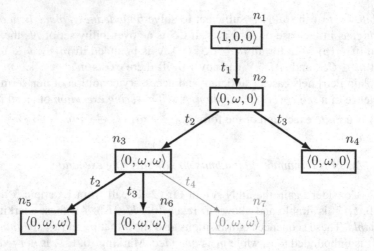

Fig. 3. The KM trees $\mathcal{T}_1$ (whole tree) and $\mathcal{T}_1'$ (bold nodes, i.e. w/o n_7) of resp. $\mathcal{N}_1$ and $\mathcal{N}_1'$

stuttering path n_1, n_2, n_3 of $\mathcal{N}_1$ summarises all the (infinitely many) possible executions obtained by firing a sequence of the form $t_1 t_2^n$. Indeed, for all $k \geq 1, \ell \geq 0$: $m_0, t_1, \langle 0, k + \ell, 0 \rangle, t_2, \langle 0, k + \ell - 1, 2 \rangle, t_2, \ldots, t_2, \langle 0, k, 2 \times \ell \rangle$ is an execution of $\mathcal{N}_1$, so, an arbitrary number of tokens can be obtained in both p_2 and p_3 by firing sequences of the form $t_1 t_2^n$. Finally, observe that a *self-covering execution* of $\mathcal{N}_1$; such as $m_0, t_1, \langle 0, 1, 0 \rangle, t_2, \langle 0, 0, 2 \rangle, t_4, \langle 0, 0, 2 \rangle$ can be detected in $\mathcal{T}_1$, by considering the path n_1, n_2, n_3, n_7, and noting that the label of (n_3, n_7) is t_4 with $\mathit{effect}(t_4) \succeq \mathbf{0}$.

The `Build`-**KM** *Algorithm.* Let us now show how to build algorithmically the **KM** of an ωPN. Recall that, *in the case of plain PNs*, the Karp & Miller tree [16] can be regarded as a *finite over-approximation of the (potentially infinite) reachability tree of the PN*. Thus, the Karp & Miller algorithm works by unfolding the transition relation of the PN, and adds two ingredients to guarantee that the tree is finite. First, a node n that has an ancestor n' *with the same label* is *not developed* (it has no children). Second, when a node n with label m has an ancestor n' with label $m' \prec m$, an *acceleration function* is applied to produce a marking m_ω s.t. $m_\omega(p) = \omega$ if $m(p) > m'(p)$ and $m_\omega(p) = m(p)$ otherwise. This acceleration is *sound* wrt to coverability since the sequence of transition that has produced the branch (n, n') can be iterated an arbitrary number of times, thus producing arbitrary large numbers of tokens in the places marked by ω in m_ω. Remark that these two constructions are not sufficient to ensure termination of the algorithm in the case of ωPN, as ωPN are *not finitely branching* (firing an ω-output-transition can produce infinitely many different successors). To cope with this difficulty, our solution unfolds the ω-*semantics* $\rightarrow_\omega$ instead of the concrete semantics $\rightarrow$. This has an important consequence: whereas the presence of a node labeled by m with $m(p) = \omega$ in the **KM** tree of a PN $\mathcal{N}$ implies that $\mathcal{N}$ *does not terminate*, this is *not true anymore* in the case of ωPN. For instance, all nodes but n_1 in $\mathcal{T}_1'$ (Fig. 3) are marked by ω, yet the corresponding ωPN $\mathcal{N}_1'$ (Fig. 2) *does terminate*.

Input an ωOPN $\mathcal{N} = \langle P, T \rangle$ and an ω-marking m_0
Output the KM of $\mathcal{N}$, starting from m_0

```
Build-KM (N, m₀):
1   T := ⟨N, E, λ, μ, n₀⟩ where N = {n₀} with λ(n₀) = m₀
2   U := {n₀}
3   while U ≠ ∅:
4     select and remove n from U
5     if ∄n̄ st (n̄, n) ∈ E⁺ and λ(n) = λ(n̄):
6       forall t in T s.t. ∀p ∈ P: I(t)(p) ≠ ω implies λ(n)(p) ≥ I(t)(p):
7         m' := Post(N, λ(n), t)
8         if nbω(m') > nbω(λ(n)):
9           T' := Build-KM(N, m')
10          add all edge and nodes of T' to T
11          let n' be the root of T'
12        else
13          n' := new node with λ(n') = m'
14          U := U ∪ {n'}
15        E := E ∪ (n, n') s.t. μ(n, n') = t.
16  return T

    Post(N,n,t):
17  m' := λ(n) − I(t) + O(t)
18  if ∃n̄ : (n̄, n) ∈ E⁺ ∧ λ(n̄) ≺ λ(n):
19    mₓ(p) := { m'(p)   if effect(n̄ ⤳ n · t)(p) ≤ 0
               { ω        otherwise
20    return mₓ
21  else:
22    return m'
```

Fig. 4. The algorithm to build the KM of an ωPN

Our version of the Karp & Miller tree adapted to ωPN is given in Fig. 4. It builds a tree $\mathcal{T} = \langle N, E, \lambda, \mu, n_0 \rangle$ where: N is a set of nodes; $E \subseteq N \times N$ is a set of edges; $\lambda : N \mapsto (\mathbb{N} \cup \{\omega\})^P$ is a function that labels nodes by ω-markings[3]; $\mu : E \mapsto T$ is a labeling function that labels arcs by transitions; and $n_0 \in N$ is the root of the tree. For each edge e, we let $effect(e) = effect(\mu(e))$. Let E^+ and E^* be respectively the transitive and the transitive reflexive closure of E. A *stuttering path* is a finite sequence $n_0, n_1, \ldots, n_\ell$ s.t. for all $1 \leq i \leq \ell$: *either* $(n_{i-1}, n_i) \in E$ or $(n_i, n_{i-1}) \in E^+$ and $\lambda(n_i) = \lambda(n_{i-1})$. A stuttering path $n_0, n_1, \ldots, n_\ell$ is a *(plain) path* iff $(n_{i-1}, n_i) \in E$ for all $1 \leq i \leq \ell$. Given two nodes n and n' s.t. $(n, n') \in E^*$, we denote by $n \leadsto n'$ the (unique path) from n to n'. Given a stuttering path $\pi = n_0, n_1, \ldots, n_\ell$, we denote by $\mu(\pi)$ the sequence $\mu(n_0, n_1)\mu(n_1, n_2) \cdots \mu(n_{\ell-1}, n_\ell)$ assuming $\mu(n_i, n_{i+1}) = \varepsilon$ when $(n_i, n_{i+1}) \notin E$; and by $effect(\pi) = \sum_{i=1}^{\ell} effect(n_{i-1}, n_i)$, letting $effect(n_{i-1}, n_i) = 0$ when $(n_i, n_{i+1}) \notin E$.

Build-KM follows the intuition given above. At all times, it maintains a frontier U of tree nodes that are candidate for development (initially, U = $\{n_0\}$, with $\lambda(n_0) =$

---

[3] We extend λ to set of nodes S in the usual way: $\lambda(S) = \{\lambda(n) \mid n \in S\}$.

m_0). Then, Build-KM iteratively picks up a node n from U (see line 4), and develops it (line 6 onwards) if n has no ancestor n' with the same label (line 5). *Developing* a node n amounts to computing all the marking m s.t. $\lambda(n) \to_\omega m$ (line 17), performing accelerations (line 19) if need be, and inserting the resulting children in the tree. Remark that Build-KM is *recursive* (see line 9): every time a marking m with an extra ω is created, it performs a recursive call to Build-KM$(\mathcal{N}, m)$, using m as initial marking[4].

The rest of the section is devoted to proving that this algorithm is correct. We start by establishing termination, then soundness (every stuttering path in the tree corresponds to an execution of the ωOPN) and finally completeness (every execution of the ωOPN corresponds to a stuttering path in the tree). To this end, we rely on the following notions. Symmetrically to *self-covering executions* we define the notion of *self-covering (stuttering) path* in a tree: a (stuttering) path π is *self-covering* iff $\pi = \pi_1 \pi_2$ with *effect*$(\pi_2) \geq \mathbf{0}$. A self-covering stuttering path $\pi = \pi_1 \pi_2$ is ω-*maximal* iff for all nodes n, n' along π_2: nbω $(n) =$ nbω (n').

Termination. Let us show that Build-KM always terminates. First observe that the depth of recursive calls is at most by $|P| + 1$, as the number of places marked by ω along a branch does not decrease, and since we perform a recursive call only when a place gets marked by ω and was not before. Moreover, the branching degree of the tree is bounded by the number $|T|$ of transitions. Thus, by König's lemma, an infinite tree would contain an infinite branch. We rule out this possibility by a classical wqo argument: if there were an infinite branch in the tree computed by Build-KM$(\mathcal{N}, m_0)$, then there would be two nodes n_1 along the branch n_2 (where n_1 is an ancestor of n_2) s.t. $\lambda(n_1) \preceq \lambda(n_2)$ and *effect*$(n_1 \rightsquigarrow n_2) \succeq \mathbf{0}$. Since the depth of recursive calls is bounded, we can assume, wlog, that n_1 and n_2 have been built during the same recursive call, hence $\lambda(n_1) \prec \lambda(n_2)$ is not possible, because this would trigger an acceleration, create an extra ω and start a new recursive call. Thus, $\lambda(n_1) = \lambda(n_2)$, but in this case the algorithm stops developing the branch (line 5). See the appendix for a full proof.

Proposition 8. *For all ωPN $\mathcal{N}$ and marking m_0,* Build-KM$(\mathcal{N}, m_0)$ *terminates.*

Then, following the intuition that we have sketched at the beginning of the section, we show that KM is *sound* (Lemma 9) and *complete* (Lemma 11). *Note that we first establish these results assuming that the ωPN $\mathcal{N}$ given as parameter is an ωOPN, then prove that the results extend to the general case of ωPN .*

Soundness. To establish *soundness* of our algorithm, we show that, for every path $n_0, \ldots, n_k$ in the tree returned by Build-KM$(\mathcal{N}, m_0)$, and for every target marking $m \in \gamma(\lambda(n_k))$, we can find an execution of $\mathcal{N}$ reaching a marking $m' \in \gamma(n_k)$ that covers m. This implies that, if $\lambda(n_k)(p) = \omega$ for some p, then, we can find a family of executions that reach a marking in $\gamma(n_k)$ with an arbitrary number of tokens in p. For instance, consider the path n_1, n_2, n_3 in $\mathcal{T}_1'$ (Fig. 3), and let $m = \langle 0, 2, 4 \rangle$. Then, a corresponding execution is $\langle 1, 0, 0 \rangle \xrightarrow{t_1} \langle 0, 4, 0 \rangle \xrightarrow{t_2} \langle 0, 3, 2 \rangle \xrightarrow{t_2} \langle 0, 2, 4 \rangle$. Remark that the execution is not necessarily the sequence of transitions labeling the path in the tree:

---

[4] Although this differs from classical presentations of the Karp & Miller technique, we have retained it because it simplifies the proofs of correctness.

in this case, we need to iterate t_2 to transfer tokens from p_2 to p_3, which is summarised in one edge (n_2, n_3) in $\mathcal{T}_1$, by the acceleration.

Lemma 9. *Let $\mathcal{N}$ be an ωOPN, let m_0 be an ω-marking and let $\mathcal{T}$ be the tree returned by* Build-KM$(\mathcal{N}, m_0)$. *Let $\pi = n_0, \dots, n_k$ be a stuttering path in $\mathcal{T}$, and let m be a marking in $\gamma(\lambda(n_k))$. Then, there exists an execution $\rho_\pi = m_0 \xrightarrow{t_1} m_1 \cdots \xrightarrow{t_\ell} m_\ell$ of $\mathcal{N}$ s.t. $m_\ell \in \gamma(\lambda(n_k))$, $m_\ell \succeq m$ and $m_0 \in \gamma(\lambda(n_0))$. Moreover, when for all $0 \leq i \leq j \leq k$:* nbw $(n_i) =$ nbw (n_j), *we have: $t_1 \cdots t_\ell = \mu(\pi)$.*

Completeness. Proving completeness amounts to showing that every execution (starting from m_0) of an ωPN $\mathcal{N}$ is witnessed by a stuttering path in Build-KM$(\mathcal{N}, m_0)$. It relies on the following property:

Lemma 10. *Let $\mathcal{N}$ be an ωOPN, let m_0 be an ω-marking, and let $\mathcal{T}$ be the tree returned by* Build-KM$(\mathcal{N}, m_0)$. *Then, for all nodes n of* Build-KM$(\mathcal{N}, m_0)$:

- *either n has no successor in the tree and has an ancestor $\overline{n}$ s.t. $\lambda(\overline{n}) = \lambda(n)$.*
- *or the set of successors of n corresponds to all the $\to_\omega$ possible successors of $\lambda(n)$, i.e.: $\{\mu(n, n') \mid (n, n') \in E\} = \{t \mid \lambda(n) \xrightarrow{t}_\omega\}$. Moreover, for each n' s.t. $(n, n') \in E$ and $\mu(n, n') = t$: $\lambda(n') \succeq \lambda(n) + \text{effect}(t)$.*

We can now state the completeness property:

Lemma 11. *Let $\mathcal{N}$ be an ωOPN with set of transitions T, let m_0 be an initial marking and let $m_0 \xrightarrow{t_1} m_1 \xrightarrow{t_2} \cdots \xrightarrow{t_n} m_n$ be an execution of $\mathcal{N}$. Then, there are a stuttering path $\pi = n_0, n_1, \dots, n_k$ in* Build-KM$(\mathcal{N}, m_0)$ *and a monotonic increasing mapping $h : \{1, \dots, n\} \mapsto \{0, \dots, k\}$ s.t.: $\mu(\pi) = t_1 t_2 \cdots t_n$ and $m_i \preceq \lambda(n_{h(i)})$ for all $0 \leq i \leq n$.*

From ωOPN to ωPN. We have shown completeness and soundness of the Build-KM algorithm for ωOPN. Let us show that each ωPN $\mathcal{N}$ can be turned into an ωOPN remlw$(\mathcal{N})$ that (i) terminates iff $\mathcal{N}$ terminates and (ii) that has the same coverability sets as $\mathcal{N}$. The ωOPN remlw$(\mathcal{N})$ is obtained from $\mathcal{N}$ by replacing each transition $t \in T$ by a transition $t' \in T'$ s.t. $O(t') = O(t)$ and $I(t') = \{I(t)(p) \otimes p \mid I(t)(p) \neq \omega\}$. Intuitively, t' is obtained from t by deleting all ω input arcs. Since t' always consumes *less tokens* than t does, the following is easy to establish:

Lemma 12. *Let $\mathcal{N}$ be an ωPN. For all executions $m_0, t'_1, m_1, \dots, t'_n, m_n$ of* remlw$(\mathcal{N})$: *$m_0, t_1, m_1, \dots, t_n, m_n$ is an execution of $\mathcal{N}$. For all finite (resp. infinite) executions $m_0, t_1, m_1, \dots, t_n, m_n$ $(m_0, t_1, m_1, \dots, t_j, m_j, \dots)$ of $\mathcal{N}$, there is an execution $m_0, t'_1, m'_1, \dots, t'_n, m'_n$ $(m_0, t_1, m'_1, \dots, t_j, m'_j, \dots)$ of* remlw$(\mathcal{N})$, *s.t. $m_i \preceq m'_i$ for all i.*

Intuitively, this means that, when solving coverability, (place) boundedness or termination on an ωPN $\mathcal{N}$, we can analyse remlw$(\mathcal{N})$ instead, because $\mathcal{N}$ terminates iff remlw$(\mathcal{N})$ terminates, and removing the ω-labeled input arcs from $\mathcal{N}$ does not allow to reach higher markings. Finally, we observe that, for all ωPN $\mathcal{N}$, and all initial marking m_0: the trees returned by Build-KM$(\mathcal{N}, m_0)$ and Build-KM (remlw$(\mathcal{N}, m_0)$)

respectively are isomorphic[5]. This is because we have defined $c - \omega$ to be equal to c: applying this rule when computing the effect of a transition t (line 17), is equivalent to computing the effect of the corresponding t' in $\mathsf{reml}\omega(\mathcal{N})$, i.e. letting $I(t')(p) = 0$ for all p s.t. $I(t)(p) = \omega$. Thus, we can lift Lemma 9 and Lemma 11 to ωPN. This establish correctness of the algorithm for the general ωPN case.

Applications of the Karp & Miller Tree. With these results we conclude that the Karp & Miller tree can be used to compute a coverability set and to decide termination of ωPN.

Theorem 13. *Let $\mathcal{N}$ be an ωPN with initial marking m_0, and let $\mathcal{T}$ be the tree returned by $\langle N, E, \lambda, \mu, n_0 \rangle = \mathtt{Build\text{-}KM}(\mathcal{N}, m_0)$. Then: (i) $\lambda(N)$ is a coverability set of $\mathcal{N}$ and (ii) $\mathcal{N}$ terminates iff $\mathcal{T}$ contains an ω-maximal self-covering stuttering path.*

Proof. Point (i) follows from Lemma 9 (lifted to ωPN). Let us now prove both directions of point (ii).

First, we show that if $\mathtt{Build\text{-}KM}(\mathcal{N}, m_0)$ contains an ω-maximal self-covering stuttering path, then $\mathcal{N}$ admits a self-covering execution from m_0. Let $n_0, \ldots, n_k$, $n_{k+1}, \ldots, n_\ell$ be an ω-maximal self-covering stuttering path, and assume $effect(n_{k+1}, \ldots, n_\ell) \geq \mathbf{0}$. Let us apply Lemma 9 (lifted to ωPN), by letting $m = \mathbf{0}$ and $\pi = \pi_2$, and let m_1 and m_2 be markings s.t. $m_1 \xrightarrow{\mu(\pi_2)} m_2$. The existence of m_1 and m_2 is guaranteed by Lemma 9 (lifted to ωPN), because all the nodes along π_2 have the same number of ω's as we are considering an ω-*maximal* self-covering stuttering path. Since $effect(\pi_2)$ is positive, so is $effect(\mu(\pi_2))$. Thus, there exists[6] m_2' s.t. $m_1 \xrightarrow{\mu(\pi_2)} m_2'$ and $m_2' \succeq m_1$. By invoking Lemma 9 (lifted to ωPN) again, letting $\pi = \pi_1$ and $m = m_1$, we conclude to the existence of a sequence of transitions σ, a marking m_0 and a marking $m_1' \succeq m_1$ s.t. $m_0 \xrightarrow{\sigma} m_1'$. Since $m_1' \succeq m_1$, $\mu(\pi_2)$ is again firable from m_1'. Let $\overline{m}_2 = m_2 + m_1' - m_1$. Clearly, $m_1' \xrightarrow{\mu(\pi_2)} \overline{m}_2$, with $\overline{m}_2 \succeq m_1'$. Hence, $m_0 \xrightarrow{\sigma} m_1' \xrightarrow{\mu(\pi_2)} \overline{m}_2$ is a self-covering execution of $\mathcal{N}$.

Second, let us show that, if $\mathcal{N}$ admits a self-covering execution from m_0, then $\mathtt{Build\text{-}KM}(\mathcal{N}, m_0)$ contains an ω-maximal self-covering stuttering path. Let $\rho = m_0 \xrightarrow{t_1} m_1 \cdots \xrightarrow{t_n} m_n$ be a self-covering execution and assume $0 \leq k < n$ is a position s.t. $m_k \preceq m_n$. Let σ_1 denote $t_1, \ldots t_k$ and σ_2 denote $t_{k+1}, \ldots t_n$. Let us consider the execution ρ', defined as follows

---

[5] That is, if $\mathtt{Build\text{-}KM}(\mathcal{N}, m_0)$ returns $\langle N, E, \lambda, \mu, n_0 \rangle$ and $\mathtt{Build\text{-}KM}(\mathsf{reml}\omega(\mathcal{N}, m_0))$ returns $\langle N', E', \lambda', \mu', n_0' \rangle$, then, there is a bijection $h : N \mapsto N'$ s.t. (i) $h(n_0) = n_0'$, (ii) for all $n \in N$: $\lambda(n) = \lambda(h(n))$, (iii) for all n_1, n_2 in N: $(n_1, n_2) \in E$ iff $(h(n_1), h(n_2)) \in E'$, (iv) for all $(n_1, n_2) \in E$: $\mu(n_1, n_2) = \mu'(h(n_1), h(n_2))$.

[6] Remark that, although $effect(\mu(\pi_2)) \succeq \mathbf{0}$, we have no guarantee that $m_2 \succeq m_1$, as we could have $effect(\mu(\pi_2)) = \omega$ for some p, and maybe the amount of tokens that has been produced in p by $\mu(\pi_2)$ to yield m_2 does not allow to have $m_2(p) \geq m_1(p)$. However, in this case, it is always possible to reach a marking with enough tokens in p to cover $m_1(p)$, since $effect(\mu(\pi_2)) = \omega$.

$$\rho' = m_0 \xrightarrow{\sigma_1} \underbrace{m_k \xrightarrow{t_{k+1}} m_{k+1} \cdots \xrightarrow{t_n} m_n}_{\sigma_2} \underbrace{\xrightarrow{t_{k+1}} m_{n+1} \cdots \xrightarrow{t_n} m_{2n-k}}_{\sigma_2} \cdots$$

$$\cdots \underbrace{\xrightarrow{t_{k+1}} m_{(|P|+1)n-|P|k+1} \cdots \xrightarrow{t_n} m_{(|P|+2)n-(|P|+1)k}}_{\sigma_2}$$

where for all $n + 1 \leq j \leq (|P| + 2)n - (|P| + 1)k$: $m_j - m_{j-1} = m_{f(j)} - m_{f(j-1)}$ with f the function defined as $f(x) = ((x - k) \mod (n - k)) + k$ for all x. Intuitively, ρ' amounts to firing $\sigma_1(\sigma_2)^{|P|+1}$ (where P is the set of places of $\mathcal{N}$) from m_0, by using, each time we fire σ_2, the same effect as the one that was used to obtain ρ (remember that the effect of σ_2 is non-deterministic when ω's are produced). It is easy to check that ρ' is indeed an execution of $\mathcal{N}$, because ρ is a self-covering execution.

Let $n_0, n_1, \ldots n_\ell$ and h be the stuttering path in $\mathtt{Build\text{-}KM}(\mathcal{N}, m_0)$ and the mapping corresponding to ρ' (and whose existence is established by Lemma 11). Since, $m_k \preceq m_n$, $\mathit{effect}(t_{k+1} \cdots t_n) \geq 0$ and by Lemma 11 (lifted to ωPN), all the following stuttering paths are self-covering:

$$n_0, \ldots, n_{h(k)}, \ldots, n_{h(n)}$$

$$n_0, \ldots, n_{h(k)}, \ldots, n_{h(n)}, \ldots, n_{h(2n-k)}$$

$$n_0, \ldots, n_{h(k)}, \ldots, n_{h(n)}, \ldots, n_{h(2n-k)}, \ldots, n_{h(3n-2k)}$$

$$\vdots$$

$$n_0, \ldots, n_{h(k)}, \ldots, n_{h(n)}, \ldots, n_{h(2n-k)}, \ldots, n_{h(3n-2k)}, \ldots, n_{h((|P|+2)n-(|P|+1)k)}$$

Let us show that one of them is ω-maximal, i.e. that there is $1 \leq j \leq |P| + 1$ s.t. $\mathrm{nb}\omega\left(n_{h(jn-(j-1)k)}\right) = \mathrm{nb}\omega\left(n_{h((j+1)n-jk)}\right)$. Assume it is not the case. Since the number of ω's can only increase along a stuttering path, this means that

$$0 \leq \mathrm{nb}\omega\left(n_{h(n)}\right) < \mathrm{nb}\omega\left(n_{h(2n-k)}\right) < \mathrm{nb}\omega\left(n_{h(3n-2k)}\right) < \mathrm{nb}\omega\left(n_{h((|P|+2)n-(|P|+1)k)}\right)$$

However, this implies that $\mathrm{nb}\omega\left(n_{h((|P|+2)n-(|P|+1)k)}\right) > |P|$, which is not possible as P is the set of places of $\mathcal{N}$. Hence, we conclude that there exists an ω-maximal self-covering stuttering path in $\mathtt{Build\text{-}KM}(\mathcal{N}, m_0)$. $\square$

4   From ωPN to Plain PN

Let us show that we can, from any ωPN $\mathcal{N}$, build a plain PN $\mathcal{N}'$ whose set of reachable markings allows to recover the reachability set of $\mathcal{N}$. This construction allows to solve reachability, coverability and (place) boundedness. The idea of the construction is depicted in Fig. 5, and can be outlined as follows. A transition t in the ωPN is simulated in *three steps* in the PN. First, t' fires, which (i) moves a token fom the *global lock* $lock_g$ to the local lock $lock_t$ and (ii) consumes the same fixed amount of tokens as t, i.e., if $I_t(p) \neq \omega$, then, t' consumes $I_t(p)$ tokens in p, for all p. Once t' has fired, all transitions are blocked but the $t_{-\omega}^{q_i}$ and $t_{+\omega}^{p_i}$ transitions, that can be fired an arbitrary

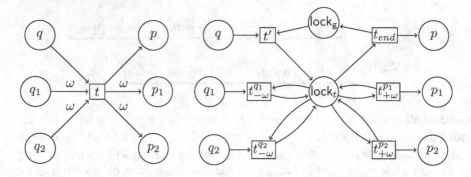

Fig. 5. Transforming an ωPN into a plain PN

number of times to simulate the ω-arcs of t. Finally, t_{end} moves the lock back to $\mathsf{lock_g}$, and produces $O_t(p)$ tokens in all p s.t. $O_t(p) \neq \omega$.

Formally, we turn the ωPN $\mathcal{N} = \langle P, T, m_0 \rangle$ into a plain PN $\mathcal{N}' = \langle P', T', m_0' \rangle$ using the following procedure. Assume that $T = T_{plain} \uplus T_\omega$, where T_ω is the set of ω-transitions of $\mathcal{N}$. Then:

1. We add to the net one place (called the *global lock*) $\mathsf{lock_g}$, and for each ω-transition t, one place lock_t. That is, $P' = P \cup \{\mathsf{lock_g}\} \cup \{\mathsf{lock}_t \mid t \in T_\omega\}$.
2. Each transition t in $\mathcal{N}$ is replaced by a set of transitions T_t in $\mathcal{N}'$. In the case where t is a plain transition, T_t contains a single transition that has the same effect as t, except that it also tests for the presence of a token in $\mathsf{lock_g}$. In the case where t is an ω-transition, T_t is a set of plain transitions that simulate the effect of t, as in Fig. 5. Formally, $T' = \cup_{t \in T} T_t$, where the T_t sets are defined as follows:
   - If t is a plain transition, then $T_t = \{t'\}$, where, $I(t') = I(t) \cup \{\mathsf{lock_g}\}$ and $O(t') = O(t) \cup \{\mathsf{lock_g}\}$.
   - If t is an ω-transition, then: $T_t = \{t', t_{end}\} \cup \{t^p_{-\omega} \mid I(t)(p) = \omega\} \cup \{t^p_{+\omega} \mid O(t)(p) = \omega\}$ where $I(t') = I(t) + \{\mathsf{lock_g}\}$; $O(t') = I(t_{end}) = \{\mathsf{lock}_t\}$; $O(t_{end}) = \{\mathsf{lock_g}\} + O(t)$. Furthermore, for all p s.t. $I(t)(p) = \omega$: $I(t^p_{-\omega}) = \{p, \mathsf{lock}_t\}$ and $O(t^p_{-\omega}) = \{\mathsf{lock}_t\}$. Finally, for all p s.t. $O(t)(p) = \omega$: $I(t^p_{+\omega}) = \{\mathsf{lock}_t\}$ and $O(t^p_{-\omega}) = \{p, \mathsf{lock}_t\}$.
3. We let f be the function that associates each marking m of $\mathcal{N}$ to the marking $f(m)$ of $\mathcal{N}'$ s.t. $m'(\mathsf{lock_g}) = 1$; for all $p \in P$: $m'(p) = m(p)$; and for all $p \notin P \cup \{\mathsf{lock_g}\}$: $m'(p) = 0$. Then, the initial marking of $\mathcal{N}'$ is $f(m_0)$.

It is straightforward to check that:

Lemma 14. *Let $\mathcal{N}$ be an ωPN and let $\mathcal{N}'$ be its corresponding PN. Then $m \in \mathrm{Reach}(\mathcal{N})$ iff $f(m) \in \mathrm{Reach}(\mathcal{N}')$.*

The above construction can be carried out in polynomial time. Thus, ωPN generalise Petri nets, the known complexities for reachability [17,18], (place) boundedness and coverability [20] carry on to ωPN:

Corollary 15. *Reachability for ωPN is decidable and* EXPSPACE*-hard. Coverability, boundedness and place boundedness for ωPN are* EXPSPACE*-c.*

This justifies the result given in Table 1 for reachability, coverability and (place) bound-edness, for ωPN. However, the above construction fails for deciding termination. For instance, assume that the leftmost part of Fig. 5 is an ωPN $\mathcal{N} = \langle P, T, m_0 \rangle$ with $m_0(q) = 1$. Clearly, all executions of $\mathcal{N}$ are finite, while $t'(t_{+\omega}^{p_1})^{\omega}$ is an infinite tran-sition sequence that is firable in $\mathcal{N}'$. Termination, however is decidable, by the KM technique of Section 3, and EXPSPACE-hard, as ωPN generalise Petri nets. In the next section, we show that the Rackoff technique [20] can be generalised to ωPN, and prove that termination is EXPSPACE-c for ωPN.

5   Extending the Rackoff Technique for ωPN

In this section, we extend the Rackoff technique to ωPN to prove the existence of short self-covering sequences. For applications of interest, such as the termination problem, it is sufficient to consider ωOPN, as proved in Lemma 12. Hence, we only consider ωOPN in this section. As in Rackoff's work [20], the idea here is to use small solutions of linear Diophantine equations to limit lengths of sequences. As in the work of Brazdil et al. [3], we modify the effect of a sequence of transitions to ensure that ω-transitions are fired at least once. But the results of [3], proved in the context of games, can not be used here directly for the termination problem.

As observed in [20], beyond some large values, it is not necessary to track the exact value of markings to solve some problems. We use threshold functions $h :\{0, \ldots, |P|\} \to \mathbb{N}$ to specify such large values. Let $\mathrm{nb}\overline{\omega}(m) = |\{p \in P \mid m(p) \in \mathbb{N}\}|$.

Definition 16. Let $h : \{0, \ldots, |P|\} \to \mathbb{N}$ be a threshold function. Given an ω-marking m, the markings $[m]_{h \to \omega}$ and $[m]_{\omega \to h}$ are defined as follows:

$$([m]_{h \to \omega})(p) = \begin{cases} m(p) & \text{if } m(p) < h(\mathrm{nb}\overline{\omega}(m)), \\ \omega & \text{otherwise.} \end{cases}$$

$$([m]_{\omega \to h})(p) = \begin{cases} m(p) & \text{if } m(p) \in \mathbb{N}, \\ h(\mathrm{nb}\overline{\omega}(m) + 1) & \text{otherwise.} \end{cases}$$

In $[m]_{h \to \omega}$, values that are too high are abstracted by ω. In $[m]_{\omega \to h}$, ω is replaced by the corresponding natural number. This kind of abstraction is formalized in the following threshold semantics.

Definition 17. Given an ωPN $\mathcal{N}$, a transition t, an ω-marking m that enables t and a threshold function h, we define the transition relation $\xrightarrow{t}_h$ as $m \xrightarrow{t}_h [m + \mathit{effect}(t)]_{h \to \omega}$.

The transition relation $\xrightarrow{t}_h$ is extended to sequences of transitions in the usual way. Note that if $m \xrightarrow{t}_h m'$, then $\omega(m) \subseteq \omega(m')$. In words, a place marked ω will stay that way along any transition in threshold semantics.

Let $R = \max\{|\mathit{effect}(t)(p)| \mid t \in T, p \in P, \mathit{effect}(t)(p) < \omega\}$. The following proposition says that ω can be replaced by large enough numbers without disabling sequences. The proof is by a routine induction on the length of sequences, using the fact that in an ωOPN, any transition can reduce at most R tokens from a place.

Proposition 18. *For some ω-markings m_1 and m_2, suppose $m_1 \xrightarrow{\sigma}_h m_2$ and $\omega(m_2) = \omega(m_1)$. If m'_1 is a marking such that $m'_1 \preceq_{\omega(m_1)} m_1$ and $m'_1(p) \geq R|\sigma|$ for all $p \in \omega(m_1)$, then $m'_1 \xrightarrow{\sigma} m'_2$ such that $m'_2 \preceq_{\omega(m_2)} m_2$ and $m'_2(p) \geq m'_1(p) - R|\sigma|$.*

Definition 19. *Given an ω-marking m_1 and a threshold function h, an ω-maximal threshold pumping sequence (h-PS) enabled at m_1 is a sequence σ of transitions such that $m_1 \xrightarrow{\sigma}_h m_2$, effect$(\sigma) \geq 0$ and $\omega(m_2) = \omega(m_1)$.*

In the above definition, note that we require *effect*$(\sigma)(p) \geq 0$ for *any* place p, irrespective of whether $m_1(p) = \omega$ or not.

Definition 20. *Suppose σ is an ω-maximal h-PS enabled at m_1 and $\sigma = \sigma_1\sigma_2\sigma_3$ such that $m_1 \xrightarrow{\sigma_1}_h m_3 \xrightarrow{\sigma_2}_h m_3 \xrightarrow{\sigma_3}_h m_2$. We call σ_2 a simple loop if all intermediate ω-markings obtained while firing σ_2 from m_3 (except the last one, which is m_3 again) are distinct from one another.*

In the above definition, since $m_3 \xrightarrow{\sigma_2}_h m_3$ and $m_1 \xrightarrow{\sigma_1\sigma_3}_h m_2$, one might be tempted to think that $\sigma_1\sigma_3$ is also an ω-maximal h-PS enabled at m_1. This is however not true in general, since there might be some $p \in \omega(m_1)$ such that *effect*$(\sigma_1\sigma_3)(p) < 0$ (which is compensated by σ_2 with *effect*$(\sigma_2)(p) > 0$). The presence of the simple loop σ_2 is required due to its compensating effect. The idea of the proof of the following lemma is that if there are a large number of loops, it is enough to retain a few to get a shorter ω-maximal h-PS.

Lemma 21. *There is a constant d such that for any ωPN $\mathcal{N}$, any threshold function h and any ω-maximal h-PS σ enabled at some ω-marking m_1, there is an ω-maximal h-PS σ' enabled at m_1, whose length is at most $(h(\mathrm{nb}\overline{\omega}(m_1))2R)^{d|P|^3}$.*

Proof (Sketch). This proof is similar to that of [20, Lemma 4.5], with some modifications to handle ω-transitions. It is organized into the following steps.

Step 1: We first associate a vector with a sequence of transitions to measure the effect of the sequence. This is the step that differs most from that of [20, Lemma 4.5]. The idea in this step is similar to the one used in [3, Lemma 7].

Step 2: Next we remove some simple loops from σ to obtain σ'' such that for every intermediate ω-marking m in the run $m_1 \xrightarrow{\sigma}_h m_2$, m also occurs in the run $m_1 \xrightarrow{\sigma''}_h m_2$.

Step 3: The sequence σ'' obtained above need not be a h-PS. With the help of the vectors defined in step 1, we formulate a set of linear Diophantine equations that encode the fact that the effects of σ'' and the simple loops that were removed in step 2 combine to give the effect of a h-PS.

Step 4: Then we use the result about existence of small solutions to linear Diophantine equations to construct a sequence σ' that meets the length constraint of the lemma.

Step 5: Finally, we prove that σ' is a h-PS enabled at m_1.

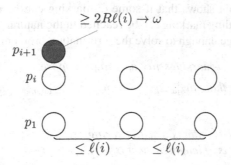

Fig. 6. Intuition for the threshold functions

Details of Step 1: Let $P_\omega \subseteq \omega(m_1)$ be the set of places p such that some transition t in σ has $effect(t)(p) = \omega$. If we ensure that for each place $p \in P_\omega$, some transition t with $effect(t)(p) = \omega$ is fired, we can ignore the effect of other transitions on p. This is formalized in the following definition of the effect of any sequence of transitions $\sigma_1 = t_1 \cdots t_r$. We define the function $\Delta_{P_\omega}[\sigma_1] : \omega(m_1) \to \mathbb{Z}$ as follows.

$$\Delta_{P_\omega}[\sigma_1](p) = \begin{cases} 1 & p \in P_\omega, \exists i \in \{1,\ldots,r\} : effect(t_i)(p) = \omega \\ 0 & p \in P_\omega, \forall i \in \{1,\ldots,r\} : effect(t_i)(p) \neq \omega \\ \sum_{1 \leq i \leq r} effect(t_i)(p) & \text{otherwise} \end{cases}$$

Applying the above definition to simple loops, it is possible to remove some of them to get shorter pumping sequences. Details about how to do it are in the remaining steps of the proof, which can be found in the technical report [13]. □

Definition 22. *Let $c = 2d$. The functions $h_1, h_2, \ell : \mathbb{N} \to \mathbb{N}$ are as follows:*

$$h_1(0) = 1 \qquad \ell(0) = (2R)^{c|P|^3} \qquad h_2(0) = R$$
$$h_1(i+1) = 2R\ell(i) \qquad \ell(i+1) = (h_1(i+1)2R)^{c|P|^3} \qquad h_2(i+1) = R\ell(i)$$

All the above functions are non-decreasing. Due to the selection of the constant c above, we have $(2xR)^{c|P|^3} \geq x^{|P|} + (2xR)^{d|P|^3}$ for all $x \in \mathbb{N}$.

The goal is to prove that if there is a self-covering execution, there is one whose length is at most $\ell(|P|)$. That proof uses the result of Lemma 21 and the definition of ℓ above reflects it. For the intuition behind the definition of h_1 and h_2, suppose that the proof of the length upper bound of $\ell(|P|)$ is by induction on $|P|$ and we have proved the result for $|P| = i$. For the case of $i + 1$, we want to decide the value beyond which it is safe to abstract by replacing numbers by ω. As shown in Fig. 6, suppose the initial prefix of a self-covering execution for i places is of length at most $\ell(i)$. Also suppose the pumping portion of the self-covering execution is of length at most $\ell(i)$. The total length is at most $2\ell(i)$. Since each transition can reduce at most R tokens from any place, it is enough to have $2R\ell(i)$ tokens in p_{i+1} to safely replace numbers by ω.

The following lemma shows that if some ω-marking can be reached in threshold semantics, a corresponding marking can be reached in the natural semantics where ω is replaced by a value large enough to solve the termination problem.

Lemma 23. *For some ω-markings m_3 and m_4, suppose $m_3 \xrightarrow{\sigma}_{h_1} m_4$. Then there is a sequence σ' such that $[m_3]_{\omega \to h_1} \xrightarrow{\sigma'} m_4'$, $m_4' \succeq_{\omega(m_4)} [m_4]_{\omega \to h_2}$ and $|\sigma'| \le h_1(\text{nb}\overline{\omega}(m_3))^{|P|}$.*

Lemma 24. *If an ωPN $\mathcal{N}$ admits a self-covering execution, then it admits one whose sequence of transitions is of length at most $\ell(|P|)$.*

Proof. Suppose $\sigma = \sigma_1 \sigma_2$ is the sequence of transitions in the given self-covering execution such that $m_0 \xrightarrow{\sigma_1} m_1 \xrightarrow{\sigma_2} m_2$ and $m_2 \succeq m_1$. A routine induction on the length of any sequence of transitions σ shows that if $m_3 \xrightarrow{\sigma} m_4$, we have $m_3 \xrightarrow{\sigma}_{h_1} m_4'$ with $m_4' - m_3 \succeq m_4 - m_3$. Hence, we have $m_0 \xrightarrow{\sigma_1}_{h_1} m_1' \xrightarrow{\sigma_2}_{h_1} m_2'$ with $m_2' \succeq m_1'$. By monotonicity, we infer that for any $i \in \mathbb{N}^+$, $m_i' \xrightarrow{\sigma_2}_{h_1} m_{i+1}'$ with $m_{i+1}' \succeq m_i'$.

Let $j \in \mathbb{N}^+$ be the first number such that $\omega(m_j') = \omega(m_{j+1}')$. We have $m_0 \xrightarrow{\sigma_1 \sigma_2^{j-1}}_{h_1} m_j' \xrightarrow{\sigma_2}_{h_1} m_{j+1}'$ and σ_2 is an ω-maximal h_1-PS enabled at m_j'.

By Lemma 21, there is a h_1-PS σ_2' enabled at m_j' whose length is at most $(h_1(\text{nb}\overline{\omega}(m_j'))2R)^{d|P|^3}$. By Lemma 23, there is a sequence σ_1' such that $m_0 \xrightarrow{\sigma_1'} m_j''$, $m_j'' \succeq_{\omega(m_j')} [m_j']_{\omega \to h_2}$ and $|\sigma_1'| \le (h_1(|P|))^{|P|}$. By Definition 22 and Definition 16, we infer that $m_j''(p) = R\ell(\text{nb}\overline{\omega}(m_j')) = R(h_1(\text{nb}\overline{\omega}(m_j'))2R)^{c|P|^3} \ge R|\sigma_2'|$ for all $p \in \omega(m_j')$. Hence, we infer from Proposition 18 that $m_0 \xrightarrow{\sigma_1'} m_j'' \xrightarrow{\sigma_2'} m_{j+1}''$. Since σ_2' is a h_1-PS, $\textit{effect}(\sigma_2') \succeq 0$, and so $m_{j+1}'' \succeq m_j''$. Therefore, firing $\sigma_1' \sigma_2'$ at m_0 results in a self-covering execution. The length of $\sigma_1' \sigma_2'$ is at most $(h_1(|P|))^{|P|} + (h_1(\text{nb}\overline{\omega}(m_j'))2R)^{d|P|^3} \le \ell(|P|)$. $\square$

Lemma 25. *Let $k = 3c$. Then $\ell(i) \le (2R)^{k^{i+1}|P|^{3(i+1)}}$ for all $i \in \mathbb{N}$.*

Theorem 26. *The termination problem for ωPN is* EXPSPACE-*c.*

The idea of the proof of the above theorem is to construct a non-deterministic Turing machine that guesses and verifies a self-covering sequence. By Lemma 24, the length of such a sequence can be limited and hence made to work in EXPSPACE.

6   Extensions with Transfer or Reset Arcs

In this section, we consider two extensions of ωPN, namely: ωPN with *transfer arcs* (ωPN+T) and ωPN with *reset arcs* (ωPN+R). These extensions have been considered in the case of plain Petri nets: Petri nets with transfer arcs (PN+T) and Petri nets with reset arcs (PN+R) have been extensively studied in the literature [7,1,8,22]. Intuitively, a *transfer arc* allows to *transfer all the tokens* from a designated place p to a given place q, while a *reset arc consumes all tokens* from a designated place p.

Formally, an *extended* ωPN is a tuple $\langle P, T \rangle$, where P is a finite set of places and T is finite set of transitions. Each transition is a pair $t = (I, O)$ where $I : P \mapsto \mathbb{N} \cup \{\omega, \mathsf{T}, \mathsf{R}\}; O : P \mapsto \mathbb{N} \cup \{\omega, \mathsf{T}\}; |\{p \mid I(p) \in \{\mathsf{T}, \mathsf{R}\}\}| \leq 1; |\{p \mid O(p) \in \{\mathsf{T}\}\}| \leq 1;$ there is p s.t. $I(p) = \mathsf{T}$ *iff* there is q s.t. $O(q) = \mathsf{T}$; and *if* there is p s.t. $I(p) = \mathsf{R}$, *then*, $O(p) \in \mathbb{N} \cup \{\omega\}$ for all p. A transition (I, O) s.t. $I(p) = \mathsf{T}$ (resp. $I(p) = \mathsf{R}$) for some p is called a *transfer* (*reset*). An ωPN with *transfer arcs* (resp. *with reset arcs*), ωPN+T (ωPN+R) for short, is an extended ωPN that contains no reset (transfer). An ωPN+T s.t. $I(t)(p) \neq \omega$ for all transitions t and places p is an ωOPN+T. The class ωIPN+T is defined symmetrically. An ωPN+T which is both an ωOPN+T *and* an ωIPN+T is a (plain) PN+T. The classes ωOPN+R, ωIPN+R and PN+R are defined accordingly.

Let $t = (I, O)$ be a transfer or a reset. t is *enabled* in a marking m iff for all p: $I(p) \notin \{\omega, \mathsf{T}, \mathsf{R}\}$ implies $m(p) \geq I(p)$. In this case firing t yields a marking $m' = m - m_I + m_O$ (denoted $m \xrightarrow{t} m'$) where for all p: $m_I(p) = m(p)$ if $I(p) \in \{\mathsf{T}, \mathsf{R}\};$ $0 \leq m_I(p) \leq m(p)$ if $I(p) = \omega; m_I(p) = I(p)$ if $I(p) \notin \{\mathsf{T}, \mathsf{R}, \omega\}; m_O(p) = m(p')$ if $O(p) = I(p') = \mathsf{T}$; $m_O(p) \geq 0$ if $O(p) = \omega$; and $m_O(p) = O(p)$ if $O(p) \notin \{\mathsf{T}, \omega\}.$ The semantics of transitions that are neither transfers nor resets is as defined for ωPN.

Let us now investigate the status of the problems listed in Section 2, in the case of ωPN+T and ωPN+R. First, since ωPN+T (ωPN+R) extend PN+T (PN+R), the lower bounds for the latters carry on: reachability and place-boundedness are undecidable [6] for ωPN+T and ωPN+R; boundedness is undecidable for ωPN+R [8]; and coverability is Ackerman-hard for ωPN+T and ωPN+R [22]. On the other hand, the construction given in Section 4 can be adapted to turn an ωPN+T (resp. ωPN+R) $\mathcal{N}$ into a PN+T (PN+R) $\mathcal{N}'$ satisfying Lemma 14 (i.e., projecting $\mathrm{Reach}(\mathcal{N}', m_0)$ on the set of places of $\mathcal{N}$ yields $\mathrm{Reach}(\mathcal{N}, m_0)$). Hence, boundedness for ωPN+T [8], and coverability for both ωPN+T and ωPN+R are decidable [1].

As far as *termination* is concerned, it is decidable [7] and Ackerman-hard [22] for PN+R and PN+T. Unfortunately, the construction presented in Section 4 does not preserve termination, so we cannot reduce termination of ωPN+T (resp. ωPN+R) to termination of PN+T (PN+R). Actually, termination becomes undecidable when considering ωOPN+R or ωOPN+T:

Theorem 27. *Termination is undecidable for* ωOPN+T *and* ωOPN+R *with one* ω-*output-arc.*

Proof. We first prove undecidability for ωOPN+T. The proof is by reduction from the *parameterised termination problem* for *Broadcast protocols* (BP) [9]. It is well-known that PN+T generalise broadcast protocols, hence the following *parameterised termination problem for PN+T* is *undecidable*: 'given a PN+T $\langle P, T \rangle$ and an ω-marking $\overline{m}_0$, does $\langle P, T, m_0 \rangle$ terminate *for all* $m_0 \in \downarrow (\overline{m}_0)$?' From a PN+T $\mathcal{N} = \langle P, T \rangle$ and an ω-marking $\overline{m}_0$, we build the ωOPN+T (with only one ω-output-arc) $\mathcal{N}' = \langle P', T', m_0' \rangle$ where $P' = P \uplus \{p_{init}\}$, $T' = T \uplus \{(I, O)\}$, $I = \{p_{init}\}$, $O = \{\omega \otimes p \mid \overline{m}_0(p) = \omega\}$, and $m_0' = \{\overline{m}_0 \otimes p \mid \overline{m}_0(p) \neq \omega\}$. Clearly, $\mathcal{N}'$ terminates iff $\langle P, T, m_0 \rangle$ terminates for all $m_0 \in \downarrow (\overline{m}_0)$. Hence, termination for ωOPN+T is *undecidable* too. Finally, we can transform an ωOPN+R $\mathcal{N} = \langle P, T, m_0 \rangle$ into an ωOPN+T $\mathcal{N}' = \langle P \uplus \{p_{trash}\}, T', m_0 \rangle$, where $t' \in T'$ iff either (i) $t' \in T$ and t' is not a reset, or (ii) there is a reset $t \in T$ and a place $p \in P$ s.t. $I(t)(p) = \mathsf{R}, I(t')(p) = \mathsf{T}$, $O(t')(p_{trash}) = \mathsf{T}$, for all $p' \neq p$: $I(t')(p') = I(t)(p')$ and for all $p'' \neq p_{trash}$:

$O(t')(p'') = O(t)(p'')$. Intuitively, the construction replaces each reset (resetting place p) in $\mathcal{N}$ by a transfer from p to p_{trash} in $\mathcal{N}'$, where p_{trash} is a fresh place from which no transition consume. Since $\mathcal{N}'$ terminates iff $\mathcal{N}$ terminates, termination is undecidable for ωPN+R too.   □

However, the construction of Section 4 can be applied to ωIPN+T and ωIPN+R to yield a corresponding PN+T (resp. PN+R) that preserves termination. Hence, termination is decidable and Ackerman-hard for those models. This justifies the results on ωPN+T and ωPN+R given in Table 1.

References

1. Abdulla, P.A., Cerans, K., Jonsson, B., Tsay, Y.-K.: General Decidability Theorems for Infinite-state Systems. In: LICS 1996. IEEE (1996)
2. Borosh, I., Treybig, L.: Bounds on positive integral solutions of linear diophantine equations. Proceedings of the American Mathematical Society 55(2), 299–304 (1976)
3. Brázdil, T., Jančar, P., Kučera, A.: Reachability games on extended vector addition systems with states. In: Abramsky, S., Gavoille, C., Kirchner, C., Meyer auf der Heide, F., Spirakis, P.G. (eds.) ICALP 2010. LNCS, vol. 6199, pp. 478–489. Springer, Heidelberg (2010)
4. Delzanno, G., Raskin, J.-F., Van Begin, L.: Towards the Automated Verification of Multi-threaded Java Programs. In: Katoen, J.-P., Stevens, P. (eds.) TACAS 2002. LNCS, vol. 2280, pp. 173–187. Springer, Heidelberg (2002)
5. Delzano, G.: Constraint-Based Verification of Parameterized Cache Coherence Protocols. FMSD 23(3) (2003)
6. Dufourd, C.: Réseaux de Petri avec reset/transfert: Décidabilité et indécidabilité. PhD thesis, ENS de Cachan (1998)
7. Dufourd, C., Finkel, A., Schnoebelen, P.: Reset Nets Between Decidability and Undecidability. In: Larsen, K.G., Skyum, S., Winskel, G. (eds.) ICALP 1998. LNCS, vol. 1443, pp. 103–115. Springer, Heidelberg (1998)
8. Dufourd, C., Jančar, P., Schnoebelen, P.: Boundedness of reset P/T nets. In: Wiedermann, J., Van Emde Boas, P., Nielsen, M. (eds.) ICALP 1999. LNCS, vol. 1644, pp. 301–310. Springer, Heidelberg (1999)
9. Esparza, J., Finkel, A., Mayr, R.: On the Verification of Broadcast Protocols. In: LICS 1999. IEEE (1999)
10. Finkel, A., Schnoebelen, P.: Well-structured transition systems everywhere! TCS 256(1-2), 63–92 (2001)
11. Finkel, A., McKenzie, P., Picaronny, C.: A well-structured framework for analysing Petri net extensions. Inf. Comput. 195(1-2), 1–29 (2004)
12. Geeraerts, G., Heußner, A., Raskin, J.F.: Queue-Dispatch Asynchronous Systems. To appear in the Proceedings of ACSD. IEEE (2013), http://arxiv.org/abs/1201.4871
13. Geeraerts, G., Heußner, A., Praveen, M., Raskin, J.F.: ω-Petri nets ArXiV.org CoRR abs/1301.6572 (2013), http://arxiv.org/abs/1301.6572
14. Geeraerts, G., Raskin, J.-F., Van Begin, L.: Expand, enlarge and check: New algorithms for the coverability problem of wsts. J. Comput. Syst. Sci. 72(1) (2006)
15. German, S., Sistla, A.: Reasoning about systems with many processes. J. ACM 39(3), 675–735 (1992)
16. Karp, R.M., Miller, R.E.: Parallel Program Schemata. JCSS 3, 147–195 (1969)

17. Lipton, R.: The reachability problem requires exponential space. Tech. Report. Yale University (1963)
18. Mayr, E.W.: An algorithm for the general Petri net reachability problem. SIAM J. of Computing 3(13), 441–460 (1984)
19. Petri, C.A.: Kommunikation mit Automaten. PhD thesis, Institut fur Instrumentelle Mathematik, Bonn (1962)
20. Rackoff, C.: The covering and boundedness problems for vector addition systems. TCS 6, 223–231 (1978)
21. Reisig, W.: Petri Nets: An Introduction. Springer (1985)
22. Schnoebelen, P.: Revisiting Ackermann-hardness for lossy counter machines and reset Petri nets. In: Hliněný, P., Kučera, A. (eds.) MFCS 2010. LNCS, vol. 6281, pp. 616–628. Springer, Heidelberg (2010)

Results on Equivalence, Boundedness, Liveness, and Covering Problems of BPP-Petri Nets

Ernst W. Mayr and Jeremias Weihmann

Technische Universität München, 85748 Garching, Germany
{mayr,weihmann}@in.tum.de
http://www14.in.tum.de/personen/index.html.en

Abstract. Yen proposed a construction for a semilinear representation of the reachability set of BPP-Petri nets which can be used to decide the equivalence problem of two BPP-PNs in doubly exponential time. We first address a gap in this construction which therefore does not always represent the reachability set. We propose a solution which is formulated in such a way that a large portion of Yen's construction and proof can be retained, preserving the size of the semilinear representation and the double exponential time bound (except for possibly larger values of some constants). In the second part of the paper, we propose very efficient algorithms for several variations of the boundedness and liveness problems of BPP-PNs. For several more complex notions of boundedness, as well as for the covering problem, we show NP-completeness. To demonstrate the contrast between BPP-PNs and a slight generalization regarding edge multiplicities, we show that the complexity of the classical boundedness problem increases from linear time to coNP-hardness. Our results also imply corresponding complexity bounds for related problems for process algebras and (commutative) context-free grammars.

1 Introduction

Basic Parallel Processes Petri nets (BPP-PNs, also known as communication-free Petri nets) are characterized by the simple topological constraint that each transition has exactly one input place (connected by an edge with multiplicity 1). There are several reasons why it is insightful to investigate this class. Studying nontrivial subclasses of Petri nets helps understanding the dynamics of Petri nets in general, which could finally lead to a primitive recursive algorithm for the reachability problem of general Petri nets (a non primitive recursive algorithm was given by Mayr [11]). Furthermore, this Petri net class is closely related to both Basic Parallel Processes, a subclass of Milner's Calculus of Communicating Systems (CCS, see, e.g., [1, 2]), as well as (commutative) context-free grammars (see, e.g., [6, 3]).

The strong topological constraint on BPP-PNs limits the computational power of these nets in the sense that they are unable to model synchronizing actions since the fireability of a transition only depends on exactly one place. Esparza [3] showed that the reachability problem of BPP-PNs is, nevertheless, NP-hard.

J.-M. Colom and J. Desel (Eds.): PETRI NETS 2013, LNCS 7927, pp. 70–89, 2013.

Furthermore, he showed that the problem is in NP. Both results together yield an alternative proof for the NP-completeness of the uniform word problem for commutative context-free grammars (as shown earlier by Huynh [6]).

Another proof for NP-membership, based on canonical firing sequences, was given by Yen [19]. In addition, he proposed an exponential time construction for a semilinear representation of the reachability set of BPP-PNs. He then used this semilinear representation to argue that the equivalence problem for BPP-PNs has a doubly exponential time bound.

In section 3, we address a gap in this construction. We show that, in general, the construction actually computes a proper superset of the reachability set. We then show how to fix the construction in such a way that most parts of Yen's argumentation can be retained while maintaining the size and running time bounds of the original construction (in the sense that all specified constants stay the same).

For some notions of boundedness and liveness of BPPs/BPP-PNs ([9, 13, 14], also see [15]), polynomial time algorithms are already known. In addition to these, we also investigate a number of other variations of the boundedness, the covering, and the liveness problem for BPP-PNs in sections 4 and 5. For two variants of the boundedness problem, and for the covering problem, we show NP-completeness. Using, among other things, results from section 3, we can decide most of the remaining problems very efficiently in linear time. (Extensions of) these algorithms are also applicable to related problems of BPPs and (commutative) context-free grammars (e.g., finiteness of context-free grammars, see [4]). These minor results can be found in the technical report [12].

Linear time algorithms not only make these problems tractable in practice but also show that BPP-PNs are too restricted if we are searching for classes of Petri nets where these problems are hard. Further variations and generalizations of BPP-PNs need to be investigated in order to mark the boundary where these problems cease to be easy. As a first example, we show that the classical boundedness problem becomes coNP-hard if we slightly weaken the restriction on the multiplicities of edges from places to transitions in BPP-PNs.

2 Preliminaries

$\mathbb{Z}$, $\mathbb{N}_0$, and $\mathbb{N}$ denote the sets of all integers, all nonnegative integers, and all positive integers, respectively, while $[a, b] = \{a, a + 1, \ldots, b\} \subsetneq \mathbb{Z}$, and $[k] = [1, k] \subsetneq \mathbb{N}$. For two vectors $u, v \in \mathbb{Z}^k$, we write $u \geq v$ if $u(i) \geq v(i)$ for all $i \in [k]$, and $u > v$ if $u \geq v$ and $u(i) > v(i)$ for some $i \in [k]$. When k is understood, a denotes, for a number $a \in \mathbb{Z}$, the k-dimensional vector with $a_i = a$ for all $i \in [k]$.

A Petri net N is a 3-tuple (P, T, F) where P is a finite set of n places, T is a finite set of m transitions with $S \cap T = \emptyset$, and $F : P \times T \cup T \times P \to \mathbb{N}_0$ is a flow function. A marking μ (of N) is a function $P \to \mathbb{N}_0$. A pair (N, μ_0) such that μ_0 is a marking of N is called a marked Petri net, and μ_0 is called its initial marking. We will omit the term "marked" if the presence of a certain initial marking is clear from the context.

Throughout this paper, n and m will always refer to the number of places and transitions of the Petri net under consideration. For a transition $t \in T$, ${}^\bullet t$ ($t^\bullet$, resp.) is the preset (postset, resp.) of t and denotes the set of all places p such that $F(p, t) > 0$ ($F(t, p) > 0$, resp.). Analogously, ${}^\bullet p$ and $p^\bullet$ are defined for the places $p \in P$.

A Petri net naturally corresponds to a directed bipartite graph with edges from P to T and vice versa such that there is an edge from $p \in P$ to $t \in T$ (from t to p, resp.) labelled with w if $0 < F(p, t) = w$ (if $0 < F(t, p) = w$, resp.). The label of an edge is called multiplicity. If a Petri net is visualized, places are usually drawn as circles and transitions as bars. If the Petri net is marked by μ, then for each place p the circle corresponding to p contains $\mu(p)$ so called tokens.

For a Petri net $N = (P, T, F)$ and a marking μ of N, a transition $t \in T$ can be applied at μ producing a vector $\mu' \in \mathbb{Z}^n$ with $\mu'(p) = \mu(p) - F(p, t) + F(t, p)$ for all $p \in P$. The transition t is enabled at μ or in (N, μ) if $\mu(p) \geq F(p, t)$ for all $p \in P$. We say that t is fired at marking μ if t is enabled and applied at μ. If t is fired at μ, then the produced vector μ' is a marking, and we write $\mu \xrightarrow{t} \mu'$.

Intuitively, if a transition is fired, it first removes $F(p, t)$ tokens from p and then adds $F(t, p)$ tokens to p. An element of T^* is called a transition sequence, an element of T^∞ is called an ∞-transition sequence. For the empty sequence $\sigma = ()$ of transitions, we define $\mu \xrightarrow{\sigma} \mu$. For a nonempty transition sequence $\sigma = (t_1, \ldots, t_k)$, we write $\mu_0 \xrightarrow{\sigma} \mu_k$ if there are markings $\mu_1, \ldots, \mu_{k-1}$ such that $\mu_0 \xrightarrow{t_1} \mu_1 \xrightarrow{t_2} \mu_2 \ldots \xrightarrow{t_k} \mu_k$.

The Parikh map $\Psi : T^* \to \mathbb{N}_0^m$ maps a transition sequence σ to its Parikh image $\Psi[\sigma]$ where $\Psi[\sigma](t) = k$ for a transition t if t appears exactly k times in σ. A Parikh vector is simply an element of $\mathbb{N}_0^m$ (hence each Parikh vector is the Parikh image of some transition sequence). For a Parikh vector Φ we write $t \in \Phi$ if $\Phi(t) > 0$, and $t \in \sigma$ if $t \in \Psi[\sigma]$.

If there is a marking μ' such that $\mu \xrightarrow{\sigma} \mu'$, then we say that σ (the Parikh vector $\Psi[\sigma]$, resp.) is enabled at μ and leads from μ to μ'. For a marked Petri net (N, μ_0), we call a transition sequence that is enabled at μ_0 a firing sequence, and we say that a marking μ is reachable if there is a firing sequence leading to μ. Analogously, an ∞-transition sequence σ is enabled at μ if each finite prefix of σ is enabled at μ. If σ is enabled at μ_0, we call σ an ∞-firing sequence. The reachability set $\mathcal{R}(N, \mu_0)$ of (N, μ_0) consists of all markings μ of N for which there is a firing sequence σ such that $\mu_0 \xrightarrow{\sigma} \mu$. We say that a marking μ can be covered or μ is coverable if there is a reachable marking $\mu' \geq \mu$.

The displacement $\Delta : \mathbb{N}_0^m \to \mathbb{Z}^n$ maps Parikh vectors $\Phi \in \mathbb{N}_0^m$ onto the change of tokens at the places $p_1, \ldots, p_n$ when applying transition sequences with Parikh image Φ. That is, we have $\Delta[\Phi](p) = \sum_{t \in T} \Phi(t) \cdot (F(t, p) - F(p, t))$ for all places p. Accordingly, we define the displacement $\Delta[\sigma]$ of a transition sequence σ by $\Delta[\sigma] := \Delta[\Psi[\sigma]]$. A Parikh vector or a transition sequence having nonnegative displacement at all places is called a nonnegative loop since, immediately after being fired, the loop is enabled again. A nonnegative loop having positive displacement at place p is a positive loop (for p).

Sometimes it is convenient to only consider those places and transitions that are relevant w.r.t. a given set of transitions or a Parikh vector. For a Petri net $\mathcal{P} = (P, T, F, \mu_0)$, and a set D of transitions, the Petri net $\mathcal{P}_D$ consists of all transitions $t \in D$, all places $p \in \bigcup_{t \in D}({}^\bullet t \cup t^\bullet)$, and the flow function F and initial marking μ_0 restricted to these subsets of transitions and places. For a Parikh vector Φ we define $\mathcal{P}_\Phi := \mathcal{P}_D$ where $D = \{t \mid t \in \Phi\}$.

In the case of BPP-PNs the strongly connected components (SCCs) are also of major interest. The directed acyclic graph obtained by shrinking all SCCs to super nodes while maintaining the edges between distinct SCC as edges between the corresponding super nodes is called the condensation (of the graph). We call an SCC C a top component (bottom component, resp.) if it has no incoming (no outgoing, resp.) edges in the condensation. For two not necessarily distinct SCCs C_1, C_2, we write $C_1 \geq C_2$ if there is a path from C_1 to C_2 in the condensation.

An important concept in the analysis of Petri nets are traps. A subset $T \subseteq P$ of places is a trap if ${}^\bullet t \cap T \neq \emptyset$ implies $t^\bullet \cap T \neq \emptyset$, i.e., every transition that removes a token from T also adds a token to T. Once a trap is marked, it cannot be unmarked by firing a transition.

Some marked Petri nets have reachability sets that are semilinear. A set $S \subseteq \mathbb{N}_0^n$ is semilinear if there is a finite number of linear sets $L_1, \ldots, L_k \subseteq \mathbb{N}_0^n$ such that $S = \bigcup_{i \in [k]} L_i$. A set $L \subseteq \mathbb{N}_0^n$ is linear if there is a finite number of vectors $b, p_1, \ldots, p_\ell \in \mathbb{N}_0^n$ such that $L = \mathcal{L}(b, \{p_1, \ldots, p_\ell\})$ where $\mathcal{L}(b, \{p_1, \ldots, p_\ell\}) := \{b + \sum_{i \in [\ell]} a_i p_i \mid a_i \in \mathbb{N}_0, i \in [\ell]\}$. The vector b is the constant vector of L while the vectors p_i are the periods of L. A semilinear representation of a semilinear set S is a set consisting of k pairs $(b_i, \{p_{i,1}, \ldots, p_{i,\ell_i}\})$, $i \in [k]$, such that $S = \bigcup_{i \in [k]} \mathcal{L}(b_i, \{p_{i,1}, \ldots, p_{i,\ell_i}\})$.

If two Petri nets allow the construction of semilinear representations of the respective reachability sets within a certain time bound, then not only many problems that are in general undecidable are decidable for this class but time bounds can be given as well. For example, the equivalence problem of (conveniently encoded representations of) semilinear sets is in Π_2^P [5, 7]. This implies that the equivalence problem for that class, i.e., the question if two Petri nets of the class have the same set of reachable markings, is decidable in exponential time w.r.t. the combined time to construct the semilinear sets.

When we talk about the input size, a "reasonable" encoding/description is assumed. We specifically assume that every number is encoded in binary representation. Furthermore, we assume for convenience that a Petri net is encoded as an enumeration of places $p_1, \ldots, p_n$ and transitions $t_1 \ldots, t_m$ followed by an enumeration of the edges with their respective edge multiplicities. A vector of $\mathbb{N}_0^k$ is encoded as a k-tuple. If we regard a tuple as an input (e.g. a marked Petri net), then it is encoded as a tuple of the encodings of the particular components. All running times given in later sections assume the random-access machine (RAM) as the model of computation.

3   The Equivalence Problem of BPP-Petri Nets Revisited

In this section we consider the equivalence problem of BPP-PNs.

Definition 1 (Equivalence problem of BPP-PNs). *Given two BPP-PNs* $\mathcal{P}$ *and* $\mathcal{P}'$, *are* $\mathcal{R}(\mathcal{P})$ *and* $\mathcal{R}(\mathcal{P}')$ *equal?*

In [19], Yen proposed a construction for a semilinear representation of the reachability set of BPP-PNs. The obtained representation has exponential size in the size of the BPP-PN. The author used the fact that the equivalence problem of semilinear sets is in Π_2^P (see [5, 7]) to show a double exponential time bound for this problem. The construction of the semilinear representation is contained in the proposed proof of the following theorem.

Theorem 1 ([19], Theorem 5). *Let* $\mathcal{P} = (P, T, F, \mu_0)$ *be a BPP-Petri net of size* s. *For some fixed constants* c_1, c_2, d_1, d_2, d_3 *independent of* s, *we can construct in* $DTIME(2^{c_2 s^3})$ *a semilinear reachability set* $\mathcal{R}(\mathcal{P}) = \bigcup_{\nu \in B} \mathcal{L}(\nu, \rho_\nu)$ *whose size is bounded by* $\mathcal{O}(2^{c_1 s^3})$, *where*

1. B *is the set of reachable markings with no component larger than* $2^{d_1 s^2}$, *and*
2. ρ_ν *is the set of all* $\vartheta \in \mathbb{N}^k$ *such that*
   (a) ϑ *has no component larger than* $2^{d_2 s^2}$, *and*
   (b) $\exists\, \sigma, \sigma_1, \sigma_2 \in T^*$, $\exists$ *marking* μ_1,
       (i) $\mu_0 \xrightarrow{\sigma_1} \mu_1 \xrightarrow{\sigma_2} \nu$,
       (ii) $\mu_1 \xrightarrow{\sigma} \mu_1 + \vartheta$,
       (iii) $|\sigma|, |\sigma_1\sigma_2| \leq 2^{d_3 s^2}$.

We show that there are BPP-PNs such that the constructed semilinear set contains markings that are not reachable. Consider the BPP-PN $\mathcal{P}$ with initial marking $\mu_0 = (1, 0, 0, 0)$ of Figure 1. The marking $\nu = (0, 0, 0, 1)$ is reachable.

In particular we have $\mu_0 \xrightarrow{t_1}$ $\mu_1 = (0, 1, 0, 0) \xrightarrow{t_2} \nu$ as well as $\mu_0 \xrightarrow{t_3} \mu_1' = (0, 0, 1, 0) \xrightarrow{t_4} \nu$. Notice that we can safely and w.l.o.g. assume $\nu \in B$ since we can blow up the size of the net by adding unrelated places. Now observe that $\mu_1 \xrightarrow{t_5} \mu_1 + \vartheta$ where $\vartheta = (0, 1, 0, 0)$, as well as $\mu_1' \xrightarrow{t_6} \mu_1' + \vartheta'$ where $\vartheta' = (0, 0, 1, 0)$. As before, we can safely assume $|t_1t_2|, |t_3t_4|, |t_5|, |t_6| \leq 2^{d_3 s^2}$. Therefore, we find $\vartheta, \vartheta' \in \rho_\nu$. But then, the unreachable marking $(0, 1, 1, 1)$ is in $\mathcal{L}(\nu, \rho_\nu)$. Hence, the constructed semilinear set $S :=$ $\bigcup_{\nu \in B} \mathcal{L}(\nu, \rho_\nu)$ cannot equal $\mathcal{R}(\mathcal{P})$.

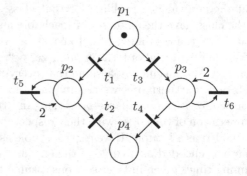

Fig. 1. A counter example for the construction proposed in Theorem 1

The inclusion $\mathcal{R}(\mathcal{P}) \subseteq S$ is proven correctly in [19]. Our goal is to repair the construction in such a way that we can more or less completely reuse the proof given for this direction. Our first step is to show that there is a certain subclass of BPP-PNs for which the other direction $S \subseteq \mathcal{R}(\mathcal{P})$ is also true. To this end, observe that the crucial property of the net of Figure 1 that makes this net a counter example is that ν is reachable by the two firing sequences $t_1 t_2$ and $t_3 t_4$ which have different Parikh images. We will later show that the restriction to those BPP-PNs having the nice property that any two firing sequences leading to the same marking have the same Parikh image yields a variation of this theorem which is correct.

Before we can prove such a theorem, we first need some observations about enabled Parikh vectors and nonnegative loops in BPP-PNs.

Lemma 1. *Let $\mathcal{P} = (N, \mu_0)$ be a BPP-PN. A Parikh vector Φ is enabled in $\mathcal{P}$ if and only if*

(a) $\mu_0 + \Delta[\Phi] \geq 0$, *and*
(b) *each top component of $\mathcal{P}_\Phi$ has a marked place.*

Proof. This Lemma is a variation of Theorem 3.1 of [3] which is better suited for our purposes. The theorem states that Φ is enabled if and only if (a) holds and if, within $\mathcal{P}_\Phi$, each place can be reached from a marked place. □

Lemma 2. *Let $\mathcal{P} = (P, T, F, \mu_0)$ be a Petri net, $\sigma = \sigma_1 \cdots \sigma_k \in T^k$ a firing sequence in $\mathcal{P}$, and let μ_i, $i \in [k]$, be defined by $\mu_0 \xrightarrow{\sigma_1} \mu_1 \xrightarrow{\sigma_2} \ldots \xrightarrow{\sigma_k} \mu_k$. Then, for each place p of $\mathcal{P}_{\Psi[\sigma]}$, there is an $i \in [0, k]$ such that p is marked in μ_i.*

Proof. Each place p of $\mathcal{P}_{\Psi[\sigma]}$ is in the pre- or postset of some transition σ_i. If $p \in {}^\bullet \sigma_i$, then p must be marked in μ_{i-1}. If $p \in \sigma_i{}^\bullet$, then p is marked in μ_i. □

Lemma 3. *Let Φ be a nonnegative loop of a BPP-PN $\mathcal{P} = (P, T, F)$, and let $C_1, \ldots, C_k$, $k \geq 1$, denote the top components of $\mathcal{P}_\Phi$. Then Φ can be split into nonnegative loops $\Phi_1, \ldots, \Phi_k$, $k \leq n$, such that*

(a) $\Phi = \sum_{i=1}^{k} \Phi_i$, *and*
(b) *the only top component of $\mathcal{P}_{\Phi_i}$ is C_i.*

Proof. For $\Phi = 0$, the lemma is obviously true, hence we assume $\Phi > 0$. We show that we can extract Φ_1 from Φ. By iteratively applying this procedure, we obtain the nonnegative loops of interest. $\mathcal{P}_{\Phi_1}$ will contain C_1 and all nodes that are reachable from C_1. Since it is possible that firing a transition t of an SCC C with $C \leq C_1, C_2$ and $C \neq C_1, C_2$ exactly $\Phi(t)$ times requires tokens coming from C_1 and C_2, $\mathcal{P}_{\Phi_1}$ and $\mathcal{P}_{\Phi - \Phi_1}$ will in general not be disjoint.

We first note that a top component of $\mathcal{P}_\Phi$ always contains a transition since otherwise there would be a transition of Φ that removes tokens from the component but no transition that adds tokens to it, a contradiction to Φ being a nonnegative loop. Let $\Phi' = \Phi$ and $\Phi_1 = 0$. By moving a transition t from Φ' to Φ_1 we mean setting $\Phi'(t) := \Phi'(t) - 1$ and $\Phi_1(t) := \Phi_1(t) + 1$.

We start with a top component C_1 of $\mathcal{P}_\Phi$, and move each transition $t \in C_1$ exactly $\Phi'(t)$ times from Φ' to Φ_1. Then we iterate the following process:

(i) If $\Delta[\Phi_1](p) > 0$ for a place p and $p = {}^\bullet t$ for a transition $t \in \Phi'$, then we move t from Φ' to Φ_1.

(ii) Otherwise, if there is a top component of $\mathcal{P}_{\Phi'}$ that is not a top component of $\mathcal{P}_\Phi$, we move each transition t of this component exactly $\Phi'(t)$ times from Φ' to Φ_1.

The procedure ends when none of these two cases is applicable.

We first prove that at each step of the procedure the only top component of $\mathcal{P}_{\Phi_1}$ is C_1. After the first step this is obviously true since we completely move C_1. Assume this holds for $\ell - 1$ steps. If the ℓ-th step is of case (i), then it holds after ℓ steps since the only nodes that are possibly added to the induced graph $\mathcal{P}_{\Phi_1}$ by moving t are t and $t^\bullet$ which can be reached by the place ${}^\bullet t$ which is already part of $\mathcal{P}_{\Phi_1}$ before moving t. If the ℓ-th step is of case (ii), then the moved component C was originally created by moving a transition t such that $t^\bullet$ and C share a place. This shows that $\mathcal{P}_{\Phi_1}$ has only one top component after the last step of the procedure.

Observe that the top components of $\mathcal{P}_{\Phi'}$ are exactly $C_2, \ldots, C_k$. The reason is that C_1 is moved, $C_2, \ldots, C_k$ remain untouched, and (ii) ensures the moving of all newly created top components.

Now, we show that $\Delta[\Phi_1] \geq \mathbf{0}$ holds *at each step of the procedure*. After the first step, i.e., after moving the top component, this holds since otherwise Φ wouldn't be a nonnegative loop. Suppose this holds after $\ell - 1$ steps. If the ℓ-th step is of case (i), then it obviously still holds after that.

Suppose, the ℓ-th step is of case (ii), where C is the new top component that is moved during this step. Consider the situation immediately *before the ℓ-th step*. Let Φ_C be defined by $\Phi_C(t) = \Phi'(t)$ if $t \in C$, and $\Phi_C(t) = 0$ otherwise. Our goal is to show that $\Delta[\Phi_C] \geq \mathbf{0}$ since this and the induction hypothesis imply $\Delta[\Phi_1 + \Phi_C] \geq \mathbf{0}$, i.e., after moving all transitions of C in the ℓ-th step, the resulting Parikh vector is a nonnegative loop.

First notice that for all places $p \notin C$ we have $\Delta[\Phi_C](p) \geq 0$. (This follows from the fact that C is a transition induced SCC, implying ${}^\bullet t \in C$ for all $t \in C$.)

Consider a place $p \in C$. By the induction hypothesis, we have $\Delta[\Phi_1](p) \geq 0$. $\Delta[\Phi_1](p) > 0$ cannot occur since otherwise the ℓ-th step would be of case (i) (applied to a transition $t \in C$ having $p = {}^\bullet t$). Thus, we have $\Delta[\Phi_1](p) = 0$.

Now, observe that for all $t \in \Phi'$ having $t^\bullet \in C$ we have $t \in \Phi_C$ since C is a top component. This implies $\Delta[\Phi_C](p) \geq \Delta[\Phi'](p)$. Combining all these observations we obtain

$$\Delta[\Phi_C](p) = \Delta[\Phi_1 + \Phi_C](p) \geq \Delta[\Phi_1 + \Phi'](p) = \Delta[\Phi](p) \geq 0.$$

Now, we show that Φ' is a nonnegative loop *at the end of the procedure*. Let p be a place, and consider the situation *after the last step*. If $\Delta[\Phi_1](p) > 0$, then there is no transition $t \in \Phi'$ having ${}^\bullet t = p$ since otherwise (i) would be applicable, and the procedure wouldn't have stopped, yet. This implies $\Delta[\Phi'](p) \geq 0$. If $\Delta[\Phi_1](p) = 0$, then $\Delta[\Phi'](p) \geq 0$ follows from $\Delta[\Phi_1 + \Phi'] = \Delta[\Phi] \geq \mathbf{0}$. As shown above, the case $\Delta[\Phi_1](p) < 0$ cannot occur.

Since $\bullet t \neq \emptyset$ for all transitions t, each top component of $\mathcal{P}_\Phi$ contains at least one place. This implies $k \leq n$, concluding the proof.   □

Lemma 4. *Let $\mathcal{P} = (P, T, F, \mu_0)$ be a BPP-PN, and Φ, ϑ Parikh vectors such that ϑ is a nonnegative loop, and Φ and $\Phi + \vartheta$ are enabled. Then, for each firing sequence α such that $\mathcal{P}_\Phi$ is a subnet of $\mathcal{P}_{\Psi[\alpha]}$, there are transition sequences $\alpha_1, \ldots, \alpha_{k+1}$ and nonnegative loops $\tau_1, \ldots, \tau_k$, $k \leq n$, such that*

(a) $\alpha = \alpha_1 \cdots \alpha_{k+1}$,
(b) $\vartheta = \tau_1 + \ldots + \tau_k$,
(c) $\mathcal{P}_{\tau_i}$, $i \in [k]$, has exactly one top component, and this top component is the i-th top component of $\mathcal{P}_\vartheta$ using a properly chosen numbering of the top components, and
(d) τ_i, $i \in [k]$, is enabled at marking μ_i where $\mu_0 \xrightarrow{\alpha_1 \cdots \alpha_i} \mu_i$.

Proof. Consider the decomposition of ϑ by Lemma 3 into nonnegative loops $\tau_1, \ldots, \tau_k$, $k \leq n$, such that $\vartheta = \sum_{i=1}^k \tau_i$, and the i-th top component C_i of $\mathcal{P}_\vartheta$ is the unique top component of $\mathcal{P}_{\tau_i}$.

Let $i \in [k]$. Assume that C_i and $\mathcal{P}_\Phi$ are disjoint. Then, C_i is a top component of $\mathcal{P}_{\Phi+\vartheta}$, and C_i is marked at μ_0 by Lemma 1 since $\Phi + \vartheta$ is enabled at μ_0. Therefore, by the same lemma, τ_i is enabled at μ_0.

Now, assume that C_i and $\mathcal{P}_\Phi$ are not disjoint, i.e., they share a place p. Since $\mathcal{P}_\Phi$ is a subnet of $\mathcal{P}_{\Psi[\alpha]}$, Lemma 2 implies that there are transition sequences α', α'' such that $\alpha = \alpha' \cdot \alpha''$ and p is marked at μ' where $\mu_0 \xrightarrow{\alpha'} \mu'$. Therefore, by Lemma 1, τ_i is enabled at μ'.

We conclude that, by splitting the sequence α at appropriate positions, there are transition sequences $\alpha_1, \ldots, \alpha_{k+1}$ such that $\alpha = \alpha_1 \alpha_2 \cdots \alpha_{k+1}$, and $\mu_0 \xrightarrow{\alpha_1} \mu_1 \cdots \xrightarrow{\alpha_k} \mu_k \xrightarrow{\alpha_{k+1}} \mu$, and τ_i is enabled at μ_i where we assume w.l.o.g. that the top components of $\mathcal{P}_\vartheta$ are conveniently numbered.   □

Having collected and proven these observations, we can show the following restricted variation of Theorem 1.

Theorem 2. *Let $\mathcal{P} = (P, T, F, \mu_0)$ be a BPP-Petri net of size s such that for all firing sequences τ, τ' leading to the same marking $\Psi(\tau) = \Psi(\tau')$ holds. For some fixed constants c_1, c_2, d_1, d_2, d_3 independent of s, we can construct in $DTIME(2^{c_2 s^3})$ a semilinear reachability set $\mathcal{R}(\mathcal{P}) = \bigcup_{\nu \in B} \mathcal{L}(\nu, \rho_\nu)$ whose size is bounded by $\mathcal{O}(2^{c_1 s^3})$, where*

1. B is the set of reachable markings with no component larger than $2^{d_1 s^2}$, and
2. ρ_ν is the set of all $\vartheta \in \mathbb{N}^k$ such that
   (a) ϑ has no component larger than $2^{d_2 s^2}$, and
   (b) $\exists \sigma, \sigma_1, \sigma_2 \in T^*$, $\exists$ marking μ_1,
       (i) $\mu_0 \xrightarrow{\sigma_1} \mu_1 \xrightarrow{\sigma_2} \nu$,
       (ii) $\mu_1 \xrightarrow{\sigma} \mu_1 + \vartheta$,
       (iii) $|\sigma|, |\sigma_1 \sigma_2| \leq 2^{d_3 s^2}$.

Proof. Assume $B \neq \emptyset$, and let $\nu \in B$, and $\mu \in \mathcal{L}(\nu, \rho_\nu)$ be arbitrarily chosen. Our goal is to show that μ is reachable. W.l.o.g. let $\rho_\nu = \{\vartheta_1, \ldots, \vartheta_\ell\}$. By definition, there are $a_1, \ldots, a_\ell \in \mathbb{N}_0$ such that $\mu = \nu + \sum_{i=1}^{\ell} a_i \vartheta_i$. For ϑ_i, $i \in [\ell]$, let $\sigma_{i,1}$ denote the sequence σ_1 as defined in the theorem. Since all firing sequences leading to ν have the same Parikh image Φ_ν, $\mathcal{P}_{\Psi[\sigma_{i,1}]}$ is a subnet of $\mathcal{P}_{\Phi_\nu}$.

Let α be some firing sequence having Parikh image Φ_ν, and let μ_j, $j \in [\|\alpha\|]$, be defined by $\mu_0 \xrightarrow{\alpha_1 \cdots \alpha_j} \mu_j$. By applying Lemma 4 to $\Psi[\sigma_{i,1}]$, θ_i, and α for all $i \in [\ell]$, we find that for any partial loop of any θ_i there is a $j \in [0, \|\alpha\|]$ such that the partial loop under consideration is enabled at μ_j. Therefore, the Parikh vector $\Psi[\alpha] + \sum_{i=1}^{\ell} a_i \vartheta_i$ leading to μ is enabled at μ_0.   □

We can use this theorem and corresponding construction in a mediate way to construct a semilinear representation of $\mathcal{R}(\mathcal{P})$ in exponential time for every BPP-PN $\mathcal{P}$. For that we need the following definition.

Definition 2 (Parikh extension). *Let $\mathcal{P} = (P, T, F, \mu_0)$, $P = \{p_1, \ldots, p_n\}$, $T = \{t_1, \ldots, t_m\}$ be a Petri net. The Parikh extension $\mathcal{P}^e = (P^e, T, F^e, \mu_0^e)$ of $\mathcal{P}$ is obtained by adding an unmarked place p_i^* for each transition t_i such that $F(t_i, p_i^*) = 1$.*

Figure 2 illustrates the Parikh extension of the net of Figure 1. If we fire a firing sequence σ, $\mu_0' \xrightarrow{\sigma} \mu_1$, in the Parikh extension $\mathcal{P}^e$, then the new place p_i^*, $i \in [m]$, counts how often the transition t_i is fired. In other words, the projection of μ_1 onto the new places equals $\Psi[\sigma]$. Hence, for each marking μ reachable in $\mathcal{P}^e$ all firing sequences leading to μ have the same Parikh image. This allows us to prove the next theorem. We remark that the concept of the Parikh extension is closely related to the concept of extended Parikh maps used in [10] for persistent Petri nets.

Theorem 3. *Let $\mathcal{P} = (P, T, F, \mu_0)$ be a BPP-Petri net of size s. For some fixed constants c_1, c_2, d_1, d_2 independent of s, we can construct in $DTIME(2^{c_2 s^3})$ a semilinear representation of the reachability set $\mathcal{R}(\mathcal{P})$ whose size is bounded by*

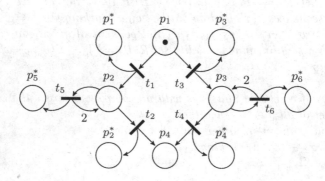

Fig. 2. The Parikh extension of the net of Figure 1

$\mathcal{O}(2^{c_1 s^3})$ where no component of any constant vector is larger than $2^{d_1 s^2}$ and no component of any period is larger than $2^{d_2 s^2}$.

Proof. We compute the Parikh extension $\mathcal{P}^e$ of $\mathcal{P}$. First notice that $\mathcal{P}^e$ is a BPP-PN. Since all firing sequences of $\mathcal{P}^e$ leading to the same marking have the same Parikh image, we can apply the construction given in [19], which is correct for $\mathcal{P}^e$ by Theorem 2, in order to obtain the semilinear representation $\mathcal{SL}(\mathcal{P}^e)$ of $\mathcal{R}(\mathcal{P}^e)$. Now notice that a marking μ is reachable in $\mathcal{P}$ if and only if there is a marking μ' that is reachable in $\mathcal{P}^e$ such that the projection of μ' onto the places of $\mathcal{P}$ equals μ (to see this, simply apply the same firing sequence). Therefore, the projection of $\mathcal{SL}(\mathcal{P}^e)$ onto the places of $\mathcal{P}$ yields the semilinear representation $\mathcal{SL}(\mathcal{P})$ of $\mathcal{R}(\mathcal{P})$.

The running time of this projection is linear in the size of $\mathcal{SL}(\mathcal{P}^e)$. In turn, the size of $\mathcal{P}^e$ is linear in the size of $\mathcal{P}$. Hence, the constants c_1, c_2, d_1, d_2 may be larger for this theorem than for Theorem 2 but all specified constants (like the cube of s^3) are not increased. $\qquad\square$

In the next sections we investigate several variations of the boundedness, and the liveness problem for BPP-PNs. In addition we show that the covering problem is NP-complete for BPP-PNs.

4   Boundedness Problems for BPP-PNs

We first define the concepts of boundedness we are interested in.

Definition 3. *Let* $\mathcal{P} = (P, T, F, \mu_0)$ *be a Petri net. A place* $p \in P$ *is*

(i) *unbounded if, for all* $k \in \mathbb{N}$, *there is a reachable marking* $\mu \in \mathcal{R}(\mathcal{P})$ *such that* $\mu(p) \geq k$.

(ii) *unbounded on an* ∞-*firing sequence* σ *if, for all* $k \in \mathbb{N}$, *there is a finite prefix of* σ *leading to a marking* μ *such that* $\mu(p) \geq k$.

(iii) *persistently unbounded if, for all reachable markings* $\mu \in \mathcal{R}(\mathcal{P})$, p *is unbounded in the Petri net* (P, T, F, μ).

A set $S \subseteq P$ *of places is*

(iv) *(placewise) unbounded if some place (all places, respectively) of* S *are unbounded.*

(v) *(∞-)unbounded if* S *contains a place that is unbounded (on an* ∞-*firing sequence, respectively).*

(vi) *placewise (∞-)unbounded if all places of* S *are unbounded (on an* ∞-*firing sequence, respectively).*

(vii) *simultaneously unbounded if, for all* $k \in \mathbb{N}$, *there is a reachable marking* $\mu \in \mathcal{R}(\mathcal{P})$ *such that* $\mu(p) \geq k$ *for all* $p \in S$.

(viii) *simultaneously* ∞-*unbounded if there is an* ∞-*firing sequence* σ *such that, for all* $k \in \mathbb{N}$, *there is a finite prefix of* σ *leading to a marking* μ *satisfying* $\mu(p) \geq k$ *for all* $p \in S$.

We remark that, for a place, "persistently unbounded" implies "unbounded on an ∞-firing sequence" which implies "unbounded". Furthermore, by Lemma 3.2 of [10] a set $S \subseteq P$ of places is simultaneously unbounded on some ∞-firing sequence if and only if there is an ∞-firing sequence σ such that all places $p \in S$ are unbounded on (the same sequence) σ. Hence, this on first sight weaker characterization yields another definition for the same concept.

4.1   Concepts of Non-simultaneously Unboundedness

In this subsection we investigate concepts of unboundedness where the places under consideration are not required to be simultaneously (∞-)unbounded, and provide efficient algorithms for the corresponding decision problems.

Lemma 5. *Let $\mathcal{P} = (P, T, F, \mu_0)$ be a BPP-PN, and $p \in P$ a place. Then the following are equivalent.*

1. *p is unbounded.*
2. *There is a firing sequence σ leading to a marking μ and a positive loop τ enabled at μ such that $\Delta[\tau](p) > 0$.*
3. *p is unbounded on some ∞-firing sequence.*
4. *There are strongly connected components C_1, C_2, C_3, C_4 of $\mathcal{P}$ such that*
   (a) *$p \in C_4$,*
   (b) *$C_1 \geq C_2 \geq C_3 \geq C_4$,*
   (c) *C_1 contains a marked place, and*
   (d) *C_2 contains a transition t with $^\bullet t \in C_2$ and $\sum_{p' \in t^\bullet \cap (C_2 \cup C_3)} F(t, p') \geq 2$.*

Proof. $1 \Rightarrow 2$: By definition, there is an infinite sequence of enabled Parikh vectors $(\Phi'_1, \Phi'_2, \ldots)$ such that $\Delta[\Phi'_i](p) < \Delta[\Phi'_{i+1}](p)$, $i \in \mathbb{N}$. It is easy to see that this sequence contains an infinite subsequence $(\Phi_1, \Phi_2, \ldots)$ such that $\Phi_i \leq \Phi_{i+1}$, $i \in \mathbb{N}$ (see, e.g., Lemma 4.1. of [8]). In particular, we have $\Delta[\Phi_1](p) < \Delta[\Phi_2](p)$, i.e., there is a positive loop ϑ such that $\Delta[\vartheta](p) > 0$ and $\Phi_1 + \vartheta = \Phi_2$. Since both Φ_1 and Φ_2 are enabled, we can apply Lemma 4 to Φ_1, ϑ and some firing sequence α having Parikh image Φ_1. Let $\alpha_1, \ldots, \alpha_{k+1}$ and $\tau_1, \ldots, \tau_k$ be defined as in the lemma. Then we have $\Delta[\tau_i](p) > 0$ for some $i \in [k]$. Let $\tau := \tau_i$. For $\mu_0 \xrightarrow{\alpha_1 \cdots \alpha_i} \mu$, τ is enabled at μ, concluding the proof.

$1 \Rightarrow 4$: We continue where the proof for $1 \Rightarrow 2$ ended. Let C'_2 be the unique top component of $\mathcal{P}_{\Psi[\tau]}$, and C'_4 the SCC of $\mathcal{P}_{\Psi[\tau]}$ containing p. Since τ is enabled at μ, by Lemma 1 there are places p_1 and p_2 such that p_1 is marked at μ_0, $\mathcal{P}$ contains a path from p_1 to p_2, p_2 is contained in C'_2, and $\mu(p_2) > 0$. Define C_1 as the SCC of $\mathcal{P}$ containing p_1.

Since τ is a nonnegative loop, C'_2 contains a transition. If there is a transition t of C'_2 such that $\sum_{p' \in t^\bullet \cap C'_2} F(t, p') \geq 2$, then simply define $C'_3 = C'_2$. Now, assume that such a transition doesn't exist. Then, we have $C'_4 \neq C'_2$ since the total number of tokens in C'_2 cannot increase by firing τ. In particular, there is a path $(p_2, t, p_3, \ldots, p)$ from C'_2 to C'_4 where $p_3 \notin C'_2$. Let C'_3 be the SCC of $\mathcal{P}_{\Psi[\tau]}$ containing p_3. If $t^\bullet \cap C'_2 = \emptyset$, then τ decreases the number of tokens at C'_2, a contradiction to τ being a nonnegative loop. Therefore, $t^{\bullet} \cap C_2' \neq \emptyset$, and we obtain $\sum_{p' \in t^{\bullet} \cap (C_2' \cup C_3')} F(t, p') \geq 2$. Now, let C_i for $i \in [2, 4]$ be the SCC of $\mathcal{P}$ containing C_i', and observe that $C_1, \ldots, C_4$ satisfy the properties (a)–(d).

$2 \Rightarrow 3$: p is unbounded on the ∞-firing sequence $\sigma \tau^{\infty}$.

$3 \Rightarrow 1$: This follows immediately from the definitions.

$4 \Rightarrow 1$: To mark $^{\bullet}t$, we first fire along a path starting at a marked place of C_1 and ending at $^{\bullet}t$. Then we fire $k \in \mathbb{N}$ times along a cycle containing t. After that, at least k tokens can be transferred to p. □

We remark that a Petri net is unbounded if and only if there is a firing sequence σ, a marking μ, and a positive loop ϑ such that $\mu_0 \xrightarrow{\sigma} \mu$, and ϑ is enabled at μ (see [8]). Lemma 5 states that the same principle holds for single places of a BPP-PN. In general, however, this is not true. We further note that (in contrast to, e.g., persistent Petri nets, see [10]) this concept doesn't hold for sets of places of BPP-PNs, i.e., a set $S \subseteq P$ of places of a BPP-PN is not necessarily simultaneously ∞-unbounded if it is simultaneously unbounded. An example is given in Figure 3. We can use the characterization provided by Lemma 5 to give efficient algorithms for certain boundedness problems.

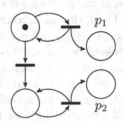

Fig. 3. $\{p_1, p_2\}$ is simultaneously unbounded but not simultaneously-∞-unbounded

Theorem 4. *Given a BPP-PN $\mathcal{P} = (P, T, F, \mu_0)$ and sets $S_1, \ldots, S_k \subseteq P$ of places, we can determine in linear time which sets are*

(a) *(∞-)unbounded in $\mathcal{P}$.*
(b) *placewise (∞-)unbounded in $\mathcal{P}$.*

Proof. Using Tarjan's modified depth-first search [18], we find the strongly connected components of $\mathcal{P}$. Then, we use four DFSs in the condensation to determine all C_1, C_2, C_3, and finally C_4-components. For (a), we simply check if each S_i, $i \in [k]$, contains a place p that is in some C_4-component. For (b), we check for each S_i, $i \in [k]$, if each place of S_i is in some C_4-component. □

Definition 4 (Boundedness problem for BPP-PNs). *Given a BPP-PN $\mathcal{P}$, are all places of $\mathcal{P}$ bounded?*

Corollary 1. *The boundedness problem for BPP-PNs is decidable in linear time.*

Proof. Apply Theorem 4 to the set of all places. □

We remark that in [9] was shown that boundedness of Basic Parallel Processes can be decided in polynomial time.

Interestingly, a slight relaxation of BPP-PNs leads to a class of nets for which the boundedness problem is coNP-hard.

Theorem 5. *Let multiplicity generalized BPP-PNs be the Petri net class consisting of all Petri nets $\mathcal{P} = (P, T, F, \mu_0)$ which satisfy $|^{\bullet}t| = 1$ for all $t \in T$. The boundedness problem for multiplicity generalized BPP-PNs is coNP-hard.*

Proof. We reduce 3-SAT in logspace to the unboundedness problem which is the complement of the boundedness problem. The reduction is illustrated in Figure 4. A similar reduction was used by Esparza [3] to show the NP-hardness of the reachability problem for BPP-PNs. Let $C_1 \wedge C_2 \wedge \ldots \wedge C_\ell$ be a formula in 3-CNF with k variables $x_1, \ldots, x_k$ and ℓ clauses $C_1, \ldots, C_\ell$.

We create a BPP-PN $\mathcal{P}$ as follows. Each variable is represented by a place containing one token, and each clause C_j is represented by a place c_j containing 3 tokens. For each variable x_i, there are two transitions x_i and $\overline{x_i}$ representing the truth assignment of x_i where x_i ($\overline{x_i}$, resp.) puts a token to place c_j if the literal x_i ($\overline{x_i}$, resp.) is contained in C_j.

Then, there is a counting place p which counts how many clauses are satisfied. If clause C_j is satisfied by an assignment, c_j contains at least 4 and at most 6 tokens which allows us to transfer exactly one token from c_j to p. If C_j is unsatisfied, c_j contains 3 tokens, and we cannot transfer any token to p.

Therefore, $\mathcal{P}$ is unbounded if and only if p is unbounded if and only if all ℓ clauses can be satisfied. This shows the NP-hardness of the unboundedness problem and therefore the coNP-hardness of the boundedness problem. □

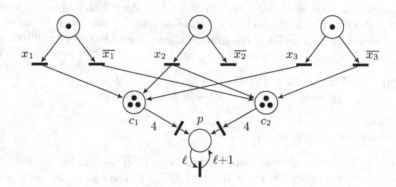

Fig. 4. The formula $C_1 \wedge C_2 = (x_1 \vee x_2 \vee x_3) \wedge (\overline{x_1} \vee x_2 \vee \overline{x_3})$ can be satisfied if and only if this BPP-PN is unbounded

In order to decide the next variation of the boundedness problem in linear time, we need some observations about traps in BPP-PNs.

Lemma 6. *Let $\mathcal{P} = (P, T, F, \mu_0)$ be a BPP-PN and $R \subseteq P$ be a set of places. The trap $Q \subseteq R$ of maximum cardinality w.r.t. R can be determined in linear time.*

Proof. Apply the following procedure. Let $Q' := R$. While there is a transition $t \in T$ such that $\bullet t \in Q'$ and $t^\bullet \cap Q' = \emptyset$, remove $\bullet t$ from Q'. Let Q denote the resulting set. Q must be a trap since otherwise the procedure wouldn't have stopped. Furthermore, Q is maximal w.r.t. inclusion since the procedure can't remove a place from a maximal trap. There is exactly one maximal trap w.r.t. inclusion which therefore is a maximum trap.

We can implement the procedure in linear time as follows. We use two arrays A and N, and a list L, as well as a collection Q. The collection Q is initialized with the set R. The array A has length $|T|$ and $A[i]$ is initialized with $|t_i{}^\bullet \cap R|$. The array N has length $|P|$ and $N[i]$ is initialized with an empty list if $p_i \notin R$, and otherwise with a list of all transitions t_j such that $t_j \in {}^\bullet p_i$. The list L is initialized with all transitions t_i having ${}^\bullet t_i \in Q$ and $A[i] = 0$. It's not hard to see that these data structures can be initialized in linear time.

Now, as long as L is not empty, we do the following. First, we pop some transition t_i from the list, and let $p_j = {}^\bullet t_i$. Then, if $p_j \in Q$, we remove p_j from Q, and, for each t_k contained in the list stored at $N[j]$, we decrease $A[k]$ by 1, and add t_k to L if $A[k] = 0$ after the decreasing step.

When L is empty, $Q \subseteq R$ is the trap of maximum cardinality w.r.t. R. The running time of this procedure is linear.   □

Lemma 7. *Let $\mathcal{P} = (P, T, F, \mu_0)$ be a BPP-PN, and $R \subseteq P$ be a subset of places such that no set $Q \subseteq R$ is a trap. Then, there is a firing sequence σ leading to a marking where no place of R is marked such that $\Delta[\sigma](p) \geq 0$ for all $p \notin R$.*

Proof. By definition, if a set $Q \subseteq P$ is not a trap, then there is a transition t with ${}^\bullet t \in Q$ and $t^\bullet \cap Q = \emptyset$. Define the transitions $t_1, \ldots, t_{|R|}$ and the sets $R_0, \ldots, R_{|R|+1}$ recursively as follows. We start with $R_1 = R$. Given R_i for $i \in [|R|]$, then t_i is a transition with ${}^\bullet t_i \in R_i$ and $t_i{}^\bullet \cap R_i = \emptyset$, and $R_{i+1} = R_i - {}^\bullet t_i$. In other words, $R_{|R|} \subsetneq \cdots \subsetneq R_1$, and we can successively empty $R_{|R|}, \ldots, R_1$ by firing the transitions $t_{|R|}, \ldots, t_1$ each an appropriate number of times. Since these transitions don't remove tokens from places outside of R, the displacement of the firing sequence at these places is nonnegative.   □

Lemma 8. *Let $\mathcal{P} = (P, T, F, \mu_0)$ be a BPP-PN, and $Q \subseteq R \subseteq P$ be the trap of maximum cardinality w.r.t. R. Then there is a firing sequence σ leading to μ with $\mu(p) = 0$ for all $p \in R$ if and only if all places of Q are unmarked.*

Proof. "$\Rightarrow$": If Q is marked, then R will always be marked, regardless of the transitions fired.

"$\Leftarrow$": Notice that $R' := R \setminus Q$ doesn't contain a trap by the maximality of Q. Consider the BPP-PN $\mathcal{P}'$ which emerges from $\mathcal{P}$ by removing Q and all transitions incident to Q. R' also doesn't contain a trap w.r.t. $\mathcal{P}'$. By Lemma 7, R' can be emptied in $\mathcal{P}'$. Therefore, R can be emptied in $\mathcal{P}$.   □

These observations enable us to prove the following theorem.

Theorem 6. *Given a BPP-PN $\mathcal{P} = (P, T, F, \mu_0)$ and a place $p \in P$, we can decide in linear time if p is persistently unbounded.*

Proof. We use the terminology of Lemma 5. Let C_4 be the SCC containing p. For the marking μ_0' having exactly one token at each place, we determine the set $R \subseteq P$ of all places contained in SCCs C_1 for which SCCs C_2 and C_3 exist such that C_1, C_2, C_3, and C_4 satisfy the properties mentioned in Lemma 5. By this lemma, p is unbounded at each marking μ such that there is a place $r \in R$ with $\mu(r) > 0$.

Therefore, p is not persistently unbounded if and only if there is a marking reachable from μ_0 where no place of R is marked. By Lemma 8, we only have to determine if the maximum trap $Q \subseteq R$ w.r.t. R is marked. By Lemma 6, this can be done in linear time.   □

4.2   Simultaneously Unboundedness and the Covering Problem

In this subsection, we consider the covering problem as well as boundedness problems where we ask if many places are simultaneously (∞-)unbounded.

Definition 5 (SU). *Given a BPP-PN $\mathcal{P} = (P, T, F, \mu_0)$ and a subset $S \subseteq P$ of places, is S simultaneously unbounded?*

Definition 6 (S-∞-U). *Given a BPP-PN $\mathcal{P} = (P, T, F, \mu_0)$ and a subset $S \subseteq P$ of places, is S simultaneously ∞-unbounded?*

Definition 7 (Covering problem for BPP-PNs (covering)). *Given a BPP-PN $\mathcal{P}$, and a marking μ of $\mathcal{P}$, is there a reachable marking $\mu' \geq \mu$?*

In order to prove the next theorems, we need the following corollary which is an immediate consequence of Lemma 2 of [19].

Corollary 2. *Let $\mathcal{P} = (P, T, F, \mu_0)$ be a BPP-PN with $m = |T|$ and largest edge multiplicity W, and let μ be a reachable marking. Then there is a firing sequence $\sigma = \pi_1\alpha_1\pi_2\alpha_2 \cdots \pi_m\alpha_m$ leading from μ_0 to μ such that, for all $i \in [m]$, π_i is a nonnegative loop, and α_i satisfies $-mW \leq \Delta[\alpha_i](p) \leq mW$ for all $p \in P$.*

Theorem 7. *SU and S-∞-U are NP-complete even if we restrict the input to BPP-PNs $\mathcal{P} = (P, T, F, \mu_0)$ with $|t^\bullet| = 1$ and $F(t, t^\bullet) \leq 2$ for all $t \in T$.*

Note that a further restriction to $F(t, t^\bullet) = 1$ leads to S-Systems, a subclass of BPP-PNs, which are always bounded.

Theorem 8. *The covering problem for BPP-PNs is NP-complete.*

Proof. For two problems A and B let $A \preceq_{\log} B$ denote the existence of a logspace many-one reduction from A to B.

We first show the NP-hardness of SU and S-∞-U by showing 3-SAT $\preceq_{\log}$ SU and 3-SAT $\preceq_{\log}$ S-∞-U.

Given a formula F in 3-CNF over the variables $x_1, \ldots, x_k$ and clauses $C_1, \ldots, C_\ell$, we construct a BPP-PN such that a certain subset $S = \{c_i \mid i \in [\ell]\}$ of places is simultaneously (∞)-unbounded if and only if F can be satisfied. An example is illustrated in Figure 5 (cf. the reduction in the proof of Theorem 5).

Next, we show covering $\in$ NP by reducing covering in logspace to the reachability problem. We modify $\mathcal{P}$ by adding, for each $p \in P$, a transition t_p having $F(p, t_p) = 1$. Call the resulting BPP-PN $\mathcal{P}'$. Notice that a marking μ can be covered in $\mathcal{P}$ if and only if μ can be covered in $\mathcal{P}'$ if and only if μ is reachable in $\mathcal{P}'$.

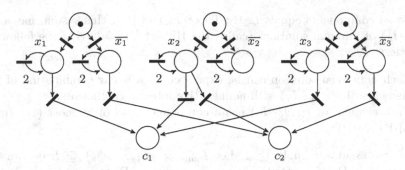

Fig. 5. The formula $C_1 \wedge C_2 = (x_1 \vee x_2 \vee x_3) \wedge (\overline{x_1} \vee x_2 \vee \overline{x_3})$ can be satisfied if and only if $\{c_1, c_2\}$ is simultaneously (∞-)unbounded

Now, we show $\mathsf{SU} \in \mathsf{NP}$ and that **covering** is NP-hard by showing $\mathsf{SU} \prec_{\log}$ **covering**. Again consider the net $\mathcal{P}'$. Let W be the largest edge multiplicity of $\mathcal{P}'$. Define the marking μ by $\mu(p) = \mu_0(p) + (n+m)^2 W + 1$ if $p \in S$, and $\mu(p) = 0$ otherwise. Notice that μ has polynomial encoding size.

Assume that S is simultaneously unbounded in $\mathcal{P}$. Then μ is coverable in $\mathcal{P}'$. Now, assume that μ is coverable in $\mathcal{P}'$, i.e., μ is reachable in $\mathcal{P}'$. In accordance with Corollary 2, let $\pi_1 \alpha_1 \cdots \pi_{n+m} \alpha_{n+m}$ be a firing sequence of $\mathcal{P}'$ leading from μ_0 to μ. Then $-(n+m)^2 W \leq \sum_{i \in [n+m]} \Delta[\alpha_i](p) \leq (n+m)^2 W$ for all $p \in P$. Hence, each place $p \in S$ has some i such that $\Delta[\pi_i](p) > 0$. Therefore, for each $k \in \mathbb{N}$, the marking μ'_k reached in $\mathcal{P}'$ by the firing sequence $\pi_1^{(n+m)^2 W + k}$. $\alpha_1 \cdots \pi_{n+m}^{(n+m)^2 W + k} \cdot \alpha_{n+m}$ satisfies $\mu'_k(p) \geq k$ for all $p \in S$. By removing all transitions from this firing sequence that are not part of $\mathcal{P}$, we obtain a firing sequence of $\mathcal{P}$ leading to a marking μ_k of $\mathcal{P}$ with $\mu_k(p) \geq k$ for all $p \in S$. Therefore, S is simultaneously unbounded in $\mathcal{P}$.

It remains to be shown that $\mathsf{S}\text{-}\infty\text{-}\mathsf{U} \in \mathsf{NP}$. To this end, we will describe a nondeterministic procedure accepting if and only if the given set $S \subseteq P$ is simultaneously ∞-unbounded. Suppose, the latter is the case. Then there is an ∞-firing sequence on which S is simultaneously unbounded. A similar argument as in the proof of Lemma 5 shows that there are transitions sequences σ, τ such that $\sigma\tau^\infty$ is an ∞-firing sequence and τ is a positive loop having $\Delta[\tau](p) \geq 1$ for all $p \in S$.

By Lemma 1, τ is enabled at exactly those markings μ where all top components of $\mathcal{P}_\tau$ are marked. Therefore, there is a marking μ^* such that each place has either zero or one tokens and such that τ is enabled at μ^*. We will use the existence of μ^* later.

Let $D \in \mathbb{Z}^{n \times m}$ be the displacement matrix of $\mathcal{P}$, i.e., the i-th column of D equals $\Delta[t_i]$. Consider the system $D\Phi \geq \mathbf{0}$ of linear diophantine inequalities. Obviously, the set L of nontrivial nonnegative integral solutions of this system equals the set of nonnegative loops having at least one transition. Now, consider the system $(D, -I_n)y = 0$ having the set L' of nontrivial solutions where I_n is the $n \times n$-identity matrix. The set of projections of the elements of L' onto the first m components equals L. By Theorem 1 of [17], this system has a set $\mathcal{H}(D, -I_n)$ of minimal solutions (called the Hilbert basis) having the following properties:

(i) Each nontrivial solution can be expressed as a linear combination of the elements of $\mathcal{H}(D, -I_n)$ with nonnegative integral coefficients.
(ii) Each element of $\mathcal{H}(D, -I_n)$ has a component sum of at most $(1 + (m + n)W)^{m+n}$.

W.l.o.g., we assume $n, m, W \geq 1$. Let $L_{\min} = \{\Phi_1, \ldots, \Phi_r\} \subseteq L$ denote the projection of $\mathcal{H}(D, -I_n)$ onto the first m components. From (ii) we immediately obtain $r \leq ((1 + (m + n)W)^{m+n} + 1)^n \leq (2nmW)^{c(m+n)n}$ for some constant $c > 0$.

Since τ is a nonnegative loop, we can write $\Psi[\tau] = \sum_{i \in [r]} a_i \Phi_i$ for suitable $a_i \in \mathbb{N}_0$. Now, define $a'_i := \min\{a_i, 1\}$ and $\Phi := \sum_{i \in [r]} a'_i \Phi_i$. For each $p \in S$, we have $\Delta[\tau](p) > 0$, implying the existence of an i with $a_i > 0$ and $\Delta[\Phi_i](p) > 0$. Therefore, Φ is a nonnegative loop with $\Delta[\Phi](p) > 0$ for all $p \in S$. Furthermore, by Lemma 1, Φ is enabled at μ since $\mathcal{P}_\Phi = \mathcal{P}_{\Psi[\tau]}$. (ii) and $r \leq (2nmW)^{c(m+n)n}$ imply that the largest component of Φ is at most $(2nmW)^{d(m+n)n}$ for some constant $d > 0$. Therefore, the encoding size of Φ is polynomial.

Now, we can describe the nondeterministic procedure which accepts if and only if S is simultaneously unbounded on some ∞-firing sequence: We guess μ^* and Φ in polynomial time and check nondeterministically and in polynomial time if μ^* can be covered and if Φ is enabled at μ^*.

This completes the proof. $\square$

We note that it can be decided in linear time if the set P of all places is simultaneously (∞-)unbounded. This is the case if and only if all top components C contain a marked place and a transition t with $\sum_{p \in t^\bullet \cap C} F(t, p) \geq 2$. Hence, the problems SU and S-∞-U are hard only if the input set S satisfies $1 < |S| < |P|$.

Slight generalisations of the results and algorithms of this subsection also imply several (minor) results for other problems for BPP-PNs, as well as for (commutative) context-free grammars. These can be found in the technical report [12].

5   Liveness Problems for BPP-PNs

Many different notions of liveness can be found in literature. We are mainly interested in the following.

Definition 8. *Let $\mathcal{P} = (P, T, F, \mu_0)$ be a Petri net. A transition t is*

- *L_0-live or* dead *if there is no firing sequence containing t.*
- *L_1-live or* potentially fireable *if it isn't dead.*
- *L_2-live or* arbitrarily often fireable *if for each $k \in \mathbb{N}$ there is a firing sequence containing t at least k times.*

- L_3-live *or* infinitely often fireable *if there is an ∞-firing sequence containing t infinitely often.*
- L_4-live *or* live *if t is potentially fireable at each reachable marking.*

A subset $S \subseteq T$ of transitions is called L_x-live, $x \in [0,4]$, if all transitions of S are L_x-live.

The concepts of $L_0, \dots, L_4$-liveness are referred to in [16]. Notice, that L_i-liveness implies L_j-liveness, where $4 \geq i \geq j \geq 1$. Using the results of Section 4, we can efficiently solve many decision problems involving these notions of liveness.

Theorem 9. *Given a BPP-PN $\mathcal{P} = (P, T, F, \mu_0)$ and sets $S_1, \dots, S_k \subseteq T$ of transitions, we can determine in linear time which sets are*

(a) L_0-live. (b) L_1-live. (c) L_2-live. (d) L_3-live.

Proof. Consider the Parikh extension $\mathcal{P}^e = (P^e, T, F^e, \mu_0^e)$ of $\mathcal{P}$ (see Definition 2). A transition t_i is not L_0-live iff t_i is L_1-live iff for the SCC C_i containing p_i^* there is a marked SCC C such that $C_i \leq C$ (see Lemma 1). Hence, we can answer (a) and (b) in linear time by computing $\mathcal{P}^e$, collecting the SCCs of $\mathcal{P}^e$ and investigating the found SCCs in a similar fashion as in Theorem 4.

For (c) and (d) notice that t_i is L_2-live iff p_i^* is unbounded iff p_i^* is unbounded on some ∞-firing sequence (see Lemma 5) iff t_i is L_3-live. Hence, we simply apply the algorithm of Theorem 4 to $\mathcal{P}^e$ and the sets $S_1^*, \dots, S_k^*$, where $S_i^* = \{p_j^* \mid t_j \in S_j\}$. □

Corollary 3. *Given a BPP-PN $\mathcal{P} = (P, T, F, \mu_0)$, we can decide in linear time, if (a transition t of) $\mathcal{P}$ is*

(a) L_0-live. (b) L_1-live. (c) L_2-live. (d) L_3-live.

Theorem 10. *Given a BPP-PN $\mathcal{P} = (P, T, F, \mu_0)$ and a transition $t \in T$, we can decide in linear time if t is L_4-live.*

Proof. As before, consider the Parikh extension $\mathcal{P}^e = (P^e, T, F^e, \mu_0^e)$ of $\mathcal{P}$. It is easy to see that a transition t_i is L_4-live iff p_i^* is persistently unbounded. □

In [13], Mayr showed that L_4-liveness is decidable in polynomial time for Basic Parallel Processes. Our results imply a quadratic time algorithm.

Corollary 4. *Given a BPP-PN $\mathcal{P} = (P, T, F, \mu_0)$, we can decide in quadratic time, if $\mathcal{P}$ is L_4-live.*

In the same paper, other interesting notions of liveness were investigated, namely the partial deadlock reachability problem and the partial livelock reachability problem. For both problems polynomial time algorithms were proposed for PA-processes in general. Using our results, linear time algorithms can be given for BPPs/BPP-PNs.

Theorem 11 (deadlock). *Given a BPP-PN $\mathcal{P} = (P, T, F, \mu_0)$ and a set S of transitions, we can decide in linear time if there is a reachable marking μ such that $\mu(^\bullet t) = 0$ for all $t \in S$.*

Proof. Let $R = \bigcup_{t \in S} {}^\bullet t$. By Lemma 6, we determine in linear time the maximum trap $Q \subseteq R$ w.r.t. R. By Lemma 8, R can be emptied if and only if Q is unmarked which can be checked in linear time. $\square$

Theorem 12 (livelock). *Given a BPP-PN $\mathcal{P} = (P, T, F, \mu_0)$ and a set S of transitions, we can decide in linear time if there is a reachable marking μ such that for all markings μ' reachable from μ, we have $\mu'({}^\bullet t) = 0$ for all $t \in S$.*

Proof. We introduce a counting place p and an edge from each transition $t \in S$ to p. A marking μ as defined in the lemma exists if and only if p is not persistently unbounded. By Theorem 6, this can be decided in linear time. $\square$

6   Conclusion

We showed in conjunction with [19] that the equivalence problem is decidable in doubly exponential time. Furthermore, we investigated several boundedness and liveness problems for BPP-PNs. For some of them, as well as for the covering problem, NP-completeness was shown. For most of the other problems, linear time could be achieved implying linear time algorithms for many problems in related areas. Open problems include:

- Is the equivalence problem complete for some known complexity class?
- Are the problems SU and S-∞-U decidable in polynomial time for sets S of constant size?
- How do the complexities of the reachability and the covering problem, and of variations of the boundedness and liveness problem behave if we consider different generalizations of BPP-PNs?

Acknowledgements. We thank Javier Esparza for many helpful discussions.

References

1. Christensen, S.: Distributed bisimilarity is decidable for a class of infinite state-space systems. In: Cleaveland, W.R. (ed.) CONCUR 1992. LNCS, vol. 630, pp. 148–161. Springer, Heidelberg (1992)
2. Christensen, S., Hirshfeld, Y., Moller, F.: Bisimulation equivalence is decidable for basic parallel processes. In: Best, E. (ed.) CONCUR 1993. LNCS, vol. 715, pp. 143–157. Springer, Heidelberg (1993)
3. Esparza, J.: Petri nets, commutative context-free grammars, and basic parallel processes. Fundamenta Informaticae 31(1), 13–25 (1997)
4. Esparza, J., Rossmanith, P., Schwoon, S.: A uniform framework for problems on context-free grammars. EATCS Bulletin 72, 169–177 (2000)
5. Huynh, D.T.: The complexity of semilinear sets. In: de Bakker, J.W., van Leeuwen, J. (eds.) ICALP 1980. LNCS, vol. 85, pp. 324–337. Springer, Heidelberg (1980)
6. Huynh, D.T.: Commutative grammars: The complexity of uniform word problems. Information and Control 57(1), 21–39 (1983)

7. Huynh, D.T.: A simple proof for the sum upper bound of the inequivalence problem for semilinear sets. Elektronische Informationsverarbeitung und Kybernetik, 147–156 (1986)
8. Karp, R.M., Miller, R.E.: Parallel program schemata. Journal of Computer and System Sciences 3(2), 147–195 (1969)
9. Kučera, A.: Regularity is decidable for normed PA processes in polynomial time. In: Chandru, V., Vinay, V. (eds.) FSTTCS 1996. LNCS, vol. 1180, pp. 111–122. Springer, Heidelberg (1996)
10. Landweber, L.H., Robertson, E.L.: Properties of conflict-free and persistent Petri nets. J. ACM 25, 352–364 (1978)
11. Mayr, E.W.: An algorithm for the general Petri net reachability problem. In: Proceedings of the Thirteenth Annual ACM Symposium on Theory of Computing, STOC 1981, pp. 238–246. ACM, New York (1981)
12. Mayr, E.W., Weihmann, J.: Results on equivalence, boundedness, liveness, and covering problems of BPP-Petri nets. Technical Report TUM-I1325, Institut für Informatik, TU München (March 2013)
13. Mayr, R.: Tableau methods for PA-processes. In: Galmiche, D. (ed.) TABLEAUX 1997. LNCS, vol. 1227, pp. 276–290. Springer, Heidelberg (1997)
14. Mayr, R.: Decidability and Complexity of Model Checking Problems for Infinite-State Systems. PhD thesis, Technische Universitt Mijnchen (1998)
15. Mayr, R.: On the complexity of bisimulation problems for basic parallel processes. In: Montanari, U., Rolim, J.D.P., Welzl, E. (eds.) ICALP 2000. LNCS, vol. 1853, pp. 329–341. Springer, Heidelberg (2000)
16. Murata, T.: Petri nets: Properties, analysis and applications. Proceedings of the IEEE 77(4), 541–580 (1989)
17. Pottier, L.: Minimal solutions of linear diophantine systems: bounds and algorithms. In: Book, R.V. (ed.) RTA 1991. LNCS, vol. 488, pp. 162–173. Springer, Heidelberg (1991)
18. Tarjan, R.: Depth-first search and linear graph algorithms. SIAM Journal on Computing 1(2), 146–160 (1972)
19. Yen, H.-C.: On reachability equivalence for BPP-nets. Theoretical Computer Science 179(1-2), 301–317 (1997)

A Semantics for Every GSPN*

Christian Eisentraut[1], Holger Hermanns[1], Joost-Pieter Katoen[2], and Lijun Zhang[3]

[1] Saarland University — Computer Science, Germany
[2] Department of Computer Science, RWTH Aachen University, Germany
[3] State Key Laboratory of Computer Science, Institute of Software,
Chinese Academy of Sciences

Abstract. Generalised Stochastic Petri Nets (*GSPN*s) are a popular modelling formalism for performance and dependability analysis. Their semantics is traditionally associated to continuous-time Markov chains (*CTMC*s), enabling the use of standard *CTMC* analysis algorithms and software tools. Due to ambiguities in the semantic interpretation of confused *GSPN*s, this analysis strand is however restricted to nets that do not exhibit non-determinism, the so-called well-defined nets. This paper defines a simple semantics for *every GSPN*. No restrictions are imposed on the presence of confusions. Immediate transitions may be weighted but are not required to be. Cycles of immediate transitions are admitted too. The semantics is defined using a non-deterministic variant of *CTMC*s, referred to as Markov automata. We prove that for well-defined bounded nets, our semantics is weak bisimulation equivalent to the existing *CTMC* semantics. Finally, we briefly indicate how every bounded *GSPN* can be quantitatively assessed.

Keywords: timed and stochastic nets, semantics, confusion, (weak) bisimulation, continuous-time Markov chains.

1 Introduction

Generalised Stochastic Petri Nets (*GSPN*s) [4,3,8] constitute a formalism to model concurrent computing systems involving stochastically governed timed behaviour. *GSPN*s are based on Petri nets, and are in wide-spread use as a modelling formalism in different engineering and scientific communities. From Petri nets they inherit the underlying bipartite graph structure, partitioned into *places* and *transitions*, but extend the formalism by distinguishing between *timed transitions* and *immediate transitions*. The latter can fire immediately and in zero time upon activation. The firing time of a timed transition is governed by a *rate*, which serves as a parameter of a negative exponential distribution. Timed transitions are usually depicted as non-solid bars, while immediate transitions are depicted as solid bars.

The precise semantics of a *GSPN* may conceptionally be considered as consisting of two stages. First, an abstract, high-level semantics describes *when* which transitions may fire, and *with what probability*. Speaking figuratively in terms of a token game,

---

* This work is supported by the EU FP7 Programme under grant agreement no. 295261 (MEALS) and 318490 (SENSATION), by the DFG as part of the SFB/TR 14 AVACS, and by DFG/NWO bilateral research programme ROCKS.

J.-M. Colom and J. Desel (Eds.): PETRI NETS 2013, LNCS 7927, pp. 90–109, 2013.
© Springer-Verlag Berlin Heidelberg 2013

this semantics determines how tokens can be moved from place to place by the firing of transitions. Then second, a lower-level mathematical description of the underlying stochastic process, typically a continuous time Markov chain (*CTMC*, for short), is derived to represent the intended stochastic behaviour captured in the first stage. This Markov chain is then subject to the analysis of steady-state or transient probabilities of markings, or more advanced analysis such as stochastic model checking.

The modelling power of *GSPNs* is particularly owed to the presence of immediate transitions [12]. Unfortunately, this characteristic strength of the formalism may lead to semantically intricate situations [9,12,13,14,15,25,32]. One of the most prominent cases is *confusion* [3,8]. In confused nets, the firing order of two concurrently enabled, non-conflicting immediate transitions determines whether two subsequent transitions are in conflict or not. The net in Fig. 1 is confused, since transitions t_1 and t_2 are not in direct conflict, but firing transition t_1 first leads to a direct conflict between t_2 and t_3, which does not occur if t_2 fires first instead. Confusion is not a problem of the high-level (token game) semantics of a net,

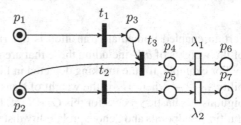

Fig. 1. Confused *GSPN*, see [3, Fig. 21]

as it is entirely clear which transition may fire, and how tokens are moved in either case. It is rather a problem of the underlying stochastic process that ought to be defined by this net. Recall that the transitions t_1 through t_3 are all immediate, and thus happen without elapse of time. Thus, their firing is basically transparent to a continuous time evolution. Places p_4 and p_5 enable two distinct timed transitions with rate λ_1 and λ_2 respectively, cf. Fig. 1. Now, depending on how the confusion between the transitions (and potentially the direct conflict between t_2 and t_3) is resolved, the underlying stochastic behaviour *either* corresponds to an exponential delay with rate λ_1, *or* to a delay with rate λ_2. Which of the two delays happens is not determined by the net structure, and as such is *non-deterministic*. Figure 2 shows a graphical representation of this phenomenon as a marking graph. States correspond to markings of the net in Fig. 1, and there is an obvious graphical correspondence with respect to the representation of the firing of timed or immediate transitions by similarly shaped edges. In state $\{p_2, p_3\}$ the direct conflict between t_2 and t_3 in the net yields a non-deterministic choice.

As the resulting process is not a *CTMC*, workarounds have been developed. To resolve (or: avoid) non-determinism, *priorities* and *weights* have been introduced [1]. Intuitively, weights are assigned to immediate transitions at the net level so as to induce a probabilistic choice instead of a non-deterministic

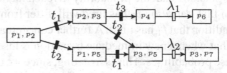

Fig. 2. Non-deterministic behaviour of the confused *GSPN* of Fig. 1

choice between (equally-prioritised) immediate transitions. Ignoring priorities, whenever more than one immediate transition is enabled, the probability of selecting a

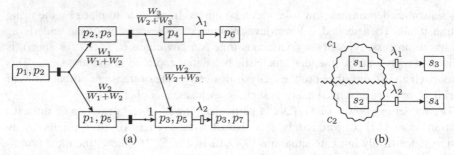

Fig. 3. (a) Probabilistic behaviour of *weighted* confused *GSPN* in Fig. 1; (b) the resulting *CTMC*

certain enabled immediate transition is determined by its weight relative to the sum of the weights of *all* –including those that are independent– enabled transitions.

For example, for the marking depicted in Fig. 1, transition t_1 is selected with probability $\frac{W_1}{W_1+W_2}$ where W_i is the weight of transition t_i. In this way, we obtain an unambiguous stochastic process for this *GSPN*, cf. Fig. 3(a). Now, the unlabelled edges have multiple endpoints and denote probability distributions over markings. We can consider this as a semi-Markov process, which has both zero-time delay and exponentially distributed time delay edges, as worked out, for instance by Balbo [8]. In order to derive a *CTMC* from this process, sequences of zero-time delay edges are fused into probability distributions over states. For our example net, we obtain the *CTMC* in Fig. 3(b) with initial distribution μ^0 with $\mu^0(s_1) = c_1$ and $\mu^0(s_2) = c_2$ where

$$c_1 = \frac{W_1}{W_1+W_2} \cdot \frac{W_3}{W_2+W_3} \text{ and } c_2 = \frac{W_2}{W_1+W_2} + \frac{W_1}{W_1+W_2} \cdot \frac{W_2}{W_2+W_3}.$$

These quantities correspond to the reachability probability of marking $\{p_4\}$ and $\{p_3, p_5\}$, respectively from the initial marking. Unfortunately, this approach has a drawback, related to the *dependence* and *independence* of transitions, an important concept in Petri net theory. In our example net of Fig. 1, the transitions t_1 and t_2 are *independent*. Their firings happen independent from each other, as the two transitions share no places. Transitions t_2 and t_3, in contrast, are dependent, as the firing of one of them influences the firing of the other (by disabling it) via the shared input place p_2. However, the expected independence between t_1 and t_2 is not reflected in our *GSPN* above after introducing weights. Instead, the probability to reach marking p_4 (and marking p_5) under the condition that transition t_2 has fired will differ from the corresponding probability under the condition that t_1 has fired. A further conceptual drawback from a modelling perspective, is that when a new immediate transition is inserted between t_1 and t_3, then this changes these probabilities. This is irritating, since we only refine one immediate transition into a sequence of two immediate transitions. Since immediate transitions do not take time, this procedure should not result in a change of the underlying stochastic model. However, it does. We can also consider this phenomenon as a problem of locality. A local change of the net has unexpected global consequences with respect to the probabilities.

To remedy this defect, several approaches to define the stochastic process at the net level have been proposed. At the core of these approaches, immediate transitions are usually partitioned according to their conflict behaviour, based on a structural analysis of the net. The standard approach is to partition them into *extended conflict sets* (shortly, *ECS*s) [1], which is a generalisation of structural conflicts in the presence of priorities (which are not treated here). Intuitively, two transitions are in structural conflict in a marking, if both are enabled in this marking, and firing any of them will disable the other. Inside an *ECS*, weights are used to decide immediate transition firings, while no choice is resolved probabilistically across *ECS*s. For confusion-free nets, the *ECS* does provide a way of resolving conflicts probabilistically with a localised interpretation of weights. Unfortunately, for confused nets, this solution approach suffers from the same problem as our initial approach: The *ECS*s for the net in Fig. 1 are given by the partition $\{\{t_1\}, \{t_2, t_3\}\}$. As transitions t_2 and t_3 are in the same *ECS*, the decision which to fire will be resolved probabilistically according to their weights. Transitions t_1 and t_2, in contrast, are in different *ECS*. Thus, the decision will still need to be resolved non-deterministically, given that they may be enabled at the same moment. Inserting immediate transition t_4 between t_1 and t_3 as mentioned above will lead to the *ECS*s $\{\{t_1\}, \{t_4\}, \{t_2, t_3\}\}$. Thus, still only the decision between transitions t_2 and t_3 is resolved probabilistically and not influenced by t_4. So, since some decisions are forced to be non-deterministic, this approach does in general not yield a mathematically well-defined stochastic process. Moreover, it is easy to see that in our example, any partition of immediate transitions will suffer from one of the semantic problems discussed.

In summary, certain nets lead to undesirable semantic problems. Due to this fact, several researchers have identified certain classes of nets as *not well-defined* (aka. *ill-defined*) [3,14,15]. Such nets are excluded both semantically and from an analysis point of view. Several different definitions have occurred in the literature. However, ill-defined nets, with confused nets being a prominent example, are not *bad* nets *per se*. As Balbo states [7]: "*this underspecification [in confused nets] could be due either to a precise modelling choice [...] or to a modelling error*". We firmly believe that the modeller should have full freedom of modelling choices, and that such choices should not be treated as errors *by definition*.

Contribution of this paper. This paper presents a semantics for *GSPN*s that is *complete* in the sense that it gives a meaning to *every GSPN*. Our semantics is *conservative* with respect to the well-established existing semantics of *well-defined* nets. More precisely, we show that for well-defined bounded *GSPN*s, our semantics is weak bisimulation equivalent to the classical *CTMC* semantics. This entails that measures of interest, such as steady-state and transient probabilities are identical. Finally, we sketch the available analysis trajectory for our semantics, including confused bounded nets.

Outline. We first recall the definition of *GSPN*s in Section 2. In Section 3 we present the *MA* semantics for *GSPN*s based on the marking graph. The bisimulation semantics will be discussed in Section 4. In Section 5 we describe quantitative analysis approaches for arbitrary (bounded) *GSPN*s, and Section 6 concludes the paper.

2   Generalised Stochastic Petri Nets

This section introduces *GSPN*s, where, for the sake of simplicity, we do not consider transition priorities. For a set X, we use $\Sigma(X)$ to denote the set of all partitions of X. For a set of places P, a *marking* m is a multi-set over P of the form $m : P \to \mathbb{N}$. We let M denote the set of all markings over P, and use m, m_0 etc to denote its elements.

Definition 1 (Generalized stochastic Petri net). *A* generalised stochastic Petri net G *(GSPN) is a tuple* $(P, T, I, O, H, m_0, W, \mathcal{D})$ *where:*

- *P is a finite set of* places,
- *$T = T_i \cup T_t$ is a finite set of transitions ($P \cap T = \emptyset$) partitioned into the sets T_t and T_i of* timed *and* immediate *transitions,*
- *$I, O, H : T \to M$ defines the transitions'* input *places,* output *places,* inhibition *places[1],*
- *$m_0 \in M$ is the initial marking,*
- *$W : T \to \mathbb{R}_{>0}$ defines the transitions'* weights, *and*
- *$\mathcal{D} : M \to \Sigma(T)$ is a marking-dependent* partition *satisfying the condition that $T_t \in \mathcal{D}(m)$ for all markings $m \in M$.*

The above definition agrees, except for the last component $\mathcal{D}$, with the classical *GSPN* definition in the literature [2,3,8]. We use the marking-dependent partition function $\mathcal{D}$ as a generalisation of the extended conflict set mentioned before. It serves to express for which immediate transitions choices are resolved probabilistically, and for which non-deterministically. This information is usually not provided in the net definition. Instead the (marking independent) *ECS* are derived based on a structural analysis of the net at hand. The reason why we include this information in an explicit form in the definition is mainly ought to formal reasons. However, it also enables (but does not enforce) a view where the choices between immediate transitions are resolved as a consequence of a conscious modelling decision, possibly decoupled from the net structure. The constraint $T_t \in \mathcal{D}(m)$ is due to the fact that all enabled timed transitions are always weighted against each other in a race. On the expense of slightly more complicated definitions in the following, we could eliminate this technicality and let $\mathcal{D} : M \to \Sigma(T_i)$.

The *input, output* and *inhibition* functions assign to each transition a mapping $P \to \mathbb{N}$, specifying the corresponding cardinalities. A transition has *concession* if sufficiently many tokens are available in all its input places, while the corresponding inhibition places do not contain sufficiently many tokens for an inhibitor arc to become effective. Firing a transition yields a (possibly) new marking, which is obtained by removing one or more tokens from each input place and adding tokens to the transition's output places. Immediate transitions execute immediately upon becoming enabled, whereas timed transitions are delayed by an exponentially distributed duration which is uniquely specified by a *transition rate* (i.e., a positive real number defined by the weights).

For notational convenience, we write cascaded function application with indexed notation of the first parameter. For example, we write I_t, O_t and H_t for $I(t)$, $O(t)$ and $H(t)$, respectively. The semantics of a *GSPN* is defined by its *marking graph*, which

---

[1] If transition t has no inhibitor places, we let $H(t) = \infty$.

is obtained by playing the "token game". Immediate transitions are fired with priority over timed transitions [2,12,3]. Accordingly, if both timed and immediate transitions have concession in a marking, only the immediate transitions become enabled. Let G be a *GSPN* with marking $m \in M$.

Definition 2 (Concession and enabled transitions).

1. The set of transitions with concession *in marking m is defined by:*

$$\mathrm{conc}(m) = \{t \in T \mid \forall p \in P.\ m(p) \geq I_t(p) \wedge m(p) < H_t(p)\}.$$

2. The set of enabled transitions *in marking m is defined by:* $\mathrm{en}_m = \mathrm{conc}(m) \cap T_i$ *if* $\mathrm{conc}(m) \cap T_i \neq \emptyset$, *and* $\mathrm{en}_m = \mathrm{conc}(m)$ *otherwise.*

A marking m is *vanishing* whenever an immediate transition is enabled in m, otherwise it is *tangible*. Given the priority of immediate transitions over timed ones, the sojourn time in vanishing markings is zero. In a vanishing marking, none of the timed transitions which have concession is enabled. In a *tangible* marking m, only timed transitions can be enabled. The residence time in tangible marking m is determined by a negative exponential distribution with rate $\sum_{t \in \mathrm{en}_m} W(t)$. The effect of executing a transition is formalised in the classical way:

Definition 3 (Transition execution). *Let the* transition execution relation $[\cdot\rangle \subseteq M \times T \times M$ *be such that for all markings $m, m' \in M$ and transitions $t \in T$ it holds:*

$$m\,[t\rangle\,m' \quad \Longleftrightarrow \quad t \in \mathrm{en}_m \wedge \forall p \in P.\ m'(p) = m(p) - I_t(p) + O_t(p).$$

We now recall the notion of *marking graph*, obtained from reachable markings:

Definition 4 (Reachable marking graph). *The* marking graph *of the GSPN G is the labelled digraph $MG(G) = (RS, E)$, where*

- *RS is the smallest set of reachable markings satisfying: $m_0 \in RS$, and $m \in RS \wedge m\,[t\rangle\,m'$ implies $m' \in RS$.*
- *The edge between m and m' is labelled by the transition t such that $m\,[t\rangle\,m'$.*

This graph describes how a net may evolve in terms of its markings. However, it fails to faithfully represent the stochastic aspects of the net. This is made more precise below.

Recall the idea that we consider certain immediate transitions probabilistically dependent from some other transitions (mainly when they are in conflict), while we consider them independent from others. Traditionally, these relations are captured by extended conflict sets (*ECS*s [1]). Here, we consider a generalisation of this concept in the form of an arbitrary immediate transitions partition $\mathcal{D}_m$. For each marking m, the partition $\mathcal{D}_m$ determines a way of resolving conflicts between immediate transitions. Each set $C \in \mathcal{D}_m$ consists of transitions whose conflicts are resolved probabilistically in m. On the other hand, transitions of different sets are considered to behave in an independent manner, i.e., we make a non-deterministic selection if several of them are enabled in m. Our semantics will be general enough that we may allow the latter even if there is a *structural* conflict between these transitions. Let us make this precise.

Assume that some transitions in the set $C \in \mathcal{D}_m$ are enabled and C is chosen to be fired. Under this condition, the probability that a specific transition fires is given as the normalised weight of the enabled transitions in C. Precisely, $\mathbf{P}_C\{t \mid m\} = 0$ if $t \notin C \cap en_m$, and otherwise:

$$\mathbf{P}_C\{t \mid m\} = \frac{W(t)}{W_C(m)} \quad \text{where} \quad W_C(m) = \sum_{t \in C \cap en_m} W(t). \tag{1}$$

If m is a vanishing marking, $W_C(m)$ denotes the cumulative weight of all enabled (i.e., immediate) transitions in C. In this case the probability $\mathbf{P}_C\{t \mid m\}$ of taking the immediate transition t in m is determined by the weight assignment W. Note that $\mathbf{P}_C\{t \mid m\}$ is 0 if t is neither enabled nor an element from C. The case that m is tangible is similar. Then only timed transitions are enabled, and recall that the set of timed transitions T_t is an element in $\mathcal{D}_m$. Thus, $C = T_t$. Accordingly,

$$W_C(m) = \sum_{t \in en_m} W(t)$$

is the exit rate from the tangible marking m. In this case, $\mathbf{P}_C\{t \mid m\}$ is the probability of taking the transition t if the tangible marking m is left.

In both cases, several distinct transition firings may lead from m to the same marking m'. These need to be accumulated. With some overload of notation we define

$$\mathbf{P}_C(m, m') = \sum_{m[t\rangle m'} \mathbf{P}_C\{t \mid m\}.$$

3   Markov Automata Semantics for GSPNs

Our aim is to provide a semantics to every *GSPN*. In particular, this includes nets in which multiple immediate transitions are enabled in a marking, nets with cycles of immediate transitions, as well as confused nets. Obviously, stochastic processes such as *CTMCs* do not suffice for this purpose, as they cannot express non-determinism. We therefore resort to an extension of *CTMCs* with non-determinism, *Markov automata* (*MAs*, for short) as introduced in [20]. This model permits to represent the concepts above, including a formulation in terms of a semi-Markov process with zero-timed delay and exponentially distributed time delays [8], while in addition supporting non-determinism between transition firings in vanishing markings. Figure 2 and 3(a) are in fact graphical representations of *MA*.

3.1   Markov Automata

We first introduce some preliminary notions that we shall use in the rest of the paper. A *subdistribution* μ over a set S is a function $\mu : S \mapsto [0, 1]$ such that $\sum_{s \in S} \mu(s) \leq 1$. Let $Supp(\mu) = \{s \in S \mid \mu(s) > 0\}$ denote the support of μ and $\mu(S') := \sum_{s \in S'} \mu(s)$ the probability of $S' \subseteq S$ with respect to μ. Let $|\mu| := \mu(S)$ denote the *size* of the subdistribution μ. We say μ is a *full distribution*, or simply distribution, if $|\mu| = 1$. Let $Dist(S)$ and $Subdist(S)$ be the set of distributions and subdistributions over S, respectively. For $s \in S$, let $\delta_s \in Dist(S)$ denote the *Dirac* distribution for s, i.e., $\delta_s(s) = 1$. Let μ and μ' be two subdistributions. We define the subdistribution $\mu'' :=$ $\mu \oplus \mu'$ by $\mu''(s) = \mu(s) + \mu'(s)$, if $|\mu''| \leq 1$. Conversely, we say that μ'' can be split into μ and μ', or that (μ, μ') is a splitting of μ''. Since $\oplus$ is associative and commutative, we use the notation $\bigoplus_{i \in I}$ for arbitrary sums over a finite index set I. Moreover, if $c \cdot |\mu| \leq 1$ and $c > 0$, we let $c\mu$ denote the subdistribution defined by: $(c\mu)(s) = c \cdot \mu(s)$. For $s \in S$ and $\mu \in Subdist(S)$ let $\mu \ominus s$ denote the subdistribution μ' with $\mu'(t) = \mu(t)$ if $t \neq s$ and $\mu'(s) = 0$.

Definition 5 (Markov automaton). *A* Markov automaton A *is a quadruple* $(S, \twoheadrightarrow, \dashrightarrow, \mu^0)$, *where*

- S *is a non-empty countable set of* states,
- $\twoheadrightarrow \subset S \times Dist(S)$ *is a set of* immediate edges,
- $\dashrightarrow \subset S \times \mathbb{R}_{>0} \times Dist(S)$ *is a set of* timed edges, *and*
- $\mu^0 \in Dist(S)$ *is an* initial distribution *over the states* S.

It is required that every state $s \in S$ *has at most one* outgoing *timed edge.*[2]

We let s, u and their variants with indices range over S, and μ over $Dist(S)$. An immediate edge $(s, \mu) \in \twoheadrightarrow$ is denoted by $s \twoheadrightarrow \mu$. The operational interpretation of edge $s \twoheadrightarrow \mu$ is that from s a next state will be probabilistically determined according to distribution μ and that in s no time elapses. Similarly, a timed edge $(s, \lambda, \mu) \in \dashrightarrow$ is denoted by $s \xrightarrow{\lambda} \mu$. We use $\lambda, r \in \mathbb{R}_{>0}$ to denote the rate of a negative exponential distribution. An edge $(s, \mu) \in \twoheadrightarrow$ is said to originate from state s.

A *state* $s \in S$ is called *tangible* if no immediate edge originates from s. A *probability distribution* over states is called *tangible* if all states in its support set are tangible. We write $s \xrightarrow{\alpha} \mu$ if either (i) $\alpha = \varepsilon$ (i.e. the edge is unlabelled) and $s \twoheadrightarrow \mu$ or (ii) $\alpha \in \mathbb{R}_{>0}$, s is tangible and $s \xrightarrow{\alpha} \mu$, or (iii) $\alpha = 0$, $\mu = \delta_s$, and s has no outgoing transition. This notation combines immediate edges (i) with timed edges (ii), but timed edges are only considered from tangible states. Clause (iii) generalizes the implicit tangibility check of clause (ii) to states without outgoing edges. The inclusion of a tangibility check inside the above clauses (ii) and (iii) of $\xrightarrow{\alpha}$ will have an interesting effect, discussed in Section 4.2. We stipulate that *non-determinism* occurs in an *MA* whenever multiple immediate edges originate from a state. In that case, it is deliberately left unspecified with which probability a particular immediate edge is taken. This represents a non-deterministic choice. Obviously, *CTMCs* can be considered as special cases of *MA*s: A *CTMC* is a *MA* with $\twoheadrightarrow = \emptyset$.

3.2 Basic Semantics of GSPNs

We are now in the position to define the semantics of every *GSPN*—including the *non* well-defined ones—by means of a *MA*. The intuition is rather simple. Basically

---

[2] This is not a restriction since the effect of two timed edges $s \xrightarrow{r} \mu$ and $s \xrightarrow{r'} \mu'$ can be combined into a single timed edge $s \xrightarrow{r+r'} \mu''$.

the semantics of a *GSPN* corresponds to its reachable marking graph, cf. Def. 4. States correspond to markings, taking an immediate edge in the *MA* is the counterpart to firing an immediate transition in the net, and likewise for timed edges and timed transitions. The marking graph can therefore directly be interpreted as a Markov automaton.

Definition 6 (Basic *MA* semantics for *GSPN*s). *The MA semantics of the GSPN* $G = (P, T, I, O, H, m_0, W, \mathcal{D})$ *is the MA* $A_G = (S, \dashrightarrow, \multimap, \mu^0)$, *where*

- $S = RS$ *is the reachable set of markings in the marking graph,*
- $\mu^0 = \delta_{m_0}$,
- *for every* $m \in RS$, *and each equivalence* $C \in \mathcal{D}_m$,

  1. *there is an edge* $m \xrightarrow{r} \mu$ *if and only if* m *is a tangible marking,* $r = W_C(m)$ *and* $\mu(m') = \mathbf{P}_C(m, m')$ *for all* $m' \in RS$,
  2. *there is an edge* $m \multimap \mu$ *if and only if* m *is a vanishing marking and* $\mu(m') = \mathbf{P}_C(m, m')$ *for all* $m' \in RS$.

So, the basic *MA* semantics is the marking graph of a *GSPN*. Every marking of the *GSPN* that is reachable by a sequence of (net) transitions from the initial marking corresponds to a state in the *MA*. As discussed before, in marking m of the net all enabled timed transitions t induce an exponentially distributed stochastic delay with a rate r that is the sum of all weights of enabled transitions. In this case, the probability to reach a marking m', say, by edge t is given as the edge's relative weight. This is reflected in clause 1 of the above *MA* semantics. If no timed transition is enabled in marking m, then no timed edge originates from state m.

In contrast, the enabled immediate transitions in a marking need to be represented by more than one immediate edge in the *MA*. Recall that each equivalence class $C \in \mathcal{D}_m$ corresponds to an *ECS* in *GSPN* terminology. For every such set C, the enabled transitions in C fire with a probability that is equal to their weight in relation to the sum of the weights of all enabled transitions in C. However, transitions that are in different sets in $\mathcal{D}_m$ are entirely independent. More precisely, transitions from different sets in $\mathcal{D}_m$ compete in a non-deterministic way. This is reflected in clause 2 of the above definition. The non-deterministic choice between transitions across different sets of $\mathcal{D}_m$ is represented by introducing an immediate edge for *every* set in the partition $\mathcal{D}_m$. The probabilistic decision among transitions *within a single* set, in turn, is reflected by the distribution over markings the corresponding immediate edge leads to.

3.3   Well-Defined *GSPN*s

The aim of this section is to formalise and generalise well-defined *GSPN*s in terms of our new semantics. A central notion for this purpose is the concept of weak edges.

Labelled trees. The notion of weak edge is defined using labelled trees. For $\sigma, \sigma' \in \mathbb{N}_{>0}^*$, let $\sigma \leq \sigma'$ if there exists a (possibly empty) $\phi \in \mathbb{N}_{>0}^*$ such that $\sigma\phi = \sigma'$. We write $\sigma < \sigma'$ whenever $\sigma \leq \sigma'$ and $\sigma \neq \sigma'$. Let L be a set of labels. An *(infinite)* *L-labelled tree* is a partial function $\mathcal{T} : \mathbb{N}_{>0}^* \to L$ satisfying

- if $\sigma \leq \sigma'$ and $\sigma' \in \text{dom}(\mathcal{T})$, then $\sigma \in \text{dom}(\mathcal{T})$,

- if $\sigma i \in dom(\mathcal{T})$ for $i \in \mathbb{N}_{>1}$, then $\sigma(i-1) \in dom(\mathcal{T})$, and
- $\varepsilon \in dom(\mathcal{T})$.

The empty word ε is called the root of $\mathcal{T}$ and $\sigma \in dom(\mathcal{T})$ is a node of $\mathcal{T}$. For node σ of tree $\mathcal{T}$, let $Children(\sigma) = \{\sigma i \mid \sigma i \in dom(\mathcal{T})\}$. Node σ is a leaf of tree $\mathcal{T}$ if there is no $\sigma' \in dom(\mathcal{T})$ with $\sigma < \sigma'$; then $Children(\sigma) = \emptyset$. We denote the set of all leaves of $\mathcal{T}$ by $Leaf_{\mathcal{T}}$ and the set of all inner nodes of $\mathcal{T}$ by $Inner_{\mathcal{T}}$. If the tree only consists of the root, then $Inner_{\mathcal{T}} = Leaf_{\mathcal{T}} = \{\varepsilon\}$. In any other case the two sets are disjoint. We consider L-labelled trees with finite branching, i.e., $|Children(\sigma)| < \infty$ for all nodes σ.

Weak edges. Weak edges for probabilistic systems have been defined in the literature via probabilistic executions in [31], trees [17], or infinite sums [16]. We adopt the tree notation here. The material presented below concerning weak edges provides no innovation over the classical treatment, it is included for the benefit of the reader. Let $L = S \times \mathbb{R}_{>0}$. A node in an L-labelled tree is labelled by a state and the (by definition non-zero) probability of reaching this node from the root of the tree. For a node σ we write $Sta_{\mathcal{T}}(\sigma)$ for the first component of $\mathcal{T}(\sigma)$ and $Prob_{\mathcal{T}}(\sigma)$ for the second component of $\mathcal{T}(\sigma)$. If $\mathcal{T}$ is clear from the context we omit the subscripts.

Definition 7 (Weak edge tree). *Let* $(S, \longrightarrow, \dashrightarrow, \mu^0)$ *be an MA. A weak edge tree* $\mathcal{T}$ *is a* $S \times \mathbb{R}_{>0}$-*labelled tree satisfying the following conditions*

1. $Prob(\varepsilon) = 1$,
2. $\forall \sigma \in Inner_{\mathcal{T}} \setminus Leaf_{\mathcal{T}} : \exists \mu : Sta(\sigma) \longrightarrow \mu$ *and* $Prob(\sigma) \cdot \mu - \xi$ *where* $\xi(Sta(\sigma')) = Prob(\sigma')$ *for all* $\sigma' \in Children(\sigma)$,
3. $\sum_{\sigma \in Leaf_{\mathcal{T}}} Prob(\sigma) = 1$.

A weak edge tree $\mathcal{T}$ corresponds to a probabilistic execution fragment: it starts from the root's state $Sta(\varepsilon)$, and resolves non-deterministic choices at every inner node of the tree, which represents the state in the MA it is labelled with. The second component of σ, $Prob(\sigma)$, is the probability of reaching the state $Sta(\sigma)$ via immediate edges in the MA, starting from the state $Sta(\varepsilon)$. The distribution associated with edge tree $\mathcal{T}$, denoted $\mu_{\mathcal{T}}$, is defined as $\mu_{\mathcal{T}} \overset{def}{=} \bigoplus_{\sigma \in Leaf_{\mathcal{T}}} \rho_\sigma$, where $\rho_\sigma \in Subdist(S)$ with $\rho_\sigma(s) = Prob(\sigma)$ if $s = Sta(\sigma)$ and $\rho_\sigma(s) = 0$ otherwise. Subdistribution $\mu_{\mathcal{T}}$ is said to be *induced* by $\mathcal{T}$. We are now in a position to define *weak edges*: For $s \in S$ and $\mu \in Dist(S)$, let $s \Longrightarrow \mu$ if μ is induced by some internal edge tree $\mathcal{T}$ with $Sta(\varepsilon) = s$.

We now generalise edges to edges originating in subdistributions over states. Let $\mu \in Dist(S)$. If for every state $s_i \in Supp(\mu)$, $s_i \Longrightarrow \mu_i'$ for some μ_i', then we write $\mu \Longrightarrow \bigoplus_{s_i \in Supp(\mu)} \mu(s_i)\mu_i'$. We apply a similar definition for $\overset{\alpha}{\longrightarrow}$ instead of $\Longrightarrow$. Finally, for $\alpha \in \mathbb{R}$, we write $s \overset{\alpha}{\Longrightarrow} \mu$ if there exist μ_1 and μ_2 such that $s \Longrightarrow \mu_1$, $\mu_1 \overset{\alpha}{\longrightarrow} \mu_2$ and $\mu_2 \Longrightarrow \mu$.

Intuitively, the weak edges in Def. 7 (referred to as *weak transitions* in the automata literature) are used to capture all possible evolutions along immediate edges starting from s. Thus, any edge itself is a weak edge, and note that from state s, there is always a weak edge $s \Longrightarrow \delta_s$, even if s is tangible.

Well-defined GSPNs. We are now ready to define well-defined GSPNs.

Definition 8 (Well-defined GSPN). *Let* $G = (P, T, I, O, H, m_0, W, \mathcal{D})$ *be a GSPN with MA semantics* A_G. *We say* G *is* well-defined, *if for every state* $m \in RS$, *and every pair* (μ, μ') *of distributions over* tangible *states it holds:* $m \Longrightarrow \mu$ *and* $m \Longrightarrow \mu'$ *implies* $\mu = \mu'$.

Different to [32], we are only interested in the probability to reach a marking, and whether it is uniquely specified, but not in the sequences of edges leading to tangible markings. Phrased differently, we are only interested in tangible state to tangible state probabilities [14,8].

It is not surprising that a well-defined *GSPN* induces a unique *CTMC*: states will correspond to those tangible markings, edge $\overset{r}{\dashrightarrow}$ is obtained by extending the weak edge until tangible states are reached. The uniqueness is guaranteed by the definition of well-defined *GSPN*s. This is summarised in the following definition:

Definition 9 (CTMC induced by a well-defined GSPN). *The well-defined GSPN* G *induces the CTMC* $C_G = (S, \longrightarrow, \dashrightarrow, \mu^0)$, *where*

- *S is the set of reachable tangible markings of G,*
- *$m \overset{r}{\dashrightarrow} \mu$ iff μ is the unique distribution over tangible markings such that a distribution μ' exists with $m \overset{r}{\dashrightarrow} \mu'$ and $\mu' \Longrightarrow \mu$ in the basic MA semantics of G,*
- *μ_0 is the unique distribution over tangible markings such that $m_0 \Longrightarrow \mu_0$.*

Lemma 1. *The induced CTMC of a well-defined GSPN is unique (up to isomorphism).*

4   Bisimulation Semantics

The basic *MA* semantics we have introduced already has several advantages. It is complete, i.e. it provides semantics for every net, and it is amenable to several analysis techniques that are being established (see Sec. 5 for further details). Nevertheless, we want to address more desirable properties the current proposal does not have: (i) the semantics should be conservative with respect to the existing standard semantics for well-defined nets, (ii) immediate edges should be disregarded as much as possible, and exponential delays should be only distinguished up to lumpability. This ensures that the actual formal semantics agrees with the intuitive behaviour of a net and semantic redundancies are avoided as much as possible. For instance, the introduction of a new immediate transition between t_1 and t_3 in Fig. 1, which should be independent of every other concurrently enabled transition, should not affect the underlying semantics.

We now will implement the above requirements by defining the semantics of a *bounded GSPN* as its basic *MA* semantics modulo a behavioural equivalence, weak bisimilarity [20]. The basic *MA* semantics modulo weak bisimilarity will exactly represent the behavioural kernel of the *GSPN*. (The setting of unbounded *GSPN*s is left for further study.)

We first need the notion of a convex combination of weak edges. Let $\mu \overset{\alpha}{\Longrightarrow}_C \gamma$ if there exists a finite index set I, and weak edges $\mu \overset{\alpha}{\Longrightarrow} \gamma_i$ and a factor $c_i \in (0, 1]$ for every $i \in I$, with $\sum_{i \in I} c_i = 1$ and $\gamma = \bigoplus_{i \in I} c_i \gamma_i$. This notion is standard for probabilistic automata, and inherited here for MA; see [31] for more details. Let the set of all splittings of immediate successor subdistributions be defined as

$$split(\mu) = \{(\mu_1, \mu_2) \mid \exists \mu' : \mu \Longrightarrow_C \mu' \wedge \mu' = \mu_1 \oplus \mu_2\}.$$

Definition 10 (Weak bisimulation [20]). *A symmetric relation $\mathcal{R}$ on subdistributions over S is called a* weak bisimulation *if and only if whenever $\mu_1 \mathcal{R} \mu_2$ then for all $\alpha \in \mathbb{R} \cup \{\varepsilon\}$: $|\mu_1| = |\mu_2|$ and for all $s \in Supp(\mu_1)$ there exist $\mu_2^{\rightarrow}, \mu_2^{\Delta}$: $(\mu_2^{\rightarrow}, \mu_2^{\Delta}) \in split(\mu_2)$ and*

(i) $\mu_1(s)\delta_s \mathcal{R} \mu_2^{\rightarrow}$ and $(\mu_1 \ominus s) \mathcal{R} \mu_2^{\Delta}$

(ii) whenever $s \xrightarrow{\alpha} \mu_1'$ for some μ_1' then $\mu_2^{\rightarrow} \xRightarrow{\alpha}_C \mu''$ and $(\mu_1(s) \cdot \mu_1') \mathcal{R} \mu''$

Two subdistributions μ and γ are weak bisimilar, *denoted by $\mu \approx \gamma$, if the pair (μ, γ) is contained in some weak bisimulation.*

Note that weak bisimilarity is a relation over *distributions*, which is a natural choice for stochastic processes. Its basic idea is that two distributions μ and γ are bisimilar, if the edge of every state in the support of μ can by matched by a weak edge of a subdistribution of γ (Condition (ii)) in the usual sense of (probabilistic) bisimulation, however, enhanced by the idea that before γ is to be split into suitable subdistributions, it may perform an arbitrary sequence of weak immediate edges (Condition (i)). As it has been shown in [19], Condition (i) is the essential difference that distinguishes weak bisimulation for MAs from weak bisimulation for Probabilistic Automata [31]. Furthermore, although not obvious from the definition, it is exactly this condition that allows to *fuse* sequences of immediate edges into their unique final goal distribution, if existing.

Bisimulation can be lifted to a relation between MAs with disjoint state space. Two MAs A, A' are bisimilar, denoted $A \approx A'$, if their initial distributions are bisimilar in the direct sum, which is the MA obtained by considering the disjoint union of states and edges respectively. This shall be used in the next section to compare the semantics of models.

4.1 Revisiting Well-Definition

To illustrate why we consider weak MA bisimilarity a semantic equivalence especially well-suited for $GSPN$ semantics, let us recall the standard procedure applied to derive a $CTMC$ from the basic MA semantics underlying a well-defined $GSPN$. We illustrate this process with the MA from Fig. 3(a) as an example. For convenience, we repeat it in Fig. 4(a) below. This figure shows the basic MA semantics of the $GSPN$ in Fig. 1 in the case that every immediate edge is weighted, and choices among immediate edges are always resolved probabilistically. For a shorter notation, we now denote edge probabilities by x_1, x_2 and so on. When we want to transform this MA into a $CTMC$, we successively remove every immediate edge by replacing a state with an outgoing immediate edge by the distribution that this immediate edge leads to. The result of this replacement is shown in Figs. 4(b) and 4(c). Finally, when no such states remain, we obtain the $CTMC$ in Fig. 4(d), where $c_1 = x_1 x_3$ and $c_2 = x_2 + x_1 x_4$. The effect of this iterative process of fusing transitions can also be formulated via matrix operations [3].

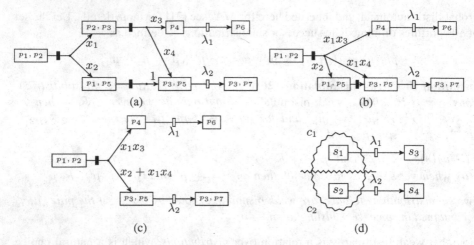

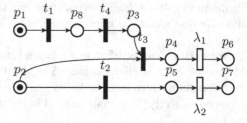

Fig. 4. From the *MA* semantics (a) a *CTMC* is obtained (d) by step-wise fusing immediate edges in (b) and (c)

In this example, this leads to a unique result, as every state has at most one outgoing immediate edge. In general, this leads to unique results whenever the net is well-defined. For nets with non-determinism, however, this approach does not lead to mathematically well-defined results.

For this purpose consider now the net in Fig. 5. Assume now that we do not resolve every choice of immediate transitions probabilistically, but only the conflict between t_2 and t_3. Hence let $\mathcal{D}_m = \{\{t_2, t_3\}, \{t_1\}, \{t_4\}\}$. Note that these are exactly the *ECS*s of the net. We then obtain the non-deterministic basic *MA* semantics in Fig. 6(a). Applying the fusing procedure as before is clearly not possible, since already in the initial state

Fig. 5. Confused *GSPN* with additional transition

of the *MA*, the marking $\{p_1, p_2\}$, we have two outgoing immediate edges, which will finally lead to two different distributions over tangible markings.

Although it is thus not possible to fully remove immediate edges here – as they are a necessary semantic component to express non-deterministic choice – we want to remove immediate edges whenever they can be fused. In our example, this would lead to the *MA* in Fig. 6(b). Only in the first state two immediate edges remain. They fully capture the non-deterministic behaviour of this *GSPN*.

Weak *MA* bisimilarity has been designed to exactly perform the task of removing immediate edges by fusion when the result is uniquely defined. In fact, the *MA* in Fig. 6(b) is the (state- and transition-wise) minimal *MA* that is weakly bisimilar to the *MA* in Fig. 6(a).

Speaking more generally, weak bisimilarity gives us a powerful means to conservatively generalise the notion of tangible and vanishing markings. Formally, a tangible

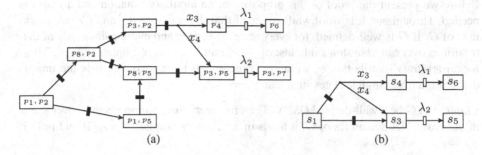

Fig. 6. A basic *MA* (a) with non-determinism and the smallest *MA* weakly bisimilar to it (b). In (b), state s_1 subsumes markings $\{p_1, p_2\}$ and $\{p_8, p_2\}$ from (a). All other markings with immediate behaviour are removed as a result of fusing them.

marking has been defined as a marking that has no outgoing immediate transitions. Markings that are not tangible are called *vanishing*. More intuitively speaking, as the words *tangible* and *vanishing* suggest, vanishing markings are semantically insignificant, while tangible markings constitute the semantic essence of a net's behaviour. Now, in the context of non-deterministic behaviour, besides of those states without immediate transitions, also those states with a non-deterministic choice between immediate transitions are semantically *tangible* in the literal sense (as long as the choice makes a behaviour difference in the end).

To make this precise, we will define the notion of *significant* markings as a conservative extension of tangible markings, and show that for well-defined nets, they coincide with tangible markings and vice versa.

Definition 11 (Significant marking). *Given a GSPN G and its basic MA semantics A_G, we call a marking m insignificant if it is vanishing and – in A_G – m is a state that has at least one outgoing immediate edge $m \longrightarrow \mu$ such that $\mu \approx \delta_m$. Otherwise we call marking m significant.*

Whereas every tangible marking is also significant, not every vanishing marking is insignificant. Only those vanishing markings are also insignificant, which have an immediate successor distribution that is semantically equivalent to the marking itself, and could thus fully replace the marking without affecting the behaviour of the net. Only in *well-defined GSPNs* significant and tangible, and vanishing and insignificant coincide respectively, as stated in the following proposition.

Proposition 1 (Preservation). *If G is a* well-defined *GSPN, then a marking m of G is tangible if and only if it is significant.*

Furthermore, the CTMC associated with a well defined GSPN enjoys a strong relation to the original net in terms of the MA semantics:

Proposition 2. *The basic MA semantics A_G of a well-defined GSPN G is weakly bisimilar to the CTMC C_G induced by G.*

Before we present the proof of this proposition, an auxiliary notation and a claim is needed. Throughout, it is worthwhile to recall that the states of A_G and C_G are markings of G. If G is well-defined, for every state $m \in RS$, and every pair (μ, μ') of distributions over *tangible states* it holds: $m \implies \mu$ and $m \implies \mu'$ implies $\mu = \mu'$. Thus, for an arbitrary distribution γ, we may write $\gamma \Longmapsto \mu$ to express that μ is the unique distribution over tangible states such that $\gamma \implies \mu$.

Claim. Let G be a well-defined *GSPN*. Then for every distribution γ and γ' over states of the basic *MA* semantics of G, it holds that $\gamma \implies \gamma'$ implies $\gamma \Longmapsto \mu$ if and only if $\gamma' \Longmapsto \mu$.

This follows immediately from the uniqueness of μ.

Proof (Proposition 2). In order to prove $A_G \approx C_G$, we will provide a bisimulation $\mathcal{R}$ and show that the pair of initial distributions of A_G and C_G is contained in $\mathcal{R}$. Let S_t be the state space of C_G, the set of all reachable tangible markings of A_G. Recall that the state space of A_G is the set RS of all reachable markings. Let $\mathcal{R}$ be the symmetric closure of the relation $\{(\gamma, \mu) \in Dist(RS) \times Dist(S_t) \mid \gamma \Longmapsto \mu\}$. The pair of initial distributions of A_G and C_G is contained in $\mathcal{R}$, which follows immediately from the definition of the initial distribution of C_G.

Recall that in C_G we have an edge $m \xrightarrow{r} \mu$ if and only if μ is the unique distribution over tangible markings such that a distribution μ' exists with $m \xrightarrow{r} \mu'$ and $\mu' \implies \mu$ in the basic *MA* semantics of G. We will refer to this fact by $(\star)$ whenever used in the sequel.

We will now check that every pair of $\mathcal{R}$ satisfies the bisimulation conditions. Consider an arbitrary pair $(\gamma, \mu) \in \mathcal{R} \cap Dist(RS) \times Dist(S_t)$. Clearly $|\gamma| = |\mu|$, as $\gamma \implies \mu$. Now consider an arbitrary state (i.e. marking) $s \in Supp(\gamma)$. By the definition of hyperedges and of $\Longmapsto$ it is easy to see that there exists a splitting $\mu^{\rightarrow} \oplus \mu^{\Delta} = \mu$, such that $\delta_s \Longmapsto \mu^{\rightarrow}$ and $\mu^{\Delta} \Longmapsto \mu^{\Delta}$, which immediately implies $\gamma(s)\delta_s \ \mathcal{R} \ \mu^{\rightarrow}$ and $\gamma \ominus s \ \mathcal{R} \ \mu^{\Delta}$. This satisfies Clause (i) of Definition 10. Now assume $s \longrightarrow \gamma'$. Then, by Claim 4.1, we see that $\gamma(s)\gamma' \Longmapsto \mu^{\rightarrow}$ and thus immediately $\gamma(s)\gamma' \ \mathcal{R} \ \mu^{\rightarrow}$. Now assume $s \xrightarrow{r} \gamma'$. Note that this implies that s is tangible, and thus $\mu^{\rightarrow} = \gamma(s)\delta_s$. But then by $(\star)$ the result follows. This finishes Clause (ii).

Now, for the symmetric case, consider an arbitrary pair $(\mu, \gamma) \in \mathcal{R} \cap Dist(S_t) \times Dist(RS)$, and let $t \in Supp(\mu)$. From the definition of $\mathcal{R}$ it follows that $\gamma \Longmapsto \mu$ and thus $\gamma \implies \mu$. Hence, $(\mu(t)\delta_t, \mu \ominus t) \in split(\gamma)$. We then choose $\gamma^{\rightarrow} = \mu(t)\delta_t$ and $\gamma^{\Delta} = \mu \ominus t$. Then for Clause (i) it suffices to note that $\mu(t)\delta_t \ \mathcal{R} \ \mu(t)\delta_t$ and $\mu \ominus t \ \mathcal{R} \ \mu \ominus t$, as for arbitrary distributions ξ over tangible states we have $\xi \Longmapsto \xi$. For Clause (ii), consider $t \xrightarrow{r} \mu'$ in the *CTMC* C_G. Note that this is the only possible transition of t (if any), as t is tangible. But then by $(\star)$, also $t \implies \mu'$ in A_G, and as before $\mu' \mathcal{R} \mu'$ follows.

Proposition 2 provides us with a kind of correctness criterion for the setup we presented. The MA weak bisimulation semantics indeed conservatively extends the classical semantics. Furthermore, many traditionally ill-defined and confused nets can still be related to a CTMC modulo weakly bisimilarity. This is linked to the fact that weak bisimilarity embodies the notion of lumpability, apart from immediate transition fusing.

4.2 Timeless Traps

Cycles of immediate transitions are an intricate problem in classical *GSPN* theory, their circular firing is often called a timeless trap [9], see Fig 7(a) for an example. *GSPN*s with timeless traps are traditionally excluded from the analysis, basically because the firing precedence of immediate over timed transitions makes the system diverge on the cycle without letting time progress. This is an awkward phenomenon, related to Zeno computations. In our *MA* reformulation, timeless traps are represented as cycles in the *MA*, and as such do not pose specific semantic problems. Furthermore, weak bisimilarity is sensitive to cycles of immediate transitions, but only to those that cannot be escaped by firing an alternative immediate transition. This is due to a built-in fairness notion in the weak bisimulation semantics, (rooted in the inclusion of a tangibility check inside the definition of the abbreviation $\xrightarrow{\alpha}$). As a consequence, if a timeless trap can be left by firing a (finite sequence of) immediate transitions leading to a tangible marking, this is equivalent to a single immediate transition firing. This implies that the net

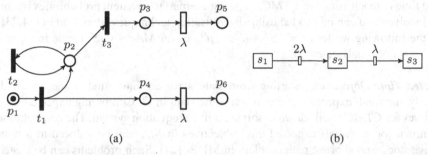

(a)   (b)

Fig. 7. A timeless trap that can be escaped by an immediate transition firing (a), and the smallest *MA* weakly bisimilar to its semantics (b). In (b), state s_1 subsumes markings $\{p_1\}$, $\{p_2, p_4\}$, and $\{p_3, p_4\}$. State s_2 subsumes markings $\{p_3, p_6\}$, and $\{p_4, p_5\}$, while state s_3 represents marking $\{p_5, p_6\}$.

in Fig. 7(a) is in fact weak bisimilar to the small chain-structured 3-state CTMC in Fig 7(b). And thus the net is analysable via the classical *CTMC* machinery. This example shows that the combination of lumping and fusing of immediate transitions as supported by weak bisimulation can have powerful effects. Variations to the definition of $\xrightarrow{\alpha}$ can induce more liberal notions of weak bisimiliarity, including the option to escape timeless traps unconditionally [27]. That option is not supported by the setup presented here, which has originally been designed to support strong compositionality properties [20]. Since compositionality is not a first-class concern in the Petri net world, this avenue seems worthwhile to be investigated further.

5   Quantitative Analysis of Markov Automata

So far, we have provided the details of a semantics of every definable *GSPN*. Thanks to Proposition 2, the steady-state and transient analysis of a well-defined *GSPN* under our semantics yields the same results as the evaluation of the induced *CTMC*. The remaining question is whether a quantitative analysis of a non well-defined *GSPN* is possible, and if so, how such analysis could be performed. Due to the possible presence of non-determinism, we can no longer consider *the* probability of a certain event. We stipulate that such probabilities depend on the resolution of non-determinism. Rather than considering, e.g., the probability to reach a state (i.e., a marking), it is common to determine the minimal and maximal reachability probabilities. These values correspond to the worst and best resolution of the non-determinism, respectively. Objectives that do not address the timing of net transitions, such as reachability, can be addressed using standard techniques for Markov decision processes (MDPs) such as linear programming, value, or policy iteration [5, Ch. 10]. Properties that involve the elapsed time are more interesting. In the following we briefly consider two such objectives: expected time and long run averages. For details we refer to [21] where Markov automata without probabilistic branching are considered. The inclusion of probabilistic branching however is rather straightforward. Long run average probabilities are the pendant to steady-state probabilities in *CTMCs*. Expected time objectives correspond to the expected time to reach a state in *CTMCs*. The counterpart to transient probabilities is a bit more involved and can be tackled using discretisation techniques advocated in [34,23].

In the following we let $A = (S, \longrightarrow, \dashrightarrow, \mu^0)$ be an *MA*, $s \in S$ a state in A, and $G \subseteq S$ a set of (goal) states.

Expected Time Objectives. Starting from state s we are interested in the maximal, or dually, minimal, expected time to reach some state in G. Computing expected time objectives for *CTMCs* boils down to solving a linear equation system. The computation of minimal (or maximal) expected time objectives in *MA* can be reduced to a non-negative stochastic shortest path problem in MDPs [21]. Such problems can be casted as a linear programming problem [10] for which efficient algorithms and tools (such as SoPLEX) exist.

Long-run average objectives. Intuitively speaking, the long-run average of being in a state in G while starting from state s is the fraction of time (on the long run) that the *MA* A will spent in states in G. We assume w.l.o.g. that G only contains tangible states, as the long-run average time spent in any vanishing state is zero. The general idea of computing the minimal long-run time spent in G is the following three-step procedure:

1. Determine the maximal end components[3] $\{A_1, \ldots, A_k\}$ of the MA at hand.
2. Determine the minimal long-run time spent in G within each end component A_j.
3. Solve a stochastic shortest path problem [10].

The first step is performed by a graph-based algorithm, whereas the last two steps boil down to solving linear programming problems. Determining the minimal expected long-run time in an end component can be reduced to a long-run ratio objective in an MDP equipped with two cost functions. Basically, it is the long-run ratio of the expected time of being in a state in G relative to the total expected time elapsed so far.

---

[3] A maximal end component is the analogue of a maximal strong component in the graph-theoretic sense, and is a standard notion for MDPs.

A prototypical implementation of our semantics is provided as part of the SCOOP tool, see: `http://wwwhome.cs.utwente.nl/timmer/scoop/webbased.html`. This is based on translating *GSPN*s to an intermediate process-algebraic formalism [33] whose operational semantics yields Markov automata. The tool also supports expected time, timed reachability, and long-run analysis as described just above.

6 Conclusion

This paper has presented a semantics of *GSPN*s in terms of a non-deterministic variant of *CTMC*s, called Markov automata [20]. We have shown that for well-defined bounded *GSPN*s, our semantics is weak bisimulation equivalent to the *CTMC* semantics existing in the literature [8,13,4,3]. This "backward compatibility" result intuitively means that our semantics is the same as the classical *GSPN* semantics up to an equivalence that preserves all quantitative measures of interest such as transient, steady-state probabilities and CSL (without next) formulae [6]. Thus, any tool based on our *MA*-semantics yields for well-defined bounded nets the same results as popular *GSPN* tools such as GreatSPN, SMART, and MARCIE.

The main contribution of this paper is that our semantics applies to *every GSPN*. That is to say, our semantic framework is not restricted to well-specified or confusion-free nets. The key to treating confused nets is (not surprisingly) the use of non-determinism. We claim that our approach can also be applied to other stochastic net formalisms such as SANs [28,30].

The semantics closes a gap in the formal treatment of *GSPN*s, which is now no longer restricted to well-defined nets. This abandons the need for any check, either syntactically or semantically, for well-definedness. This gap was particularly disturbing because several published semantics for higher-level modelling formalisms—e.g., UML, AADL, WSDL—map onto *GSPN*s without ensuring the mapping to be free of confusion, thereby inducing ill-defined models. Our Markov automata semantics provides the basis to also cover the confused and ill-specified semantic fragments of these formalisms. Indeed, we were able to relax both notions by considering the Markov automata semantics modulo weak bisimulation. To proceed this way seemed like a natural way forward for quite some time to us, but to arrive there was an astonishingly difficult notational and technical endeavour.

Possible Extensions. This paper does not consider the preservation (by the notion of weak bisimulation) of more detailed marking information such as the exact token occupancy of a place. Our notion of weak bisimulation is rather coarse and abstracts from this information. It is however straightforward to include this information by a simple extension of weak bisimulation that respects a certain state labelling, and this is fairly routine [17,6]. The same is true for other reward structures—except rewards attached to immediate transitions, which are more involved to handle. The proof for "backward compatibility" of our semantics for unbounded (but e.g., finitely branching) *GSPN*s is left for further study.

References

1. Ajmone Marsan, M., Balbo, G., Chiola, G., Conte, G.: Generalized stochastic Petri nets revisited: Random switches and priorities. In: Petri Nets and Performance Models (PNPM), pp. 44–53. IEEE CS Press (1987)
2. Ajmone Marsan, M., Balbo, G., Chiola, G., Conte, G., Donatelli, S., Franceschinis, G.: An introduction to Generalized Stochastic Petri Nets. Microel. and Rel. 31(4), 699–725 (1991)
3. Ajmone Marsan, M., Balbo, G., Conte, G., Donatelli, S., Franceschinis, G.: Modelling with Generalized Stochastic Petri Nets. John Wiley & Sons (1995)
4. Ajmone Marsan, M., Conte, G., Balbo, G.: A class of generalized stochastic Petri nets for the performance evaluation of multiprocessor systems. ACM TOCS 2(2), 93–122 (1984)
5. Baier, C., Katoen, J.-P.: Principles of Model Checking. MIT Press (2008)
6. Baier, C., Katoen, J.-P., Hermanns, H., Wolf, V.: Comparative branching-time semantics for Markov chains. Inf. Comput 200(2), 149–214 (2005)
7. Balbo, G.: Introduction to stochastic Petri nets. In: Brinksma, E., Hermanns, H., Katoen, J.-P. (eds.) FMPA 2000, LNCS, vol. 2090, pp. 84–155. Springer, Heidelberg (2001)
8. Balbo, G.: Introduction to Generalized Stochastic Petri Nets. In: Bernardo, M., Hillston, J. (eds.) SFM 2007. LNCS, vol. 4486, pp. 83–131. Springer, Heidelberg (2007)
9. Bause, F.: No way out ∞ The timeless trap. Petri Net Newsletter 37, 4–8 (1990)
10. Bertsekas, D.P., Tsitsiklis, J.N.: An analysis of stochastic shortest path problems. Mathematics of Operations Research 16(3), 580–595 (1991)
11. Buchholz, P.: Exact and ordinary lumpability in finite Markov chains. J. Applied Probability 31, 59–75 (1994)
12. Chiola, G., Donatelli, S., Franceschinis, G.: GSPNs versus SPNs: What is the actual role of immediate transitions? In: Petri Nets and Performance Models (PNPM), pp. 20–31. IEEE CS Press (1991)
13. Chiola, G., Ajmone Marsan, M., Balbo, G., Conte, G.: Generalized stochastic Petri nets: A definition at the net level and its implications. IEEE TSE 19(2), 89–107 (1993)
14. Ciardo, G., Zijal, R.: Well-defined stochastic Petri nets. In: MASCOTS, pp. 278–284 (1996)
15. Deavours, D.D., Sanders, W.H.: An efficient well-specified check. In: Petri Nets and Performance Models (PNPM), pp. 124–133. IEEE CS Press (1999)
16. Deng, Y., van Glabbeek, R., Hennessy, M., Morgan, C.: Testing finitary probabilistic processes. In: Bravetti, M., Zavattaro, G. (eds.) CONCUR 2009. LNCS, vol. 5710, pp. 274–288. Springer, Heidelberg (2009)
17. Desharnais, J., Gupta, V., Jagadeesan, R., Panangaden, P.: Weak bisimulation is sound and complete for PCTL*. Inf. Comput. 208(2), 203–219 (2010)
18. Hillston, J.: A Compositional Approach to Performance Modelling. PhD thesis, University of Edinburgh (1994)
19. Eisenraut, C., Hermanns, H., Zhang, L.: Concurrency and composition in a stochastic world. In: Gastin, P., Laroussinie, F. (eds.) CONCUR 2010. LNCS, vol. 6269, pp. 21–39. Springer, Heidelberg (2010)
20. Eisenraut, C., Hermanns, H., Zhang, L.: On probabilistic automata in continuous time. In: LICS, pp. 342–351. IEEE (2010)
21. Guck, D., Han, T., Katoen, J.-P., Neuhäußer, M.R.: Quantitative timed analysis of interactive markov chains. In: Goodloe, A.E., Person, S. (eds.) NFM 2012. LNCS, vol. 7226, pp. 8–23. Springer, Heidelberg (2012)
22. Hermanns, H. (ed.): Interactive Markov Chains. LNCS, vol. 2428. Springer, Heidelberg (2002)
23. Hatefi, H., Hermanns, H.: Model Checking Algorithms for Markov Automata. ECEASST 53, 1–15 (2012)

24. Hermanns, H., Herzog, U., Mertsiotakis, V., Rettelbach, M.: Exploiting stochastic process algebra achievements for Generalized Stochastic Petri Nets. In: Petri Nets and Performance Models (PNPM), pp. 183–192. IEEE CS Press (1997)
25. Katoen, J.-P.: GSPNs revisited: Simple semantics and new analysis algorithms. In: ACSD, pp. 6–11. IEEE (2012)
26. Kemeny, J.G., Snell, J.L., Knapp, A.W.: Denumerable Markov Chains, 2nd edn. Springer (1976)
27. Lohrey, M., D'Argenio, P.R., Hermanns, H.: Axiomatising divergence. Inf. Comput. 203(2), 115–144 (2005)
28. Meyer, J.F., Movaghar, A., Sanders, W.H.: Stochastic activity networks: Structure, behavior, and application. In: Petri Nets and Performance Models (PNPM), pp. 106–115. IEEE CS Press (1985)
29. Milner, R.: Communication and Concurrency. Prentice-Hall (1989)
30. Sanders, W.H., Meyer, J.F.: Stochastic Activity Networks: Formal definitions and concepts. In: Brinksma, E., Hermanns, H., Katoen, J.-P. (eds.) FMPA 2000, LNCS, vol. 2090, pp. 315–343. Springer, Heidelberg (2001)
31. Segala, R.: Modeling and Verification of Randomized Distributed Real-Time Systems. PhD thesis, Laboratory for Computer Science, Massachusetts Institute of Technology (1995)
32. Teruel, E., Franceschinis, G., De Pierro, M.: Well-defined Generalized Stochastic Petri Nets: A net-level method to specify priorities. IEEE TSE 29(11), 962–973 (2003)
33. Timmer, M., Katoen, J.-P., van de Pol, J., Stoelinga, M.I.A.: Efficient modelling and generation of markov automata. In: Koutny, M., Ulidowski, I. (eds.) CONCUR 2012. LNCS, vol. 7454, pp. 364–379. Springer, Heidelberg (2012)
34. Zhang, L., Neuhäußer, M.R.: Model checking interactive Markov chains. In: Esparza, J., Majumdar, R. (eds.) TACAS 2010. LNCS, vol. 6015, pp. 53–68. Springer, Heidelberg (2010)

Expressing and Computing Passage Time Measures of GSPN Models with HASL

Elvio Gilberto Amparore[1], Paolo Ballarini[2], Marco Beccuti[1],
Susanna Donatelli[1], and Giuliana Franceschinis[3]

[1] Università di Torino, Dipartimento di Informatica
{beccuti,susi,amparore}@di.unito.it
[2] Ecole Centrale Paris, Laboratoire MAS
paolo.ballarini@ecp.fr
[3] Università Piemonte Orientale, Dipartimento di Informatica
giuliana.franceschinis@di.unipmn.it

Abstract. Passage time measures specification and computation for Generalized Stochastic Petri Net models have been faced in the literature from different points of view. In particular three aspects have been developed: (1) how to select a specific token (called the tagged token) and measure the distribution of the time employed from an entry to an exit point in a subnet; (2) how to specify in a flexible way any condition on the paths of interest to be measured, (3) how to efficiently compute the required distribution. In this paper we focus on the last two points: the specification and computation of complex passage time measures in (Tagged) GSPNs using the Hybrid Automata Stochastic Logic (HASL) and the statistical model checker COSMOS. By considering GSPN models of two different systems (a flexible manufacturing system and a workflow), we identify a number of relevant performance measures (mainly passage-time distributions), formally express them in HASL terms and assess them by means of simulation in the COSMOS tool. The interest from the measures specification point of view is provided by the possibility of setting one or more timers along the paths, and setting the conditions for the paths selection, based on the measured values of such timers. With respect to other specification languages allowing to use timers in the specification of performance measures, HASL provides timers suspension, reactivation, and rate change along a path.

1 Introduction

Performance analysis of systems through Generalized Stochastic Petri Net (GSPN) models or similar formalisms has evolved significantly since their introduction, in particular an interesting research direction concerns how to express and compute relevant performance measures. The languages that have been proposed to this purpose span from classical reward based ones to logics like the Continuous Stochastic Logic (CSL) and its extensions (action based, timed-automata based and reward based), to automata based languages. In this paper the focus is on the automata based languages, allowing to analyze a measure of interest on a *selected set of paths* through the model state space. Such paths are executions of a stochastic process (quite often a Continuous Time Markov Chain (CTMC) for which efficient analysis techniques exist) usually described

J.-M. Colom and J. Desel (Eds.): PETRI NETS 2013, LNCS 7927, pp. 110–129, 2013.
© Springer-Verlag Berlin Heidelberg 2013

by means of a high level formalism like GSPN or Stochastic Process Algebra (SPA) of various sorts.

The *passage time distribution* is a specific type of performance index which is particularly useful when reasoning about properties related with Service Level Agreements (SLA) or safety requirements. In these cases classical performance measures based on mean values, like the average response time, are not sufficient and an estimate of the probability distribution for the time to complete a specific model activity (for example a recovery process) is needed instead. Moreover often only specific behaviors should be accounted for in the computation of such probability distributions, which brings us back to the need of estimating the passage time distribution on a subset of model evolution paths.

A passage time measure specification for CTMC is usually based on the definition of entry, goal and forbidden states: the distribution of the time required to reach a goal state from any entry state without hitting any forbidden state can be computed with different methods and tools [18,13]. This typically requires the (automatic) manipulation of either the CTMC or of the high level model used to generate the CTMC. When specific paths must be isolated however, the CTMC may undergo a transformation, often obtained by synchronizing it with an automata describing the paths of interest. Examples of languages proposed in the literature to express these type of measures are the Extended Stochastic Probes (XSP [11], operating on PEPA models [17]), Path Automata (PA, operating on Stochastic Activity Networks [19]), Probe Automata (PrA [3], operating on GSPN or on Tagged GSPN [7]). More recently the HASL (Hybrid Automata Stochastic Logic) [8] language has been introduced, operating on any Discrete Event Stochastic Process (but up to now experimented only with an extended version of GSPNs): we shall exploit its expressive power in the present paper, with reference precisely to GSPN models.

Another issue when expressing passage time measures on high level models is the possibility to identify entities in the model (customers), and measure the time required for a selected entity to go through a number of steps (activities), corresponding to the movement of the entity through a sequence of components in the high level model (possibly conditioned on some state-based predicate being true). In GSPN terms this often leads to the need to follow a specific token through the net. This issue is not trivial if the token to be followed may go through places containing other tokens, since they are indistinguishable: this aspect has been tackled in the literature by introducing a formalism extension called Tagged GSPNs [11].

This paper introduces the use of the HASL logic for the specification of passage time measures over GSPN models extended with general firing time distributions. The measure computation is performed using the statistical model checker COSMOS: stochastic discrete event simulation of the GSPN stochastic process synchronized with a Linear Hybrid Automaton (LHA) is performed. Cosmos generates a (statistically significant) sample of GSPN paths conforming to the HASL specification, and estimates the measures of interest from such sample (a confidence interval is provided for each measure).

The conclusion is that the HASL expressive power is adequate to specify (passage time) performance measure as can be specified with PrA and with the TGSPN passage time measure specification language, and that the COSMOS statistical model checker is an appropriate and useful tool to estimate such measures, even in presence of non exponential transition firing times and on models with very large state spaces. In addition more complex paths selection is possible, in particular those characterized by conditions on the duration of specific phases along the path (requiring to set one or more timers during the evolution, with the possibility of suspending, resuming and proceeding with different rates).

The paper is organized as follows: Section 2 describes the stochastic logic HASL, the associated LHA and the statistical model checker COSMOS. The use of LHA for the specification of passage time of paths of a stochastic Petri net model is then illustrated in Section 3, based on a simple workflow model taken from the literature; in this section we also report results for the passage time computation using COSMOS. The literature on the definition of passage time for tagged GSPN, based on entry, exit and forbidden condition for subnet identification or based on Probe Automata, is then recalled in Section 4. The difference and similarities of HASL based specification with respect to Probe Automata specification of passage times is discussed in Section 5, supported by a classical FMS example. Finally some conclusive remarks are given in Section 6.

2   Background: HASL

The Hybrid Automata Stochastic Logic (HASL) [9] is a recently introduced, automata-based, formalism for statistical model checking of discrete event stochastic processes (DESP). It enjoys two main features: generality and expressiveness. HASL is general with respect to modelling capabilities as it addresses a class of models (i.e. DESPs) which includes, but (unlike most stochastic logics) is not limited to, CTMCs. With respect to expressiveness HASL turns out to be a powerful language through which temporal reasoning is naturally blended with elaborate reward-based analysis. In that respect HASL unifies the expressiveness of CSL[4] and its action-based variant [5], timed-automata [15,10] and reward-based [16] extensions, in a single powerful formalism. The HASL model checking method belongs to the family of statistical model checking approaches (i.e. those that employ stochastic simulation as a means to estimate a model's property) and, more specifically, it employs confidence-interval methods to estimate the expected value of the target measure (i.e. either a measure of probability or a generic real-valued measure). Finally a prototype software tool for HASL model checking, named COSMOS [8], gives the possibility to actually apply HASL to real case studies (see [8] and [14] for a comparison of COSMOSwith other tools implementing statistical model checking). In the following we informally introduce the basic elements of the HASL methodology referring the reader to the literature [9] for formal details.

2.1   HASL Models: DESPs as GSPNs

The HASL logic refers to DESP models. Informally a DESP is a stochastic process consisting of a (possibly infinite) set S of states and whose dynamic is triggered by a (finite) set E of (time-consuming) discrete events. For reasons of generality no restrictions are considered on the nature of the delay distributions associated with events, thus any distribution with non-negative support may be considered. In practice the HASL framework [9] has been formalized referring to a high-level representation of DESP, namely:

an HASL model consists of an (extended) GSPN whose timed-transitions may be associated with any delay distribution with non-negative support (e.g. Exponential, Deterministic, Uniform, LogNormal, etc.). The choice of GSPNs as modeling formalisms is due to two factors: (1) they allow a flexible modeling w.r.t. the policies defining the process (choice, service and memory) and (2) they provide an efficient path generation (due the simplicity of the *firing rule* which drives their dynamics). For the sake of space in this paper we omit the formal definition of DESP but we rather introduce the rationale behind it through description of some examples. Also, in the remainder we will simply use the notation term GSPN referring, in facts, to a DESP model in GSPN form.

2.2 HASL Formulas: A Linear Hybrid Automaton and a Target Expression

A HASL formula is a pair $(\mathscr{A}, Z)$ where $\mathscr{A}$ is Linear Hybrid Automaton (i.e. a restriction of hybrid automata [2]) and Z is an expression involving *data variables* of $\mathscr{A}$. The goal of HASL model checking is to estimate the value of Z by synchronization of a GSPN $\mathscr{N}$ with the automaton $\mathscr{A}$. This is achieved through stochastic simulation of the synchronized process $(\mathscr{N} \times \mathscr{A})$, a procedure by means of which infinite timed executions of process $\mathscr{N}$ are selected through automaton $\mathscr{A}$ until some final state is reached or the synchronization fails. During such synchronization, data variables evolve and the values they assume condition the successive evolution of the synchronization. The synchronization stops as soon as either: a final location of $\mathscr{A}$ is reached (in which case the values of the variables are considered in the estimate of Z), or the considered trace of $\mathscr{N}$ is rejected by $\mathscr{A}$ (in which case variables' values are discarded).

LHA: Again here we only provide an informal description of LHA referring the reader to [9] for formal definitions. Simply speaking an LHA is an automaton consisting of the following elements: a finite set of locations L (some of which are *Final* and some other are *Initial*); a finite set of events E (corresponding to the GSPN transition labels); a finite non-empty set X of n real variables; a location labeling function $(\Lambda : L \to Prop)$ which associates each location with a (boolean) property that refers to GSPN states (i.e. markings); a flow function $(flow : L \mapsto Ind^n)$ which associates with each location an n-tuple of GSPN indicators (conditions referring to GSPN markings) expressing the rate (i.e. the first derivative) at which each data variable in X changes in that location; variable's flow can be simple constants or functions of the current state of the GSPN (hence they are expressed by means of *GSPN state indicators*, denote *Ind*). Finally a transition from a source location l to a target location l' has the following form $l \xrightarrow{E', \gamma, U} l'$, where: I) $E' \subseteq E \cup \{\sharp\}$ is a set of labels of synchronizing events (including the extra label $\sharp$ denoting autonomous edges); II) γ is a constraint (i.e. a boolean combination of inequalities of the form $\sum_{1 \leq i \leq n} \alpha_i x_i + c \prec 0$ where $\alpha_i, c \in Ind$ are GSPN state indicators and $\prec \in \{=, <, >, \leq, \geq\}$ and $x_i \in X$). III) U is a set of *updates* (i.e. an n-tuple of functions $u_1, ..., u_n$ where each u_k is of the form $x_k = \sum_{1 \leq i \leq n} \alpha_i x_i + c$ where the $\alpha_i, c \in Ind$ are GSPN indicators) by means of which new values are assigned to variables of X on traversing of the edge.

Example of LHA: Figure 2 depicts two variants of a simple two locations LHA characterizing path measures of the toy GSPN model of Figure 1 (such a GSPN may represent

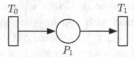

Fig. 1. A toy GSPN model

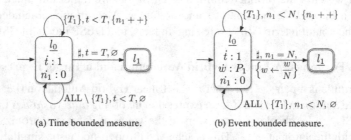

(a) Time bounded measure.   (b) Event bounded measure.

Fig. 2. Example of LHA for simple properties of the GSPN model in Fig. 1

a simple unbounded queue with service/arrival represented by T_1 and T_0 respectively, while tokens in place P_1 represent the customers in queue and in service). The locations named l_0 and l_1 are the initial and the final locations for the two automata. Both automata employ two data-variables: an integer variable n_1 (hence with flow $\dot{n}_1 = 0$ in every location), counting the occurrences of transition T_1, and a real-valued variable t which is used as a timer (hence with flow $\dot{t} = 1$) to record the simulation-time along the path; moreover the LHA of Figure 2(b) has a variable w with flow equal to the marking of place P_1 (hence it measures the integral of the number of waiting customers in the queue) along the observed path. Both automata have two synchronizing edges (the self-loops on l_0) and one autonomous edge (from l_0 to l_1). The topmost synchronizing edge synchronizes with occurrences of T_1 and increments the value of n_1 thus counting the occurrences of T_1. The bottommost synchronizing edge, on the other hand, synchronizes with any transitions (denoted ALL) of the GSPN except T_1, without performing any update. The autonomous edge $l_0 \rightarrow l_1$ instead leads to the final location as soon as its constraint is fulfilled. Note that the the condition leading to the final location in the LHA of Figure 2 (b) represents a time-bounded constraint on the simulation time t, i.e.: as soon as $t = T$ the processed path is accepted (and, by that time, n_1 will be equal to the number of firings of T_1 up to time $t = T$). On the other hand, the condition that leads to the final location in the LHA of Figure 2 (b), is an event-bounded constraint, as it accepts paths as soon as T_1 has occurred $n_1 = N$ times. More specifically, variable n_1 is incremented every time T_1 fires in location l_0, until n_1 becomes equal to N; in the same location variable w grows with rate equal to the marking of P_1. Just before taking the transition from l_0 to l_1, w contains therefore a quantity that corresponds to the summation of the residence times of all tokens observed in $P1$ along the simulation; when the transition is taken, it is updated to $w = w/n_1$, hence, on acceptance, w will be equal to the average residence time of customers in place P_1 along the observed path, that is to say after the first N occurrences of T_1.

HASL Expression: The second component of an HASL formula is an expression, denoted as Z and defined by the following grammar:

$$Z ::= E(Y) \mid Z + Z \mid Z \times Z \mid CDF_I(Y) \mid PDF_I(Y) \mid PROB()$$
$$Y ::= c \mid Y + Y \mid Y \times Y \mid Y/Y \mid last(y) \mid min(y) \mid max(y) \mid int(y) \mid avg(y) \tag{1}$$
$$y ::= c \mid x \mid y + y \mid y \times y \mid y/y$$

y is an arithmetic expression built on top of LHA data variables (x) and constants (c). Y is a path dependent expression built on top of basic path random variables such as $last(y)$ (i.e. the last value of y along a synchronizing path), $min(y)/max(y)$ (the minimum/maximum value of y along a synchronizing path), $int(y)$ (the integral over time along a path) and $avg(y)$ (the average value of y along a path). Finally Z, the target measure of an HASL experiment, is an arithmetic expression built on top of the first moment of Y $(E[Y])$, and thus allowing to consider more complex measures including, e.g. $Var(Y) \equiv E[Y^2] - E[Y]^2$, $Covar(Y_1, Y_2) \equiv E[Y_1 \cdot Y_2] - E[Y_1] \cdot E[Y_2]$.

The expressions $CDF_I(Y)$ and $PDF_I(Y)$ compute a sequence of cumulative/instantaneous values, subdivided in discrete samples of uniform size described by a *sampling interval* $I = \langle t_0, t_{Final}, \Delta t \rangle$, where t_0 and t_{Final} are the extremes of the sampling and Δt is the size of the uniform step at which the samples are taken. Usually, the target data variable of CDF/PDF is a time counter, in order to compute a density or a distribution function. The target expression $PROB()$ measures the mean number of paths accepted by the LHA automaton over the total number of simulated paths. It is the only operator that is influenced by rejected paths, since $E(Y)$, $CDF_I(Y)$ and $PDF_I(Y)$ takes samples from accepted paths only.

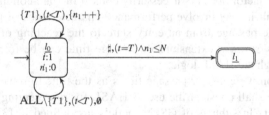

Fig. 3. LHA to compute the probability of observing less than N firings of $T1$ within time T

Measures of probability with HASL

With reference to the two LHAs of Figure 2, and considering the quantities accumulated along the accepted paths in the variables t, n_1 and w, we can define the expressions $E(last(n_1))$ for the LHA (a) to estimate the mean number of $T1$ firing up to time T, while for the LHA ((b) we could define $E(last(w))$ to estimate the mean average waiting time for behaviors up to the N-th firing of $T1$ and $CDF_I(t)$ for a given interval, to estimate the (normalized) distribution of the time to complete.

Note that, in the definition provided by HASL, the CDF is actually normalized, so as to asymptotically reach 1. This is motivated by the fact that the CDF of certain quantities may be "defective" in HASL, since it is computed only on accepted paths: if only a subset of the paths is accepted, the CDF may not tend to 1 but to an asymptotic value which is the probability of accepted paths.

Figure 3 shows another simple example of LHA (a small variant of the LHA in Figure 2(a)) for measuring ϕ_1: *the probability that the number of occurrences of transition T_1 within time T does not exceed N.* This is achieved by using the condition $(t = T) \wedge (n_1 < N)$ on the autonomous edge from l_0 to l_1 and $t < T$ on the other two edges departing from l_0. If a path counts more than N occurrences of $T1$ within time T, the edge $l_0 \to l_1$ will not be triggered and the LHA will not be able to synchronize with any successive GSPN transitions. Such a path is therefore rejected. The rate of acceptance is measured by defining the target measure $Z = PROB()$.

3 HASL and Passage Time of Selected Paths

Given a discrete-state stochastic process with state space S, and sets $E, G, D \subset S$ the *passage-time $Prob(E, G, D, t)$* is a CDF measure expressing the probability of reaching any goal state in G starting from any enter state in E and avoiding forbidden states in D and with a delay no greater than t [18,13]. The measure $Prob(E, G, D, t)$ is defined in function of the state-dependent random variable $P_x^{G,D}$, denoting the probability to reach a goal state in G starting from state x and avoiding states in D.

Here we discuss how to take advantage of an expressive property specification formalism, such as HASL, from the point of view of passage-time related measures. More specifically we are going to show how the HASL formalism can be used to express "standard" passage time measures as well as more complex ones. In other words we consider the possibility of an extended characterisation of passage-time measure where the constraining factor does not necessarily consist of state-conditions (i.e. the forbidden states D) but it may involve performance characteristics of the model (i.e. measured during the passage from an entry state to the reaching of a goal state). Such *performance-constrained* extensions of passage time can be formally expressed and measured through the HASL logic.

In this section we consider passage times as the time to traverse a set of selected paths, while we shall consider the use of HASL for representing the passage time of tagged customers in subnets of GSPN models, as defined in [3], in Section 4. The presentation in this section is supported by a specific GSPN model taken from the literature [1], which represents a business workflow. This will allow us to illustrate the richness of HASL in characterizing paths based on the actions performed and on the states visited along the path, as in any stochastic logic, but also based on accumulated discrete and continuous measures, including accumulated performance indices. The generality of the model considered will also allow us to bring evidence of the usefulness of such a rich specification language.

3.1 Business Workflow Model

The model considered is a case of an order-handling business process model, taken from [1]. Fig. 4 shows the GSPN that models the processing of a single order. The workflow involves two separate tasks of preparing and sending the bill to the client, and to ship

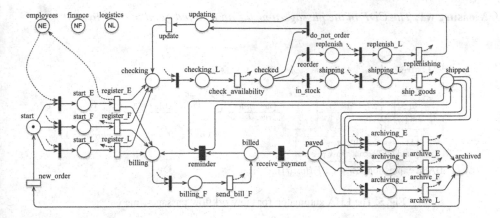

Fig. 4. The workflow model for an order-handling process

the requested goods. The company reckons on three types of employees: those who manage accounting (F), logistics (L) and generic employees (E). Different tasks are carried out by different employees. The Petri net is made of some subnets consisting of an immediate transition, a place and a timed transition. Such subnets first allocate one of these staff resources, execute the specified task and then release the resource. The staff is represented by three places *finance*, *logistics* and *employees*. Arrows from and to these three places are drawn only for the case of the activity represented by the *register_E* transition, done by a generic employee, which is reserved just before the *register_E* transition becomes enabled, and released upon firing. All the other subnets whose timed transitions have labels with suffixes "_E", "_F" and "_L" use a similar schema, acquiring and releasing the appropriate employee resources.

The Petri net represents the lifetime of an order from the *start* place, when it is received, to the *archived* place, when it has been served. Upon receiving of an order, one employee prepares the request to the warehouse and to the accounting department. A logistic personnel checks if the requested item is available: if it is not, a reorder is issued (*replenish*), and the shipping is delayed until the items are available (*update*). In the meanwhile, the bill is sent to the client. If after some time the payment has not been received (*receive_payment*), the billing is resent (*reminder*). When the item has been shipped and payed, the request is archived. The actual model contains several replicas of the model in Figure 4, all sharing the three resource places. Since replicas are kept separate, we can easily follow the possible paths followed by each single order.

3.2 HASL Based Passage-Time Measures

Referring to the GSPN model of the business workflow we illustrate a number of passage-time measures expressed in HASL terms. For each measure we provide first an informal specification then the corresponding formal characterization as an HASL formula (i.e. an LHA paired with an HASL expression as by grammar (1)). Finally for each such measures we provide numerical results obtained by running experiments with the COSMOS tool.

Measure w1: *The CDF of the passage-time for an ordered good to be delivered.*

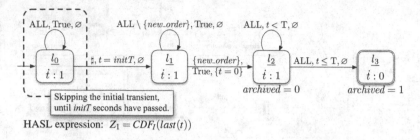

HASL expression: $Z_1 = CDF_I(last(t))$

Fig. 5. The LHA automaton for the workflow passage-time w_1

The probability measure w1 can be encoded by the HASL formula $\phi_{w1} = (\mathscr{A}_1, Z_1)$ where $\mathscr{A}_1$ is the LHA depicted in Figure 5 and Z_1 is the HASL expression $Z_1 = CDF_I(last(t))$. The automaton $\mathscr{A}_1$ employs one data variable: a timer t that records the simulation time of the synchronizing paths. $\mathscr{A}_1$ works as follows: the initial location l_0 is used only to emulate a transient window, letting a random trajectory of duration $initT$ being simulated before the actual analysis starts in location l_1. The automaton then remains in l_1 for as long as the first occurrence of (the GSPN) transition *new_order*, whose firing takes $\mathscr{A}_1$ into location l_2: note that on traversing the $l_1 \rightarrow l_2$ edge the timer t is reset, corresponding to the beginning of the passage-time measuring. From l_2 a path is accepted (i.e. by reaching final location l_3) as soon as the GSPN place *archived* is filled in with a token (representing the delivering of previous incoming order). The edge from l_2 to the final locations l_3 is taken only when the condition *archived*=1 is met. If at the instant when place *archived* is filled in with one token the passage-time is $t < T$ then the path is accepted. On the contrary, if $t > T$ the LHA becomes unable to synchronize with further transitions of the GSPN, and the path is rejected.

Measure w2: *The CDF of the passage-time for an ordered good to be delivered given that it was out-of-stock.*

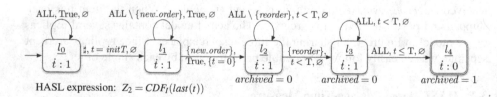

HASL expression: $Z_2 = CDF_I(last(t))$

Fig. 6. The LHA automaton for the workflow property w_2

The probability measure w2 can be encoded by the HASL formula $\phi_{w2} = (\mathscr{A}_2, Z_2)$ where $\mathscr{A}_2$ is the LHA depicted in Figure 6 and $Z_2 = CDF_I(Last(t))$. Automaton $\mathscr{A}_2$ is a variant of $\mathscr{A}_1$ and employs the same data variable t for the elapsed time. As in $\mathscr{A}_1$

the initial location l_0 is used only to let the transient window pass before entering in l_1 where the actual analysis of the synchronized path begins. From l_1 the LHA moves to l_2 on firing of a *new_order* transition (which also trigger the timer t reset). From l_2, l_3 is reached only on occurrence of a *reorder* transition (i.e. the ordered good was out-of-stock). From l_3 the final location l_4 is then reached under exactly the same conditions as of $\mathscr{A}_1$. As a result the GSPN paths leading to the final location are those which contain an occurrence of *new_order* followed by an occurrence of *reorder* and that finally lead to a marking $M(archived) = 1$.

Measure w3: *The CDF of the passage-time for an ordered good to be delivered given that it is not out-of-stock and that the total delay for checking its availability and shipping it does not exceed K .*

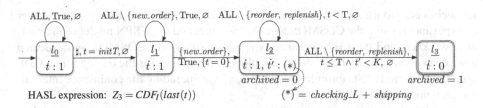

HASL expression: $Z_3 = CDF_I(last(t))$ $(*) = checking\_L + shipping$

Fig. 7. The LHA automaton for the workflow property w_3

The probability measure w3 can be encoded by the HASL formula $\phi_{w3} = (\mathscr{A}_3, Z_3)$ where $\mathscr{A}_3$ is the LHA depicted in Figure 7 and $Z_3 = CDF_I(last(t))$. Automaton $\mathscr{A}_3$ employs two data variables: a timer t, (as $\mathscr{A}_1$ and $\mathscr{A}_2$) plus an extra real-valued variable t' which is used to further conditioning the accepted paths. More specifically t' is used to measure the total time that the system spends in either checking the availability of the ordered good or in shipping it while assuming that the ordered good does not require a *reorder* (i.e. it is present in stock). For this the rate of variation of t' (i.e. its flow) corresponds to the sum of the marking of places *checking_L* and *shipping* (location l_2). Furthermore note that the exclusion of paths containing an occurrence of either *reorder* or *replenish* is obtained by using $ALL \setminus \{reorder, replenish\}$ in the on all arcs departing from l_2 as synchronization (constraint) set: i.e. if the currently simulated GSPN path has lead to l_2 then the occurrence of either *reorder* or *replenish* causes the path to be rejected (no LHA transition is enabled). Finally, from l_2 a path is accepted as soon as place *archived* is filled in with one token (delivery of the ordered good) and the condition on t' is satisfied (i.e. if $t' < K$, where K is a constant parameter of the $\mathscr{A}_3$).

Measure w4: *The CDF of the passage-time for an ordered good to be delivered given that the total delay for reordering and updating the stock does not exceed K .*

The probability measure w4 can be encoded by the HASL formula $\phi_{w4} = (\mathscr{A}_4, Z_4)$ where $\mathscr{A}_4$ is the LHA depicted in Figure 8 and $Z_4 = CDF_I(Last(t))$. Automaton $\mathscr{A}_4$ is a variant of $\mathscr{A}_3$ and uses the same two data variables only that now t' is used to measure

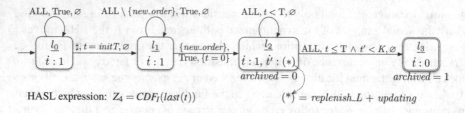

ALL, True, ∅   ALL \ {new_order}, True, ∅   ALL, $t <$ T, ∅

$\sharp, t = initT, \varnothing$ {new_order},
                                          True, {t = 0}   ALL, $t \leq$ T $\wedge\ t' < K, \varnothing$

l_0 : $i : 1$ l_1 : $i : 1$ l_2 : $i : 1,\ \dot{t}' : (*)$ archived $= 0$ l_3 : $i : 0$ archived $= 1$

HASL expression: $Z_4 = CDF_I(last(t))$ $(*) = replenish\_L + updating$

Fig. 8. The LHA automaton for the workflow property w_4

the total time that the system spends in either replenishing after a re-order or updating the stock. Thus the rate of variation of t' corresponds to the sum of the marking of places *replenish_L* and *updating* (location l_2).

Experiments: To assess the value of w_1, w_2, w_3 and w_4 we have performed a number of experiments using the COSMOS tool: i..e. we encoded the GSPN model of the business workflow and the LHA formulae $\phi_{w1}, \phi_{w2}, \phi_{w3}, \phi_{w4}$ into COSMOS and executed a set of experiments. Figure 9 shows the plots of the results for the passage-time CDF corresponding to HASL formulae $\phi_{w1}, \phi_{w2}, \phi_{w3}, \phi_{w4}$, including the confidence intervals for each time sample.

The plot of ϕ_{w1} represents the CDF of the average passage time in the workflow net, from the arrival of a new order to its archiving. The workflow has three replicas of the model, and has 1 generic employee, 1 accounting employee and 2 persons at the logistics. The transition names appearing on LHA edges refer to the first of the three replicas (due to the system symmetry the result is independent on the replica chosen for the measure). The other three curves represent the passage time distributions of a selected set of paths, where only paths that respect the additional constraints described in the LHAs are considered. For instance, ϕ_{w3} selects only paths that avoid reorders and also perform checking and shipping within a given time bound, so the average passage time of ϕ_{w3}-accepted paths will be less that of ϕ_{w1}. On the contrary, ϕ_{w2} consider

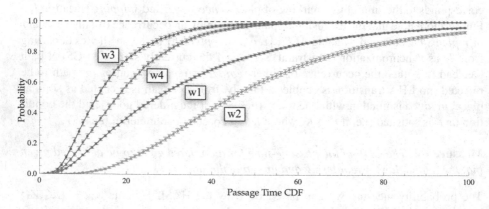

Fig. 9. Passage times for the workflow properties w_1 to w_4, with *sampling interval* $I = \langle 0, 100, 1 \rangle$

only paths that do at least one reorder, so their average passage time will be greater than that of ϕ_{w1}. Finally ϕ_{w4} lays in between ϕ_{w1} and ϕ_{w3} because it admits paths that require a reorder, but impose a limit on the total delay for reordering and updating the stock.

The workflow model keeps the status of each ordered good separated, by using a distinct replication of the subnet where the order is circulating. In general, keeping tracks of specific tokens in a net can be derived by a proper tagging of a selected token through the subnets where the tagged token flows. This can be done automatically by using *tagged GSPNs*. In the next sections we shall consider the use of HASL for the specification and computation of passage times, as defined for queueing networks and tagged GSPN.

Table 1. COSMOS runtime in function of confidence-level, interval-width and number of cores

measure	conf-level	width	num-cores	build-time	runtime	gen-paths
w1			1	2.91	56.07	6.7 e03
w1		0.01	2	2.91	32.86	6.7 e03
w1	99%		4	2.91	25.66	6.7 e03
w1			1	2.91	5569.42	6.635 e06
w1		0.001	2	2.91	3307.99	6.635 e06
w1			4	3.50	2521.05	6.635 e06
w1			1	2.87	33.37	3.9 e03
w1		0.01	2	2.85	20.15	3.9 e03
w1	95%		4	2.84	13.07	3.9 e03
w1			1	3.21	3262.36	3.842 e06
w1		0.001	2	2.82	1917.90	3.842 e06
w1			4	3.33	1365.10	3.842 e06

COSMOS performances. Table 1 reports the performances of the COSMOS tool on an Apple MacBook Pro, processor Intel dual Core i7 2.8GHz, 8GB 1333 MHZ DDR3 RAM, 256KB L2 cache, and 4MB L3 cache. when assessing measure **w1** of the workflow model. The table shows how the simulation-time (i.e. the runtime for sampling trajectories in a quantity sufficient to match the required accuracy of estimation) varies in function of the chosen accuracy (i.e. the confidence level and width of the estimated interval) of an experiment. Since Cosmos also allows for the parallelization of trajectory simulation, we have considered different levels of parallelization (number of cores over which an experiment is distributed). Observe that, quite sensibly, the runtime gain is less than linear with respect to the level of parallelization and in particular the gain drastically decreases when the chosen number of cores chosen for the computation is bigger than the actual CPU cores (in our machine this value is two). Results for more than 4 cores are not shown since they remains practically equal to the case of 4 cores. Finally observe that Cosmos adopts a model-driven code-generation scheme, i.e. a customized C++ instance of the HASL simulator is generated for each input pair (GSPN, HASL-formula) and compiled before the actual computation starts. The associated computation time is reported in the "build-time" column of Table 1, whereas the "gen-paths" column indicates the number of trajectories(paths) generated by each experiment.

4   Background: Tagged GSPN and Probe Automata

GSPN have been widely used in research and applications to compute classical performance measures, as the mean number of tokens in places or the throughput of transitions. One type of performance index which is not straightforward to define on GSPN models is the distribution of the time required for a (specific) token to pass through a given sub-net , since it requires "token-centric" [6,12] view of the system where the tokens represent system entities/customers moving through the net. However, in general, tokens in GSPNs cannot be interpreted as entities which travel throughout the models, but they are indistinguishable quantities consumed and generated by transition firing. Hence, the idea of selecting one token to make it "tagged" and of following it throughout the net is not a trivial task, unless exploiting certain structural properties of the net.

The work in [6] proposes to exploit invariant properties (i.e. p-invariant) to identify places where tokens with indistinguishable behavior are preserved so that such tokens can be treated similarly to the customers of Queuing Network models and can thus be tagged by the modeler to compute the distribution of the time required by one specific token to travel between points of the net. This has led to the introduction of the Tagged Generalized Stochastic Petri Net (TGSPN) formalism which extends classical GSPN with primitives to specify the subnet on which the passage time should be computed. In particular the counterpart of the observed (tagged) customer is the identification of a p-semiflow that leads to a partial unfolding of the subnet identified by the semiflow: transitions and places of the subnet are replicated: each place P_i has a replica P_i^{tag} (same for transitions), and the token in the tagged subnet represents the tagged customer. The condition upon which to start, finish or stop the computation of the passage time are specified by the *Entry*, *Exit* and *Forbid* conditions, identifying respectively the transitions corresponding to the start and stop of passage time count, and those causing the abort of the measurement (i.e. allowing to discard the paths where any forbidden transition fires). Actually the conditions are specified as triplets $\{\langle t, C_{in}, C_{out} \rangle\}$, where C_{in} is the marking condition that has to be satisfied in the tangible marking where t is fired, and C_{out} is the marking condition which has to be satisfied in the tangible marking reached after firing it. Any of the three elements may be omitted. The TGSPN computation of passage time is based on the tangible reachability graph. Therefore, the semantics is slightly more complicated, since arcs in a TRG are labeled with *extended firing sequences*: a timed transition followed by zero or more immediate, leading to a tangible marking. So $Entry = \{\langle t, C_{in}, C_{out} \rangle\}$ means that the computation of the passage time starts when we reach a tangible marking satisfying C_{out} from a tangible marking satisfying C_{in} with an extended firing sequence containing t.

To make this more concrete we introduce another example taken from the literature. This is the model of a Flexible Manufacturing System (FMS), a model that was already used in [6] to explain tagged GSPN, and that we report in Fig. 10 for ease of reference. This FMS comprises four manufacturing stations, from M_1 through M_4, where two of them, M_2 and M_3, can fail. Raw parts are loaded on suitable pallets at the Load/Unload (L/U) station represented by the pair – place *Pallets* and single server transition *load* – and are then manufactured, being sequentially brought to the four machines. The model then cycle back (pallet reused on a new raw part). Machine M_2 and M_3 can fail, but while machine M_3 has no spares, so when it breaks down it has to undergo a reparation

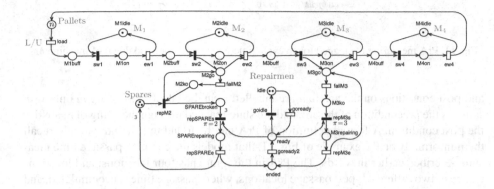

Fig. 10. A FMS with machines breakdown and repair

by a repairman (initially located in place *REPMANidle*), machine M_2 has a set of spares available in place *Spares*. When all spares have been used, a repairman will intervene to repair all spares, and make them available again. Repairmen are not always available: they cycle between vacation and repair periods. Upon return from a vacation (*goready*) a repairman checks if a machine has failed (*repSPAREs*, *repM₃s*) and in such case starts a repair activity (*repSPAREe*, *repM₃e*), otherwise goes back to vacation (*goidle*). After the repairman ends working on a failed machine, he takes a rest (*goready2*) before starting a new cycle. If both M_2 and M_3 require the intervention of a repairman, priority is given to machine M_3.

The modeler could be interested in several passage time measures on such model, for instance the distribution of the waiting time in place M_2ko, that is the time spent by a part in machine M_2 awaiting for the spare parts to be replaced by a repairman. This could be simply expressed as: ($\mathscr{P}_1$) the first passage time from a state where M_2ko becomes marked (due to the firing of transition *failM₂* when there are no tokens representing spare tools in place *Spares*) to a state where the same place becomes empty (due to the firing of transition *repM₂*). This passage time is specified in TGSPN as $Entry = \{\langle failM_2, Spares = 0, -\rangle\}$ and $Exit = \{\langle repM_2, -, -\rangle\}$. Note that the transition *repM₂* in the triplet means "an extended firing sequence that contains *repM₂*", which could match, for instance, the firing of the timed transition *repSPAREe* followed by the two immediates *repM₂* and *repSPAREs*.

However, it has been recognized that the "triplet-based" specification language of TGSPN is not flexible enough to express passage time measures in presence of more elaborate customer behaviour. For example we cannot express the requirement of a passage time distribution of the full cycle in the system (time between two successive firings of transition *load*^{tag}), taking into account only those paths which have experienced at least one "real" breakdown of machine M_2 (a breakdown when no spare is available, which is equivalent to reaching a tangible marking satisfying the condition $M_2ko > 0$), before starting the computation of the passage time. To cope with this limitation Probe Automata (PrA) have been introduced in [3]. PrA uses a path automaton [19], which recognizes paths based on the transitions that fire along the path, enriched with the pre

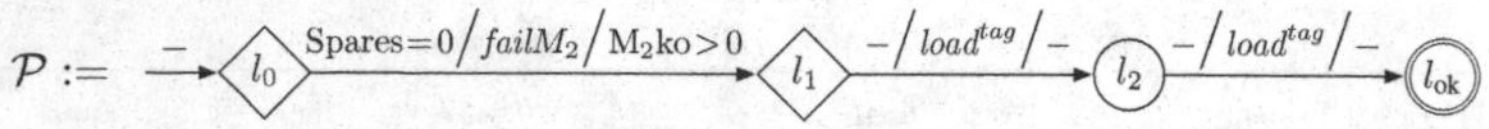

$$\mathcal{P} := \quad \xrightarrow{\quad} \langle l_0 \rangle \xrightarrow{\text{Spares}=0\,/\,failM_2\,/\,M_2\text{ko}>0} \langle l_1 \rangle \xrightarrow{-\,/\,load^{tag}\,/\,-} \bigcirc l_2 \xrightarrow{-\,/\,load^{tag}\,/\,-} \bigcirc\bigcirc l_{ok}$$

Fig. 11. PrA measuring the tagged token cycle time after a stop for breakdown of machine M_2

and post conditions on the arcs: an arc labelled t can be taken in marking m only if m satisfies the precondition of the arc, and the state reached through the firing of t satisfies the post condition. A formal definition of PrA can be found in [3], and we only recall them informally on the example of Fig. 11, that models the complex passage time measure described earlier in words. The PrA in the figure has four locations, and locations can be of two different types: passage locations, where passage time is accumulated, and non-passage location, where time is not accumulated. Non-passage locations (l_0 and l_1) are drawn as rhombuses and passage (l_2) locations are drawn as circles. As usual, initial locations (only l_0) are identified by an entering edge, while final locations (l_{ok}) have a double border. Finally, edges are labeled with constraints written as $C_{in}/t/C_{out}$.

5   HASL and Probe Automata

When using HASL for the specification of passage time properties of PrA type, there are some peculiarities of PrA that require specific attention. A very general difference is that PrA have been explicitly designed to specify passage time over GSPN subnets, so we cannot expect LHA to be as compact as PrA in representing the same property.

There are indeed two peculiar features of PrA that make them more compact than LHA for passage time specification. The first one is that PrA have implicit loops over locations, accepting all transition firings different from the one present on outgoing arcs. In LHA instead, all events have to be specified, otherwise a path is rejected. The second one is that pre and post conditions over arcs in PrA have no direct counterpart in LHA. An equivalent LHA automaton will have a multiplication of locations to model the same behavior of the PrA, to distinguish locations that do and do not satisfy the conditions. This may lead to a significant increase in the number of locations of the LHA. These differences are not really limitations in the expressiveness, it is only an issue of how easy it is to model a property.

On the other side LHA allow for multiple data variables with flow that can depend on the location. PrA instead has a single, implicit, clock variable, which is equivalent of having a single data variable t with flow of 1 in the counting locations, and 0 otherwise. LHA variables allow to store state informations, that in the PrA would require a multiplication of locations.

Another fundamental difference between PrA and LHA is that PrA are defined over the TRG of the GSPN, while LHA observe the full RG, which includes also vanishing states, so what can be observed with PrA differs in nature from what can be observed by an LHA on the same GSPN. An automatic translation from PrA to LHA would require, for each arc of the PrA labeled with a transition t, the expansion into a subnet that accept all sub-paths from tangible to a tangible marking, passing only through vanishing states and having at least a firing of t along the path.

A difference that requires some caution is that measuring of passage-time through Probe-Automata [3] is based on the idea that a probe may be plugged in (hence becoming operative) at any moment in time of the process described by the considered GSPN, which basically amounts to saying that probes compute passage time assuming that the initial state(s) for the probe have a distribution that is equivalent to the steady state probability. Since HASL does not (inherently) support steady-state measures, then, in order to cope with the *probe plugging in* approach, it is necessary that an LHA for passage-time measures is equipped with a so-called *transient emulator*, i.e. a (unique) initial location whose goal is simply to simulate the GSPN model for a given delay *initT* (assuming that at *initT* the GSPN has reached steady state).

Finally, PrA have been carefully designed to allow a numerical solution, while LHA have instead been designed with simulation in mind. This last difference involves two main aspects: the PrA assumes that the underlying state model is finite and known (this is necessary if we want to allow for an initial distribution, but especially if we want to use the steady state distribution as the initial distribution, as discussed above) and a PrA has a single timer, so as to allow for a reuse of classical Markov renewal theory results in the computation of the passage time.

Using again the FMS model, we consider three PrA passage time specifications $\mathcal{P}_1$ to $\mathcal{P}_3$ (taken from [3]), provide the corresponding LHA (F_1 to F_3) and add a new FMS property expressed through LHA F_4, which cannot be expressed by PrA.

The first example (Figure 12) shows a simple passage time CDF specification that leads to the computation of the distribution of a piece of the cycle (time to load the pallet plus the work time of M_1 and M_2) for the tagged customer. $\mathcal{P}_1$ shows the PrA, while F_1 is the LHA. As expected there are some more arcs, and more annotations over arcs. Note the use of the initial additional location l_0 to skip the initial transient. The passage location l_1 states that time should be accumulated starting in l_2, which is rendered in the LHA by a reset of the clock t while entering location l_2.

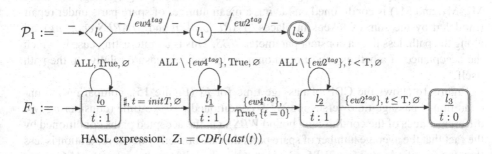

HASL expression: $Z_1 = CDF_l(last(t))$

Fig. 12. The Probe Automaton $\mathcal{P}_1$ and the LHA automaton F_1 for the FMS model

Figure 13 is the cycle time for a tagged token conditioned on at most one breakdown occurring during the cycle (to either the tagged customer itself, or to any other, untagged, customer). The specification of "at most one breakdown" cycle in PrA requires to keep memory of the failures occurred in the past, which leads to the introduction of location l_2 The corresponding LHA appears to be more complex, at least in terms

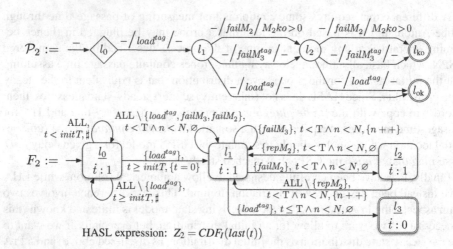

HASL expression: $Z_2 = CDF_l(last(t))$

Fig. 13. The Probe Automaton $\mathscr{P}_2$ and the LHA automaton F_2 for the FMS model

of arc annotations, but it is actually more general: the number of breakdowns is accumulated in the discrete variable n, and paths are accepted if there have been at most N breakdowns.

Property $\mathscr{P}_3$ of Figure 14 shows an example of PrA arcs with pre and post conditions. By requiring that $failM_2$ is pre conditioned on place $Spares$ being equal to zero and post conditioned on M_2ko being greater than zero, we are catching the first breakdown of M_2 that finds no spare parts available. In the LHA this is translated expanding the arc from l_0 to l_1 of the PrA into the three locations l_1, l_2, l_3 of the LHA.

Finally, Figure 15 shows an FMS property that is not present in [3]. It is an example in which the acceptance of a passage time (in this case the time to go through machines M_1, M_3, and M_3) is conditioned on having a mean number of spare parts under repair (modeled by the sum of tokens in places $SPARESbroken$ and $SPARESrepairing$), all along the path, less than a constant parameter WBS. This is an interesting case in which the acceptance of a path is conditioned on a performance measure relative to the path itself.

Figure 16 shows the CDF of passage time for F_4 (Figure 15), comparing the unconditioned passage-time versus the WBS conditioned passage-time (considering two different values of the conditioning bound WBS), i.e. the accepted paths conditioned by the fact that the average number of spare parts that are broken or under reparation is less than the specified value (i.e. WBS). Note that, for this plot, we have instructed Cosmos to report "unnormalized", defective, passage time CDF.

All the measures shown in this section refer to the initial marking of 9 pallets in the load/unload station, that can be compared with the results of [3]. Note that the state space of the FMS model may be quite large as the number of pallets grows (for instance, with 20 pallets it has more than 3 million states, or for 300 pallets reaches 150 billion states). With such large state spaces, the simulation approach becomes virtually the only applicable method.

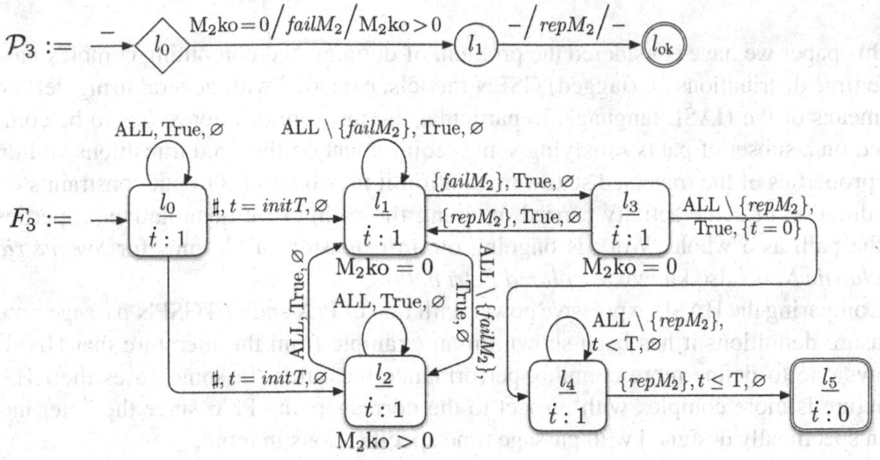

$$\mathcal{P}_3 := \quad \xrightarrow{\quad -\quad} \boxed{l_0} \xrightarrow{\quad M_2\mathrm{ko}=0\,/\,failM_2\,/\,M_2\mathrm{ko}>0\quad} \boxed{l_1} \xrightarrow{\quad -\,/\,repM_2\,/\,-\quad} \boxed{l_{ok}}$$

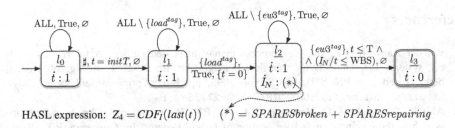

HASL expression: $Z_3 = CDF_I(last(t))$

Fig. 14. The Probe Automaton $\mathcal{P}_3$ and the LHA automaton F_3 for the FMS model

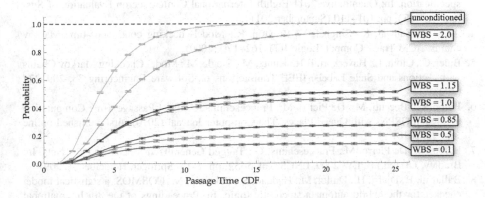

HASL expression: $Z_4 = CDF_I(last(t))$ $(*) = SPARESbroken + SPARESrepairing$

Fig. 15. The LHA automaton for the FMS property F_4

Fig. 16. Passage time property F_4 of the FMS model

6   Conclusion and Future Work

In this paper we have considered the problem of defining and computing complex passage time distributions on (tagged) GSPN models, extended with general firing delays, by means of the HASL language. In particular the measure of interest has to be computed on a subset of paths satisfying some requirement on the fired transitions and on the properties of the traversed states. In addition it may be useful to put constraints on the duration of some activity carried on along the path, or on quantitative properties of the path as a whole. Work is ongoing on an extension of Cosmos for *Symmetric Stochastic Nets* (also known as *colored Petri nets*).

Comparing the HASL expressive power with that of PrA and of TGSPN passage time measure definitions it has been shown on an example from the literature that HASL allows one to define more complex performance measures. In some cases the LHA structure is more complex with respect to the corresponding PrA, since the latter has been specifically designed with passage time specifications in mind.

Finally, the experimental results performed on two different models (i.e. FMS and Business workflow) have shown that the COSMOS statistical model checker is an adequate tool to estimate such measures, even in presence of non exponential transition firing times and on models with very large state spaces.

References

1. van der Aalst, W.M.P.: Business process management demystified: A tutorial on models, systems and standards for workflow management. In: Desel, J., Reisig, W., Rozenberg, G. (eds.) ACPN 2003, LNCS, vol. 3098, pp. 1–65. Springer, Heidelberg (2004), http://dx.doi.org/10.1007/978-3-540-27755-2_1
2. Alur, R., Courcoubetis, C., Henzinger, T.A., Ho, P.H.: Hybrid automata: An algorithmic approach to the specification and verification of hybrid systems. In: Grossman, R.L., Ravn, A.P., Rischel, H., Nerode, A. (eds.) HS 1991 and HS 1992. LNCS, vol. 736, pp. 209–229. Springer, Heidelberg (1993)
3. Amparore, E., Beccuti, M., Donatelli, S., Franceschinis, G.: Probe automata for passage time specification. In: Quantitative 2011 Eighth International Conference on Evaluation of Systems (QEST), pp. 101–110 (September 2011)
4. Aziz, A., Sanwal, K., Singhal, V., Brayton, R.: Model-checking continuous-time Markov chains. ACM Trans. Comput. Logic 1(1), 162–170 (2000)
5. Baier, C., Cloth, L., Haverkort, B.R., Kuntz, M., Siegle, M.: Model Checking Markov Chains with Actions and State Labels. IEEE Transactions on Software Engineering 33, 209–224 (2007)
6. Balbo, G., Beccuti, M., De Pierro, M., Franceschinis, G.: First Passage Time Computation in Tagged GSPNs with Queue Places. The Computer Journal (2010) (first published online July 22, 2010)
7. Balbo, G., De Pierro, M., Franceschinis, G.: Tagged Generalized Stochastic Petri Nets. In: Bradley, J.T. (ed.) EPEW 2009. LNCS, vol. 5652, pp. 1–15. Springer, Heidelberg (2009)
8. Ballarini, P., Djafri, H., Duflot, M., Haddad, S., Pekergin, N.: COSMOS: a statistical model checker for the hybrid automata stochastic logic. In: Proceedings of the 8th International Conference on Quantitative Evaluation of Systems (QEST 2011), pp. 143–144. IEEE Computer Society Press (September 2011)

9. Ballarini, P., Djafri, H., Duflot, M., Haddad, S., Pekergin, N.: HASL: an expressive language for statistical verification of stochastic models. In: Proc. Valuetools (2011)
10. Chen, T., Han, T., Katoen, J.P., Mereacre, A.: Quantitative Model Checking of Continuous-Time Markov Chains Against Timed Automata Specifications. In: Symposium on Logic in Computer Science, pp. 309–318 (2009)
11. Clark, A., Gilmore, S.: State-aware performance analysis with extended stochastic probes. In: Thomas, N., Juiz, C. (eds.) EPEW 2008. LNCS, vol. 5261, pp. 125–140. Springer, Heidelberg (2008), http://dx.doi.org/10.1007/978-3-540-87412-6_10
12. Dingle, N.J., Knottenbelt, W.J.: Automated Customer-Centric Performance Analysis of Generalised Stochastic Petri Nets Using Tagged Tokens. Electron. Notes Theor. Comput. Sci. 232, 75–88 (2009)
13. Dingle, N.J., Harrison, P.G., Knottenbelt, W.J.: Uniformisation and Hypergraph Partitioning for the Distributed Computation of Response Time Densities in Very Large Markov Models. Journal of Parallel and Distributed Computing 64(8), 309–920 (2004)
14. Djafri, H.: Numerical and Statistical Approaches for Model Checking of Stochastic Processes. Ph.D. thesis, ENS Cachan (June 2012)
15. Donatelli, S., Haddad, S., Sproston, J.: Model checking timed and stochastic properties with CSL^{TA}. IEEE Trans. Softw. Eng. 35(2), 224–240 (2009)
16. Haverkort, B., Cloth, L., Hermanns, H., Katoen, J.P., Baier, C.: Model checking performability properties. In: Proc. DSN 2002 (2002)
17. Hillston, J.: Process algebras for quantitative analysis. In: Proceedings of the 20th Annual IEEE Symposium on Logic in Computer Science, pp. 239–248. IEEE Computer Society, Washington, DC (2005), http://portal.acm.org/citation.cfm?id=1078035.1079698
18. Kulkarni, V.: Modeling and Analysis of Stochastic Systems. Chapman & Hall, London (1995)
19. Obal II, W.D., Sanders, W.H.: State-space support for path-based reward variables. Perform. Eval. 35, 233–251 (1999), http://dx.doi.org/10.1016/S0166-53169900010-3

On Multi-enabledness in Time Petri Nets[*]

Hanifa Boucheneb[1], Didier Lime[2], and Olivier H. Roux[2]

[1] Department of Computer Engineering and Software Engineering,
École Polytechnique de Montréal,
P.O. Box 6079, Station Centre-ville, Montréal, Québec, Canada, H3C 3A7
hanifa.boucheneb@polymtl.ca
[2] LUNAM Université, École Centrale de Nantes, IRCCyN
1, rue de la Noë, 44 300 Nantes, France
{didier.lime,olivier-h.roux}@irccyn.ec-nantes.fr

Abstract. We consider time Petri nets with multiple-server semantics. We first prove that this setting is strictly more expressive, in terms of timed bisimulation, than its single-server counterpart. We then focus on two choices for the firing of multiple instances of the same transition: the more conservative safety-wise non deterministic choice, where all firable instances may fire in any order, and a simpler alternative, First Enabled First Fired (FEFF), where only the oldest instance may fire, obviously leading to a much more compact state-space. We prove that both semantics are not bisimilar but actually simulate each other with strong timed simulations, which in particular implies that they generate the same timed traces. FEFF is then very appropriate to deal with linear timed properties of time Petri nets.

Keywords: Time Petri nets, multiple/single-server semantics, firing choice policies, strong/weak timed simulations.

1   Introduction

The theory of Petri Nets provides a general framework to specify the behavior of real-time reactive systems, including their time constraints. Time constraints may be expressed in terms of stochastic delays of transitions (stochastic Petri nets), fixed values associated with places or transitions [13], or intervals labelling places, transitions or arcs [7, 10–12, 15]. Among these time extensions of Petri nets, we consider here time Petri nets [11] "à la Merlin" (threshold semantics) in both single-server and multiple-server semantics. This model associates with each transition a static firing interval constraining their firing dates. The multi-enabledness appears as soon as we consider non-safe Petri nets and is consequently both a theoretical and a practical problems. As an example, in a production line, a multi-enabled transition can either model a queue for a machine able to process one peace at a time or a conveyer belt able to move more than one object. The multiple-server semantics allows to handle, at the same time, several enabling instances of the same transition, whereas it is not allowed

---

[*] This work has been partially funded by project ImpRo ANR-2010-BLAN-0317.

in the single-server semantics. In [6], the authors discussed and showed the benefits of multiple-server semantics over single-server semantics such as scaling and
conciseness power. Multi-enabledness allows very compact representations for
some systems [6], where system resources are represented by tokens. In such a
case, adding new resources to a system consists in simply adding new tokens in
the marking, without changing the structure of the Petri net [6].

Timing information is either associated with tokens (age semantics) [8] or enabled transitions (threshold semantics) [4, 6]. For the multiple-server threshold
semantics, each enabling instance of the same transition has its own timing information characterized by means of a clock (used to measure the time elapsing
since its enabling) or a firing interval (indicating its firing delays). In [4], two
firing choice policies have been proposed to manage the different enabling instances of the same transition: non deterministic (NDF) and First Enabled First
Fired (FEFF) firing policies. For the NDF firing choice, all possible firing orders
of these instances are considered, whereas, for the FEFF firing choice, only one
firing order, corresponding to firing the oldest one first, is considered. The NDF
firing choice includes the FEFF firing choice. Consequently, it strongly simulates the FEFF firing choice. However, as it considers all possible firing orders
of instances of the same transitions, it may cause a blow-up of the state space,
compared to the FEFF firing choice.

In this paper, we first give an overview of the different semantics of multi-
enabledness for time Petri nets. Then, we show that the multiple-server semantics
adds expressiveness relatively to the single-server semantics. We also prove that
the FEFF firing choice strongly simulates the NDF firing choice. As, the NDF
firing choice strongly timed simulates the FEFF firing choice, it follows that both
firing choice policies (NDF and FEFF) strongly simulate each other and then
generate the same timed language.

This paper is organized as follows. Section 2 defines formalisms and notions
used in the paper such as timed transition systems, strong (weak) timed simulation relations and time Petri nets. Then, it discusses the different semantics
of multi-enabledness of time Petri nets, proposed in the literature. Section 3 is
devoted to the threshold semantics of time Petri nets, in the context of multiple-
server policy and the comparison of the expressiveness relatively to the single-
server semantics. Section 4 compares two firing choice policies: NDF and FEFF.
The conclusion is presented in Section 5.

2   Preliminaries

Let $\mathbb{N}$, $\mathbb{Q}^+$ and $\mathbb{R}^+$ be the sets of natural, non-negative rational and non-negative
real numbers, respectively.

2.1   Timed Transition Systems and Timed (bi)Simulation

As usual, we shall define the operational semantics of our time Petri nets by
means of timed transition systems (TTS) combining both discrete (actions) and
continuous (time elpasing) transitions [6]:

Definition 1 (Timed Transition System). *A TTS is a 4-tuple* $\mathcal{S} =< Q, q_0, \Sigma, \rightarrow>$ *where Q is a set of states, $q_0 \in Q$ is the initial state, Σ is the set of discrete actions (disjoint from the time domain $\mathbb{R}^+$ of the continuous actions), and $\rightarrow \in Q \times (\Sigma \cup \mathbb{R}^+) \times Q$ is the transition relation.*

A tuple $(q, a, q') \in \rightarrow$, also denoted $q \xrightarrow{a} q'$, represents the transition from state q to state q' by the discrete or continuous action a.

In the sequel, we assume that all TTSs satisfy the classical time-related conditions where $d, d' \in \mathbb{R}^+$:

- time determinism: if $q \xrightarrow{d} q'$ and $q \xrightarrow{d} q''$ then $q' = q''$;
- time additivity: if $q \xrightarrow{d} q'$ and $q' \xrightarrow{d'} q''$ then $q \xrightarrow{d+d'} q''$;
- null delay: $\forall q : q \xrightarrow{0} q$;
- time continuity: if $q \xrightarrow{d} q'$ then $\forall d' \leq d, \exists q'', q \xrightarrow{d'} q''$ and $q'' \xrightarrow{d-d'} q'$.

Definition 2 (Run in a TTS). *A run ρ in a TTS $\mathcal{S} =< Q, q_0, \Sigma, \rightarrow>$ is a (possibly infinite) sequence $q_0 a_0 q_1 a_1 \ldots a_{n-1} q_n \ldots$ such that $\forall i, (q_i, a_i, q_{i+1}) \in \rightarrow$.*

We assume w.l.o.g. that in a run, discrete and continuous transitions are strictly alternating. The case of a run ending in infinite delay raises no theoretical issue but we omit it for the sake of readability. Any run ρ can therefore be written as: $\rho = q_0 \xrightarrow{d_1} q_1 \xrightarrow{t_1} q_2 \xrightarrow{d_2} q_3 \xrightarrow{t_2} \cdots$, where d_i and t_i for $i > 0$ are continuous and discrete actions, respectively.

Definition 3 (Timed and Untimed Traces). *For any run $\rho = q_0 \xrightarrow{d_1} q_1 \xrightarrow{t_1} q_2 \xrightarrow{d_2} q_3 \xrightarrow{t_2} \cdots$, the timed trace (timed word) of ρ is the sequence $d_1 t_1 d_2 t_2 \ldots$. The untimed trace (also called firing sequence) of ρ is the sequence $t_1 t_2 \ldots$.*

Definition 4 (Timed Language). *The timed language of $\mathcal{S}$, denoted $L(\mathcal{S})$, is the set of its timed traces.*

In order to compare the different semantic choices we shall introduce the notion of simulation:

Definition 5 (Strong timed (bi)simulation). *Let $\mathcal{S}_1 =< Q_1, q_{10}, \Sigma, \rightarrow_1>$ and $\mathcal{S}_2 =< Q_2, q_{20}, \Sigma, \rightarrow_2>$ be two timed transition systems.*
A binary relation $\prec_S \subseteq Q_1 \times Q_2$ is a (strong) timed simulation iff $\forall (q_1, q_2) \in \prec_S$, $\forall a \in \Sigma \cup \mathbb{R}^+, (\exists q_1' \in Q_1, q_1 \xrightarrow{a}_1 q_1') \Rightarrow (\exists q_2' \in Q_2, q_2 \xrightarrow{a}_2 q_2'$ and $(q_1', q_2') \in \prec_S)$.
A strong timed simulation $\simeq_S \subseteq Q_1 \times Q_2$ is a (strong) timed bisimulation if $\simeq_S^{-1} \subseteq Q_2 \times Q_1$ is also a strong timed simulation.

We say that transition system $\mathcal{S}_1$ is strongly simulated by $\mathcal{S}_2$ (i.e., $\mathcal{S}_2$ strongly simulates $\mathcal{S}_1$), if there exists a strong timed simulation relation $\prec_S \subseteq Q_1 \times Q_2$ s.t. $(q_{10}, q_{20}) \in \prec_S$. Note that such a simulation implies that $L(\mathcal{S}_1) \subseteq L(\mathcal{S}_2)$.

Similarly, transition systems $\mathcal{S}_1$ and $\mathcal{S}_2$ are strongly timed bisimilar if there exists a strong timed bisimulation $\simeq_S \subseteq Q_1 \times Q_2$ and $(q_{10}, q_{20}) \in \simeq_S$.

An invisible action is any action which does not belong to $\Sigma \cup \mathbb{R}^+$. We denote all invisible actions by $\epsilon \notin \Sigma \cup \mathbb{R}^+$.

Let $\sigma \in (\Sigma \cup \{\epsilon\} \cup \mathbb{R}^+)^+$ be a timed trace and $vis(\sigma)$ the timed trace obtained by eliminating invisible actions (ϵ) of σ and grouping continuous actions.

Definition 6 (Weak timed simulation). *Let* $S_1 = < Q_1, q_{10}, \Sigma \cup \{\epsilon\}, \rightarrow_1 >$
and $S_2 = < Q_2, q_{20}, \Sigma \cup \{\epsilon\}, \rightarrow_2 >$ *be two timed transition systems and*
$\preceq_W \subseteq Q_1 \times Q_2$ *a binary relation. Relation* $\preceq_W$ *is a* weak timed simulation *iff*
$\forall (q_1, q_2) \in \preceq_W, \forall a \in \Sigma \cup \mathbb{R}^+, (\exists q'_1 \in Q_1, q_1 \xrightarrow{a}_1 q'_1) \Rightarrow (\exists q'_2 \in Q_2, \exists \sigma \ s.t.$
$vis(\sigma) = a, q_2 \xrightarrow{\sigma}_2 q'_2 \ and \ (q'_1, q'_2) \in \preceq_W)$.

We derive the notions of weak timed bisimulation and weak timed (bi)similarity
of transitions systems in exactly the same way as for strong simulations.

2.2   Time Petri Nets

We now introduce the formalisms considered in this article:

Definition 7 (Petri net). *A Petri net is defined by a 4-tuple:*
$< P, T, \mathsf{Pre}, \mathsf{Post}, M_0 >$, *where:*

- *P and T are finite sets of places and transitions, respectively (s.t. $P \cap T = \emptyset$);*
- *$\mathsf{Pre}, \mathsf{Post} \in [P \times T \longrightarrow \mathbb{N}]$ are the backward incidence and the forward incidence functions, respectively. They indicate, for each transition, the tokens
  needed for its firing and those produced;*
- *$M_0 \in [P \longrightarrow \mathbb{N}]$ is the initial distribution of tokens in places, called the
  initial marking.*

A marking M of a Petri net is a function from P to $\mathbb{N}$. Let $M \in [P \longrightarrow \mathbb{N}]$ be a
marking and t a transition. Transition t is k-enabled for $k > 0$ iff $M \geq k \times \mathsf{Pre}(., t)$
and $M \not\geq (k+1) \times \mathsf{Pre}(., t)$. In this case, k is the enabling degree of t in M.
If t is k-enabled for some $k > 0$, we simply say it is enabled. When we say t is
multi-enabled we emphasize the fact that $k > 1$. By convention, t is said to be
0-enabled in M if it is not enabled in M. Note that in case, a transition has at
least an input place, the set of its enabling instances in M is finite. We suppose
here that each transition has at least an input place.

If t is enabled in M, it may *fire*, leading to the marking M' s.t.
$\forall p \in P, M'(p) = M(p) - \mathsf{Pre}(p, t) + \mathsf{Post}(p, t)$.

Let INT be the set of intervals of $\mathbb{R}^+$ of the form $[a, b]$ or $[a, \infty[$, where
$a, b \in \mathbb{Q}^+$. For any interval $I \in INT$, the lower and the upper bounds of I are
denoted $\downarrow I$ and $\uparrow I$, respectively.

Definition 8 (Time Petri Net). *A time Petri net (TPN) is defined by a
7-tuple: $< P, T, \mathsf{Pre}, \mathsf{Post}, Is, M_0 >$ where:*

- *$< P, T, \mathsf{Pre}, \mathsf{Post}, M_0 >$ is a Petri net;*
- *$Is \in [T \rightarrow INT]$ is a function which associates with each transition t a static
  firing interval $Is(t)$.*

Intuitively, a transition t is firable if it is maintained enabled during a time
inside its static firing interval. It must be fired without any additional delay,
if it is maintained enabled $\uparrow Is(t)$ time units, unless it is immediately disabled
by a conflicting transition. Firing a transition takes no time but leads to a new
marking.

2.3   TPN Semantics

Several semantics are proposed in the literature for TPN that can be classified according to four policies (service, granularity, memory and choice) [1–3, 9] and the characterization of timing information (clock state vs interval state).

Server Policy. For time Petri Nets, time is used to model a non instantaneous service represented by a timed transition. In this context, the service policy specifies whether several enabling instances of the same transition may be considered simultaneously (multiple-server semantics) or not (single-server semantics). The multiple-server semantics allows to handle, at the same time, several services per transition whereas it is not allowed in the single-server semantics. For single-server semantics, the multi-enabledness is not ambiguous since a transition can do only one thing at the same time (only one enabling instance of each transition is considered at each state), whereas different interpretations can be defined for multiple-server semantics.

Granularity Policy. This policy indicates which objects timing information is associated with. Timing information is either associated with tokens (age semantics) [8] or enabled transitions (threshold semantics) [6]. In [5], the authors considered the age semantics where tokens are managed FIFO based on their ages. As tokens are handled FIFO, an enabled transition will always use the oldest tokens from each place. The difference between age and threshold semantics is highlighted in [9] by the difference between the *individual token interpretation* and the *collective token interpretation*. It is particularly significant when two or more tokens are needed, in a given place, to fire a transition. Let us consider the example of [6] depicted in Fig. 1 showing the difference between age and threshold semantics. In this example, a server answers to requests, in a delay between 2 and 3 time units for each request and we want to detect a too heavy load of the system. More precisely, we want to detect the presence of more than 40 requests during a period of 30 time units. The modeling of such a system is given in Fig. 1. Using the age semantics, the transition *Loaded* will never be fired. Indeed, each token will stay at most 3 time units in place *Running*: no multi-set of 40 tokens will exist with a token older than 3 time units. Using the threshold semantics, the transition *Loaded* will be enabled once 40 tokens will be in place *Running*, and it will fire 30 time units after, as long as at least 40 tokens are in the place, independently of their ages.

In this paper, we focus on the threshold semantics.

Memory Policy. This policy specifies when the timing information is set or reinitialized. For the age semantics, the timing information of tokens is set at their creation. In the context of threshold semantics, the memory policy relies on the notion of newly enabled transitions. In the classical semantics (also called intermediate semantics), this notion is defined using intermediate markings (markings resulting from the consumption of tokens): when a transition is fired, all transitions not enabled in the intermediate marking but enabled in the successor marking are considered as newly enabled. The firing of a transition

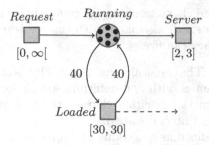

Fig. 1. A TPN illustrating the difference between age and threshold semantics [6]

is then not atomic w.r.t. markings. In [1, 2], the authors have discussed other semantics where the firing of transitions is considered atomic: atomic and persistent atomic semantics. In such semantics, all transitions not enabled before firing a transition t but enabled after its firing are newly enabled. Another difference between the intermediate, atomic and persistent atomic semantics lies in the particular case of the fired transition. If the fired transition enables again itself, it is considered as newly enabled in the intermediate and atomic semantics but not newly enabled in the persistent atomic semantics. For the single-server and threshold semantics, the intermediate, atomic and persistent atomic semantics are equivalent w.r.t. weak timed bisimulation if the intervals are all right-closed and persistent atomic is more expressive otherwise [2, 3, 14].

Firing Choice Policy. This policy specifies the enabled transitions to fire first and those to disable first, in case of conflicts. In the context of TPNs, the choice of transition to fire first is non deterministic for different transitions. For the age semantics, in [5], tokens are managed First in First Out (FIFO semantics). When a transition is fired, it consumes the oldest tokens first. In the case of the threshold and multiple-server semantics, the multi-enabledness of a transition t can be considered as different transitions, which we call *enabling instances*, and which are either totally independent (non deterministic firing choice (NDF)) or managed so as to fire the oldest one first (First Enabled First Fired (FEFF) policy).

Disabling Choice Policy. This policy specifies which enabling instances of transitions to disable first: the most recent ones first (Last Enabled First Disabled (LEFD)) or the oldest ones first (First Enabled First Disabled (FEFD)) are possible policies. As for the firing choice policy we can also take into account all possible choices non-deterministically (NDD).

Clock vs Interval States. Besides these policies, there are, in the setting of threshold semantics, two known characterizations of timing information. The first one is based on clocks. A clock is either associated with each enabled transition to measure its enabling time (time elapsed since it became enabled most recently) or associated with each token to measure its age (time elapsed since its creation). The second characterization of timing information is based on dynamically decreasing intervals associated with enabled transitions indicating the time remaining until they can fire.

TPN " à la Merlin". The classical semantics of TPN is single-server, threshold and intermediate semantics with non deterministic choice of transitions to fire first. The timing information is either characterized by means of clocks (clock states) or time intervals (interval states).

The clock state is defined as a marking and a function which associates with each enabled transition the value of its clock. The clock of a transition t is set to 0, when it is newly enabled. Afterwards, its value increases synchronously with time until it is fired or disabled by firing a conflicting transition. It is firable if its clock value reaches its static firing interval. It must be fired without any additional delay transition when its clock reaches $\uparrow Is(t)$, unless it is disabled.

The interval state is defined as a marking and a function which associates with each enabled transition the time interval in which the transition can be fired. When a transition t is newly enabled, its firing interval is set to its static firing interval. The bounds of this interval decrease synchronously with time, until t is fired or disabled by another firing. t is firable, if the lower bound of its firing interval reaches 0. It must be fired, without any additional delay, if the upper bound of its firing interval reaches 0, unless it is disabled.

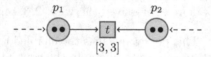

Fig. 2. A simple TPN with multiple enabledness

Illustrative Example. Let us point out, by means of the simple example of Fig. 2, some subtle differences between the semantics discussed above. Assume that initially, the marking is $\binom{p_1}{p_2} = \binom{1}{1}$. At date 1, a token arrives in p_1 leading to the marking $\binom{2}{1}$. Then, another token arrives in p_2 at date 2 leading to the marking $\binom{p_1}{p_2} = \binom{2}{2}$. The transition t is 2-enabled in the marking $\binom{2}{2}$.

For all semantics, the transition t will be fired twice. However, the firing dates of the transitions vary according to the service, granularity and memory policies.

For the single-server and threshold semantics, only one enabling instance of transition t is considered from each state. This instance is enabled since date 0. When this instance is fired, transition t is again enabled. The timing information of this enabling instance of t may be either reinitialized or not, according to the memory policy (intermediate, atomic or persistent atomic semantics). For both intermediate and atomic semantics, the transition t will be fired at dates 3 and 6. Indeed, t is enabled since date 0 and then is fired at date 3. When t is fired, it is again enabled in the resulting marking, its timing information is reinitialized.

Therefore, it will be fired again 3 time units later (i.e., at date 6). For persistent atomic semantics, the transition t will be fired at dates 3 and 3, since, in this case, when t is fired for the first time, its timing information is not reinitialized.

For the age semantics, there are 2 tokens in place P_1 created at dates 0 and 1, respectively. There are also 2 tokens in place P_2 created at dates 0 and 2, respectively. In case, the tokens are managed FIFO, the dates of the first and the second firing of t are 3 and 5. The first firing of t uses the tokens of p_1 and p_2 with age 0. The second firing of t uses the remaining tokens.

For the multiple-server (multi-enabledness) semantics, the two enabling instances of transition t can be fired at dates:

- 3 and 5 for the threshold semantics, since the first and the second instances of t are enabled since dates 0 and 2, respectively.
- 3 and 5 for the age semantics with FEFF discipline. t is considered to be mutli-enabled as soon as the marking is $\binom{2}{1}$.
- 3 and 5 or 4 and 5 for the age semantics with non deterministic choice of tokens to be used. Indeed, the dates of the first and the second firing of t depend on the ages of tokens used. There are two possibilities. The first firing of t may use either the tokens of p_1 and p_2 with age 0 or the token of p_1 with age 1 and the token of p_2 with age 0. The second firing of t uses the remaining tokens. So, the dates of the first and the second firing of t are either 3 and 5 or 4 and 5.

In [6], the authors discussed and showed the benefits of multiple-server semantics over single-server semantics such as scaling and conciseness power. However, does the multiple-server semantics increase the expressive power of time Petri nets or not? So, the first aim of this paper is to investigate this question for the threshold semantics. The other aim is the comparison of two firing choice policies: NDF and FEFF.

3   Threshold Semantics in the Context of Multiple-Server Policy

In the threshold and multiple-server semantics, a clock or a firing interval is associated with each enabling instance of a transition. The choice of transition to fire first is non deterministic for different transitions. The enabling instances of the same transition are either considered totally independent (non deterministic choice) or managed so as to fire the oldest one first (First Enabled First Fired (FEFF) policy). In the sequel, we consider the case of non deterministic choice of the transition to fire first.

Let M be a marking. For economy of notations, we suppose that the set of transitions is strictly ordered ($t < t'$ or $t' < t$, for any pair of transitions of T) and the enabling instances of transitions of M are managed in an ordered list, denoted en. In en, the enabling instances of the same transition are ordered from the oldest to the newest one and transitions appear in increasing order. An enabling instance of this list is referred to as t^i where $t \in T$ is its transition and i is its position in the list en.

3.1   Clock Based Timing Information

For the clock based timing information, the TPN state is defined by the triplet $q = (M, \mathsf{en}, \nu)$, where $M \in [P \longrightarrow \mathbb{N}]$ is a marking, en is the list of enabling instances of transitions in M, where the enabling instances of the same transition are ordered from the oldest to the newest one and transitions appear in increasing order, and $\nu \in [\mathsf{en} \longrightarrow \mathbb{R}^+]$ is a clock valuation over en. For each $t^i \in \mathsf{en}$, $\nu(t^i)$ is the clock value of the ith enabling instance. The initial state is $q_0 = (M_0, \mathsf{en}_0, \nu_0)$, where M_0 is the initial marking, en_0 is the appropriately ordered list of enabling instances of transitions in M_0 and $\forall t^i \in \mathsf{en}$, $\nu_0(t^i) = 0$.

All clocks of transitions evolve uniformly with time. Let $q = (M, \mathsf{en}, \nu)$ be a state, $d \in \mathbb{R}^+$. We denote $\nu + d$ the function ν' defined by $\forall t^i \in \mathsf{en}, \nu'(t^i) = \nu(t^i) + d$. It specifies the evolution of time by d units. $(M, \mathsf{en}, \nu) \xrightarrow{d} (M, \mathsf{en}, \nu')$ iff

$$\nu' = \nu + d \text{ and } \forall t^i \in \mathsf{en}, \nu(t^i) + d \leq\uparrow Is(t)$$

Let $q = (M, \mathsf{en}, \nu)$ be a state and $t \in T$. Transition t is firable at state $q = (M, \mathsf{en}, \nu)$ iff there is at least an enabling instance t^i of t in en s.t. its clock has reached its firing interval (i.e., $\nu(t^i) \geq\downarrow Is(t)$). In case transition t^i is firable at state $q = (M, \mathsf{en}, \nu)$, its firing will consume $\mathsf{Pre}(p, t)$ tokens from each place p and produce $\mathsf{Post}(p, t)$ tokens in each place p. Consequently, it will disable transitions that are in conflict and enable new instances of transitions. We denote $\mathsf{CF}(M, \mathsf{en}, t^i)$ the set of enabling instances of transitions of en in conflict in M with t^i. The set of newly enabled instances in the marking reached from M by firing t^i is denoted $\mathsf{Nw}(M, \mathsf{en}, t^i)$.

If t^i is firable at $q = (M, \mathsf{en}, \nu)$, its firing leads to the state $q' = (M', \mathsf{en}', \nu')$ s.t.

- $M' = M - \mathsf{Pre}(., t) + \mathsf{Post}(., t)$,
- en' is computed from en by eliminating enabling instances of $\mathsf{CF}(M, \mathsf{en}, t^i)$ and then inserting the enabling instances of $\mathsf{Nw}(M, \mathsf{en}, t^i)$ and
- ν' is computed from ν by eliminating clock values of enabling instances of $\mathsf{CF}(M, \mathsf{en}, t^i)$ and inserting value 0 for each enabling instance of $\mathsf{Nw}(M, \mathsf{en}, t^i)$.

We write $(M, \mathsf{en}, \nu) \xrightarrow{t^i} (M', \mathsf{en}', \nu')$, for $t^i \in \mathsf{en}$ iff t^i is firable at (M, en, ν), i.e., $t^i \in \mathsf{en}$ and its firing leads to the state (M', en', ν').

Example 1. As an example, consider the time Petri net at Fig. 3. and the state $q = (M, \mathsf{en}, \nu)$, where $M(P_1) = 3$, $\mathsf{en} = \{S^1, S^2, S^3, L^1, L^2, L^3\}$, $\nu(S^1) = \nu(L^1) = 2.5$, $\nu(S^2) = \nu(L^2) = 2.1$ and $\nu(S^3) = \nu(L^3) = 1.3$.

Let us consider the firing of the instance S^1 in the context of FEFF semantics, atomic memory policy and FEFD disabling choice policy, meaning that $\mathsf{CF}(M, \mathsf{en}, S^1) = \{S^1, L^1\}$. Thus, the firing of the transition S^1 denoted by $(M, \mathsf{en}, \nu) \xrightarrow{S^1} (M', \mathsf{en}', \nu')$ leads to the deletion of S^1 and L^1 and thus the other

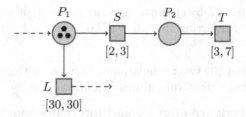

Fig. 3. A TPN illustrating CF and Nw

instances are left shifted in the list en'. Then, since $\mathsf{Nw}(M, \mathsf{en}, S^1) = \{T^1\}$, we
have en' $= \{S^1, S^2, L^1, L^2, T^1\}$ with $\nu'(S^1) = \nu'(L^1) = 2.1$, $\nu'(S^2) = \nu'(L^2) = 1.3$ and $\nu'(T^1) = 0$. □

Using clock based timing information, the behavior of a time Petri net is defined
by means of the timed transition system $< Q, q_0, \Sigma, \longrightarrow >$, where Q is the set
of clock states of the time Petri net, $q_0 = (M_0, \mathsf{en}_0, \nu_0)$ is its initial clock state,
$\Sigma = T$, and $\longrightarrow$ is composed of continuous and discrete transitions defined as
follows:
Let $q = (M, \mathsf{en}, \nu)$ and $q' = (M', \mathsf{en}', \nu')$ be two clock states, $d \in \mathbb{R}^+$ and $t \in T$.

$(M, \mathsf{en}, \nu) \xrightarrow{d} (M, \mathsf{en}, \nu + d)$ iff $(M, \mathsf{en}, \nu) \xrightarrow{d} (M, \mathsf{en}, \nu + d)$

$(M, \mathsf{en}, \nu) \xrightarrow{t} (M', \mathsf{en}', \nu')$ iff $\exists t^i \in \mathsf{en}, (M, \mathsf{en}, \nu) \xrightarrow{t^i} (M', \mathsf{en}', \nu')$.

3.2 Interval Based Timing Information

For the interval based timing information, the interval TPN state is defined by
the triplet (M, en, I), where M and en are defined as for the clock state and $I \in$
$[\mathsf{en} \longrightarrow INT]$ is an interval function over en. For each $t^i \in \mathsf{en}$, $I(t^i)$ is the firing
interval of the ith enabling instance of t. The initial state is $q_0 = (M_0, \mathsf{en}_0, I_0)$,
where $\forall t^i \in \mathsf{en}_0, I_0(t^i) = Is(t)$.

In this case, the behavior of TPN is defined by means of the timed transition
system $< Q_I, (M_0, \mathsf{en}_0, I_0), \Sigma, \longrightarrow_I >$, where Q_I is the set of interval states of
the TPN, $(M_0, \mathsf{en}_0, I_0)$ is its initial interval state, $\Sigma = T$, and $\longrightarrow_I$ is composed
of continuous and discrete transitions defined as follows:

Let (M, en, I) and (M', en', I') be two interval states, $d \in \mathbb{R}^+$ and $t \in T$.
$(M, \mathsf{en}, I) \xrightarrow{d}_I (M, \mathsf{en}, I')$ iff

$$\forall t^i \in \mathsf{en}, d \leq \uparrow I(t^i) \text{ and } I'(t^i) = [\max(0, \downarrow I(t^i) - d), \uparrow I(t^i) - d].$$

$(M, \mathsf{en}, I) \xrightarrow{t}_I (M', \mathsf{en}', I')$ iff $\exists t^i \in \mathsf{en}$ s.t.

- $\downarrow I(t^i) = 0$,
- $M' = M - \mathsf{Pre}(., t) + \mathsf{Post}(., t)$,
- en' is computed from en by eliminating enabling instances of $\mathsf{CF}(M, \mathsf{en}, t^i)$
  and then inserting the enabling instances of $\mathsf{Nw}(M, \mathsf{en}, t^i)$ and

- I' is computed from I by eliminating firing intervals of enabling instances of $\mathsf{CF}(M,\mathsf{en},t^i)$ and inserting the interval $Is(t')$ for each enabling instance t'^j of $\mathsf{Nw}(M,\mathsf{en},t^i)$.

Theorem 1 proves that the two formulations of the semantics, clocks or intervals, are actually equivalent wrt. strong timed bisimulation.

Theorem 1 (Equivalence of clock and interval semantics). *All other semantic choices being the same, the transition systems, obtained for both clock and interval timing information, are strongly timed bisimilar.*

Proof. Let $\simeq$ be the binary relation over $\mathcal{Q}$ defined by:
$\forall (M,\mathsf{en},\nu) \in \mathcal{Q}, \forall (M',\mathsf{en}',I) \in \mathcal{Q}_I, \ (M,\mathsf{en},\nu) \simeq_S (M',\mathsf{en}',I)$ iff
$M = M', \mathsf{en} = \mathsf{en}'$, and $\forall t^i \in \mathsf{en}, I(t^i) = [\max(0, \downarrow Is(t) - \nu(t^i)), \uparrow Is(t) - \nu(t^i)]$.
It is easy to verify that $\simeq$ is a strong timed bisimulation. □

3.3   Conflicting and Newly Enabled Transitions

The notions of conflicting and newly enabled transitions are not dependent of the characterization of timing information. They mainly depend on the marking M, the list en and the memory (intermediate, atomic or persistent atomic semantics) and disabling choice (FEFD or LEFD) policies.

Conflicting Transitions. Let M be a marking, en its list of enabling instances ordered appropriately as explained previously, t and t' two enabled transitions in M, $k > 0$ and $k' > 0$ their enabling degrees in M.

For the intermediate semantics, an enabling instance of t is in conflict with some enabling instances of t' in M iff the enabling degree of t' is decreased in the intermediate marking $M - \mathsf{Pre}(.,t)$ (i.e., $M - \mathsf{Pre}(.,t) \not\geq k' \times \mathsf{Pre}(.,t')$). Let k'' be the enabling degree of t' in the intermediate marking $M - \mathsf{Pre}(.,t)$. The firing of an enabling instance of t will disable the $k' - k''$ oldest or youngest enabling instances of t', dependently of the discipline LEFD or FEFD used to manage conflicting transitions. Moreover, the fired instance t^i is supposed to be in conflict with itself (i.e., $t^i \in \mathsf{CF}(M,\mathsf{en},t^i)$).

For the persistent atomic semantics, an enabling instance of t is in conflict with some enabling instances t' in M iff the enabling degree of t' is decreased in the successor marking of M by t (i.e., $M - \mathsf{Pre}(.,t) + \mathsf{Post}(.,t) \not\geq k' \times \mathsf{Pre}(.,t')$). Let k'' be the enabling degree of t' in the successor marking of M by t. The firing of an enabling instance of t will disable the $k' - k''$ oldest or youngest enabling instances of t', dependently of the discipline LEFD or FEFD used to manage conflicting transitions.

For the atomic semantics, the set $\mathsf{CF}(M,\mathsf{en},t^i)$ is computed in the same manner as for the persistent atomic semantics, except that t^i is supposed to be in conflict with itself (i.e., $t^i \in \mathsf{CF}(M,\mathsf{en},t^i)$).

Newly Enabled Transitions. Let M be a marking, en its list of enabling instances ordered appropriately as explained previously, t^i an enabling instance of transition t in M and $M' = M - \mathsf{Pre}(.,t) + \mathsf{Post}(.,t)$ the successor marking of M by any enabling instance of t. The set $\mathsf{Nw}(M, \mathsf{en}, t^i)$ relies to the memory policy (intermediate, atomic or persistent atomic semantics) used.

For the intermediate semantics, there are new enabling instances of a transition $t' \in T$ in M', if its enabling degree in M' is greater than its enabling degree in the intermediate marking $M - \mathsf{Pre}(.,t)$. In other words, if a transition t' is k-enabled (for some $k \geq 0$) in $M - \mathsf{Pre}(.,t)$ but k'-enabled in M' with $k' > k$, then there are $k' - k$ new enabling instances of t' in M'.

For the atomic semantics, the firing of a transition is atomic. So, there are new enabling instances of a transition $t' \in T$ in M', if its enabling degree in M' is greater than its enabling degree in M. In other words, if a transition t' is k-enabled (for some $k \geq 0$) in M but k'-enabled in M' with $k' > k$, then there are $k' - k$ new enabling instances of t' in M'.

For the persistent atomic semantics, the set $\mathsf{Nw}(M, \mathsf{en}, t^i)$ is computed in the same manner as for the atomic semantics, except that if there are new enabling instances of the fired transition t in M', one of these enabling instances inherits the timing information of the fired transition.

For the rest of the paper, we fix a TPN $\mathcal{N}$ with multiple-server and threshold semantics. Moreover, we focus on the clock based timing information. Theorem 1 implies that the results shown here are also valid for the interval based timing information.

Property 1 follows from the definitions of CF and Nw: Note that according to these definitions, it holds that:

Property 1. Let $q = (M, \mathsf{en}, \nu)$ be a clock state of $\mathcal{N}$. For any pair (t^i, t^j) of enabling instances of the same transition t in en:

1. $\mathsf{CF}(M, \mathsf{en}, t^i) - \{t^i\} = \mathsf{CF}(M, \mathsf{en}, t^j) - \{t^j\}$
2. $\mathsf{Nw}(M, \mathsf{en}, t^i) = \mathsf{Nw}(M, \mathsf{en}, t^j)$
3. $t^i \in \mathsf{CF}(M, \mathsf{en}, t^i)$ iff $t^j \in \mathsf{CF}(M, \mathsf{en}, t^j)$

3.4 Multiple-Server Semantics Adds Expressiveness

In the context of threshold semantics, we establish in theorems 2 and 3 that the multiple-server semantics adds expressiveness relatively to the single-server policy.

Theorem 2. *Every TPN $\mathcal{N}$ can be translated into a TPN $\mathcal{N}'$ s.t. $\mathcal{N}$ in the single-server and threshold semantics is strongly or weakly timed bisimilar to $\mathcal{N}'$ in the multiple-server and threshold semantics.*

Proof. (sketch of the proof) In the single-server and threshold semantics, there is one clock (one firing interval) per transition, even if it is multi-enabled.

For the intermediate semantics, to achieve the translation, it suffices to add a place p_t for each transition t of $\mathcal{N}$ with $\mathsf{Pre}(p_t, t) = \mathsf{Post}(p_t, t) = 1$ and $M_0(p_t) = 1$. Doing so, we eliminate the multi-enabledness of t. Moreover, if t is again enabled after its firing, it is newly enabled in $\mathcal{N}'$, because t is not enabled in the intermediate marking (p_t is empty). Therefore, $\mathcal{N}'$ under the multiple-server, threshold and intermediate semantics is strongly timed bisimilar to $\mathcal{N}$ under the single-server, threshold and intermediate semantics.

This translation works also for the persistent atomic semantics, since the tokens of the added places are present before and after any firing. Therefore, if a transition t is again enabled after its firing in $\mathcal{N}$, it is also enabled again in $\mathcal{N}'$ after its firing. Under multiple-server, threshold and persistent atomic semantics, $\mathcal{N}'$ is strongly timed bisimilar to $\mathcal{N}$ under single-server, threshold and persistent atomic semantics.

For the atomic semantics, if a transition t is again enabled after its firing, it is considered as newly enabled. To deal with this case, the translation needs to add two places $p_{t_{in}}$ and $p_{t_{out}}$, and a transition t_t for each transition t of $\mathcal{N}$ with $\mathsf{Pre}(p_{t_{in}}, t) = 1, \mathsf{Post}(p_{t_{out}}, t) = 1, \mathsf{Pre}(p_{t_{out}}, t_t) = 1, \mathsf{Post}(p_{t_{in}}, t_t) = 1$, $M_0(p_{t_{in}}) = 1, M_0(p_{t_{out}}) = 0$ and $Is(t_t) = [0, 0]$. If a transition t is again enabled after its firing in $\mathcal{N}$, its firing in $\mathcal{N}'$ will empty place $p_{t_{in}}$, enable transition t_t and then disable t. As there is no delay between firings of t and t_t and the unique role of t_t is to allow the enabling of t (i.e., invisible transition), it follows that $\mathcal{N}$ under single-server, threshold and atomic semantics is weakly timed bisimilar to $\mathcal{N}'$ under multiple-server, threshold and atomic semantics.   □

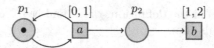

Fig. 4. Multiple-server threshold semantics with no equivalent single-server semantics

Theorem 3. *There is no TPN under single-server and threshold semantics equivalent to the TPN at Fig. 4 under multiple-server and threshold semantics (neither w.r.t. timed bisimulation nor w.r.t. timed language acceptance).*

Proof. Let $n_a(d)$ and $n_b(d)$ be the numbers of firings of transitions a and b at date d, respectively. Necessarily, $n_a(d) \geq n_b(d)$ and at date d, it remains $n_a(d) - n_b(d)$ occurrences of b in less than 2 time units. For each of these occurrences of b, we need a clock to ensure that it occurs in the interval $[1, 2]$, relatively to the corresponding occurrence of a. Under single-server and threshold semantics, the number of clocks is finite, since we have one clock per transition. Since $n_a(d) - n_b(d)$ can grow to infinity, the translation into an equivalent TPN under single-server and threshold semantics should have an infinite number of transitions. This translation is then impossible.   □

4   NDF vs FEFF Firing Choice Policies

Consider a TPN $\mathcal{N}$ under multiple-server and threshold semantics. There are two main firing choice policies: non deterministic (NDF) and FEFF. We compare these two possibilities by supposing that the memory and disabling choice policies are the same in both cases. The results are however valid whatever these choices for memory and disabling policies are, thanks to the parametrization using CF.

For NDF firing choice, if several enabling instances of the same transition t are firable from a state $q = (M, \nu)$, then all these instances will be fired from q and the firing of one of them will not disable the others. Since NDF firing choice includes the FEFF one, we have the following obvious lemma.

Lemma 1. *The NDF firing choice strongly timed simulates the FEFF firing choice.*

Proof. Let $\mathcal{S}_1 = < Q_1, q_0, \Sigma, \to_1 >$ and $\mathcal{S}_2 = < Q_2, q_0, \Sigma, \to_2 >$ be the transition systems of $\mathcal{N}$ under multiple-server and threshold semantics for NDF and FEFF firing choice policies, respectively. Since NDF firing choice includes the FEFF one, it follows that $Q_2 \subseteq Q_1$ and $\to_2 \subseteq \to_1$. Therefore, the NDF firing choice strongly simulates the FEFF firing choice.   □

However, for the NDF firing choice policy, all enabling instances of the same transition will be fired in different orders, which may cause a blow-up of the state space. It would be interesting if we can consider only one firing order without loosing properties of the model. In this sense, we show in the following that under FEFF or NDF firing choice semantics, $\mathcal{N}$ has the same timed language. The proof of this claim is based on the strong timed simulation relation over states Q of $\mathcal{N}$ defined in subsection 4.1.

4.1   FEFF Simulates NDF Firing Choice Policy

We first consider a TPN $\mathcal{N}$ under multiple-server and threshold semantics with NDF firing choice policy.

Let $\preceq$ be the relation over states Q of the TPN $\mathcal{N}$ defined by:
$\forall (M, \mathsf{en}, \nu), (M', \mathsf{en}', \nu') \in Q, (M, \mathsf{en}, \nu) \preceq (M', \mathsf{en}', \nu')$ iff $M = M'$, $\mathsf{en} = \mathsf{en}'$ and

$$\forall t^i \in \mathsf{en}, \nu(t^i) = \nu'(t^i) \quad \text{or} \quad \downarrow Is(t) \leq \nu'(t^i) < \nu(t^i).$$

Lemma 2. *The relation $\preceq$ is a strong timed simulation.*

Proof. It suffices to show that:
$\forall (M, \mathsf{en}, \nu) \in Q, \forall (M', \mathsf{en}', \nu') \in Q$ s.t. $(M, \mathsf{en}, \nu) \preceq (M', \mathsf{en}', \nu')$,
$\forall d \in \mathbb{R}^+, \forall t^f \in \mathsf{en}$,
(i) $(M, \mathsf{en}, \nu) \xrightarrow{d} (M, \mathsf{en}, \nu + d) \Rightarrow$

$$((M', \mathsf{en}', \nu') \xrightarrow{d} (M', \mathsf{en}', \nu' + d) \text{ and } (M, \mathsf{en}, \nu + d) \preceq (M', \mathsf{en}', \nu' + d))$$

(ii) $\exists (M_1, \mathsf{en}_1, \nu_1) \in Q, (M, \mathsf{en}, \nu) \xrightarrow{t^f} (M_1, \mathsf{en}_1, \nu_1) \Rightarrow$

$$(\exists (M_1', \mathsf{en}_1', \nu_1') \in Q, (M', \mathsf{en}', \nu') \xrightarrow{t^f} (M_1', \mathsf{en}_1', \nu_1') \text{ and }$$

$$(M_1, \mathsf{en}_1, \nu_1) \preceq (M_1', \mathsf{en}_1', \nu_1'))$$

Proof of (i): $(M, \mathsf{en}, \nu) \xrightarrow{d} (M, \mathsf{en}, \nu + d)$ iff $\forall t^i \in \mathsf{en}, \nu(t^i) + d \leq\uparrow Is(t)$. By assumption, $(M, \mathsf{en}, \nu) \preceq (M', \mathsf{en}', \nu')$, which means that:
(1) $M = M'$, $\mathsf{en} = \mathsf{en}'$ and

$$\forall t^i \in \mathsf{en}, \nu'(t^i) = \nu(t^i) \ \text{ or } \downarrow Is(t) \leq \nu'(t^i) < \nu(t^i).$$

Therefore, $\forall t^i \in \mathsf{en}, \nu'(t^i) \leq \nu(t^i)$. It follows that $M = M'$, $\mathsf{en} = \mathsf{en}'$ and $\forall t^i \in \mathsf{en}, \nu'(t^i) + d \leq \nu(t^i) + d \leq\uparrow Is(t)$ and then

$$(M', \mathsf{en}', \nu') \xrightarrow{d} (M', \mathsf{en}', \nu' + d).$$

Moreover, (1) implies that $M = M'$, $\mathsf{en} = \mathsf{en}'$,

$$\forall t^i \in \mathsf{en}, \nu'(t^i) + d = \nu(t^i) + d \ \text{ or } \downarrow Is(t) \leq\downarrow Is(t) + d \leq \nu'(t^i) + d < \nu(t^i) + d.$$

and then $(M, \mathsf{en}, \nu + d) \preceq (M', \mathsf{en}', \nu' + d)$.

Proof of (ii): $(M, \mathsf{en}, \nu) \xrightarrow{t^f} (M_1, \mathsf{en}_1, \nu_1)$ and $(M, \mathsf{en}, \nu) \preceq (M', \mathsf{en}', \nu')$ imply that $M = M'$, $\mathsf{en} = \mathsf{en}'$ and $t^f \in \mathsf{en}, \nu(t^f) \geq \downarrow Is(t)$ and $(\nu'(t^f) = \nu(t^f) \text{ or } \downarrow Is(t) \leq \nu'(t^f) < \nu(t^f))$. Therefore, $t^f \in \mathsf{en}', \nu'(t^f) \geq \downarrow Is(t)$ and then $\exists (M'_1, \mathsf{en}'_1, \nu'_1) \in \mathcal{Q}, (M', \mathsf{en}', \nu') \xrightarrow{t^f} (M'_1, \mathsf{en}'_1, \nu'_1)$.
Moreover, it holds that $M_1 = M'_1$, $\mathsf{en}_1 = \mathsf{en}'_1$ and $\forall t'^i \in \mathsf{en}_1$, if $t'^i \in \mathsf{Nw}(M, \mathsf{en}, t^f), \nu_1(t'^i) = 0, \nu'_1(t'^i) = 0$.
Otherwise, $\nu_1(t'^i) = \nu(t'^{i_o}), \nu'_1(t'^i) = \nu'(t'^{i_o})$, t'^{i_o} being the reference in en to t'^i.
Therefore, $\forall t'^i \in \mathsf{en}_1$,

$$\nu_1(t'^i) = \nu'_1(t'^i) = 0 \text{ or } \downarrow Is(t) \leq \nu'_1(t'^i) = \nu'(t'^{i_o}) < \nu(t'^{i_o}) = \nu_1(t'^i).$$

The relation $\preceq$ is then a strong timed simulation over states of the model, whatever the intermediate/atomic/persistent atomic semantics. $\qquad\square$

Let us now show that from the same state (M, en, ν), the states reached by firing two enabling instances of the same transition are s.t. the one reached by the older enabling instance strongly simulates the other one. It means that applying FEFF firing choice will preserve the timed traces of (M, en, ν) obtained by NDF firing choice, whatever the intermediate/atomic/persistent atomic semantics.

Lemma 3. *Let (M, en, ν), $(M_f, \mathsf{en}_f, \nu_f)$ and $(M_g, \mathsf{en}_g, \nu_g)$ be three states, t^f and t^g two distinct enabling instances of the same transition t in M s.t.*

$$(M, \mathsf{en}, \nu) \xrightarrow{t^f} (M_f, \mathsf{en}_f, \nu_f), \ (M, \mathsf{en}, \nu) \xrightarrow{t^g} (M_g, \mathsf{en}_g, \nu_g) \ \text{ and } \ \nu(t^f) \geq \nu(t^g).$$

Then $(M_f, \mathsf{en}_f, \nu_f) \preceq (M_g, \mathsf{en}_g, \nu_g)$.

Proof. Since t^f and t^g are two enabling instances of the same transition t, it follows that (see Property 1): $M_f = M_g$, $\mathsf{CF}(M, \mathsf{en}, t^f) - \{t^f\} = \mathsf{CF}(M, \mathsf{en}, t^g) - \{t^g\}$, $\mathsf{Nw}(M, \mathsf{en}, t^f) = \mathsf{Nw}(M, \mathsf{en}, t^g)$, and $t^f \in \mathsf{CF}(M, \mathsf{en}, t^f) \Leftrightarrow t^g \in \mathsf{CF}(M, \mathsf{en}, t^g)$.

The list en_f is obtained from en by eliminating enabling instances of $CF(M, en, t^f)$ and adding enabling instances of $Nw(M, en, t^f)$. Similarly, the list en_g is obtained from en by eliminating instances of $CF(M, en, t^g)$ and adding instances of $Nw(M, en, t^g)$. Let us consider two cases: $t^f \notin CF(M, en, t^f)$ (which may hold for the persistent atomic semantics) and $t^f \in CF(M, en, t^f)$ (which always holds for the intermediate and atomic semantics).

1) If $t^f \notin CF(M, en, t^f)$ (i.e., $t^g \notin CF(M, en, t^g)$) then $CF(M, en, t^f) = CF(M, en, t^g)$ and $Nw(M, en, t^f) = Nw(M, en, t^g)$. It follows that $en_f = en_g$ and $\nu_f = \nu_g$.

2) If $t^f \in CF(M, en, t^f)$ (i.e., $t^g \in CF(M, en, t^g)$) then $CF(M, en, t^f) - \{t^f\} = CF(M, en, t^g) - \{t^g\}$ and $Nw(M, en, t^f) = Nw(M, en, t^g)$. It follows that the lists en_f and en_g are equal. In addition, both ν_f and ν_g are obtained by eliminating the same set of clocks values, except those of the fired instances. In ν_f, the clock value of the fired instance t^f is eliminated but the one of t^g is kept. Similarly, in ν_g, the clock value of the fired instance t^g is eliminated but the one of t^f is kept. In lists en_f and en_g, transitions appears in increasing order and the enabling instances of the same transition are ordered from the oldest to the newest one. Let g' and f', with $f' < g'$, be the positions in en_f and en_g of the enabling instances t^g and t^f of en, respectively. The enabling instances between positions f' and g' are all firable instances of t. Then:

$$\forall t'^i \in en_f, \nu_f(t'^i) = \begin{cases} \nu_g(t'^i) & \text{if } i < f' \\ \nu_g(t'^{i+1}) \leq \nu_g(t'^i) & \text{if } f' \leq i < g' \\ \nu_g(t'^i) & \text{otherwise} \end{cases}.$$

As t^f and t^g are firable from (M, en, ν), it follows that:
$\forall i \in [f', g'[, \downarrow Is(t) \leq \nu_f(t^i) \leq \nu_g(t^i)$.
Consequently, $\forall t'^i \in en_f, \nu_f(t'^i) = \nu_g(t'^i) \lor \downarrow Is(t') \leq \nu_f(t'^i) \leq \nu_g(t'^i)$.
Then: $(M_f, en_f, \nu_f) \preceq (M_g, en_g, \nu_g)$.   □

We can now state the main result of this section:

Theorem 4. *Let $\mathcal{N}$ be a TPN under multiple-server and threshold semantics. The FEFF firing choice policy of $\mathcal{N}$ strongly timed simulates the NDF firing choice policy.*

Proof. Let $\mathcal{S}_1 = < Q_1, q_0, \Sigma, \rightarrow_1 >$ and $\mathcal{S}_2 = < Q_2, q_0, \Sigma, \rightarrow_2 >$ be the transition systems of $\mathcal{N}$ under multiple-server and threshold semantics for NDF and FEFF firing choice policies, respectively. $\mathcal{S}_1$ and $\mathcal{S}_2$ have the same initial state $q_0 = (M_0, en_0, , \nu_0)$. The FEFF firing choice means that the oldest enabling instance of the same transition is fired first. From the initial state $q_0 = (M_0, en_0, \nu_0)$, we have the following relationships between $\mathcal{S}_1$ and $\mathcal{S}_2$:

1) $\forall d \geq 0, (M_0, en_0, \nu_0) \xrightarrow{d}_1 (M_0, en_0, \nu_0 + d) \Leftrightarrow (M_0, en_0, \nu_0) \xrightarrow{d}_2 (M_0, en_0, \nu_0 + d)$.

2) $\forall t^g \in en_0, \quad (M_0, en_0, \nu_0) \xrightarrow{t^g}_1 (M_g, en_g, \nu_g) \Rightarrow$

$$\exists t^f \in en_0 \text{ s.t. } \nu(t^f) \geq \nu(t^g), (M_0, en_0, \nu_0) \xrightarrow{t^f}_2 (M_f, en_f, \nu_f).$$

Using Lemma 3, we can state that $(M_f, en_f, \nu_f) \preceq (M_g, en_g, \nu_g)$.

Inductively and thanks to Lemma 2 and Lemma 3, from any states $(M, \mathsf{en}, \nu_1) \in Q_1$ and $(M, \mathsf{en}, \nu_2) \in Q_2$ such that $(M, \mathsf{en}, \nu_1) \preceq (M, \mathsf{en}, \nu_2)$, we have:

1) $\forall d \geq 0, \quad (M, \mathsf{en}, \nu_1) \xrightarrow{d}_1 (M, \mathsf{en}, \nu_1 + d) \Rightarrow (M, \mathsf{en}, \nu_2) \xrightarrow{d}_2 (M, \mathsf{en}, \nu_2 + d)$.

2) $\forall t^g \in \mathsf{en}, \quad (M, \mathsf{en}, \nu_1) \xrightarrow{t^g}_1 (M_g, \mathsf{en}, \nu_g) \Rightarrow \exists t^f \in \mathsf{en}$ s.t.

$\nu(t^f) \geq \nu(t^g), (M, \mathsf{en}, \nu_2) \xrightarrow{t^f}_2 (M_f, \mathsf{en}_f, \nu_f)$ and $(M_f, \mathsf{en}_f, \nu_f) \preceq (M_g, \mathsf{en}_g, \nu_g)$.

Therefore, $\mathcal{S}_2$ strongly timed simulates $\mathcal{S}_1$.   □

4.2   FEFF vs NDF wrt Timed Language

We can now extend the previous results to timed language acceptance consideration.

Theorem 5. *Let $\mathcal{N}$ be a TPN under multiple-server and threshold semantics. The timed language of $\mathcal{N}$ is the same for both NDF and FEFF firing choice policies.*

Proof. Let $\mathcal{S}_1 = < Q_1, q_0, \Sigma, \rightarrow_1 >$ and $\mathcal{S}_2 = < Q_2, q_0, \Sigma, \rightarrow_2 >$ be the transition systems of $\mathcal{N}$ under multiple-server and threshold semantics for NDF and FEFF choice policies, respectively. Lemma 1 states that $\mathcal{S}_1$ strongly timed simulates $\mathcal{S}_2$. Theorem 4 states that $\mathcal{S}_2$ strongly timed simulates $\mathcal{S}_1$. Therefore, $\mathcal{S}_1$ and $\mathcal{S}_2$ have the same timed language.   □

4.3   FEFF vs NDF wrt Timed Bisimulation

The previous subsections prove that for the multiple-server and threshold semantics, there is a strong timed co-simulation between the NDF and FEFF firing choice policies. We will now show that this strong timed co-simulation is not a timed bisimulation.

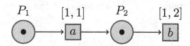

Fig. 5. A TPN $\mathcal{N}$

Let us consider the TPN $\mathcal{N}$ of Fig. 5. Let $\mathcal{S}_1 = < Q_1, q_0, \Sigma, \rightarrow_1 >$ and $\mathcal{S}_2 = < Q_2, q_0, \Sigma, \rightarrow_2 >$ be the transition systems of $\mathcal{N}$ under multiple-server and threshold semantics for NDF and FEFF choice policies, respectively.

We consider the following run. At date 1, we fire the transition a and then we let 1 more time unit elapse. For both firing choice policies, this run is possible and leads to the same state q_3, where there are two tokens in place P_2 with different ages 1 and 2: $q_0 \xrightarrow{1} q_1 \xrightarrow{a} q_2 \xrightarrow{1} q_3$ with $\rightarrow \in \{\rightarrow_1, \rightarrow_2\}$. From this state, with the NDF firing choice policy, both enabling instances of b are fired. The firing of the instance of b with the youngest token of P_2 leads to the state q_4 ($q_3 \xrightarrow{b}_1 q_4$). To mime this behavior, from q_3, the FEFF choice policy has to fire the transition b with the oldest token of P_2 leading to the state q_5 ($q_3 \xrightarrow{b}_2 q_5$). Now from state q_5, the FEFF firing choice policy allows to wait 1 time unit ($q_5 \xrightarrow{1}_2 q_6$), whereas the transition b must fire immediately with the NDF firing choice policy from q_4 ($q_4 \xrightarrow{b}_1 q_7$). Then, q_4 and q_5 are not strongly timed bisimilar.

4.4   Discussion

The NDF firing choice policy is the more conservative safety-wise choice and corresponds in general to the semantics we want to use for model checking or control purposes. However, since all firable instances may fire in any order, the number of runs may lead to a state space explosion. To avoid to compute all runs, the simpler alternative, First Enabled First Fired (FEFF), where only the oldest instance may fire, leads to a much more compact state-space. We show with the following example of Fig. 6 how FEFF generates fewer runs than NDF but preserves all the timed traces of NDF.

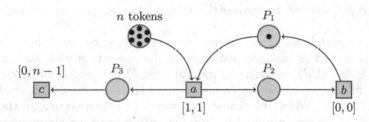

Fig. 6. A TPN illustrating the interest of the FEFF choice policy

We consider only the runs producing the trace $(ab)^n\omega$ of duration n. Since the prefix $(ab)^n$ takes also n time units, the duration of ω is 0. In all runs, the only state reachable after this prefix is q_1 with a marking with one token in P_1, n tokens in P_3 and no token in other places. As a shorthand and since the markings will now change only in place P_3, we can denote the state by the age of each n tokens enabling c. For q_1, the oldest token is $c^1 = n-1$ and the newest is $c^n = 0$ and $q_1 = \{c^1 = n-1, c^2 = n-2, c^3 = n-3 \ldots, c^n = 0\}$. All the instances of c are firable from q_1 since the timed interval of c is $[0, n-1]$. For FEFF, we obtain one only run by firing c^1 then c^2 then $c^3 \ldots$ then c^n whereas for NDF we obtain all the $n!$ combinations of all the instances leading to the same state q_f. We illustrate these runs with $n = 3$ in Figure 7. All these $n!$ runs of NDF produce the same trace $(ab)^n(c)^n$ where $\omega = (c)^n$ is in null time, which is also the trace produced by the only run of FEFF.

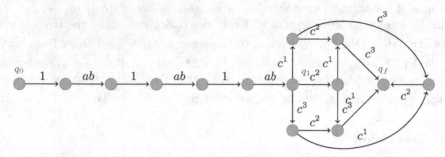

Fig. 7. Runs with NDF

5 Conclusion

In this paper, we first presented and discussed different semantics of multi-enabledness for time Petri nets. Then, we considered the threshold semantics in both contexts single-server and multiple-server semantics and investigated some questions relative to the expressiveness. We showed that multiple-server semantics adds expressiveness, in terms of timed bisimulation, relatively to single-server semantics.

For the multiple-server semantics, different firing choice policies may be used to manage the multiple instances of the same transition such as Non-Deterministic (NDF) choice and First Enabled First Fired (FEFF) choice. In the NDF firing choice, the firable instances of the same transition are fired in all possible orders, which may cause a blow-up of the state space. In the FEFF choice, these instances are fired in only one order: from the oldest ones to the more recent ones. We proved that both semantics are not bisimilar but actually simulate each other with strong timed simulations, which in particular implies that they generate the same timed traces. Consequently, a TPN under NDF or FEFF policies has the same linear (timed) properties but FEFF leads to more compact state space. FEFF is then very appropriate to deal with linear timed properties of time Petri nets.

As immediate perspective, we will investigate the extension of the results established in this paper, to the case of age-semantics.

References

1. Aura, T., Lilius, J.: A causal semantics for time petri nets. Theoretical Computer Science 243(2), 409–447 (2000)
2. Bérard, B., Cassez, F., Haddad, S., Lime, D., Roux, O.H.: Comparison of different semantics for time Petri nets. In: Peled, D.A., Tsay, Y.-K. (eds.) ATVA 2005. LNCS, vol. 3707, pp. 293–307. Springer, Heidelberg (2005)
3. Bérard, B., Cassez, F., Haddad, S., Lime, D., Roux, O.H.: The expressive power of time Petri nets. Theoretical Computer Science 474, 1–20 (2013)

4. Berthomieu, B.: La méthode des classes d'états pour l'analyse ·des réseaux temporels - mise en oeuvre, extension à la multi-sensibilisation. In: Modélisation des Systèmes Réactifs (MSR 2001), pp. 275–290. Hermes, Toulouse (2001)
5. Boucheneb, H., Bullich, A., Roux, O.H.: FIFO Time Petri Nets for conflicts handling. In: 11th International Workshop on Discrete Event Systems (WODES 2012), IFAC, Guadalajara (2012)
6. Boyer, M., Diaz, M.: Multiple enabledness of transitions in time Petri nets. In: Proc. of the 9th IEEE International Workshop on Petri Nets and Performance Models, Aachen, Germany, September 11–14. IEEE Computer Society (2001)
7. Boyer, M., Roux, O.H.: On the compared expressiveness of arc, place and transition time Petri nets. Fundamenta Informaticae 88(3), 225–249 (2008)
8. Cerone, A., Maggiolo-Schettini, A.: Timed based expressivity of time Petri nets for system specification. Theoretical Computer Science 216, 1–53 (1999)
9. Glabbeek, R., Plotkin, G.: Configuration structures. In: Proceedings of the 10th Annual IEEE Symposium on Logic in Computer Science (LICS 1995), pp. 199–209. IEEE Computer Society Press, San Diego (1995)
10. Khansa, W., Denat, J.-P., Collart-Dutilleul, S.: P-time Petri nets for manufacturing systems. In: International Workshop on Discrete Event Systems (WODES 1996), Edinburgh (U.K.), pp. 94–102 (August 1996)
11. Merlin, P.M.: A study of the recoverability of computing systems. PhD thesis, Department of Information and Computer Science, University of California, Irvine, CA (1974)
12. Myers, C.J., Rokicki, T.G., Meng, T.H.-Y.: Poset timing and its application to the timed cirucits. IEEE Transactions on CAD 18(6) (1999)
13. Ramchandani, C.: Analysis of asynchronous concurrent systems by timed Petri nets. PhD thesis, Massachusetts Institute of Technology, Cambridge, MA, Project MAC Report MAC-TR-120 (1974)
14. Reynier, P.-A., Sangnier, A.: Weak time petri nets strike back! In: Bravetti, M., Zavattaro, G. (eds.) CONCUR 2009. LNCS, vol. 5710, pp. 557–571. Springer, Heidelberg (2009)
15. Walter, B.: Timed net for modeling and analysing protocols with time. In: Proceedings of the IFIP Conference on Protocol Specification Testing and Verification, North Holland (1983)

Complexity Results for Elementary Hornets

Michael Köhler-Bußmeier and Frank Heitmann

University of Hamburg, Department for Informatics
Vogt-Kölln-Str. 30, D-22527 Hamburg
koehler@informatik.uni-hamburg.de

Abstract. In this paper we study the complexity of HORNETS, an algebraic extension of object nets. We define a restricted class: safe, elementary HORNETS, to guarantee finite state spaces.

We have shown previously that the reachability problem for this class
requires at least exponential space, which is a major increase when compared to safe, elementary object nets, which require polynomial space.

Here, we show how a safe, elementary HORNETS can be simulated by a
safe EOS, which establishes an upper bound for the complexity, since we
know that that the reachability problem for safe EOS is PSpace-complete.

Keywords: Hornets, nets-within-nets, object nets, reachability,
safeness.

1 Hornets: Higher-Order Object Nets

In this paper we study the algebraic extension of object nets, called HORNETS.
HORNETS are a generalisation of object nets [1,2,3], which follow the *nets-within-
nets* paradigm as proposed by Valk [4,5]. With object nets we study Petri nets
where the tokens are nets again, i.e. we have a nested marking. Events are also
nested. We have three different kinds of events – as illustrated by the following
example:

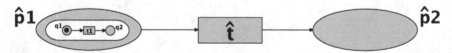

1. System-autonomous: The system net transition $\hat{t}$ fires autonomously, which
   moves the net-token from $\hat{p}_1$ to $\hat{p}_2$ without changing its marking.
2. Object-autonomous: The object net fires transition t_1 "moving" the black
   token from q_1 to q_2. The object net remains at its location $\hat{p}_1$.
3. Synchronisation: Whenever we add matching synchronisation inscriptions at
   the system net transition $\hat{t}$ and the object net transition t_1, then both must
   fire synchronously: The object net is moved to $\hat{p}_2$ and the black token moves
   from q_1 to q_2 inside. Whenever synchronisation is specified, autonomous
   actions are forbidden.

J.-M. Colom and J. Desel (Eds.): PETRI NETS 2013, LNCS 7927, pp. 150–169, 2013.

For HORNETS we extend object-nets with algebraic concepts that allow to modify the structure of the net-tokens as a result of a firing transition. This is a generalisation of the approach of algebraic nets [6], where algebraic data types replace the anonymous black tokens.

Example 1. We consider a HORNET with two workflow nets N_1 and N_2 as tokens – cf. Figure 1. To model a run-time adaption, we combine N_1 and N_2 resulting in the net $N_3 = (N_1 \| N_2)$. This modification is modelled by transition t of the HORNETS in Fig. 1. In a binding α with $x \mapsto N_1$ and $y \mapsto N_2$ the transition t is enabled. Assume that $(x \| y)$ evaluates to N_3 for α. If t fires it removes the two net-tokens from p and q and generates one new net-token on place r. The net-token on r has the structure of N_3 and its marking is obtained as a transfer from the token on v in N_1 and the token on s in N_2 into N_3. This transfer is possible since all the places of N_1 and N_2 are also places in N_3 and tokens can be transferred in the obvious way.

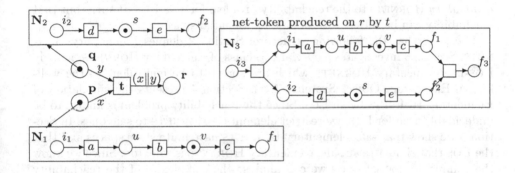

Fig. 1. Modification of the net-token's structure

The use of algebraic operations in HORNETS relates them to *algebraic higher-order (AHO) systems* [7], which are restricted to two-levelled systems but have a greater flexibility for the operations on net-tokens, since each net transformation is allowed. There is also a relationship to Nested Nets [8], which are used for adaptive systems.

It is not hard to prove that the general HORNET formalism is Turing-complete. In [9] we have proven that there are several possibilities to simulate counter programs: One could use the nesting to encode counters. Another possibility is to encode counters in the algebraic structure of the net operators.

In this paper we like to study the *complexity* that arises due the algebraic structure. Here, we restrict HORNETS to guarantee that the system has a finite state space. First, we allow at most one token on each place, which results in the class of *safe* HORNETS. However this restriction does not guarantee finite state spaces, since we have the nesting depth as a second source of undecidability [3]. Second, we restrict the universe of object nets to finite sets. Finally, we restrict the nesting depth and introduce the class of *elementary* HORNETS, which have a two-levelled nesting structure. This is done in analogy to the class of *elemtary object net systems* (Eos) [2], which are the two-level specialisation of general object nets [2,3].

If we rule out these sources of complexity the main origin of complexity is the use of algebraic transformations, which are still allowed for safe, elementary HORNETS. As a result we obtain the class of safe, elementary HORNETS – in analogy to the class of *safe* Eos [10].

We have shown in [10,11] that most problems for *safe* Eos are PSPACE-complete. More precisely: All problems that are expressible in LTL or CTL, which includes reachability and liveness, are PSPACE-complete. This means that with respect to these problems *safe* Eos are no more complex than p/t nets.

In a previous publication [12] we have shown that *safe, elementary* HORNETS are beyond PSPACE. We have shown a lower bound, i.e. that "the reachability problem requires exponential space" for safe, elementary HORNETS – similarily to well known result of for *bounded* p/t nets [13]. Here we establish an upper bound for the reachability problem giving a reduction from the problem for *safe, elementary* HORNETS to the reachability problem for *safe* Eos. It turns out that reachability can be solved with double-exponential space.

The paper has the following structure: Section 2 defines Elementary HOR-NETS. Section 3 investigates a special sub-class of elementary HORNETS, namely Eos-like elementary HORNETS which are shown to be equivalent to the well-known Elementary Object Systems, Eos. Section 4 presents the simulation of Elementary HORNETS by an Eos. Since the reachability problem is known to be undecidable even for Eos, we restrict elementary HORNETS to safe ones. In Section 5 we show that safe, elementary HORNETS have finite state spaces and that the Eos that simulates an safe, elementary HORNETS is safe, too. Since we know the complexity for safe Eos we can analyse the complexity of the reachability problem for safe, elementary HORNETS by studying the size of the simulating Eos. The work ends with a conclusion.

2   Definition of Elementary Hornets

A multiset $\mathbf{m}$ on the set D is a mapping $\mathbf{m} : D \to \mathbb{N}$. Multisets can also be represented as a formal sum in the form $\mathbf{m} = \sum_{i=1}^{n} x_i$, where $x_i \in D$.

Multiset addition is defined component-wise: $(\mathbf{m}_1 + \mathbf{m}_2)(d) := \mathbf{m}_1(d) + \mathbf{m}_2(d)$. The empty multiset $\mathbf{0}$ is defined as $\mathbf{0}(d) = 0$ for all $d \in D$. Multiset-difference $\mathbf{m}_1 - \mathbf{m}_2$ is defined by $(\mathbf{m}_1 - \mathbf{m}_2)(d) := \max(\mathbf{m}_1(d) - \mathbf{m}_2(d), 0)$.

The cardinality of a multiset is $|\mathbf{m}| := \sum_{d \in D} \mathbf{m}(d)$. A multiset $\mathbf{m}$ is finite if $|\mathbf{m}| < \infty$. The set of all finite multisets over the set D is denoted $MS(D)$.

Multiset notations are used for sets as well. The meaning will be apparent from its use.

Any mapping $f : D \to D'$ extends to a multiset-homomorphism $f^\sharp : MS(D) \to MS(D')$ by $f^\sharp \left(\sum_{i=1}^{n} x_i \right) = \sum_{i=1}^{n} f(x_i)$.

A *p/t net* N is a tuple $N = (P, T, \mathbf{pre}, \mathbf{post})$, such that P is a set of places, T is a set of transitions, with $P \cap T = \emptyset$, and $\mathbf{pre}, \mathbf{post} : T \to MS(P)$ are the pre- and post-condition functions. A marking of N is a multiset of places: $\mathbf{m} \in MS(P)$. We denote the enabling of t in marking $\mathbf{m}$ by $\mathbf{m} \xrightarrow{t}$. Firing of t is denoted by $\mathbf{m} \xrightarrow{t} \mathbf{m}'$.

2.1   Elementary Hornets

Net-Algebras. We define the algebraic structure of object nets. For a general introduction of algebraic specifications cf. [14].

Let K be a set of net-types (kinds). A (many-sorted) *specification* (Σ, X, E) consists of a signature Σ, a family of variables $X = (X_k)_{k \in K}$, and a family of axioms $E = (E_k)_{k \in K}$.

A signature is a disjoint family $\Sigma = (\Sigma_{k_1 \cdots k_n, k})_{k_1, \cdots, k_n, k \in K}$ of operators. The set of terms of type k over a signature Σ and variables X is denoted $\mathbb{T}^k_\Sigma(X)$.

We use (many-sorted) predicate logic, where the terms are generated by a signature Σ and formulae are defined by a family of predicates $\Psi = (\Psi_n)_{n \in \mathbb{N}}$. The set of formulae is denoted PL_Γ, where $\Gamma = (\Sigma, X, E, \Psi)$ is the *logic structure*.

Let Σ be a signature over K. A *net-algebra* assigns to each type $k \in K$ a set $\mathcal{U}_k$ of object nets. Each object net $N \in \mathcal{U}_k, k \in K$ net is a p/t net $N = (P_N, T_N, \mathbf{pre}_N, \mathbf{post}_N)$. We identify $\mathcal{U}$ with $\bigcup_{k \in K} \mathcal{U}_k$ in the following. We assume the family $\mathcal{U} = (\mathcal{U}_k)_{k \in K}$ to be disjoint.

The nodes of the object nets in $\mathcal{U}_k$ are not disjoint, since the firing rule allows to transfer tokens between net tokens within the same set $\mathcal{U}_k$. Such a transfer is possible, if we assume that all nets $N \in \mathcal{U}_k$ have the same set of places P_k. P_k is the place universe for all object nets of kind k. In the example of Fig. 1 the object nets N_1, N_2, and N_3 must belong to the same type since otherwise it would be impossible to transfer the markings in N_1 and N_2 to the generated N_3.

In general, P_k is not finite. Since we like each object net to be finite in some sense, we require that the transitions T_N of each $N \in \mathcal{U}_k$ use only a finite subset of P_k, i.e. $\forall N \in \mathcal{U} : |{}^\bullet T_N \cup T_N{}^\bullet| < \infty$.

The family of object nets $\mathcal{U}$ is the universe of the algebra. A net-algebra $(\mathcal{U}, \mathcal{I})$ assigns to each constant $\sigma \in \Sigma_{\lambda, k}$ an object net $\sigma^\mathcal{I} \in \mathcal{U}_k$ and to each operator $\sigma \in \Sigma_{k_1 \cdots k_n, k}$ with $n > 0$ a mapping $\sigma^\mathcal{I} : (\mathcal{U}_{k_1} \times \cdots \times \mathcal{U}_{k_n}) \to \mathcal{U}_k$.

A net-algebra is called *finite* if P_k is a finite set for each $k \in K$.

A variable assignment $\alpha = (\alpha_k : X_k \to \mathcal{U}_k)_{k \in K}$ maps each variable onto an element of the algebra. For a variable assignment α the evaluation of a term $t \in \mathbb{T}^k_\Sigma(X)$ is uniquely defined and will be denoted as $\alpha(t)$.

A net-algebra, such that all axioms of (Σ, X, E) are valid, is called *net-theory*.

Nested Markings. A marking of an elementary HORNET assigns to each system net place one or many net-tokens. The places of the system net are typed by the function $k : \widehat{P} \to K$, meaning that a place $\widehat{p}$ contains net-tokens of kind $k(\widehat{p})$. Since the net-tokens are instances of object nets, a *marking* is a *nested* multiset of the form:

$$\mu = \sum_{i=1}^{n} \widehat{p}_i[N_i, M_i] \quad \text{where} \quad \widehat{p}_i \in \widehat{P}, N_i \in \mathcal{U}_{k(\widehat{p}_i)}, M_i \in MS(P_{N_i}), n \in \mathbb{N}$$

Each addend $\widehat{p}_i[N_i, M_i]$ denotes a net-token on the place $\widehat{p}_i$ that has the structure of the object net N_i and the marking $M_i \in MS(P_{N_i})$. The set of all nested multisets is denoted as $\mathcal{M}_H$. We define the partial order $\sqsubseteq$ on nested multisets by setting $\mu_1 \sqsubseteq \mu_2$ iff $\exists \mu : \mu_2 = \mu_1 + \mu$.

The projection $\Pi_N^{1,H}(\mu)$ is the multiset of all system-net places that contain the object-net N:[1]

$$\Pi_N^{1,H}\left(\sum_{i=1}^{n} \widehat{p}_i[N_i, M_i]\right) := \sum_{i=1}^{n} \mathbf{1}_N(N_i) \cdot \widehat{p}_i \tag{1}$$

where the indicator function $\mathbf{1}_N$ is defined as: $\mathbf{1}_N(N_i) = 1$ iff $N_i = N$.

The projection $\Pi_{\mathcal{U}}^{2,H}(\mu)$ is the multiset of all object nets in the marking:

$$\Pi_{\mathcal{U}}^{2,H}\left(\sum_{i=1}^{n} \widehat{p}_i[N_i, M_i]\right) := \sum_{i=1}^{n} N_i \tag{2}$$

Analogously, the projection $\Pi_N^{2,H}(\mu)$ is the multiset of all net-tokens' markings (that belong to the object-net N):

$$\Pi_N^{2,H}\left(\sum_{i=1}^{n} \widehat{p}_i[N_i, M_i]\right) := \sum_{i=1}^{n} \mathbf{1}_k(N_i) \cdot M_i \tag{3}$$

The projection $\Pi_k^{2,H}(\mu)$ is the sum of all net-tokens' markings belonging to the same type $k \in K$:

$$\Pi_k^{2,H}(\mu) := \sum_{N \in \mathcal{U}_k} \Pi_N^{2,H}(\mu) \tag{4}$$

Synchronisation. The transitions in an HORNET are labelled with synchronisation inscriptions. We assume a fixed set of channels $C = (C_k)_{k \in K}$.

– The function family $\widehat{l}_\alpha = (\widehat{l}_\alpha^k)_{k \in K}$ defines the synchronisation constraints. Each transition of the system net is labelled with a multiset $\widehat{l}^k(\widehat{t}) = (e_1, c_1) + \cdots + (e_n, c_n)$, where the expression $e_i \in \mathbb{T}_\Sigma^k(X)$ describes the called object net and $c_i \in C_k$ is a channel. The intention is that $\widehat{t}$ fires synchronously with a multiset of object net transitions with the same multiset of labels. Each variable assignment α generates the function $\widehat{l}_\alpha^k(\widehat{t})$ defined as:

$$\widehat{l}_\alpha^k(\widehat{t})(N) := \sum_{\substack{1 \le i \le n \\ \alpha(e_i) = N}} c_i \quad \text{for} \quad \widehat{l}^k(\widehat{t}) = \sum_{1 \le i \le n} (e_i, c_i) \tag{5}$$

Each function $\widehat{l}_\alpha^k(\widehat{t})$ assigns to each object net N a multiset of channels.
– For each $N \in \mathcal{U}_k$ the function l_N assigns to each transition $t \in T_N$ either a channel $c \in C_k$ or $\perp_k$, whenever t fires without synchronisation, i.e. autonously.

---

[1] The superscript H indicates that the function is used for HORNETS.

System Net. Assume we have a fixed logic $\Gamma = (\Sigma, X, E, \Psi)$ and a net-theory $(\mathcal{U}, \mathcal{I})$. An *elementary higher-order object net* (HORNET) is composed of a system net $\widehat{N}$ and the set of object nets $\mathcal{U}$. W.l.o.g. we assume $\widehat{N} \notin \mathcal{U}$. To guarantee finite algebras for elementary HORNETS, we require that the net-theory $(\mathcal{U}, \mathcal{I})$ is finite, i.e. each place universe P_k is finite.

The system net is a net $\widehat{N} = (\widehat{P}, \widehat{T}, \mathbf{pre}, \mathbf{post}, \widehat{G})$, where each arc is labelled with a multiset of terms: $\mathbf{pre}, \mathbf{post} : \widehat{T} \to (\widehat{P} \to MS(\mathbb{T}_\Sigma(X)))$. Each transition is labelled by a guard predicate $\widehat{G} : \widehat{T} \to PL_\Gamma$. The places of the system net are typed by the function $k : \widehat{P} \to K$. As a typing constraint we have that each arc inscription has to be a multiset of terms that are all of the kind that is assigned to the arc's place:

$$\mathbf{pre}(\widehat{t})(\widehat{p}), \quad \mathbf{post}(\widehat{t})(\widehat{p}) \quad \in \quad MS(\mathbb{T}_\Sigma^{k(\widehat{p})}(X)) \tag{6}$$

For each variable binding α we obtain the evaluated functions $\mathbf{pre}_\alpha, \mathbf{post}_\alpha : \widehat{T} \to (\widehat{P} \to MS(\mathcal{U}))$ in the obvious way.

Definition 1 (Elementary Hornet). *Assume a fixed many-sorted predicate logic* $\Gamma = (\Sigma, X, E, \Psi)$. *An elementary* HORNET *is a tuple* $EH = (\widehat{N}, \mathcal{U}, \mathcal{I}, k, l, \mu_0)$ *such that:*

1. $\widehat{N}$ *is an algebraic net, called the* system net.
2. $(\mathcal{U}, \mathcal{I})$ *is a finite net-theory for the logic* Γ.
3. $k : \widehat{P} \to K$ *is the typing of the system net places.*
4. $l = (\widehat{l}, l_N)_{N \in \mathcal{U}}$ *is the labelling.*
5. $\mu_0 \in \mathcal{M}_H$ *is the initial marking.*

2.2   Events and Firing Rule

The synchronisation labelling generates the set of system events Θ:
    The labelling introduces three cases of events:

1. Synchronised firing: There is at least one object net that has to be synchronised, i.e. there is a N such that $\widehat{l}(\widehat{t})(N)$ is not empty.

    Such an event is a pair $\theta = \widehat{t}^\alpha[\vartheta]$, where $\widehat{t}$ is a system net transition, α is a variable binding, and ϑ is a function that maps each object net to a multiset of its transitions, i.e. $\vartheta \in MS(T_N)$. It is required that $\widehat{t}$ and $\vartheta(N)$ have matching multisets of labels, i.e. $\widehat{l}(\widehat{t})(N) = l_N^\sharp(\vartheta(N))$ for all $N \in \mathcal{U}$. (Remember that $l_N^\sharp$ denotes the multiset extension of l_N.)

    The intended meaning is that $\widehat{t}$ fires synchronously with all the object net transitions $\vartheta(N), N \in \mathcal{U}$.

2. System-autonomous firing: The transition $\widehat{t}$ of the system net fires autonomously, whenever $\widehat{l}(\widehat{t})$ is the empty multiset $\mathbf{0}$.

    We consider system-autonomous firing as a special case of synchronised firing generated by the function ϑ_{id}, defined as $\vartheta_{id}(N) = \mathbf{0}$ for all $N \in \mathcal{U}$.

3. Object-autonomous firing: An object net transition t in N fires autonomously, whenever $l_N(t) = \bot_k$.

Object-autonomous events are denoted as $id_{\widehat{p},N}[\vartheta_t]$, where $\vartheta_t(N') = \{t\}$ if $N = N'$ and $\mathbf{0}$ otherwise. The meaning is that in object net N fires t autonomously within the place $\widehat{p}$.

For the sake of uniformity we define for an arbitrary binding α:

$$\mathbf{pre}_\alpha(id_{\widehat{p},N})(\widehat{p}')(N') = \mathbf{post}_\alpha(id_{\widehat{p},N})(\widehat{p}')(N') = \begin{cases} 1 & \text{if } \widehat{p}' = \widehat{p} \wedge N' = N \\ 0 & \text{otherwise.} \end{cases}$$

The set of all events generated by the labelling l is $\Theta_l := \Theta_1 \cup \Theta_2$, where Θ_1 contains synchronous events (including system-autonomous events as a special case) and Θ_2 contains the object-autonomous events:

$$\begin{aligned} \Theta_1 &:= \left\{ \widehat{\tau}^\alpha[\vartheta] \quad \mid \forall N \in \mathcal{U} : \widehat{l}_\alpha(\widehat{t})(N) = l_N^\sharp(\vartheta(N)) \right\} \\ \Theta_2 &:= \left\{ id_{\widehat{p},N}[\vartheta_t] \mid \widehat{p} \in \widehat{P}, N \in \mathcal{U}_{k(\widehat{p})}, t \in T_N \right\} \end{aligned} \tag{7}$$

Firing Rule. A system event $\theta = \widehat{\tau}^\alpha[\vartheta]$ removes net-tokens together with their individual internal markings. Firing the event replaces a nested multiset $\lambda \in \mathcal{M}_H$ that is part of the current marking μ, i.e. $\lambda \sqsubseteq \mu$, by the nested multiset ρ. The enabling condition is expressed by the *enabling predicate* ϕ_{EH} (or just ϕ whenever EH is clear from the context):

$$\begin{aligned} \phi_{EH}(\widehat{\tau}^\alpha[\vartheta], \lambda, \rho) &\iff \forall k \in K : \\ &\forall \widehat{p} \in k^{-1}(k) : \forall N \in \mathcal{U}_k : \Pi_N^{1,H}(\lambda)(\widehat{p}) = \mathbf{pre}_\alpha(\widehat{\tau})(\widehat{p})(N) \wedge \\ &\forall \widehat{p} \in k^{-1}(k) : \forall N \in \mathcal{U}_k : \Pi_N^{1,H}(\rho)(\widehat{p}) = \mathbf{post}_\alpha(\widehat{\tau})(\widehat{p})(N) \wedge \\ &\Pi_k^{2,H}(\lambda) \geq \sum_{N \in \mathcal{U}_k} \mathbf{pre}_N^\sharp(\vartheta(N)) \wedge \\ &\Pi_k^{2,H}(\rho) = \Pi_k^{2,H}(\lambda) + \sum_{N \in \mathcal{U}_k} \mathbf{post}_N^\sharp(\vartheta(N)) - \mathbf{pre}_N^\sharp(\vartheta(N)) \end{aligned} \tag{8}$$

The predicate ϕ_{EH} has the following meaning: Conjunct (1) states that the removed sub-marking λ contains on $\widehat{p}$ the right number of net-tokens, that are removed by $\widehat{\tau}$. Conjunct (2) states that generated sub-marking ρ contains on $\widehat{p}$ the right number of net-tokens, that are generated by $\widehat{\tau}$. Conjunct (3) states that the sub-marking λ enables all synchronised transitions $\vartheta(N)$ in the object N. Conjunct (4) states that the marking of each object net N is changed according to the firing of the synchronised transitions $\vartheta(N)$.

Note, that conjunct (1) and (2) assures that only net-tokens relevant for the firing are included in λ and ρ. Conditions (3) and (4) allow for additonal tokens in the net-tokens.

For system-autonomous events $\widehat{t}^\alpha[\vartheta_{id}]$ the enabling predicate ϕ_{EH} can be simplified further: Conjunct (3) is always true since $\mathbf{pre}_N(\vartheta_{id}(N)) = \mathbf{0}$. Conjunct (4) simplifies to $\Pi_k^{2,H}(\rho) = \Pi_k^{2,H}(\lambda)$, which means that no token of the object nets get lost when a system-autonomous events fires.

Analogously, for an object-autonomous event $\widehat{\tau}[\vartheta_t]$ we have an idle-transition $\widehat{\tau} = id_{\widehat{p},N}$ and $\vartheta = \vartheta_t$ for some t. Conjunct (1) and (2) simplify to $\Pi_{N'}^{1,H}(\lambda) =$

$\widehat{p} = \Pi_{N'}^{1,H}(\rho)$ for $N' = N$ and to $\Pi_{N'}^{1,H}(\lambda) = \mathbf{0} = \Pi_{N'}^{1,H}(\rho)$ otherwise. This means that $\lambda = \widehat{p}[M]$, M enables t, and $\rho = \widehat{p}[M - \mathbf{pre}_N(\widehat{t}) + \mathbf{post}_N(\widehat{t})]$.

Definition 2 (Firing Rule). *Let EH be an elementary* HORNET *and* $\mu, \mu' \in \mathcal{M}_H$ *markings.*

- *The event* $\widehat{\tau}^\alpha[\vartheta]$ *is enabled in* μ *for the mode* $(\lambda, \rho) \in \mathcal{M}_H^2$ *iff* $\lambda \sqsubseteq \mu \wedge \phi_{EH}(\widehat{\tau}[\vartheta], \lambda, \rho)$ *holds and the guard* $\widehat{G}(\widehat{t})$ *holds, i.e.* $E \models_{\mathcal{I}}^\alpha \widehat{G}(\widehat{\tau})$.
- *An event* $\widehat{\tau}^\alpha[\vartheta]$ *that is enabled in* μ *can fire – denoted* $\mu \xrightarrow[EH]{\widehat{\tau}^\alpha[\vartheta](\lambda, \rho)} \mu'$.
- *The resulting successor marking is defined as* $\mu' = \mu - \lambda + \rho$.

Note, that the firing rule has no a-priori decision how to distribute the marking on the generated net-tokens. Therefore we need the mode (λ, ρ) to formulate the firing of $\widehat{\tau}^\alpha[\vartheta]$ in a functional way.

2.3   Unfolding of Elementary Hornets

Since all nets $N \in \mathcal{U}_k$ have the same set of places P_k, which is finite for elementary HORNETS, there is an upper bound for the cardinality of $\mathcal{U}_k$.

Lemma 1. *For each* $k \in K$ *we have* $|\mathcal{U}_k| \leq 2^{\left(2^{4|P_k|}\right)}$.

Proof. Note, that each common set of places P_k is finite for elementary HORNETS. Each possible transition t chooses a subset of P_k for the preset ${}^\bullet t$ and another subset for the postset $t^\bullet$. We identify t with the pair $({}^\bullet t, t^\bullet)$. A transition is an element from $T_k := \mathcal{P}(P_k) \times \mathcal{P}(P_k)$. We have $|T_k| = 2^{|P_k|} \cdot 2^{|P_k|} = (2^{|P_k|})^2 = 2^{2|P_k|}$ different transitions.

To each transition $t \in T_N$, $N \in \mathcal{U}_k$ the partial function l_N assigns either a channel $c \in C_k$ or $\perp_k$.

The set of labelled transitions is $LT_k := T_k \times (C_k \cup \{\perp_k\}))$ and we have $|LT_k| = |T_k \times (C_k \cup \{\perp_k\})| = 2^{2|P_k|} \cdot (|C_k| + 1)$ different labelled transitions.

We cannot use more channels than we have transitions in the object net, i.e. we could use at most $|T_k| \leq 2^{2|P_k|}$ different channels from $C_k \cup \{\perp_k\}$. Thus, we have:

$$|LT_k| = 2^{2|P_k|} \cdot (|C_k| + 1) \leq 2^{2|P_k|} \cdot 2^{2|P_k|} \leq 2^{4|P_k|}$$

Since each object net N in $\mathcal{U}_k$ is characterised by its set of labelled transitions and there are $|\mathcal{P}(LT_k)| = 2^{|LT_k|}$ subsets of LT_k, we have at most $2^{\left(2^{4|P_k|}\right)}$ different object nets.   qed.

In the following we identify an object net with a subset of LT_k.

An elementary HORNET is called *grounded* whenever all arc inscriptions $\mathbf{pre}(\widehat{t})(\widehat{p})$ and $\mathbf{post}(\widehat{t})(\widehat{p})$ as well as the synchronisation family $\widehat{l}_\alpha = (\widehat{l}_\alpha^k)_{k \in K}$ are grounded terms, i.e. they do not contain variables.

For an elementary HORNET EH we can replace each system net transition $\widehat{t}$ by a family of transitions which is obtained by pairing $\widehat{t}$ with a variable binding α. The constructing is very similar to the well known construction for coloured Petri nets. Note, that the set of variable bindings is finite, since due to Lemma 1 there are only finitely many assignments for each variable.

Proposition 1. *Each elementary* HORNET *EH can be bisimulated by a grounded* HORNET $G(EH)$.

Proof. We first observe, that each system net transition $\hat{t}$ has only a finite set of different variable bindings, since there are only finitely many values in $\mathcal{U}_k$ for each variable.

For each variable binding α we generate one copy of the system net transition $\hat{t}_\alpha$. We replace each arc inscription and each channel label by the evaluation of the original inscription.

Obviously, the resulting HORNET $G(EH)$ is grounded and has the same behaviour as the original one.   qed.

3   Sub-classes of Elementary Hornets

In the following we will show that the formalism of *elementary object systems* (EOS) of [2] is a special sub-class of Elementary Hornets.

We first recall the definition (EOS) and then characterise the sub-class of EOS-like Hornets.

3.1   Elementary Object Systems, EOS

An elementary object system (EOS) is composed of a system net, which is a p/t net $\hat{N} = (\hat{P}, \hat{T}, \mathbf{pre}, \mathbf{post})$ and a set of object nets $\mathcal{N} = \{N_1, \ldots, N_n\}$, which are p/t nets given as $N = (P_N, T_N, \mathbf{pre}_N, \mathbf{post}_N)$, where $N \in \mathcal{N}$. In extension we assume that all sets of nodes (places and transitions) are pairwise disjoint. Moreover we assume $\hat{N} \notin \mathcal{N}$ and the existence of the object net $\bullet \in \mathcal{N}$, which has no places and no transitions and is used to model anonymous, so called black tokens.

The system net places are typed by the mapping $d : \hat{P} \to \mathcal{N}$ with the meaning, that a place $\hat{p} \in \hat{P}$ of the system net with $d(\hat{p}) = N$ may contain only net-tokens of the object net type N.

Nested Markings. Since the tokens of an EOS are instances of object nets, a *marking* of an EOS is a *nested* multiset. A marking of an EOS OS is denoted $\mu = \sum_{k=1}^{|\mu|} (\hat{p}_k, M_k)$, where $\hat{p}_k$ is a place of the system net and M_k is the marking of a net-token with type $d(\hat{p}_k)$. To emphasise the nesting, markings are also denoted as $\mu = \sum_{k=1}^{|\mu|} \hat{p}_k[M_k]$. Markings of the form $\hat{p}[0]$ with $d(\hat{p}) = \bullet$ are abbreviated as $\hat{p}[].$[2]

The set of all markings which are syntactically consistent with the typing d is denoted $\mathcal{M}$, where $d^{-1}(N) \subseteq \hat{P}$ is the set of system net places of the type N:

$$\mathcal{M} := MS \left(\bigcup_{N \in \mathcal{N}} \left(d^{-1}(N) \times MS(P_N) \right) \right) \qquad (9)$$

---

[2] For EOS the net structure of a net-token is uniquely determined by the place. So, we do not include the object net N_i in μ for EOS.

We define the partial order $\sqsubseteq$ on nested multisets by setting $\mu_1 \sqsubseteq \mu_2$ iff $\exists \mu$: $\mu_2 = \mu_1 + \mu$.

Analogously to HORNETS, the events of an EOS are also nested and we have three different kinds of events system-autonomous, object-autonomous, and synchronisation events.

Each system net transition $\widehat{t}$ is labelled with a multiset of channels: $\widehat{l}(\widehat{t})(N) = ch_1 + \cdots + ch_n \in MS(\bigcup_{N \in \mathcal{N}} C_N)$, depicted as $\langle N{:}ch_1, N{:}ch_2, \ldots \rangle$. Similarily, an object net transition t may be labelled with a channel $l_N(t) \in C_N$ or with $\perp$, whenever t fires autonomously.

Definition 3 (EOS). *An elementary object system (EOS) is a tuple $OS = (\widehat{N}, \mathcal{N}, d, \Theta, \mu_0)$, where:*

1. *$\widehat{N}$ is a p/t net, called the* system net.
2. *$\mathcal{N}$ is a finite set of disjoint p/t nets, called* object nets.
3. *$d : \widehat{P} \to \mathcal{N}$ is the* typing *of the system net places.*
4. *$l = (\widehat{l}, l_N)_{N \in \mathcal{N}}$ is the* synchronisation labelling.
5. *$\mu_0 \in \mathcal{M}$ is the* initial marking.

The set of all events generated as in (7) as $\Theta_l := \Theta_1 \cup \Theta_2$, where Θ_1 contains the synchronisation events and – as a special sub-case – the system-autonomous events, while Θ_2 contains the object-autonomous events. The only difference is that evcent do not have variable bindings α, since there are no variables in the inscriptions, i.e. we have events in the form $\widehat{t}[\vartheta]$ instead of $\widehat{t}^\alpha[\vartheta]$.

3.2 Firing Rule

The projection Π^1 on the first component abstracts from the substructure of all net-tokens for a marking of an EOS:

$$\Pi^1 \left(\sum\nolimits_{k=1}^n \widehat{p}_k[M_k] \right) := \sum\nolimits_{k=1}^n \widehat{p}_k \tag{10}$$

The projection Π_N^2 on the second component is the sum of all net-tokens' markings M_k of the type $N \in \mathcal{N}$, ignoring their local distribution within the system net:

$$\Pi_N^2 \left(\sum\nolimits_{k=1}^n \widehat{p}_k[M_k] \right) := \sum\nolimits_{k=1}^n \mathbf{1}_N(\widehat{p}_k) \cdot M_k \tag{11}$$

where the indicator function $\mathbf{1}_N : \widehat{P} \to \{0, 1\}$ is $\mathbf{1}_N(\widehat{p}) = 1$ iff $d(\widehat{p}) = N$. Note that $\Pi_N^2(\mu)$ results in a marking of the object net N.

A system event $\widehat{\tau}[\vartheta]$ removes net-tokens together with their individual internal markings. Firing the event replaces a nested multiset $\lambda \in \mathcal{M}$ that is part of the current marking μ, i.e. $\lambda \sqsubseteq \mu$, by the nested multiset ρ. Therefore the successor marking is $\mu' := (\mu - \lambda) + \rho$. The enabling condition is expressed by the *enabling predicate* ϕ_{OS} (or just ϕ whenever OS is clear from the context):

$$\phi_{OS}(\widehat{\tau}[\vartheta], \lambda, \rho) \iff \Pi^1(\lambda) = \mathbf{pre}(\widehat{\tau}) \wedge \Pi^1(\rho) = \mathbf{post}(\widehat{\tau}) \wedge$$
$$\forall N \in \mathcal{N} : \Pi_N^2(\lambda) \geq \mathbf{pre}_N(\vartheta(N)) \wedge$$
$$\forall N \in \mathcal{N} : \Pi_N^2(\rho) = \Pi_N^2(\lambda) - \mathbf{pre}_N^\sharp(\vartheta(N)) + \mathbf{post}_N^\sharp(\vartheta(N)) \tag{12}$$

With $\widehat{M} := \Pi^1(\lambda)$ and $\widehat{M}' := \Pi^1(\rho)$ as well as $M_N := \Pi_N^2(\lambda)$ and $M_N' := \Pi_N^2(\rho)$ for all $N \in \mathcal{N}$ the predicate ϕ_{OS} has the following meaning:

1. The first conjunct expresses that the system net multiset $\widehat{M}$ corresponds to the pre-condition of the system net transition $\widehat{\tau}$, i.e. $\widehat{M} = \mathbf{pre}(\widehat{\tau})$.
2. In turn, a multiset $\widehat{M}'$ is produced, that corresponds to the post-set of $\widehat{\tau}$.
3. A multi-set $\vartheta(N)$ of object net transitions is enabled if the sum M_N of the net-tokens' markings (of type N) enable it, i.e. $M_N \geq \mathbf{pre}_N(\vartheta(N))$.
4. The firing of $\widehat{\tau}[\vartheta]$ must also obey the *object marking distribution condition*: $M_N' = M_N - \mathbf{pre}_N(\vartheta(N)) + \mathbf{post}_N(\vartheta(N))$, where $\mathbf{post}_N(\vartheta(N)) - \mathbf{pre}_N(\vartheta(N))$ is the effect of the object net's transitions on the net-tokens.

Note that conditions 1. and 2. assure that only net-tokens relevant for the firing are included in λ and ρ. Conditions 3. and 4. allow for additional tokens in the net-tokens.

Definition 4 (Firing Rule). *Let OS be an* Eos *and $\mu, \mu' \in \mathcal{M}$ markings. The event $\widehat{\tau}[\vartheta]$ is enabled in μ for the mode $(\lambda, \rho) \in \mathcal{M}^2$ iff $\lambda \sqsubseteq \mu \wedge \phi_{OS}(\widehat{\tau}[\vartheta], \lambda, \rho)$ holds.*

An event $\widehat{\tau}[\vartheta]$ that is enabled in μ for the mode (λ, ρ) can fire: $\mu \xrightarrow[OS]{\widehat{\tau}[\vartheta](\lambda, \rho)} \mu'$. The resulting successor marking is defined as $\mu' = \mu - \lambda + \rho$.

3.3 Eos-Like Elementary HORNETS

A special case arises whenever $|\mathcal{U}_k| = 1$ for all $k \in K$, i.e. we have exactly one object net N_k for each kind k. We can drop all arc inscriptions, since they are completely determined by the incident place. In this case no real transformation is possible since all expression evaluate to this special object net N_k.

Since this is exactly the behaviour that is defined by an Eos, we call an elementary HORNET with this property Eos-*like*.

For Eos-like HORNETS we have that the object net N_i in a net-token $\widehat{p}_i[N_i, M_i]$ is uniquely determined by the type $k = d(\widehat{p}_i)$ of the system net place $\widehat{p}_i$ as $N_i = N_k$. This object-net is denoted $N(k)$.

To translate the notation from HORNETS to Eos, we drop the object net in the notation of net-tokens (which can be done without loss of information for Eos-like Hornets) and write $\widehat{p}_i[M_i]$ instead of $\widehat{p}_i[N_i, M_i]$. Formally: π is the opration that "forgets" the object net N_i:

$$\pi([N_i, M_i]) := [M_i] \tag{13}$$

This notation extends to marked places $\widehat{p}_i[M_i]$ and multisets of marked places.

Analously, π forgets the variable assignment for events, i.e. we have:

$$\pi(\widehat{t}^\alpha[\vartheta]) := \widehat{t}[\vartheta] \tag{14}$$

Note, that for Eos-like HORNETS the mapping π has a unique inverse π^{-1}.

Each Eos is a special Eos-like HORNET.

Proposition 2. *For a given* EOS *OS we construct an* EOS-*like elementary* HORNET *EH(OS) with the property:*

$$\pi(\mu) \xrightarrow[OS]{\pi(w)} \pi(\mu') \iff \mu \xrightarrow[EH(OS)]{w} \mu'$$

Proof. Assume a given EOS: $OS = (\widehat{N}, \mathcal{N}, d, \Theta, \mu_0)$. We construct the corresponding elementary HORNET tuple $EH(OS) = (\widehat{N}_{EH}, \mathcal{U}, \mathcal{I}, k_{EH}, l_{EH}, \mu_{EH})$ as follows.

- We use the set of object nets as kinds: $K := \{k_N \mid N \in \mathcal{N}\}$ and define $\mathcal{U}_{k_N} := \{N\}$.
- We set the typing of the HORNET to $k_{EH}(\widehat{p}) := k_{d(\widehat{p})}$.
- Each arc from and to a system-net place $\widehat{p}$ is inscibed with the variable x_k of kind $k = k_{EH}(\widehat{p})$. with the multiplicity as given in the EOS:
  $\mathbf{pre}_{EH}(\widehat{t})(\widehat{p}) := (\mathbf{pre}(\widehat{t})(\widehat{p}) \cdot x_{k_{EH}(\widehat{p})})$
- The the guard predicate $\widehat{G}(\widehat{t})$ is true.
- Assume that the system net transition $\widehat{t}$ is labelled $\widehat{l}(\widehat{t}) = \sum_{i=1}^{n} ch_i$.
  We define the HORNET labelling as $\widehat{l}_{EH}^k(\widehat{t}) = \sum_{i=1}^{n}(x_{k(c_i)}, c_i)$ where $k(c) = k \iff c \in C_k$.
- The EOS marking $\mu_0 = \sum_{i=1}^{n} \widehat{p}_i[M_i]$ translates to: $\mu_{EH} = \sum_{i=1}^{n} \widehat{p}_i[N_i, M_i]$ for $N_i := d(\widehat{p}_i)$.

Obviously, the resulting HORNET is EOS-like.

Next, we show the equivalences for the enabling predicates ϕ_{EH} and ϕ_{OS} in (8) and (12):

- Note, that for a EOS-like HORNET we have $|\mathcal{U}_k| = 1$ and therefore we have for the first conjunct:

$$\begin{aligned}
&\forall k \in K : \forall \widehat{p} \in k^{-1}(k) : \forall N \in \mathcal{U}_k : \Pi_N^{1,H}(\lambda)(\widehat{p}) = \mathbf{pre}_\alpha(\widehat{\tau})(\widehat{p})(N) \\
\iff &\forall k \in K : \forall \widehat{p} \in k^{-1}(k) : \quad\quad \Pi_N^{1,H}(\lambda)(\widehat{p}) = \mathbf{pre}_\alpha(\widehat{\tau})(\widehat{p})(N(k)) \\
\iff &\forall k \in K : \forall \widehat{p} \in k^{-1}(k) : \quad\quad \Pi_{N(k)}^1(\pi(\lambda))(\widehat{p}) = \mathbf{pre}_{OS}(\widehat{\tau})(\widehat{p}) \\
\iff &\forall k \in K : \quad\quad\quad\quad\quad\quad\quad \Pi_{N(k)}^1(\pi(\lambda)) = \mathbf{pre}_{OS}(\widehat{\tau})_{|k^{-1}(k)} \\
\iff &\quad\quad\quad\quad\quad\quad \sum_{k \in K} \Pi_N^1(\pi(\lambda)) = \sum_{k \in K} \mathbf{pre}_{OS}(\widehat{\tau})_{|k^{-1}(k)} \\
\iff &\quad\quad\quad\quad\quad\quad\quad\quad \Pi^1(\pi(\lambda)) = \mathbf{pre}_{OS}(\widehat{\tau})
\end{aligned}$$

Analogously for the postset (second conjunct).
- Additionally, we have at the level of object-nets:

$$\Pi_k^{2,H}(\mu) = \sum_{N \in \mathcal{U}_k} \Pi_N^{2,H}(\mu) = \Pi_{N(k)}^{2,H}(\mu) = \Pi_{N(k)}^2(\pi(\mu))$$

This implies the equivalence for the third and fourth conjunct:

$$\Pi_k^{2,H}(\lambda) \geq \sum_{N \in \mathcal{U}_k} \mathbf{pre}_N^\sharp(\vartheta(N)) \iff \Pi_{N(k)}^2(\pi(\lambda)) \geq \mathbf{pre}_{OS}(\vartheta(N(k)))$$

– For the fourth conjunct we obtain:

$$\Pi_k^{2,H}(\rho) = \Pi_k^{2,H}(\lambda) + \sum_{N \in \mathcal{U}_k} \mathbf{post}_N^\sharp(\vartheta(N)) - \mathbf{pre}_N^\sharp(\vartheta(N))$$
$$\Longleftrightarrow \quad \Pi_k^{2,H}(\rho) = \Pi_k^{2,H}(\lambda) + \mathbf{post}_{N(k)}^\sharp(\vartheta(N(k))) - \mathbf{pre}_{N(k)}^\sharp(\vartheta(N(k)))$$
$$\Longleftrightarrow \quad \Pi_N^2(\pi(\rho)) = \Pi_N^2(\pi(\lambda)) - \mathbf{pre}_{OS}^\sharp(\vartheta(N(k))) + \mathbf{post}_{OS}^\sharp(\vartheta(N(k)))$$

This established the equivalence for the enabling predicates:

$$\phi_{EH}(\widehat{\mathcal{T}}^\alpha[\vartheta], \lambda, \rho) \quad \Longleftrightarrow \quad \phi_{OS}(\pi(\widehat{\mathcal{T}}[\vartheta]), \pi(\lambda), \pi(\rho))$$

Since the mapping π has a unique inverse π^{-1} for EOS-like HORNETS we obtain:

$$\phi_{OS}(\widehat{\mathcal{T}}[\vartheta]), \lambda, \rho) \quad \Longleftrightarrow \quad \phi_{EH}(\pi^{-1}(\widehat{\mathcal{T}}[\vartheta]), \pi^{-1}(\lambda), \pi^{-1}(\rho))$$

Since $\lambda \sqsubseteq \mu \iff \pi^{-1}(\lambda) \sqsubseteq \pi^{-1}(\mu)$ and $G(\widehat{t})$ holds, we have shown that the behaviour is isomorphic.   qed.

Conversely, EOS-like HORNETS create that sub-class that mimics EOS. Each EOS-like HORNET is a special EOS.

Proposition 3. *For a given* EOS-*like* HORNET *EH we construct an* EOS *OS(EH) with the property:*

$$\mu \xrightarrow[EH]{w} \mu' \quad \Longleftrightarrow \quad \pi(\mu) \xrightarrow[OS(EH)]{\pi(w)} \pi(\mu')$$

Proof. Assume that the EOS-like HORNET $EH = (\widehat{N}, \mathcal{U}, \mathcal{I}, k, l, \mu_0)$ is given. We know that $\mathcal{U}_k = \{N(k)\}$ for each $k \in K$.

The EOS $OS(EH) = (\widehat{N}_{OS}, \mathcal{N}_{OS}, d_{OS}, l_{OS}, \mu_{OS})$ that simulates EH is constructed the following way:

– The set of object nets $\mathcal{N}_{OS} := \{N(k) \mid k \in K\}$.
– We set the typing of the EOS to $d_{OS}(\widehat{p}) := N(k(\widehat{p}))$.
– In the system net we adapt the arc inscriptions. Whenever the HORNET's system net transition $\widehat{t}$ removes $\mathbf{pre}_\alpha(\widehat{t})(\widehat{p})(N)$ net-tokens of shape N, the EOS removes the same number of net-tokens.
  For a fixed variable assignment α we define:

$$\mathbf{pre}_{OS}(\widehat{t})(\widehat{p}) := \mathbf{pre}_\alpha(\widehat{t})(\widehat{p})(N(k))$$

Analogously for the postset.
    This notion is well-defined, since there is only "one" variable binding, i.e. the one that maps all variables in X_k to the unique object net N_k.
– Analogously for the synchronisation labels. For $\widehat{l}^k(\widehat{t}) = \sum_{i=1}^n (e_i, c_i)$ in the HORNET we define $\widehat{l}_{OS}(\widehat{t})(N) = \sum_{i=1}^n c_i$ in the EOS.

- We can assume that the the guard predicate $\widehat{G}(\widehat{t})$ is true for all bindings, since $\widehat{G}(\widehat{t})$ is false for some binding it is false for all bindings and the transition could be deleted.
- The Eos marking is $\mu_{OS} = \pi(\mu_0)$.

Analogously to the previous proposition, we obtain for the enabling predicates:
$$\phi_{EH}(\widehat{\tau}^\alpha[\vartheta], \lambda, \rho) \iff \phi_{OS}(\pi(\widehat{\tau}[\vartheta]), \pi(\lambda), \pi(\rho)).$$
Since $\lambda \sqsubseteq \mu \iff \pi(\lambda) \sqsubseteq \pi(\mu)$ and $G(\widehat{t})$ always holds, we have shown that the behaviour is isomorphic.   qed.

4   Eos-Simulation of Elementary HORNETS

In the following we show that each elementary HORNET can be simulated by an Eos-like HORNET (or as we have seen before: by an Eos), i.e. elementary HORNET are a conservative extension of Eos:

Theorem 1. *For each elementary* HORNET *EH there exists an* Eos *OS(EH) with the property:*
$$\mu \xrightarrow[EH]{*} \mu' \iff \overline{\mu} \xrightarrow[OS(EH)]{*} \overline{\mu'}$$

Proof. W.l.o.g. we can assume that the HORNET *EH* is grounded. The Eos
$$OS(EH) = (\widehat{N}_{OS}, \mathcal{N}_{OS}, d_{OS}, l_{OS}, \mu_{OS})$$
that simulates *EH* is defined as follows:

- We simulate all the object nets in $\mathcal{U}_k$ by one single object net $N_{OS,k}$ (cf. Fig. 2). This single object net $N_{OS,k}$ contains all the object net places P_k and the disjoint union of all transitions $T_{OS,k} := \biguplus_{N \in \mathcal{U}_k} T_N$, which is a finite set.

    Additionally, we have one place run_N for each object net $N \in \mathcal{U}_k$. We add the place run_N as the side condition to a transition t iff $t \in T_N$ and t fires object-autonomous, i.e. $l_N(t) = \perp_k$. The intuitive meaning is that $N_{OS,k}$ represents N iff run_N is marked. The simulation will guaratee that no two run places are marked simultaneously.

    For each place run_N we have a transition with the label enable_N to add a token on it and another transition with the label disable_N to remove a token from it. The set of all object nets is then $\mathcal{N}_{OS} = \{N_{OS,k} \mid k \in K\}$.

    For each net-token $[N, M]$ the corresponding Eos-marking is defined as:
$$\overline{[N, M]} := [M + \text{run}_N] \tag{15}$$

- Each system net transition $\widehat{t}$ is replaced by a subnet (cf. Figure 3). Each subnet has the places $\{\widehat{q}_i^{\widehat{p},N}, \text{begin}_i^{\widehat{p},N}, \widehat{r}_j^{\widehat{p},N}, \text{end}_j^{\widehat{p},N} \mid k \in K, \widehat{p} \in \widehat{P}_k, N \in \mathcal{U}_k, 1 \le i \le |\mathbf{pre}_\alpha(\widehat{t})(\widehat{p})(N)|, 1 \le j \le |\mathbf{post}_\alpha(\widehat{t})(\widehat{p'})(N)|\}$.
    We have the transitions $\{\text{start}, \text{finish}\} \cup \{t_N \mid k \in K, N \in \mathcal{U}_k\} \cup \{\text{off}_i^{\widehat{p},N}, \text{on}_j^{\widehat{p},N} \mid k \in K, \widehat{p} \in \widehat{P}_k N \in \mathcal{U}_k, 1 \le i \le |\mathbf{pre}_\alpha(\widehat{t})(\widehat{p})(N)|, 1 \le j \le |\mathbf{post}_\alpha(\widehat{t})(\widehat{p'})(N)|\}$. The flow relation is sketched in Figure 3.

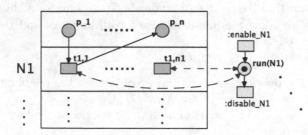

Fig. 2. Eos simulation of Hornets: The Object Net $N_{OS,k}$

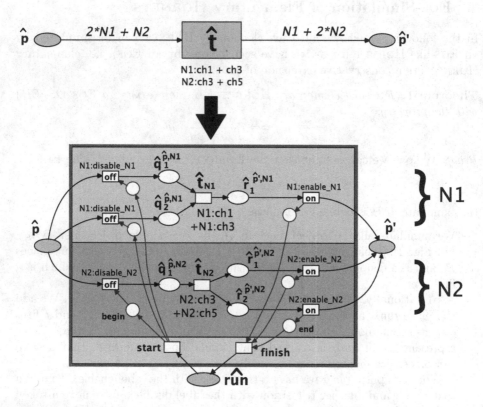

Fig. 3. Eos simulation of Hornets: The Translation of $\widehat{t}$ in the System Net $\widehat{N}_{OS}$

- The type is $d_{OS}(\widehat{p}) = N_{OS,k(\widehat{p})}$.
- The labelling is given as described above.
- A HORNET marking $\mu = \sum_{i=1}^{n} \widehat{p}_i[N_i, M_i]$ translates to $\overline{\mu}$:

$$\overline{\mu} := \widehat{\text{run}} + \sum_{i=1}^{n} \widehat{p}\overline{[N_i, M_i]} \tag{16}$$

The initial marking is $\mu_{OS} = \overline{\mu}_0$

We now show the simulation $\mu \xrightarrow[EH]{*} \mu' \iff \overline{\mu} \xrightarrow[OS(EH)]{*} \overline{\mu'}$.

This holds initially, since the initial marking is $\mu_{OS} = \overline{\mu}_0$.

It also holds for each firing step: (1) Assume that an event θ is enabled in the HORNET. If θ is an object-autonomous event for the object-net N, then this is enabled in $OS(EH)$, too, since run_N is marked.

Assume that θ is a synchronisation, which includes system-autonomous events as a special case. Then, θ is simulated by the EOS in the following way:

- First, the global run-place is cleared by start, which disables all other system-net transitions outside the subnet.
- Assume that $\widehat{t}$ removes $n := |\mathbf{pre}_\alpha(\widehat{t})(\widehat{p})(N)|$ net-tokens N from $\widehat{p}$. Then we have n transition $\mathrm{off}_i^{\widehat{p},N}$ for $1 \leq i \leq n$ (labelled with $\mathsf{disable}_N$) for each N and $\widehat{p}$ in the simulating subnet and each moves exactly one net-token from $\widehat{p}$ to one of the places $\widehat{q}_i^{\widehat{p},N}$. Note, that after firing each $\widehat{q}_i^{\widehat{p},N}$ is marked with exactly one net-token.
- As a result of the synchronisation via the channel $\mathsf{disable}_N$ all the run_N places in the transported net-tokens on $\widehat{q}_i^{\widehat{p},N}$ are now unmarked and no object-autonomous firing is possible anymore.
- For each N we join all the net-tokens of this type and synchronise with the combination with the transitions $\widehat{t}_N$. The synchronisation is possible since it is in the HORNET. Whenever $\widehat{t}$ generates $m := |\mathbf{post}_\alpha(\widehat{t})(\widehat{p}')(N)|$ tokens of the net N on $\widehat{p}'$, then we have the places $\widehat{r}_i^{\widehat{p}',N}$ for $1 \leq i \leq m$ and each transition $\mathrm{on}_i^{\widehat{p}',N}$ generates exactly one net-token on each $\widehat{r}_i^{\widehat{p}',N}$.
- In each generated net-token on $\widehat{r}_i^{\widehat{p}',N}$ we activate the run_N place via the channel enable_N. Here, we make use of the fact, that each $\widehat{r}_i^{\widehat{p}',N}$ contains exactly one net-token, so each net-token is enabled exactly once.
- The now "activated" net-tokens are then moved to the places $\widehat{p}'$ in the postset.
- Finally, we mark the global run place again (by finish) and other events can be simulated.

(2) Conversely, assume that an event θ is enabled in the simulating EOS. If θ is an object-autonomous event, then run_N is marked. Clearly, θ is enabled in the HORNET EH, too.

Assume that θ is a synchronisation. Once the subnet for $\widehat{t}$ is entered all other original events are disabled. If the global run-place ever gets marked again, than the generated marking corresponds to the marking that $\widehat{t}$ would have generated in the HORNET. However, it is possible that we enter the subnet and try to simulate an event that is not enabled in the HORNET. In this case the simulation blocks somewhere inside and the the global run-place never gets marked again. Therefore, we never get marking of the form $\overline{\mu}$ again.   qed.

Note that the construction above results in a quite drastic increase in the size of the EOS: Since we add new elements for each object net in the HORNET, by Lemma 1 the size of the EOS grows double exponentially in P_k.

But, the simulation is good for showing undecidability by giving a reduction from elementary HORNETS to EOS.

Corollary 1. *The reachability problem is undecidable for elementary* HORNETS.

Proof. Since we know that the reachability problem for EOS is undecidable [15], we obtain another undecidability proof for HORNETS by Theorem 1.   qed.

5   Boundedness for Safe, Elementary Hornets

A HORNET is safe iff each place $\widehat{p}$ in the system net carries at most one token and each net-token carries at most one token on each place p in all reachable markings μ. Since we are interested in safe HORNETS in the following, we do not use arc weights greater than 1.

Lemma 2. *A safe, elementary* HORNET *has a finite reachability set.*

Proof. Each reachable marking is of the form $\mu = \sum_{i=1}^{n} \widehat{p}_i[N_i, M_i]$. Since each each object net $N \in \mathcal{U}_k$ is safe, we know that each net-token has at most $2^{|P_k|}$ different markings M_i. Furthermore, we know $|\mathcal{U}_k| \leq 2^{\left(2^{4|P_k|}\right)}$. Assume that m is the maximum of all $|P_k|$. Then we have at most $2^{\left(2^{4m}\right)} \cdot 2^m$ different net-tokens $[N, M]$.

Each system net place $\widehat{p}$ is either unmarked or marked with one of these net-tokens. Therefore, we have at most $\left(1 + 2^{\left(2^{4m}\right)} \cdot 2^m\right)^{|\widehat{P}|}$ different markings in the safe Hornet.   qed.

In the following we study the complexity of the reachability problem for safe, elementary Hornets.

Theorem 2 ([12]). *The reachability problem for safe, elementary* HORNETS *requires exponential space.*

The proof we have given in [12] is similar to the heart of Lipton's famous result. Lipton has proven that acceptance of a counter program C translates into a reachability question for a p/t net $N(C)$. Lipton gives a construction for $N(C)$, which is of size $O(|C|^2)$. This proves that the reachability problem requires at least $2^{\sqrt{|C|}}$ space, i.e. exponential space.

In [12] we simulate a counter program C by a safe, elementary HORNET EH of size $O(poly(|C|))$. Thus we know that a question about a counter program C translates into a question about EH. When C needs $2^{|A|}$ space for the counters, we have that EH needs $2^{\sqrt[n]{|A|}}$ space for the corresponding question, where n is the order of the polynom $poly(|C|)$.

5.1   Simulation of Safe, Elementary Hornets

Since we have shown that the reachability problem Reach$_{seH}$ for the safe elementary HORNETS requires exponential space we have established a lower bound.

In the following we give a reduction of the reachability problem for safe, elementary HORNETS to that of safe EOS.

Lemma 3. *For each safe, elementary* HORNET *EH the simulating* EOS *$OS(EH)$ given in Theorem 1 is safe.*

Proof. In the simulating EOS $OS(EH)$ we have arranged the places $\widehat{p}_{N,i}$, where $i = 1..m$ on a circle. But since we consider a safe HORNET EH, we can assume $m = 1$, i.e. no net-tokens move along the circle.

The subnet of Figure 3, which replaces the system net transition $\widehat{t}$, uses a global run place (which is safe). The original system net places $\widehat{p}$ are still safe with these subnets. The same is true for the inner places that internal to the subnets.

The object net of Fig. 2 has only safe places, since the original places are still safe and the run places are never enabled more than once.   qed.

The reduction technique constructs the EOS $OS(G(EH))$ for the elemntary HORNET EH. We know that the reachability problem requires polynomial space for EOS calculated in the size of the EOS, where the size of an EOS is the sum of the sizes of the system-net and all object-nets. The natural question that arises now is: How big is the simulation gap? Or, equivalently: What is the size of $OS(G(EH))$ when compared to EH?

Theorem 3. *The reachability problem* Reach$_{seH}$ *for safe, elementary* HORNETS *can be solved with double-exponential space:* Reach$_{seH} \in 2^{O\left(2^{4m}\right)}$, *where m is the maximum of all $|P_k|$, i.e. $m := \max\{|P_k| : k \in K\}$.*

Proof. By Prop. 1 we can unfold each elementary HORNET into an equivalent grounded HORNET $G(EH)$. We obtain $G(EH)$ by replacing each system-net transition $\widehat{t}$ by a copy for each possible variable assignment. By Lemma 1 we have $|\mathcal{U}_k| \leq 2^{\left(2^{4m}\right)}$. Whenever the inscriptions of a system-net transition $\widehat{t}$ contains $|var(t)|$ variables, then we obtain $\left(2^{\left(2^{4m}\right)}\right)^{|var(t)|} = 2^{\left(|var(t)| \cdot 2^{4m}\right)}$ different copies in the grounded unfolding. We obtain that the size of the system net increases by the factor $2^{O\left(2^{4m}\right)}$, i.e. when compared to EH the grounded HORNET $G(EH)$ has a system-net with an double-exponential blow-up.

We then construct the EOS $OS(G(EH))$ for the grounded HORNET $G(EH)$.

The construction of Theorem 1 modifies the system net, i.e. each system net transition $\widehat{t}$ is replaced by the subnet of Figure 3. However, the size of the subnet contains the object nets from the arc inscriptions $\mathbf{pre}_\alpha(\widehat{t})(\widehat{p})(N)$ and $\mathbf{post}_\alpha(\widehat{t})(\widehat{p})(N)$. Therefore, the increase is bounded.

This is different for the object nets. All the object nets N in $\mathcal{U}_k$ are simulated by a single object net $N_{OS,k}$ in the EOS. The object net $N_{OS,k}$ contains all the subsets of the set of labelled transitions as transitions in $T_{OS,k}$ Since we have $2^{4|P_k|}$ labelled transitions, we have $|T_{OS,k}| \geq 2^{2^{4|P_k|}}$. Additionally, each object net in $\mathcal{U}_k$ has one run-place, i.e. $2^{\left(2^{4|P_k|}\right)}$ many, and two transitions for en-/disabling this run-place. Thus, the increase of the object-net is of size $2^{\left(c \cdot 2^{4|P_k|}\right)}$ for some constant c.

Compared to EH the simulating EOS $OS(G(EH))$ has an double-exponential blow-up for the system- and the object net.

Since we know that the reachability problem for safe EOS is PSpace-complete [10], we can decide the reachability problem for EH with a space complexity of $2^{\left(c \cdot 2^{4|P_k|}\right)} \simeq 2^{\left(Pol \cdot 2^{4|P_k|}\right)} \simeq 2^{O\left(2^{4|P_k|}\right)}$.   qed.

6   Conclusion

In this paper we studied the subclass of safe, elementary HORNETS. While all properties expressible in temporal logic are PSPACE-complete for safe EOS, we obtain that the reachability problem for safe, elementary HORNETS needs at least exponential space, which is also a lower bound for LTL model checking since the reachability problem can be expressed as a LTL property.

We like to emphasise that the power of safe, elementary HORNETS is only due to the transformations of the net-algebra, since all other sources of complexity (nesting, multiple tokens on a place, etc.) have been ruled out by the restriction of safeness and elementariness.

In this paper we have also given a reduction of the reachability problem for safe, elementary HORNETS to safe EOS. This reduction leads to the two following simulation-chains:

$$\text{EOS} \rightarrow \text{eHORNET} \rightarrow \text{grounded eHORNET}$$

and

$$\text{safe eHORNET} \rightarrow \text{safe, grounded eHORNET} \rightarrow \text{safe, EOS-like HORNET} \rightarrow \text{EOS}$$

However, we have seen that the lower and upper bounds are far from optimal, since we know that the reachability problem requires a least exponential space for safe, elemntary HORNETS, while the reduction technique leads to a *double* exponential space complexity.

Therefore, our current research focusses on the analysis of the exact complexity of the reachability problem without relying on a reduction to EOS.

References

1. Köhler, M., Rölke, H.: Concurrency for mobile object-net systems. Fundamenta Informaticae 54 (2003)
2. Köhler, M., Rölke, H.: Properties of Object Petri Nets. In: Cortadella, J., Reisig, W. (eds.) ICATPN 2004. LNCS, vol. 3099, pp. 278–297. Springer, Heidelberg (2004)
3. Köhler-Bußmeier, M., Heitmann, F.: On the expressiveness of communication channels for object nets. Fundamenta Informaticae 93, 205–219 (2009)
4. Valk, R.: Modelling concurrency by task/flow EN systems. In: 3rd Workshop on Concurrency and Compositionality, St. Augustin, Bonn, Gesellschaft für Mathematik und Datenverarbeitung. GMD-Studien, vol. 191 (1991)
5. Valk, R.: Object petri nets. In: Desel, J., Reisig, W., Rozenberg, G. (eds.) ACPN 2003, LNCS, vol. 3098, pp. 819–848. Springer, Heidelberg (2004)
6. Reisig, W.: Petri nets and algebraic specifications. Theoretical Computer Science 80, 1–34 (1991)

7. Hoffmann, K., Ehrig, H., Mossakowski, T.: High-level nets with nets and rules as tokens. In: Ciardo, G., Darondeau, P. (eds.) ICATPN 2005. LNCS, vol. 3536, pp. 268–288. Springer, Heidelberg (2005)
8. Lomazova, I.A.: Nested petri nets for adaptive process modeling. In: Avron, A., Dershowitz, N., Rabinovich, A. (eds.) Trakhtenbrot/Festschrift, LNCS, vol. 4800, pp. 460–474. Springer, Heidelberg (2008)
9. Köhler-Bußmeier, M.: Hornets: Nets within nets combined with net algebra. In: Franceschinis, G., Wolf, K. (eds.) PETRI NETS 2009. LNCS, vol. 5606, pp. 243–262. Springer, Heidelberg (2009)
10. Köhler-Bußmeier, M., Heitmann, F.: Safeness for object nets. Fundamenta Informaticae 101, 29–43 (2010)
11. Köhler-Bußmeier, M., Heitmann, F.: Liveness of safe object nets. Fundamenta Informaticae 112, 73–87 (2011)
12. Köhler-Bußmeier, M.: On the complexity of safe, elementary Hornets. In: Proceedings of the International Workshop on Concurrency, Specification, and Programming (CS&P 2012). CEUR Workshop Proceedings, vol. 928 (2012)
13. Lipton, R.J.: The reachability problem requires exponential space. Research Report 62, Dept. of Computer science (1976)
14. Ehrig, H., Mahr, B.: Fundamentals of algebraic Specification. In: EATCS Monographs on TCS. Springer (1985)
15. Köhler, M.: Reachable markings of object Petri nets. Fundamenta Informaticae 79, 401–413 (2007)

Complexity Analysis of Continuous Petri Nets*

Estíbaliz Fraca[1] and Serge Haddad[2]

[1] Instituto de Investigación en Ingeniería de Aragón (I3A)
Universidad de Zaragoza, Zaragoza, Spain
efraca@unizar.es
[2] Ecole Normale Supérieure de Cachan,
LSV, CNRS UMR 8643, INRIA, Cachan, France
haddad@lsv.ens-cachan.fr

Abstract. At the end of the eighties, continuous Petri nets were introduced for: (1) alleviating the combinatory explosion triggered by discrete Petri nets and, (2) modelling the behaviour of physical systems whose state is composed of continuous variables. Since then several works have established that the computational complexity of deciding some standard behavioural properties of Petri nets is reduced in this framework. Here we first establish the decidability of additional properties like boundedness and reachability set inclusion. We also design new decision procedures for the reachability and lim-reachability problems with a better computational complexity. Finally we provide lower bounds characterising the exact complexity class of the boundedness, the reachability, the deadlock freeness and the liveness problems.

1 Introduction

From Petri Nets to Continuous Petri Nets. Continuous Petri nets (CPN) were introduced in [5] by considering continuous states (specified by a non negative real number of tokens in places) where the dynamics of the system is triggered either by discrete events or by a continuous evolution ruled by speed of firings. In the former case such nets are called autonomous CPNs while in the latter they are called timed CPNs. In both cases, the evolution is due to a *fractional* transition firing (infinitesimal and simultaneous in the case of timed CPNs).

Modelling with CPNs. CPNs have been used in several significant application fields. In [3], a method based on CPNs is proposed for the fault diagnosis of manufacturing systems that manage systems intractable with discrete Petri nets (for modelling of manufacturing systems see also [17]). In [15], the authors introduce a bottom-up modelling methodology based on CPNs to represent cell metabolism and solve in this framework the regulation control problem. Combining discrete and continuous Petri nets yields hybrid Petri nets with applications

---

* This work has been partially supported by CICYT - FEDER project DPI2010-20413, by a pre-doctoral grant of the Gobierno de Aragón and by Fundación Aragón I+D and by ImpRo (ANR-2010-BLAN-0317).

J.-M. Colom and J. Desel (Eds.): PETRI NETS 2013, LNCS 7927, pp. 170–189, 2013.

to modelling and simulation of water distribution systems [9] and to the analysis of traffic in urban networks [16].

Analysis of CPNs. While several analysis methods have been developed for timed CPNs there is no hope for fully automatic techniques in the general case since standard problems of dynamic systems are known to be undecidable even for bounded nets [13].

Due to the semantics of autonomous CPNs, a marking can be the limit of the markings visited along an infinite firing sequence. Thus most of the usual properties are duplicated depending on whether these markings are considered or not. When considering these markings, reachability (resp. liveness, deadlock-freeness) becomes lim-reachability (resp. lim-liveness, lim-deadlock-freeness).

Contrary to the timed case, the analysis of autonomous CPNs (that we simply call CPNs in the sequel) appears to be less complex than the one of discrete Petri nets. In [10], exponential time decision procedures are proposed for the reachability and lim-reachability problems for general CPNs. In [14] assuming additional hypotheses on the net, the authors design polynomial time decision procedures for (lim-)reachability and boundedness. In [13], (lim-)deadlock-freeness and (lim-)liveness are shown to belong in coNP. These procedures are based on "simple" characterisations of the properties.

Our Contributions. First we revisit characterisations of properties in CPN establishing an alternative characterisation for reachability and the first characterisation for boundedness. Then based on these characterisations, we show that (lim-)reachability and boundedness are decidable in polynomial time. We also establish that the (lim-)reachability set inclusion problem is decidable in exponential time. Finally we prove that (lim-)reachability and boundedness are PTIME-hard and that (lim-)deadlock-freeness, (lim-)liveness and (lim-)reachability set inclusion problems are coNP-hard. We establish these lower bounds even when considering restricted cases of these problems.

Organisation. In Section 2, we introduce CPNs and the properties that we are analysing. In Section 3, we develop the characterisations of reachability and boundedness. Afterwards in Section 4, we design the decision procedures. Then, we provide complexity lower bounds in Section 5. Finally in Section 6, we summarise our results and give perspectives to this work. All missing proofs can be found in [8].

2   Continuous Petri Nets: Definitions and Properties

2.1   Continuous Petri Nets

Notations. $\mathbb{N}$ (resp. $\mathbb{Q}$, $\mathbb{R}$) is the set of non negative integers (resp. rational, real numbers). Given a set of numbers E, $E_{\geq 0}$ (resp. $E_{>0}$) denotes the subset of non negative (resp. positive) numbers of E. Given an $E \times F$ matrix $\mathbf{M}$ with E and F sets of indices, $E' \subseteq E$ and $F' \subseteq F$, the $E' \times F'$ submatrix $\mathbf{M}_{E' \times F'}$ denotes the restriction of $\mathbf{M}$ to rows indexed by E' and columns indexed by F'. The support of a vector $\mathbf{v} \in \mathbb{R}^E$, denoted $[\![\mathbf{v}]\!]$, is defined by $[\![\mathbf{v}]\!] \overset{\text{def}}{=} \{e \in E \mid \mathbf{v}[e] \neq 0\}$. $\mathbf{0}$

denotes the null vector. One writes $\mathbf{v} \geq \mathbf{w}$ when $\mathbf{v}$ is componentwise greater or equal than $\mathbf{w}$ and $\mathbf{v} \gneq \mathbf{w}$ when $\mathbf{v} \geq \mathbf{w}$ and $\mathbf{v} \neq \mathbf{w}$. One writes $\mathbf{v} > \mathbf{w}$ when $\mathbf{v}$ is componentwise strictly greater than $\mathbf{w}$. $\|\mathbf{v}\|_1$ is the 1-norm of $\mathbf{v}$ defined by $\|\mathbf{v}\|_1 \overset{\text{def}}{=} \sum_{e \in E} |\mathbf{v}[e]|$. Let $E' \subseteq E$, then $\mathbf{v}[E']$ denotes the restriction of $\mathbf{v}$ to components of E'.

Here, we adopt the following terminology: a *net* denotes the structure without initial marking while a *net system* denotes a net with an initial marking. The structure of CPNs and discrete nets are identical.

Definition 1. *A Petri net (PN) is a tuple $\mathcal{N} = \langle P, T, \mathbf{Pre}, \mathbf{Post} \rangle$ where:*

- *P is a finite set of places;*
- *T is a finite set of transitions, with $P \cap T = \emptyset$;*
- *$\mathbf{Pre}$ (resp. $\mathbf{Post}$), is the backward (resp. forward) $P \times T$ incidence matrix, whose items belong to $\mathbb{N}$.*

The incidence matrix $\mathbf{C}$ is defined by $\mathbf{C} \overset{\text{def}}{=} \mathbf{Post} - \mathbf{Pre}$.

Given a place (resp. transition) v in P (resp. in T), its *preset*, ${}^\bullet v$, is defined as the set of its input transitions (resp. places): ${}^\bullet v \overset{\text{def}}{=} \{t \in T \mid \mathbf{Post}[v,t] > 0\}$ (resp. ${}^\bullet v \overset{\text{def}}{=} \{p \in P \mid \mathbf{Pre}[p,v] > 0\}$). Its *postset* $v^\bullet$ is defined as the set of its output transitions (resp. places): $v^\bullet \overset{\text{def}}{=} \{t \in T \mid \mathbf{Pre}[v,t] > 0\}$ (resp. $v^\bullet \overset{\text{def}}{=} \{p \in P \mid \mathbf{Post}[p,v] > 0\}$). This notion generalizes to a subset V of places (resp. transitions) by: ${}^\bullet V \overset{\text{def}}{=} \bigcup_{v \in V} {}^\bullet v$ and $V^\bullet \overset{\text{def}}{=} \bigcup_{v \in V} v^\bullet$. In addition, ${}^\bullet V^\bullet \overset{\text{def}}{=} {}^\bullet V \cup V^\bullet$.

Given $T' \subseteq T$, $\mathcal{N}_{T'}$ is the subnet of $\mathcal{N}$ such that its set of transitions is T' and its set of places is ${}^\bullet T'^\bullet$, and its backward and forward incidence matrices are respectively $\mathbf{Pre}_{{}^\bullet T'^\bullet \times T'}$ and $\mathbf{Post}_{{}^\bullet T'^\bullet \times T'}$.

We define $\mathcal{N}^{-1}$ as the "reverse" net of $\mathcal{N}$, in which the places and transitions coincide, and its arcs are inverted.

Definition 2. *Given a PN $\mathcal{N} = \langle P, T, \mathbf{Pre}, \mathbf{Post} \rangle$, its reverse net $\mathcal{N}^{-1}$ is defined by $\mathcal{N}^{-1} \overset{\text{def}}{=} \langle P, T, \mathbf{Post}, \mathbf{Pre} \rangle$.*

A *continuous* PN system consists of a net and a non negative real marking.

Definition 3. *A CPN system is a tuple $\langle \mathcal{N}, \mathbf{m}_0 \rangle$ where $\mathcal{N}$ is a PN and $\mathbf{m}_0 \in \mathbb{R}_{\geq 0}^P$ is the initial marking.*

When a CPN system is an input of a decision problem, the items of $\mathbf{m}_0$ are rational numbers in order to characterise the complexity of the problem.

In discrete PNs the firing rule of a transition requires tokens specified by $\mathbf{Pre}$ to be present in the corresponding places. In continuous PNs a non negative real *amount* of transition firing is allowed and this amount scales the requirement expressed by $\mathbf{Pre}$ and $\mathbf{Post}$.

Definition 4. *Let $\mathcal{N}$ be a CPN, t be a transition and $\mathbf{m} \in \mathbb{R}_{\geq 0}^P$ be a marking.*

- *The* enabling degree *of t w.r.t. m, $enab(t, m) \in \mathbb{R}_{\geq 0} \cup \infty$, is defined by:*
  $enab(t, m) \overset{def}{=} \min\{\frac{m[p]}{Pre[p,t]} \mid p \in {}^\bullet t\}$ *($enab(t, m) = \infty$ iff ${}^\bullet t = \emptyset$).*
- *t is enabled in m if $enab(t, m) > 0$.*
- *t can be fired by any amount $\alpha \in \mathbb{R}$ such that[1] $0 \leq \alpha \leq enab(t, m)$, and its firing leads to marking m' defined by: for all $p \in P$, $m'[p] = m[p] + \alpha C[p, t]$.*

The firing of t from m by an amount α leading to m' is denoted as $m \overset{\alpha t}{\longrightarrow} m'$. We illustrate the firing rule of a CPN with the system in Fig. 1(a) (example taken from [10]). In the initial marking $m_0 = (1, 0, 1, 0)$, only transition t_1 is enabled and its enabling degree is 1. Hence, it can be fired by any real amount α s.t. $0 \leq \alpha \leq 1$. If t_1 is fired by an amount of 0.5, marking $m_1 = (0.5, 0.5, 1, 0)$ is reached. In m_1, transitions t_1 and t_2 are enabled, with enabling degree both equal to 0.5.

Let $\sigma = \alpha_1 t_1 \ldots \alpha_n t_n$ be a finite sequence with for all i, $t_i \in T$ and $\alpha_i \in \mathbb{R}_{\geq 0}$. σ is firable from m_0 if for all $1 \leq i \leq n$ there exist m_i such that $m_{i-1} \overset{\alpha_i t_i}{\longrightarrow} m_i$. This firing is denoted by $m_0 \overset{\sigma}{\longrightarrow} m_n$. When the destination marking is irrelevant we omit it and simply write $m_0 \overset{\sigma}{\longrightarrow}$. Let $\sigma = \alpha_1 t_1 \ldots \alpha_n t_n \ldots$ be an infinite sequence then σ is firable from m_0 if for all n, $\alpha_1 t_1 \ldots \alpha_n t_n$ is firable from m_0. This firing is denoted as $m_0 \overset{\sigma}{\longrightarrow}_\infty$.

Given a finite or infinite sequence $\sigma = \alpha_1 t_1 \ldots \alpha_i t_i \ldots$ and $\alpha \in \mathbb{R}_{\geq 0}$, the sequence $\alpha \sigma$ is defined by $\sigma \overset{def}{=} \alpha \alpha_1 t_1 \ldots \alpha \alpha_i t_i \ldots$. Given two infinite sequences $\sigma = \alpha_1 t_1 \ldots \alpha_i t_i \ldots$ and $\sigma' = \alpha'_1 t'_1 \ldots \alpha'_i t'_i \ldots$, the (non commutative) sum $\sigma + \sigma'$ is defined by: $\sigma + \sigma' \overset{def}{=} \alpha_1 t_1 \alpha'_1 t'_1 \ldots \alpha_i t_i \alpha'_i t'_i \ldots$. This notion generalises to arbitrary sequences by extending them to infinite sequences with null amounts of firings (the selected transitions are irrelevant).

Let $\sigma = \alpha_1 t_1 \ldots \alpha_n t_n$ be a finite sequence and denote $\sigma^{-1} = \alpha_n t_n \ldots \alpha_1 t_1$. By definition of the reverse net, $m \overset{\sigma}{\longrightarrow} m'$ in $\mathcal{N}$ iff $m' \overset{\sigma^{-1}}{\longrightarrow} m$ in $\mathcal{N}^{-1}$.

The Parikh image (also called firing count vector) of a (finite or infinite) firing sequence $\sigma = \alpha_1 t_1 \ldots \alpha_n t_n \ldots$ denoted $\overrightarrow{\sigma} \in (\mathbb{R}_{\geq 0} \cup \{\infty\})^T$ is defined by: $\overrightarrow{\sigma}[t] \overset{def}{=} \sum_{i|t_i=t} \alpha_i$. As in discrete PNs, when $m \overset{\sigma}{\longrightarrow} m'$, $m' = m + C\overrightarrow{\sigma}$ and this equation is called the *state equation*.

A set of places P' is a *siphon* if ${}^\bullet P' \subseteq P'^\bullet$. When a siphon does not contain tokens in some marking, it will never contain tokens after any firing sequence starting from this marking. One call it an *empty siphon*.

An interesting difference between discrete and continuous PN systems is that the sequence of markings visited by an infinite firing sequence may converge to a given marking. For example, let us consider again the CPN of Fig. 1(a), and the marking $m_1 = (0.5, 0.5, 1, 0)$. From m_1, $0.5t_2$ can be fired, reaching $m_2 = (0.5, 0.5, 0, 0.5)$. From m_2 transition t_3 can be fired by an amount of 0.5, leading to $m_3 = (0.5, 0.5, 0.5, 0)$. Iterating this process leads to the infinite firing sequence $\sigma = 2^{-1} t_2 2^{-1} t_3 \ldots 2^{-n} t_2 2^{-n} t_3 \ldots$ whose visited markings converge

---

[1] So from every marking, any (even disabled) transition can fire by a null amount without modifying the marking.

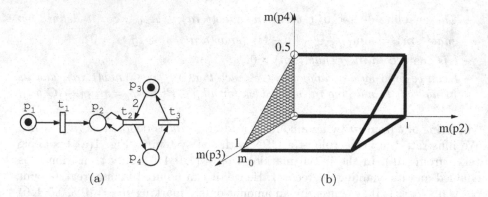

Fig. 1. (a) A CPN system (b) its lim-reachability set [10]

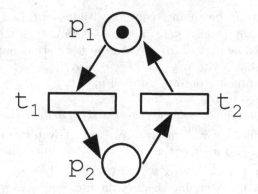

Fig. 2. A simple CPN system

toward $(0.5, 0.5, 0, 0)$. Observe that the Parikh image $\vec{\sigma} = \vec{t_2} + \vec{t_3}$ does not correspond to any finite firing sequence starting from m_1.

Consider now the PN in Fig. 2 with initial marking $m_0 = (1, 0)$. Let $\sigma = 1t_1 \frac{1}{2} t_2 \frac{1}{3} t_1 \frac{1}{4} t_2 \cdots \frac{1}{2i-1} t_1 \frac{1}{2i} t_2 \ldots$ The sequence σ is infinite and its sequence of visited markings converges toward marking m defined by: $m \overset{\text{def}}{=} (1 - \log(2), \log(2))$. Here $\vec{\sigma} = \infty \vec{t_1} + \infty \vec{t_2}$.

Let σ be an infinite firing sequence starting from m whose sequence of visited markings converges toward m', one says that m' is *limit reachable* from m which is denoted by: $m \overset{\sigma}{\longrightarrow}_\infty m'$. Thus in CPNs, two sets of reachable markings are defined.

Definition 5. *Given a CPN system* $\langle \mathcal{N}, m_0 \rangle$,

– *Its reachability set* $\text{RS}(\mathcal{N}, m_0)$ *is defined by:*
  $\text{RS}(\mathcal{N}, m_0) \overset{\text{def}}{=} \{ m \mid \text{there exists a finite sequence } m_0 \overset{\sigma}{\longrightarrow} m \}$.
– *Its lim-reachability set,* $\text{lim}-\text{RS}(\mathcal{N}, m_0)$, *is defined by:*
  $\text{lim}-\text{RS}(\mathcal{N}, m_0) \overset{\text{def}}{=} \{ m \mid \text{there exists an infinite sequence } m_0 \overset{\sigma}{\longrightarrow}_\infty m \}$.

RS or lim−RS are convex sets (see Section 3) but not necessarily topologically closed. In Fig. 1, marking $m = (1, 0, 0, 0)$ belongs to the closure of RS or lim−RS, but it does not belong to these sets. Since an infinite sequence can include null amounts of firings, $\mathrm{RS}(\mathcal{N}, m_0) \subseteq \mathrm{lim-RS}(\mathcal{N}, m_0)$. More interestingly, for all $m \in \mathrm{lim-RS}(\mathcal{N}, m_0)$, $\mathrm{lim-RS}(\mathcal{N}, m) \subseteq \mathrm{lim-RS}(\mathcal{N}, m_0)$ (see the proof in appendix of [8]). So there is no need to consider iterations of lim-reachability.

2.2   CPN Properties

Here we introduce the standard properties that a modeller wants to check on a net. In the framework of CPNs, every property its defined either w.r.t. to the reachability set or w.r.t. to the lim-reachability set.

Reachability is the main property as it is the core of safeness properties.

Definition 6 (reachability). *Given a system $\langle \mathcal{N}, m_0 \rangle$ and a marking m, m is (lim-)reachable in $\langle \mathcal{N}, m_0 \rangle$ if $m \in (\mathrm{lim-})\mathrm{RS}(\mathcal{N}, m_0)$.*

Boundedness is often related to the resources needed by the system. For CPN, boundedness and lim-boundedness coincide [14].

Definition 7 (boundedness). *A system $\langle \mathcal{N}, m_0 \rangle$ is (lim-)bounded if there exists $b \in \mathbb{R}_{\geq 0}$ such that for all $m \in (\mathrm{lim-})\mathrm{RS}(\mathcal{N}, m_0)$ and all $p \in P$, $m[p] \leq b$.*

Deadlock-freeness ensures that a system will never reach a marking where no transition is enabled, i.e a *dead marking*.

Definition 8 (deadlock-freeness). *A system $\langle \mathcal{N}, m_0 \rangle$ is (lim-)deadlock-free if for all $m \in (\mathrm{lim-})\mathrm{RS}(\mathcal{N}, m_0)$, there exists $t \in T$ such that t is enabled at m.*

The net of Fig. 1 is deadlock-free but not lim-deadlock-free: $m \overset{\mathrm{def}}{=} (0, 1, 0, 0)$ is a *dead* marking which is limit-reachable but not reachable and no reachable marking is dead.

Liveness ensures that whatever the reachable state, any transition will be fireable in some future. So the system never "looses its capacities".

Definition 9 (liveness). *A system $\langle \mathcal{N}, m_0 \rangle$ is (lim-)live if for all transition t and for all marking $m \in (\mathrm{lim-})\mathrm{RS}(\mathcal{N}, m_0)$ there exists $m' \in (\mathrm{lim-})\mathrm{RS}(\mathcal{N}, m)$ such that t is enabled at m'.*

The net of Fig. 1 is neither live nor lim-live: once t_1 becomes disabled, it will remain so whatever the finite or infinite firing sequence considered.

A home state is a marking that can be reached whatever the current state. This property can express for instance that recovering from faults is always possible. A net is *reversible* if its initial marking is an home state. Both properties are particular cases of the reachability set inclusion problem.

Definition 10 (reachability set inclusion).
Given systems $\langle \mathcal{N}, m_0 \rangle$ and $\langle \mathcal{N}', m_0' \rangle$ with $P = P'$, $\langle \mathcal{N}, m_0 \rangle$ is (lim-)reachable included in $\langle \mathcal{N}', m_0' \rangle$ if $(\mathrm{lim-})\mathrm{RS}(\mathcal{N}, m_0) \subseteq (\mathrm{lim-})\mathrm{RS}(\mathcal{N}', m_0')$.
A marking m is a home state if $\mathrm{RS}(\mathcal{N}, m_0) \subseteq \mathrm{RS}(\mathcal{N}^{-1}, m)$.
When $m = m_0$, one says that $\langle \mathcal{N}, m_0 \rangle$ is reversible.

The following table summarises the results already known about the complexity of the associated decision problems. A net is *consistent* if there exists a vector $\mathbf{v} \in \mathbb{R}_{\geq 0}$ with $[\![v]\!] = T$ and $C\mathbf{v} = 0$. No lower bounds have been established.

Table 1. Complexity bounds: previous results

Problems	Upper bounds
(lim-)reachability	in EXPTIME [10] in PTIME for lim-reachability when all transitions are fireable at least once and the net is consistent [14]
(lim-)boundedness	in PTIME when all transitions are fireable at least once [14]
(lim-)deadlock-freeness	in coNP [13]
(lim-)liveness	in coNP [13]
(lim-)reachability set inclusion	no result

3   Properties Characterisations

3.1   Preliminary Results about Reachability and Firing Sequences

Most of the results of this subsection are generalisations of results given in [14,10].

The following lemma is an almost immediate consequence of firing definition and has for corollary the convexity of the (lim-)reachability set. In this lemma depending on the sequences $\longrightarrow_{(\infty)}$ denotes either $\longrightarrow$ or $\longrightarrow_\infty$.

Lemma 11. *Given a CPN system $\langle \mathcal{N}, \mathbf{m}_0 \rangle$, (finite or infinite) sequences $\sigma, \sigma_1, \sigma_2$ markings $\mathbf{m}, \mathbf{m}', \mathbf{m}_1, \mathbf{m}_2, \mathbf{m}'_1, \mathbf{m}'_2$ and $\alpha, \alpha_1, \alpha_2 \in \mathbb{R}_{>0}$:*

(0) $\mathbf{m}_1 \xrightarrow{\sigma} \mathbf{m}'_1$ and $\mathbf{m}_1 \leq \mathbf{m}_2$ implies $\mathbf{m}_2 \xrightarrow{\sigma} \mathbf{m}'_2$ with $\mathbf{m}'_1 \leq \mathbf{m}'_2$

(1) $\mathbf{m} \xrightarrow{\sigma}_{(\infty)} \mathbf{m}$ iff $\alpha \mathbf{m} \xrightarrow{\alpha\sigma}_{(\infty)} \alpha \mathbf{m}'$

(2) $\mathbf{m} \xrightarrow{\sigma}_\infty$ iff $\alpha \mathbf{m} \xrightarrow{\alpha\sigma}_\infty$

(3) $\mathbf{m}_1 \xrightarrow{\sigma_1}_{(\infty)} \mathbf{m}'_1$ and $\mathbf{m}_2 \xrightarrow{\sigma_2}_{(\infty)} \mathbf{m}'_2$ implies $\mathbf{m}_1 + \mathbf{m}_2 \xrightarrow{\sigma_1+\sigma_2}_{(\infty)} \mathbf{m}'_1 + \mathbf{m}'_2$

(4) $\mathbf{m}_1 \xrightarrow{\sigma_1}_\infty$ and $\mathbf{m}_2 \xrightarrow{\sigma_2}_\infty$ implies $\mathbf{m}_1 + \mathbf{m}_2 \xrightarrow{\sigma_1+\sigma_2}_\infty$

(5) $\mathbf{m}_1 \xrightarrow{\alpha_1\sigma}_{(\infty)} \mathbf{m}'_1$ and $\mathbf{m}_2 \xrightarrow{\alpha_2\sigma}_{(\infty)} \mathbf{m}'_2$ implies $\mathbf{m}_1 + \mathbf{m}_2 \xrightarrow{(\alpha_1+\alpha_2)\sigma}_{(\infty)} \mathbf{m}'_1 + \mathbf{m}'_2$

(6) $\mathbf{m}_1 \xrightarrow{\alpha_1\sigma}_\infty$ and $\mathbf{m}_2 \xrightarrow{\alpha_2\sigma}_\infty$ implies $\mathbf{m}_1 + \mathbf{m}_2 \xrightarrow{(\alpha_1+\alpha_2)\sigma}_\infty$

The two next lemmas constitute a first step for the characterisation of reachability since they provide sufficient conditions for reachability and lim-reachability in particular cases.

Lemma 12. *Let $\langle \mathcal{N}, \mathbf{m}_0 \rangle$ be a continuous system, $\mathbf{m}$ be a marking and $\mathbf{v} \in \mathbb{R}_{\geq 0}^T$ that fulfill:*

- $m = m_0 + C\mathbf{v}$;
- $\forall p \in {}^\bullet[\![\mathbf{v}]\!]\ m_0[p] > 0$;
- $\forall p \in [\![\mathbf{v}]\!]^\bullet\ m[p] > 0$.

Then there exists a finite sequence σ such that $m_0 \overset{\sigma}{\longrightarrow} m$ and $\vec{\sigma} = \mathbf{v}$.

Proof. Define $\alpha_1 \overset{\text{def}}{=} \min(\frac{m_0[p]}{\sum_{t \in [\![v]\!]} Pre[p,t]\mathbf{v}[t]} \mid p \in {}^\bullet[\![\mathbf{v}]\!])$

and $\alpha_2 \overset{\text{def}}{=} \min(\frac{m[p]}{\sum_{t \in [\![v]\!]} Post[p,t]\mathbf{v}[t]} \mid p \in [\![\mathbf{v}]\!]^\bullet)$ with the convention that $\alpha_1 \overset{\text{def}}{=} 1$

(resp. $\alpha_2 \overset{\text{def}}{=} 1$) if ${}^\bullet[\![\mathbf{v}]\!]$ (resp. $[\![\mathbf{v}]\!]^\bullet$) is empty.

Due to the second and the third hypotheses α_1 and α_2 are positive.

Let $n \overset{\text{def}}{=} \max(\lceil \frac{1}{\min(\alpha_1, \alpha_2)} \rceil, 2)$.

Denote $[\![v]\!] \overset{\text{def}}{=} \{t_1, \ldots, t_k\}$ and define $\sigma' \overset{\text{def}}{=} \frac{\mathbf{v}[t_1]}{n} t_1 \ldots \frac{\mathbf{v}[t_k]}{n} t_k$ and $\sigma \overset{\text{def}}{=} \sigma'^n$.

We claim that σ is the required firing sequence.

Let us denote $m_i \overset{\text{def}}{=} m_0 + \frac{i}{n} C\mathbf{v}$. Thus $m = m_n$.

By definition of α_1 and n, in $\mathcal{N}\ m_0 \overset{\sigma'}{\longrightarrow} m_1$ and by definition of α_2, $m_n \overset{\sigma'^{-1}}{\longrightarrow}$

m_{n-1} in $\mathcal{N}^{-1}$. So in $\mathcal{N}\ m_{n-1} \overset{\sigma'}{\longrightarrow} m_n$.

Let $1 < i < n - 1$.

Using lemma 11, $\frac{n-1-i}{n-1} m_0 \overset{\frac{n-1-i}{n-1}\sigma'}{\longrightarrow} \frac{n-1-i}{n-1} m_1$ and $\frac{i}{n-1} m_{n-1} \overset{\frac{i}{n-1}\sigma'}{\longrightarrow} \frac{i}{n-1} m_n$.

Using lemma 11 again and summing, one gets: $m - m_i \overset{\sigma'}{\longrightarrow} m_{i+1}$. ∎

Lemma 13. *Let $\langle \mathcal{N}, m_0 \rangle$ be a continuous system, m be a marking and $\mathbf{v} \in \mathbb{R}_{\geq 0}^T$ that fulfill:*

- $m = m_0 + C\mathbf{v}$;
- $\forall p \in {}^\bullet[\![\mathbf{v}]\!]^\bullet\ m_0[p] > 0$.

Then there exists an infinite sequence σ such that $m_0 \overset{\sigma}{\longrightarrow}_\infty m$ and $\vec{\sigma} = \mathbf{v}$.

Proof. Let m_i be inductively defined by $m_{i+1} = \frac{1}{2} m_i + \frac{1}{2} m$. and for $i \geq 1$, let $\mathbf{v}_i = \frac{1}{2^i} \mathbf{v}$ (thus $[\![\mathbf{v}_i]\!] = [\![\mathbf{v}]\!]$). Observe that $m_i = \frac{1}{2^i} m_0 + (1 - \frac{1}{2^i}) m$. So:

- $m_{i+1} = m_i + C\mathbf{v}_i$;
- $\forall p \in {}^\bullet[\![\mathbf{v}_i]\!]^\bullet\ m_i[p] > 0$ and $m_{i+1}[p] > 0$.

Applying lemma 12, for all $i \geq 1$ there exists σ_i such that $m_i \overset{\sigma_i}{\longrightarrow} m_{i+1}$. Since $\lim_{i \to \infty} m_i = m$, the sequence $\sigma = \sigma_1 \sigma_2 \ldots$ is the required sequence. ∎

The key concept in order to get characterisation of properties, is the notion of *firing set* of a CPN system [10].

Definition 14. *Let $\langle \mathcal{N}, m_0 \rangle$ be a CPN system. Then its firing set $FS(\mathcal{N}, m_0) \subseteq 2^T$ is defined by:*

$$FS(\mathcal{N}, m_0) = \{[\![\vec{\sigma}]\!] \mid m_0 \overset{\sigma}{\longrightarrow}\}$$

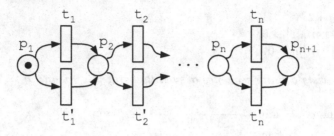

Fig. 3. A CPN system with an exponentially sized firing set

Due to the empty sequence, $\emptyset \in FS(\mathcal{N}, m_0)$. The size of a firing set may be exponential w.r.t. the number of transitions of the net. For example, consider the CPN system of Fig. 3. Its firing set is:

$$\{T' \mid \forall 1 \leq j < i \leq n \ \{t_i, t'_i\} \cap T' \neq \emptyset \Rightarrow \{t_j, t'_j\} \neq \emptyset\}$$

Thus its size is at least $2^{\frac{|T|}{2}}$.

The next two lemmas establish elementary properties of the firing set and leads to new notions.

Lemma 15. *Let $\mathcal{N}$ be a CPN and m, m' be two markings such that $[\![m]\!] = [\![m']\!]$. Then $FS(\mathcal{N}, m) = FS(\mathcal{N}, m')$.*

Proof. Since $[\![m]\!] = [\![m']\!]$, there exists $\alpha > 0$ such that $\alpha m \leq m'$.
Let $m \xrightarrow{\sigma}$. Using lemma 11 $\alpha m \xrightarrow{\alpha \sigma}$. Since $\alpha m \leq m'$, $m' \xrightarrow{\alpha \sigma}$.
Thus $FS(\mathcal{N}, m) \subseteq FS(\mathcal{N}, m')$. By symmetry, $FS(\mathcal{N}, m) = FS(\mathcal{N}, m')$. ∎
So given $P' \subseteq P$, without ambiguity we define $FS(\mathcal{N}, P')$ by:
$$FS(\mathcal{N}, P') \stackrel{\text{def}}{=} FS(\mathcal{N}, m) \text{ for any } m \text{ such that } P' = [\![m]\!]$$

Lemma 16. *Let $\langle \mathcal{N}, m_0 \rangle$ be a CPN system. Then $FS(\mathcal{N}, m_0)$ is closed by union.*

Proof. Let $m_0 \xrightarrow{\sigma}$ and $m_0 \xrightarrow{\sigma'}$.
Then using three times lemma 11, $0.5m_0 \xrightarrow{0.5\sigma}$, $0.5m_0 \xrightarrow{0.5\sigma'}$ and $m_0 \xrightarrow{0.5\sigma + 0.5\sigma'}$.
Since $[\![\overrightarrow{0.5\sigma + 0.5\sigma'}]\!] = [\![\overrightarrow{\sigma}]\!] \cup [\![\overrightarrow{\sigma'}]\!]$, the conclusion follows. ∎

Notation. We denote $\mathtt{maxFS}(\mathcal{N}, m_0)$ the maximal set of $FS(\mathcal{N}, m_0)$ that is the union of all members of $FS(\mathcal{N}, m_0)$.

The next proposition is a structural characterisation for a subset of transitions to belong to the firing set. In addition, it shows that in the positive case, a "useful" corresponding sequence always exists and furthermore one may build this sequence in polynomial time.

Proposition 17. *Let* $\langle \mathcal{N}, m_0 \rangle$ *be a CPN system and* T' *be a subset of transitions. Then:*
$$T' \in FS(\mathcal{N}, m_0) \text{ iff } \mathcal{N}_{T'} \text{ has no empty siphon in } m_0.$$
Furthermore if $T' \in FS(\mathcal{N}, m_0)$ *then there exists* $\sigma = \alpha_1 t_1 \ldots \alpha_k t_k$ *with* $\alpha_i > 0$ *for all* i, $T' = \{t_1, \ldots, t_k\}$ *and a marking* m *such that:*

- $m_0 \overset{\sigma}{\longrightarrow} m$;
- *for all place* p, $m(p) > 0$ *iff* $m_0(p) > 0$ *or* $p \in {}^\bullet T'^\bullet$.

Proof.

Necessity. Suppose $\mathcal{N}_{T'}$ contains an empty siphon Σ in m_0. Then none of the transitions belonging $\Sigma^\bullet$ can be fired in the future. Since $\mathcal{N}_{T'}$ does not contain isolated places $\Sigma^\bullet (= {}^\bullet \Sigma^\bullet) \neq \emptyset$ and so $T' \notin FS(\mathcal{N}, m_0)$.

Sufficiency. Suppose that $\mathcal{N}_{T'}$ has no empty siphon in m_0. We build by induction the sequence σ of the proposition. More precisely, we inductively prove for increasing values of i that:

- for every $j < i$ there exists a non empty set of transitions $T_j \subseteq T'$ that fulfill for all $j \neq j'$, $T_j \cap T_{j'} = \emptyset$;
- for every $j \leq i$ there exists a marking m_j with $m_j(p) > 0$ iff $m_0(p) > 0$ or $p \in {}^\bullet T_k{}^\bullet$ for some $k < j$;
- for every $j < i$ there exists a sequence $\sigma_j = \alpha_{j,1} t_{j,1} \ldots \alpha_{j,k_j} t_{j,k_j}$ with $T_j = \{t_{j,1} \ldots t_{j,k_j}\}$ and $m_j \overset{\sigma}{\longrightarrow} m_{j+1}$.

There is nothing to prove for the basis case $i = 0$.

Suppose that the assertion holds until i. If $T' = T_1 \cup \ldots \cup T_{i-1}$ then we are done.

Otherwise define $T'' = T' \setminus (T_1 \cup \ldots \cup T_{i-1})$ and $T_i = \{t \text{ enabled in } m_i \mid t \in T''\}$. We claim that T_i is not empty. Otherwise for all $t \in T''$, there exists an empty place p_t in m_i. Due to the inductive hypothesis, $m_0(p_t) = 0$ and ${}^\bullet p_t \cap (T_1 \cup \ldots \cup T_{i-1}) = \emptyset$. So the union of places p_t is an empty siphon of $\langle \mathcal{N}_{T'}, m_0 \rangle$ which contradicts our hypothesis.

Let us denote $T_i = \{t_{i,1} \ldots t_{i,k_i}\}$. Define $\alpha = \min(\frac{m_i(p)}{2k_i} \mid p \in {}^\bullet T_i)$ with the convention that $\alpha = 1$ if ${}^\bullet T_i = \emptyset$. The sequence $\sigma_i = \alpha t_{i,1} \ldots \alpha t_{i,k_i}$ is fireable from m_i and leads to a marking m_{i+1} fulfilling the inductive hypothesis.

Since T'' is finite the procedure terminates. ∎

We include the complexity result below since its proof relies in a straightforward manner on the sufficiency proof of the previous proposition.

Corollary 18. *Let* $\langle \mathcal{N}, m_0 \rangle$ *be a CPN system and* T' *be a subset of transitions. Then algorithm 1 checks in polynomial time whether* $T' \in FS(\mathcal{N}, m_0)$ *and in the negative case returns the maximal firing set included in* T' *(when called with* $T = T'$, *it returns* maxFS$(\mathcal{N}, m_0)$*).*

3.2 Characterisation of Reachability and Boundedness

In [10] a characterisation of reachability was presented. The theorem below is an alternative characterisation that only relies on the state equation and firing sets.

Algorithm 1. Decision algorithm for membership of $FS(\mathcal{N}, m_0)$

Fireable($\langle \mathcal{N}, m_0 \rangle, T'$): status
Input: a CPN system $\langle \mathcal{N}, m_0 \rangle$, a subset of transitions T'
Output: the membership status of T' w.r.t. $FS(\mathcal{N}, m_0)$
Output: in the negative case the maximal firing set included in T'
Data: *new*: boolean; P': subset of places; T'': subset of transitions
1 $T'' \leftarrow \emptyset; P' \leftarrow [\![m_0]\!]$
2 **while** $T'' \neq T'$ **do**
3 $\quad$ *new* $\leftarrow$ **false**
4 $\quad$ **for** $t \in T' \setminus T''$ **do**
5 $\quad\quad$ **if** ${}^\bullet t \subseteq P'$ **then** $T'' \leftarrow T'' \cup \{t\}; P' \leftarrow P' \cup t^\bullet;$ *new* $\leftarrow$ **true**
6 $\quad$ **end**
7 $\quad$ **if not** *new* **then return** (false, T'')
8 **end**
9 **return true**

Theorem 19. *Let $\langle \mathcal{N}, m_0 \rangle$ be a CPN system and m be a marking.*
Then $m \in \mathrm{RS}(\mathcal{N}, m_0)$ iff there exists $\mathbf{v} \in \mathbb{R}_{\geq 0}^{|T|}$ such that:

1. $m = m_0 + C\mathbf{v}$
2. $[\![\mathbf{v}]\!] \in FS(\mathcal{N}, m_0)$
3. $[\![\mathbf{v}]\!] \in FS(\mathcal{N}^{-1}, m)$

Proof.
Necessity. Let $m \in \mathrm{RS}(\mathcal{N}, m_0)$. So there exists a finite firing sequence σ such that $m_0 \xrightarrow{\sigma} m$. Let $\mathbf{v} = \vec{\sigma}$, then $m = m_0 + C\mathbf{v}$.

Since σ is fireable from m_o in $\mathcal{N}$, $[\![\mathbf{v}]\!] \in FS(\mathcal{N}, m_0)$. In $\mathcal{N}^{-1}$, $m \xrightarrow{\sigma^{-1}} m_0$. Since $\mathbf{v} = \overrightarrow{\sigma^{-1}}$, $[\![\mathbf{v}]\!] \in FS(\mathcal{N}^{-1}, m)$.

Sufficiency. Since $[\![\mathbf{v}]\!] \in FS(\mathcal{N}, m_0)$, using Proposition 17 and Lemma 11 there exists a sequence σ_1 such that $[\![\mathbf{v}]\!] = [\![\vec{\sigma_1}]\!]$, for all $0 < \alpha_1 \leq 1$, $m_0 \xrightarrow{\alpha_1 \sigma_1} m_1$ with $m_1(p) > 0$ for $p \in {}^\bullet[\![\mathbf{v}]\!]^\bullet$.

Since $[\![\mathbf{v}]\!] \in FS(\mathcal{N}^{-1}, m)$, using Proposition 17 and Lemma 11 there exists a sequence σ_2 such that $[\![\mathbf{v}]\!] = [\![\vec{\sigma_2}]\!]$, for all $0 < \alpha_2 \leq 1$, $m \xrightarrow{\alpha_2 \sigma_2} m_2$ in $\mathcal{N}^{-1}$ with $m_2(p) > 0$ for $p \in {}^\bullet[\![\mathbf{v}]\!]^\bullet$.

Choose α_1 and α_2 enough small such that the vector $\mathbf{v}' = \mathbf{v} - \alpha_1\vec{\sigma_1} - \alpha_2\vec{\sigma_2}$ is non negative and $[\![\mathbf{v}']\!] = [\![\mathbf{v}]\!]$. This is possible since $[\![\mathbf{v}]\!] = [\![\vec{\sigma_1}]\!] = [\![\vec{\sigma_2}]\!]$.

Since $m_2 = m_1 + C\mathbf{v}'$ and m_1, m_2 fulfill the hypotheses of Lemma 12, there exists a sequence σ_3 such that $\mathbf{v}' = \vec{\sigma_3}$ and $m_1 \xrightarrow{\sigma_3} m_2$.
Let $\sigma = (\alpha_1\sigma_1)\sigma_3(\alpha_2\sigma_2)^{-1}$ then $m_0 \xrightarrow{\sigma} m$. ∎

The following characterisation has been stated in [10]. We include the proof here since in that paper, the proof of necessity was not developed.

Theorem 20. *Let $\langle \mathcal{N}, m_0 \rangle$ be a CPN system and m be a marking.*
Then $m \in \mathtt{lim-RS}(\mathcal{N}, m_0)$ iff there exists $\mathbf{v} \in \mathbb{R}_{\geq 0}^{|T|}$ such that:

1. $m = m_0 + Cv$
2. $[\![v]\!] \in FS(\mathcal{N}, m_0)$

Proof.

Necessity. Let $m \in \mathrm{lim\text{-}RS}(\mathcal{N}, m_0)$. So there exists a firing sequence $\sigma = \alpha_1 t_1 \ldots \alpha_n t_n \ldots$ such that $m = \lim_{n \to \infty} m_n$, where $m_n \overset{\alpha_{n+1} t_{n+1}}{\longrightarrow} m_{n+1}$.

Thus there exists $B \in \mathbb{N}$ such that for all $p \in P$ and all $n \in \mathbb{N}$, $m_n[p] \leq B$.

Let $T' \overset{\mathrm{def}}{=} \{t \mid \exists i \in \mathbb{N} \; t = t_i\}$. There exists n_0 such that $T' = \{t \mid \exists i \leq n_0 \; t = t_i\}$ and so $T' \in FS(\mathcal{N}, m_0)$.

Let $\alpha \in \mathbb{Q}_{>0}$ such that $\alpha \leq \min(\sum_{i \leq n_0, t_i = t} \alpha_i \mid t \in T')$.

Let us define LP_n an existential linear program where $v \in \mathbb{R}^T$ is the vector of variables by:

1. $m_n - m_0 = Cv$
2. $\forall t \in T' \; v[t] \geq \alpha$
3. $\forall t \in T \setminus T' \; v[t] = 0$

Due to the existence of the firing sequence σ, for all $n \geq n_0$ LP_n admits a solution. Using linear programming theory (see [12]), since $m_n[p] \leq B$ for all n and all p, there exists B' such that for all $n \geq n_0$, LP_n admits a solution v_n whose items are bounded by B'.

So the sequence $\{v_n\}_{n \geq n_0}$ admits a subsequence that converges to some v. By continuity, v fulfills $m - m_0 = Cv$, $\forall t \subset T' \; v[t] \geq \alpha$ and $\forall t \in T \setminus T' \; v[t] - 0$. So $[\![v]\!] = T'$ and v is the desired vector.

Sufficiency. Since $[\![v]\!] \in FS(\mathcal{N}, m_0)$, using Proposition 17 and Lemma 11 there exists a sequence σ_1 such that $[\![v]\!] = [\![\overrightarrow{\sigma_1}]\!]$, for all $0 < \alpha_1 \leq 1$, $m_0 \overset{\alpha_1 \sigma_1}{\longrightarrow} m_1$ with $m_1(p) > 0$ for $p \in {}^\bullet[\![v]\!]^\bullet$.

Choose α_1 enough small such that the vector $v' = v - \alpha_1 \overrightarrow{\sigma_1}$ is non negative and $[\![v']\!] = [\![v]\!]$. This is possible since $[\![v]\!] = [\![\overrightarrow{\sigma_1}]\!]$.

Since $m = m_1 + Cv'$ and m_1 fulfills the hypotheses of lemma 13, there exists an infinite sequence σ_2 such that $v' = \overrightarrow{\sigma_2}$ and $m_1 \overset{\sigma_2}{\longrightarrow}_\infty m$.

Let $\sigma = (\alpha_1 \sigma_1)\sigma_2$ then $m_0 \overset{\sigma}{\longrightarrow}_\infty m$. ∎

We present below the first characterisation of boundedness for CPN systems.

Theorem 21. *Given a CPN system* $\langle \mathcal{N}, m_0 \rangle$. *Then* $\langle \mathcal{N}, m_0 \rangle$ *is unbounded iff:*
There exists $v \in \mathbb{R}_{\geq 0}^T$ *such that* $Cv \gneq 0$ *and* $[\![v]\!] \subseteq \mathrm{maxFS}(\mathcal{N}, m_0)$.

Proof.

Sufficiency. Assume there exists $v \in \mathbb{R}_{\geq 0}^T$ such that $Cv \gneq 0$ and $[\![v]\!] \subseteq \mathrm{maxFS}(\mathcal{N}, m_0)$. Denote $T' \overset{\mathrm{def}}{=} \mathrm{maxFS}(\mathcal{N}, m_0)$. Using proposition 17, there exists $m_1 \in RS(\mathcal{N}, m_0)$ such that for all $p \in {}^\bullet T'^\bullet$, $m_1(p) > 0$. Define $m_2 \overset{\mathrm{def}}{=} m_1 + Cv$, thus $m_2 \gneq m_1$. Since $[\![v]\!] \subseteq T'$, m_1 and m_2 fulfill the hypotheses of lemma 12. Applying it, yields a firing sequence $m_1 \overset{\sigma}{\longrightarrow} m_2$. Iterating this sequence establishes the unboundedness of $\langle \mathcal{N}, m_0 \rangle$.

Necessity. Assume $\langle \mathcal{N}, \boldsymbol{m}_0 \rangle$ is unbounded. Then there exists $p \in P$ and a family of firing sequences $\{\sigma_n\}_{n \in \mathbb{N}}$ such that $\boldsymbol{m}_0 \xrightarrow{\sigma_n} \boldsymbol{m}_n$ and $\boldsymbol{m}_n(p) \geq n$. Since $\{[\![\vec{\sigma}_n]\!]\}_{n \in \mathcal{N}}$ is finite by extracting a subsequence w.l.o.g. we can assume that all these sequences have the same support, say $T' \subseteq \mathrm{maxFS}(\mathcal{N}, \boldsymbol{m}_0)$.

Let $\mathbf{v}_n \overset{\mathrm{def}}{=} C\vec{\sigma}_n$. Define $\mathbf{w}_n = \frac{\mathbf{v}_n}{\|\mathbf{v}_n\|_1}$. Since $\{\mathbf{w}_n\}_{n \in \mathbb{N}}$ belongs to a compact set, there exists a convergent subsequence $\{\mathbf{w}_{\alpha(n)}\}_{n \in \mathbb{N}}$. Denote $\mathbf{w}$ its limit. Since $\|\mathbf{w}\|_1 = 1$, $\mathbf{w}$ is non null. We claim that $\mathbf{w}$ is a non negative vector. Since $\boldsymbol{m}_n(p) \geq n$, $\|\mathbf{v}_n\|_1 \geq \mathbf{v}_n[p] \geq n - \boldsymbol{m}_0[p]$. On the other hand, for all $p' \in P$, $\mathbf{w}_n[p'] \geq \frac{-\boldsymbol{m}_0[p']}{\|\mathbf{v}_n\|_1}$. Combining the two inequalities, for $n > \boldsymbol{m}_0[p]$, $\mathbf{w}_n[p'] \geq \frac{-\boldsymbol{m}_0[p']}{n - \boldsymbol{m}_0[p]}$. Applying this inequality to $\alpha(n)$ and letting n go to infinity yields $\mathbf{w}[p'] \geq 0$.

Due to standard results of polyhedra theory (see [1] for instance), the set $\{C_{P \times T'} \mathbf{u} \mid \mathbf{u} \in \mathbb{R}_{\geq 0}^{T'}\}$ is closed. So there exists $\mathbf{u} \in \mathbb{R}_{\geq 0}^{T'}$ such that $\mathbf{w} = C\mathbf{u}$. Considering $\mathbf{u}$ as a vector of $\mathbb{R}_{\geq 0}^{T}$ by adding null components for $T \setminus T'$ yields the required vector. ∎

4   Decision Procedures

Naively implementing the characterisation of reachability would lead to an exponential procedure since it would require to enumerate the items of $FS(\mathcal{N}, \boldsymbol{m}_0)$ (whose size is possibly exponential). For each item, say T', the algorithm would check in polynomial time (1) whether T' belongs to $FS(\mathcal{N}^{-1}, \boldsymbol{m})$ and (2) whether the associated linear program $\mathbf{v} > 0 \wedge C_{P \times T'} \mathbf{v} = \boldsymbol{m} - \boldsymbol{m}_0$ admits a solution. Guessing T' shows that the reachability problem belongs to NP.

In fact, we improve this upper bound with the help of Algorithm 2. When $\boldsymbol{m} \neq \boldsymbol{m}_0$, this algorithm maintains a subset of transitions T' which fulfills $[\![\vec{\sigma}]\!] \subseteq T'$ for any $\boldsymbol{m}_0 \xrightarrow{\sigma} \boldsymbol{m}$ (as will be proven in proposition 22). Initially T' is set to T. Then lines 4-9 build a solution to the state equation restricted to transitions of T' with a maximal support (if there is at least one). If there is no solution then the algorithm returns false. Otherwise T' is successively restricted to (1) the support of this maximal solution (line 10), (2) the maximal firing set in $\mathrm{maxFS}(\mathcal{N}_{T'}, \boldsymbol{m}_0[^\bullet T'^\bullet])$ (line 11) and, (3) the maximal firing set in $\mathrm{maxFS}(\mathcal{N}_{T'}^{-1}, \boldsymbol{m}[^\bullet T'^\bullet])$ (line 12). If the two last restrictions do not modify T' then the algorithm returns true. If T' becomes empty then the algorithm returns false.

Omitting line 12, Algorithm 2 decides the lim-reachability problem.

Proposition 22. *Algorithm 2 returns true iff $\boldsymbol{m}$ is reachable in $\langle \mathcal{N}, \boldsymbol{m}_0 \rangle$. Algorithm 2 without line 12 returns true iff $\boldsymbol{m}$ is lim-reachable in $\langle \mathcal{N}, \boldsymbol{m}_0 \rangle$.*

Proof. We only consider the non trivial case $\boldsymbol{m} \neq \boldsymbol{m}_0$.

Soundness. Assume that the algorithm returns true at line 13.

By definition, vector **sol** which is a barycenter of solutions is also a solution with maximal support and so fulfils the first statement of Theorem 19. Since $T' = [\![\mathbf{sol}]\!]$ at line 13, $[\![\mathbf{sol}]\!] \in FS(\mathcal{N}, \boldsymbol{m}_0)$ due to line 11 and $[\![\mathbf{sol}]\!] \in FS(\mathcal{N}^{-1}, \boldsymbol{m})$

Algorithm 2. Decision algorithm for reachability

Reachable($\langle \mathcal{N}, m_0 \rangle, m$): status
Input: a CPN system $\langle \mathcal{N}, m_0 \rangle$, a marking m
Output: the reachability status of m
Output: the Parikh image of a witness in the positive case
Data: $nbsol$: integer; $\mathbf{v}, \mathbf{sol}$: vectors; T': subset of transitions

1 **if** $m = m_0$ **then return** (**true**,0)
2 $T' \leftarrow T$
3 **while** $T' \neq \emptyset$ **do**
4 $\quad$ $nbsol \leftarrow 0$; $\mathbf{sol} \leftarrow 0$
5 $\quad$ **for** $t \in T'$ **do**
6 $\quad\quad$ solve $\exists? \mathbf{v}\ \mathbf{v} \geq 0 \wedge \mathbf{v}[t] > 0 \wedge C_{P \times T'} \mathbf{v} = m - m_0$
7 $\quad\quad$ **if** $\exists \mathbf{v}$ **then** $nbsol \leftarrow nbsol + 1$; $\mathbf{sol} \leftarrow \mathbf{sol} + \mathbf{v}$
8 $\quad$ **end**
9 $\quad$ **if** $nbsol = 0$ **then return false else** $\mathbf{sol} \leftarrow \frac{1}{nbsol} \mathbf{sol}$
10 $\quad$ $T' \leftarrow [\![\mathbf{sol}]\!]$
11 $\quad$ $T' \leftarrow T' \cap \mathrm{maxFS}(\mathcal{N}_{T'}, m_0[{}^\bullet T'^\bullet])$
12 $\quad$ $T' \leftarrow T' \cap \mathrm{maxFS}(\mathcal{N}_{T'}^{-1}, m[{}^\bullet T'^\bullet])$ /* deleted for lim-reachability   */
13 $\quad$ **if** $T' = [\![\mathbf{sol}]\!]$ **then return** (**true**,$\mathbf{sol}$)
14 **end**
15 **return false**

---

due to line 12. Thus m is reachable in $\langle \mathcal{N}, m_0 \rangle$ since it fulfills the assertions of Theorem 19. In case of lim-reachability, line 12 is omitted. So the assertions of Theorem 20 are fulfilled and m is lim-reachable in $\langle \mathcal{N}, m_0 \rangle$.

Completeness. Assume the algorithm returns false.

We claim that at any time the algorithm fulfils the following invariant: for any $m_0 \xrightarrow{\sigma} m$, $[\![\vec{\sigma}]\!] \subseteq T'$.

This invariant initially holds since $T' = T$. At line 10 due to the first assertion of Theorem 19, for any such σ, $[\![\vec{\sigma}]\!] \subseteq [\![\mathbf{sol}]\!]$ since $\mathbf{sol}$ is a solution with maximal support. So the assignment of line 10 lets true the invariant. Due to the second assertion of Theorem 19 and the invariant, any σ fulfils $[\![\vec{\sigma}]\!] \subseteq \mathrm{maxFS}(\mathcal{N}_{T'}, m_0[{}^\bullet T'^\bullet])$. So the assignment of line 11 lets true the invariant. Due to the third assertion of Theorem 19 and the invariant, any σ fulfils $[\![\vec{\sigma}]\!] \subseteq \mathrm{maxFS}(\mathcal{N}_{T'}^{-1}, m[{}^\bullet T'^\bullet])$. So the assignment of line 12 lets true the invariant.

If the algorithm returns false at line 9 due to the invariant the first assertion of Theorem 19 cannot be satisfied. If the algorithm returns false at line 15 then $T' = \emptyset$. So due to the invariant and since $m \neq m_0$, m is not reachable from m_0.

The case of lim-reachability is similarly handled with the following invariant: for any $m_0 \xrightarrow{\sigma}_\infty m$, $[\![\vec{\sigma}]\!] \subseteq T'$. ∎

Proposition 23. *The reachability and the lim-reachability problems for CPN systems are decidable in polynomial time.*

Proof. Let us analyse the time complexity of Algorithm 2. Since T' must be modified in lines 11 or 12 in order to start a new iteration of the main loop, there are at most $|T|$ iterations of this loop. The number of iterations of the inner loop is also bounded by $|T|$. Finally solving a linear program can be performed in polynomial time [12] as well as computing the maximal item of a firing set (see corollary 18). ∎

In [10], it is proven that the lim-reachability problem for consistent CPN systems with no empty siphons in the initial marking is decidable in polynomial time. We improve this result by showing that this problem and a similar one belong to NC $\subseteq$ PTIME (a complexity class of problems that can take advantage of parallel computations, see [11]).

Proposition 24. *The reachability problem for consistent CPN systems with no empty siphons in the initial marking and no empty siphons in the final marking for the reverse net belongs to NC.*
The lim-reachability problem for consistent CPN systems with no empty siphons in the initial marking belongs to NC.

Proof. Due to the assumptions on siphons and proposition 17 only the first assertion of Theorems 19 and 20 needs to be checked. Due to consistency, there exists $\mathbf{w} > \mathbf{0}$ such that $\boldsymbol{C}\mathbf{w} = \mathbf{0}$. Assume there is some $\mathbf{v} \in \mathbb{R}^T$ such that $\boldsymbol{m} - \boldsymbol{m}_0 = \boldsymbol{C}\mathbf{v}$. For some $n \in \mathbb{N}$ large enough, $\mathbf{v}' \stackrel{\text{def}}{=} \mathbf{v} + n\mathbf{w} \in \mathbb{R}^T_{\geq 0}$ and still fulfils $\boldsymbol{m} - \boldsymbol{m}_0 = \boldsymbol{C}\mathbf{v}'$.

Now the decision problem $\exists ?\mathbf{v} \in \mathbb{R}^T \; \boldsymbol{m} - \boldsymbol{m}_0 = \boldsymbol{C}\mathbf{v}$ belongs to NC [4]. ∎

Proposition 25. *The boundedness problem for CPN systems is decidable in polynomial time.*

Proof. Using the characterisation of Theorem 21, one first computes in polynomial time $T' = \texttt{maxFS}(\mathcal{N}, \boldsymbol{m}_0)$ (see corollary 18). Then for all $p \in P$, one solves the existential linear program $\exists ?\mathbf{v} \geq \mathbf{0} \; \boldsymbol{C}_{P \times T'}\mathbf{v} \geq \mathbf{0} \wedge (\boldsymbol{C}_{P \times T'}\mathbf{v})[p] > 0$. The CPN system is unbounded if some of these linear programs admits a solution. ∎

In discrete Petri nets, the reachability set inclusion problem is undecidable, while the restricted problem of home state is decidable (see [7] for a detailed survey about decidability results in PNs). In CPN systems, this problem is decidable thanks to the special structure of the (lim-)reachability sets.

Proposition 26. *The reachability set inclusion and the lim-reachability set inclusion problems for CPN systems are decidable in exponential time.*

Proof. Let us define $TP \stackrel{\text{def}}{=} \{(T', P') \mid T' \in FS(\mathcal{N}, \boldsymbol{m}_0) \wedge P' \subseteq P \wedge T' \in FS(\mathcal{N}^{-1}, P')\}$. For every pair $(T', P') \in TP$, define the polyhedron $E_{T', P'}$ over $\mathbb{R}^P \times \mathbb{R}^{T'}$ by:

$$E_{T', P'} \stackrel{\text{def}}{=} \{(\boldsymbol{m}, \mathbf{v}) \mid \boldsymbol{m}[P'] > \mathbf{0} \wedge \boldsymbol{m}[P \setminus P'] = \mathbf{0} \wedge \mathbf{v} > \mathbf{0} \wedge \boldsymbol{m} = \boldsymbol{C}_{P \times T'}\mathbf{v}\}$$

and $R_{T', P'}$ by: $R_{T', P'} \stackrel{\text{def}}{=} \{\boldsymbol{m} \mid \exists \mathbf{v} \; (\boldsymbol{m}, \mathbf{v}) \in E_{T', P'}\}$

Using the characterisation of Theorem 19 and Lemma 15,
$RS(\mathcal{N}, \boldsymbol{m}_0) = \bigcup_{(T',P') \in TP} R_{T',P'}$.

Due to Lemma 11, the reachability set of a CPN system is convex. So $RS(\mathcal{N}, \boldsymbol{m}_0)$ can be rewritten as:

$$RS(\mathcal{N}, \boldsymbol{m}_0) = \{ \sum_{(T',P') \in TP} \lambda_{T',P'} \boldsymbol{m}_{T',P'} \mid$$

$$\sum_{(T',P') \in TP} \lambda_{T',P'} = 1 \wedge \forall (T', P') \in TP \; \lambda_{T',P'} \geq 0 \wedge \boldsymbol{m}_{T',P'} \in R_{T',P'} \}$$

Observe that this representation is exponential w.r.t. the size of the CPN system.

Let $\langle \mathcal{N}, \boldsymbol{m}_0 \rangle$ and $\langle \mathcal{N}', \boldsymbol{m}_0' \rangle$ be two CPN systems for which one wants to check whether $RS(\mathcal{N}, \boldsymbol{m}_0) \subseteq RS(\mathcal{N}', \boldsymbol{m}_0')$. One builds the representation above for $RS(\mathcal{N}, \boldsymbol{m}_0)$ and $RS(\mathcal{N}', \boldsymbol{m}_0')$. Then one transforms the representation of the set $RS(\mathcal{N}', \boldsymbol{m}_0')$ as a system of linear constraints. This can be done in polynomial time w.r.t. the original representation [2]. So the number of constraints is still exponential w.r.t. the size of $\langle \mathcal{N}', \boldsymbol{m}_0' \rangle$.

Afterwards for every constraint of this new representation, one adds its negation to the representation of $RS(\mathcal{N}, \boldsymbol{m}_0)$ and check for a solution of such a system. $RS(\mathcal{N}, \boldsymbol{m}_0) \not\subseteq RS(\mathcal{N}', \boldsymbol{m}_0')$ iff at least one of these linear programs admits a solution. The overall complexity of this procedure is still exponential w.r.t. the size of the problem. The procedure for lim-reachability set inclusion is similar.   ■

5   Hardness Results

We now provide matching lower bounds for almost all problems analysed in the previous sections.

Proposition 27. *The reachability, lim-reachability and boundedness problems for CPN systems are PTIME-complete.*

We want to prove that the lower bounds are robust. To this aim, we recall free-choice CPNs.

Definition 28. *A CPN $\mathcal{N}$ is free-choice if:*

- $\forall p \in P \; \forall t \in T \{ \boldsymbol{Pre}[p, t], \boldsymbol{Post}[p, t] \} \subseteq \{0, 1\};$
- $\forall t, t' \in T \; {}^{\bullet}t \cap {}^{\bullet}t' \neq \emptyset \Rightarrow {}^{\bullet}t = {}^{\bullet}t'.$

Proposition 29. *The (lim-)deadlock-freeness and (lim-)liveness problems in free-choice CPN systems are coNP-hard.*

Proof. We use almost the same reduction from the 3SAT problem as the one proposed for free-choice Petri nets in [6]. However the proof of correctness is specific to continuous nets.

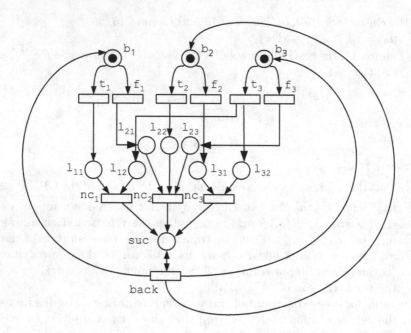

Fig. 4. The CPN corresponding to formula $(\neg x_1 \vee \neg x_3) \wedge (x_1 \vee \neg x_2 \vee x_3) \wedge (x_2 \vee \neg x_3)$

Let $\{x_1, x_2, \ldots, x_n\}$ denote the set of propositions and $\{c_1, c_2, \ldots, c_m\}$ denote the set of clauses. Every clause c_j is defined by $c_j \stackrel{\text{def}}{=} lit_{j1} \vee lit_{j2} \vee lit_{j3}$ where for all j, k, $lit_{jk} \in \{x_1, \ldots, x_n, \neg x_1, \ldots, \neg x_n\}$. The satisfiability problem consists in the existence of an interpretation $\nu : \{x_1, x_2, \ldots, x_n\} \longrightarrow \{\textbf{false}, \textbf{true}\}$, such that for all clause c_j, $\nu(c_j) = \textbf{true}$.

Every proposition x_i yields a place b_i initially marked with a token (all other places are unmarked) and input of two transitions t_i, f_i corresponding to the assignment associated with an interpretation. Every of literal lit_{jk} yields a place l_{jk} which is the output of transition t_i if $lit_{jk} = x_i$ or transition f_i if $lit_{jk} = \neg x_i$ Every clause c_j yields a transition nc_j with three input "literal" places corresponding to literals $\neg lit_{j1}, \neg lit_{j2}, \neg lit_{j3}$. An additional place suc is the output of every transition nc_j. Finally, transition $back$ has suc as a loop place and b_i for all i as output places. The reduction is illustrated in Fig. 4.

Assume that there exists ν such that for all clause c_j, $\nu(c_j) = \textbf{true}$. Then fire the following sequence $\sigma = 1t_1^* \ldots 1t_n^*$ where $t_i^* = t_i$ when $\nu(x_i) = \textbf{true}$ and $t_i^* = f_i$ when $\nu(x_i) = \textbf{false}$. Consider m the reached marking. Since $\nu(c_j) = \textbf{true}$, at least one input place of nc_j is empty in m. Moreover $m(suc) = m(b_i) = 0$ for all i. So m is dead.

Assume that there does not exist ν such that for all clause c_j, $\nu(c_j) = \textbf{true}$. Observe that given a marking m such that $m(suc) > 0$ all transitions will be fireable in the future and suc will never decrease (thus $m(suc) > 0$ for a lim-reachable marking m as well).

Table 2. Complexity bounds

Problems	Upper and lower bounds
(lim-)reachability	PTIME-complete in NC for lim-reachability (resp. reachability) when all transitions are fireable at least once (resp. and also in the reverse CPN) and the net is consistent
(lim-)boundedness	PTIME-complete
(lim-)deadlock-freeness and (lim-)liveness	coNP-complete coNP-hard even for free-choice CPNs or for CPNs when all transitions are fireable at least once and the net is consistent
(lim-)reachability set inclusion	in EXPTIME coNP-hard even for reversibility in CPNs when all transitions are fireable at least once and the net is consistent

So we only consider reachable marking m such that $m(suc) = 0$, i.e. when no transitions nc_j have been fired. Our goal is to prove that from such marking there is a sequence that produces tokens in suc. Examining the remaining transitions, the following invariants hold. For all atomic proposition x_i, and reachable marking m, one has

$$\forall i \; m[b_i] + \sum_{l_{jk} \in \{x_i, \neg x_i\}} m[l_{jk}] \geq 1$$

$$\forall j, k, j', k' \; lit_{jk} = lit_{j'k'} \Rightarrow m[l_{jk}] = m[l_{j'k'}]$$

If for some i, $m[b_i] > 0$, we fire t_i in order to empty b_i. Thus the invariants become:

$$\forall i \; \sum_{l_{jk} \in \{x_i, \neg x_i\}} m[l_{jk}] \geq 1$$

$$\forall j, k, j', k' \; lit_{jk} = lit_{j'k'} \Rightarrow m[l_{jk}] = m[l_{j'k'}]$$

Now define ν by $\nu(x_i) = $ **true** if for some $lit_{jk} = x_i$, $m(l_{jk}) > 0$. Due to the hypothesis, there is a clause c_j such that $\nu(c_j) = $ **false**. Due to our choice of ν and the invariants, all inputs of nc_j are marked. So firing nc_j marks suc. ∎

We show that even the hypotheses that allow the lim-reachability to belong in NC do not reduce the complexity of other problems.

Proposition 30. *The (lim-)deadlock-freeness, (lim-)liveness and reversibility problems in consistent CPN systems with no initially empty siphons are coNP-hard.*

6   Conclusions

In this work we have analysed the complexity of the most standard problems for continuous Petri nets. For almost all these problems, we have characterised their complexity class by designing new decision procedures and/or providing reductions to complete problems. We have also shown that the reachability set inclusion, undecidable for Petri nets, becomes decidable in the continuous framework. These results are summarised in Table 2.

There are three fruitful possible extensions of this work. Other properties like coverability could be studied. A temporal logic provides a specification language for expressing properties. In Petri nets, the model checking problem lies on the boundary of decidability depending on the type of logics (branching versus linear, propositional versus evenemential). We want to investigate this problem for continuous Petri nets. Hybrid Petri nets encompass both discrete and continuous Petri nets. So it would be interesting to examine the complexity and decidability of standard problems for the whole class or some appropriate subclasses of this formalism.

Acknowledgments. The authors would like to thank Jorge Júlvez and Manuel Silva for fruitful discussions on CPNs.

References

1. Avis, D., Fukuda, K., Picozzi, S.: On canonical representations of convex polyhedra. In: Cohen, A.M., Gao, X.-S., Takayama, N. (eds.) Mathematical Software, Proceedings of the First International Congress of Mathematical Software, pp. 350–360. World Scientific Publishing (2002)
2. Bagnara, R., Hill, P.M., Zaffanella, E.: Not necessarily closed convex polyhedra and the double description method. Formal Aspects of Computing 17(2), 222–257 (2005)
3. Cabasino, M.P., Seatzu, C., Mahulea, C., Silva, M.: Fault diagnosis of manufacturing systems using continuous Petri nets. In: Proceedings of the IEEE International Conference on Systems, Man and Cybernetics, Istanbul, Turkey, pp. 534–539. IEEE (2010)
4. Codenotti, B., Leoncini, M., Preparata, F.P.: The role of arithmetic in fast parallel matrix inversion. Algorithmica 30(4), 685–707 (2001)
5. David, R., Alla, H.: Continuous Petri nets. In: Proc. of the 8th European Workshop on Application and Theory of Petri Nets, Zaragoza, Spain, pp. 275–294 (1987)
6. Desel, J., Esparza, J.: Free Choice Petri Nets. Cambridge Tracts in Theoretical Computer Science 40 (1995)
7. Esparza, J., Nielsen, M.: Decidability issues for Petri nets - a survey. Elektronische Informationsverarbeitung und Kybernetik 30(3), 143–160 (1994)
8. Fraca, E., Haddad, S.: Complexity analysis of continuous Petri nets. Research Report LSV-13-01, LSV, ENS Cachan, France (2013)
9. Gudiño-Mendoza, B., López-Mellado, E., Alla, H.: Modeling and simulation of water distribution systems using timed hybrid Petri nets. Simulation 88(3), 329–347 (2012)

10. Júlvez, J., Recalde, L., Silva, M.: On reachability in autonomous continuous petri net systems. In: van der Aalst, W.M.P., Best, E. (eds.) ICATPN 2003. LNCS, vol. 2679, pp. 221–240. Springer, Heidelberg (2003)
11. Papadimitriou, C.H.: Computational complexity. Addison-Wesley (1994)
12. Papadimitriou, C.H., Steigliz, K.: Combinatorial Optimization. Algorithms and Complexity, 2nd edn. Dover publications (1998)
13. Recalde, L., Haddad, S., Silva, M.: Continuous Petri nets: Expressive power and decidability issues. Int. Journal of Foundations of Computer Science 21(2), 235–256 (2010)
14. Recalde, L., Teruel, E., Silva, M.: Autonomous continuous P/T systems. In: Donatelli, S., Kleijn, J. (eds.) ICATPN 1999. LNCS, vol. 1639, pp. 107–126. Springer, Heidelberg (1999)
15. Ross-Leon, R., Ramirez-Trevino, A., Morales, J.A., Ruiz-Leon, J.: Control of metabolic systems modeled with timed continuous Petri nets. In: ACSD/Petri Nets Workshops. CEUR Workshop Proceedings, vol. 827, pp. 87–102 (2010)
16. Renato Vázquez, C., Sutarto, H.Y., Boel, R.K., Silva, M.: Hybrid Petri net model of a traffic intersection in an urban network. In: Proceedings of the IEEE International Conference on Control Applications, CCA 2010, Yokohama, Japan, pp. 658–664 (2010)
17. Zerhouni, N., Alla, H.: Dynamic analysis of manufacturing systems using continuous Petri nets. In: Proceedings of the IEEE International Conference on Robotics and Automation, Los Alamitos, CA, USA, vol. 2, pp. 1070–1075 (1990)

Step Persistence in the Design of GALS Systems

Johnson Fernandes[2], Maciej Koutny[1], Marta Pietkiewicz-Koutny[1],
Danil Sokolov[2], and Alex Yakovlev[2]

[1] School of Computing Science
[2] School of Electrical & Electronic Engineering
Newcastle University, Newcastle upon Tyne, NE1 7RU, U.K.

Abstract. In this paper we investigate the behaviour of GALS (Globally Asynchronous Locally Synchronous) systems in the context of VLSI circuits. The specification of a system is given in the form of a Petri net. Our aim is to re-design the system to optimise signal management, by grouping together concurrent events. Looking at the concurrent reachability graph of the given Petri net, we are interested in discovering events that appear in 'bundles', so that they all can be executed in one clock tick. The best candidates for bundles are sets of events that appear and re-appear over and over again in the same configurations, forming 'persistent' sets of events. Persistence was considered so far only in the context of sequential semantics. Here we introduce a notion of persistent steps and discuss their basic properties. We then introduce a formal definition of a bundle and propose an algorithm to prune the behaviour of a system, so that only bundle steps remain. The pruned reachability graph represents the behaviour of a re-engineered system, which in turn can be implemented in a new Petri net using the standard techniques of net synthesis. The proposed algorithm prunes reachability graphs of persistent and safe nets leaving bundles that represent maximally concurrent steps.

Keywords: asynchronous and synchronous circuit, GALS system, persistence, step transition system, Petri net.

1 Introduction

Traditional circuit design styles have been following one of the two main strands, namely synchronous and asynchronous. In a nutshell, these two approaches differ in their techniques of synchronising interaction between circuit elements. Asynchronous designs adopt *'on request'* synchronisation where interaction is regulated by means of handshake control signals. They are designed to be adaptive to delays of signal propagation. Synchronous designs, on the other hand, assume worst case delay between circuit elements and determine a global periodic control signal for synchronisation called the *clock*. The clock signal limits the many sequencing options considered in asynchronous control. Thus synchronous circuits are considered to be a proper subset of asynchronous circuits [6].

Asynchronous logic was the dominant design style with most early computers. In particular, David Muller's speed-independent circuits, dating back to the

J.-M. Colom and J. Desel (Eds.): PETRI NETS 2013, LNCS 7927, pp. 190–209, 2013.
© Springer-Verlag Berlin Heidelberg 2013

late 1950s, have served many interesting applications such as the ILLIAC I and ILLIAC II computers [15]. However, since 1960, an era when fabrication of integrated circuits (ICs) became a feasible business, synchronous design became the mainstream technique as it met the market needs with its shorter design cycle. Today, majority of designs are synchronous, well etched in the heart of semiconductor industry together with superior CAD tools and EDA flows.

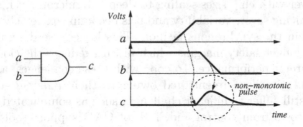

Fig. 1. Hazardous switching of an AND gate

One of the main issues with the complexity of asynchronous circuits was the handling of hazards. Hazards are manifestations of undesirable switching activity called glitches. In the asynchronous style of synchronisation, the output of each circuit element is potentially sensitive to its inputs. This can give rise to non-monotonic pulses (or glitches) when transitioning between output states, as illustrated in the waveform of Figure 1 taking the case of an AND gate. Due to tight timing between the rising edge of input a and falling edge of input b, the output c produces a non-monotonic pulse before stabilising to a low. This behaviour is hazardous as it is uncertain how the fanout of the AND gate will interpret the glitch; the output c temporary switching to logical 1 or staying at logical 0 all the time.

As shown, for instance, in paper [19], the phenomenon described in the above example can be conveniently interpreted in terms of formal models such as Keller's named transition systems [11] or Petri nets [6]. In particular, what we see in this circuit is the effect where a signal that is enabled (rising edge of c) in a certain state of the circuit may become disabled without firing after the occurrence of another signal (falling edge of b). Such an effect corresponds to the violation of *persistence* [1] property at the level of signal transitions if the latter are used to label the corresponding named transition system. Furthermore, when such a circuit is modelled by a labelled Petri net following the technique of [19], the Petri net would also be classified as a non-persistent one. Thus, it was shown in [19] that the modelling and analysis of an asynchronous circuit with respect to hazard-freedom is effectively reduced to the analysis of persistence of its corresponding Petri net model.

---

[1] Informally, persistence means that an action of a system that is enabled at some point of system's execution cannot be disabled by another enabled action. It can only be delayed.

Synchronous circuits, on the other hand, do not require persistence satisfaction as they are intrinsically immune to hazardous behaviour. The principle reason being that the clock, set at worst-case latency period, filters out undesirable circuit switching. This greatly simplifies circuit design compared to asynchronous methods wherein the same circuit had to be analysed for persistence and redesigned to ensure glitch-free operation. Clocked circuits are thus preferred over asynchronous circuits for designing functionally correct (hazard-immune) ICs efficiently. However with chip sizes scaling to deep sub-micron level, semiconductors are experiencing severe variability and it is becoming extremely complicated to design chips in the synchronous fashion. This is because designing for variability requires longer safety margins which in turn reduces the clock frequency and degrades circuit performance. To cope with these challenges, asynchronous design methodologies have re-emerged owing to their *inherent* adaptiveness. However, they still suffer significant challenges such as complicated design flow, high overhead costs from control and, lack of CAD support tools and legacy design reuse. Therefore attempts are being made to find a compromise.

An on-trend intermediate solution is mixed synchronous-asynchronous design, chiefly acting in the form of Globally Asynchronous Locally Synchronous (GALS) methodology; its benefits well known in literature [10,9,17]. GALS system design, introduced in [5], can exploit the advantages of asynchrony and at the same time maximally reuse the products of synchronous design flow. This design technique divides a digital system into synchronous islands which communicate asynchronously by handshake mechanism. Each island has its own local clock which can be activated on demand by means of a handshake control signal. Such systems comprise a mixed temporal behaviour. Asynchronous handshakes handle switching between components where adaptability can significantly improve performance, while clocking is applied to components where worst case performance is tolerable. However, it is worthy of note that modelling GALS systems would involve detection of potential hazardous states due to presence of asynchronous components, making their design and verification a significant research challenge.

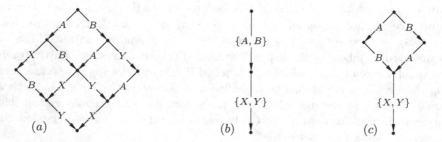

Fig. 2. Temporal representations of systems having concurrent, parallel and mixed concurrent-parallel behaviours: (*a*) interleaving model for asynchronous behaviour; (*b*) step model for synchronous behaviour; and (*c*) mixed model for GALS behaviour

Being a recent trend, there is a lack of formal models that describe correctness of GALS designs. The complexity in modelling them begins with the investigation of persistence. It should be noted that the standard notion of persistence has been defined at the level of single actions, which is also known as interleaving semantics of concurrency. This notion has been adequate for representing the correctness of the behaviour of circuits that are fully asynchronous. In asynchronous circuits, there is concurrency between independent actions and sequential order between causally related actions. This notion is well represented by Keller's named transition systems [11]. Figure 2(a) depicts such a model capturing the asynchronous behaviour of a system with four events: A, B, X and Y. Now, in synchronous circuits, the clock signal would trigger a single action or several actions. These circuits exhibit parallelism between actions in the same clock cycle and sequential order between groups of actions in adjacent clock cycles. To represent this group execution of actions, we will use *steps*, and therefore we need step transition systems to represent such a behaviour. A step represents a single action or a group of actions that are triggered simultaneously from a particular state by the clock signal. Figure 2(b) shows such a transition system model capturing the temporal behaviour of a synchronous system with the help of steps. For the case of GALS, there is a mixture of synchrony and asynchrony and hence both concurrent and parallel behaviour have to be represented. Figure 2(c) illustrates the mixed temporal behaviour seen in such systems. In all three cases, step transition systems provide a suitable behavioral model, as a single transition can be treated as a singleton step.

Synchronous and asynchronous systems have distinct techniques to guarantee functionally correct behaviour. However, for GALS systems, it is not so straightforward as correctness should be accounted from both angles. We would like to find an adequate representation of the correct behaviour of GALS systems. Here, it would be natural to define such a behaviour in analogous way as it was done for asynchronous circuits, i.e., with the use of the notion of persistence. However, when modelling GALS systems we have to consider complex actions, namely steps, and corresponding transition systems. This paper is hence centred around extending the notion of persistence to steps.

The main motivation for studying persistent steps in this paper is as follows. Digital system design based on formal models is normally associated with two main tasks: one is the verification of a system's behavioural specification or checking the model of the system implementation, while the other is the synthesis of the circuit implementation from its specification. In the context of verification we would like, for example, to check if the Petri net model of a GALS system satisfies the requirement of hazard-freedom under a particular form of synchronisation of actions (in steps). In the context of synthesis, we would like to find the optimal partitioning of actions into synchronous steps so that the complexity of control of these steps is minimised. For example, the intuitive complexity of handling synchronisations safely in the three scenarios of Figure 2 varies between them, from the most intricate in the fully asynchronous one (case (a)) to the simplest in fully synchronous case of (case (b)), placing the GALS version in the middle (case (c)). With this varying complexity, one can design systems that may exhibit hazards if they are treated as fully asynchronous, but when actions are synchronised into steps the system would behave safely. Amongst the methods for synchronising actions into steps, we can consider those that are based on the insertion of additional control circuits to physically 'bundle' actions together, or based on ensuring the appropriate 'bundling' constraints based on timing, or delays. Traditional globally clocked systems, self-timed systems working under fundamental mode assumptions, and asynchronous systems with relative timing [6] are all of the latter category.

It is this idea of bundling those steps of actions that are 'hazard-free' or persistent that motivated our notion of *bundles*, introduced in this paper. In terms of nets and corresponding transition systems, bundles are informally sets of transitions that can be executed synchronously and therefore be treated as some kind of 'atomic actions', giving rise to new 'bigger' transitions. Section 4 provides a more formal treatment for bundles and shows a constructive procedure for deriving them by pruning reachability graphs or transitions systems, depending on whether we are solving the verification or synthesis problem. For example, in the process of synthesis of the control policy for a GALS system, such a 'pruned' transition system would represent the desired behaviour, which then we would like to implement in a form of a Petri net.

We hope that the reader will find the theory presented in this paper as a necessary first step in paving the way towards automating the design of GALS systems. Right now, we are not trying to answer how this theory can be applied in the above-sketched scenarios of verification and synthesis. This will be done in our subsequent papers, which will have to answer many new questions arising on the way, including, for example, what a rigorous metric for the complexity of bundle control is, how the notions of maximal steps (global and local) affect such a complexity, or what the different forms of step persistence (A, B and C) imply in terms of hazard-avoidance in the system.

The paper is organised as follows. Section 2 recalls the basic definitions and notations concerning step transition systems and PT-nets. Section 3 introduces the notion of persistent steps and discusses their basic properties. Section 4 presents the main result of the paper, an algorithm that prunes the concurrent reachability graph of a net, which serves as an initial system specification, to obtain a representation of a desired 'GALS' behavior. Finally, section 5 contains conclusions and presents directions for future work.

2   Preliminaries

In this section we recall definitions and notations concerned with step transition systems and Petri nets used in the rest of this paper.

2.1   Step Transition Systems

Let T be a finite set of net transitions representing actions of a concurrent system. A set of transitions will be called a *step*, and we will use $\alpha, \beta, \gamma, \ldots$ to range over all steps $\mathcal{P}(T)$. Sometimes we will identify a step α with its characteristic function $\alpha : T \to \{0, 1\}$. We will also write $\alpha = \sum_{t \in T} \alpha(t) \cdot t$. The size of α will be defined by the number of its elements and denoted by $|\alpha|$.

Definition 1 (st-system). *A* step transition system *(or st-system) over* T *is a triple*

$$STS = (Q, A, q_0)$$

consisting of a set of states Q *including the initial state* $q_0 \in Q$, *and a set of* labelled arcs $A \subseteq Q \times \mathcal{P}(T) \times Q$. *It is assumed that:*

- *the transition relation is deterministic, i.e., if* $(q, \alpha, q') \in A$ *and* $(q, \alpha, q'') \in A$ *then* $q' = q''$
- *each state is reachable, i.e., if* $q \in Q$ *then there are steps* $\alpha_1, \ldots, \alpha_n$ $(n \geq 0)$ *and states* $q_1, \ldots, q_n = q$ *such that* $(q_{i-1}, \alpha_i, q_i) \in A$ *for* $1 \leq i \leq n$.

We introduce the following notations:

- $q \xrightarrow{\alpha} q'$ and $q \xrightarrow{\alpha}$ whenever $(q, \alpha, q') \in A$.
- $En_{STS}(q) = \{\alpha \mid q \xrightarrow{\alpha}\}$ is the set of all steps enabled at a state q.
- $active_{STS}(q) = \bigcup\{\alpha \mid q \xrightarrow{\alpha}\}$ is the set of all transitions active at a state q (the transitions that feature in the steps enabled at q).
- $En_{STS} = \{\alpha \mid \exists q \in Q : q \xrightarrow{\alpha}\}$ is the set of all the enabled steps of STS.
- $max(q) = \{\alpha \in En_{STS}(q) \mid \forall \beta \in En_{STS}(q) : \alpha \not\subset \beta\}$ is the set of all maximal steps enabled at a state q.

2.2 PT-nets

A *PT-net* is a tuple $\mathcal{N} = (P, T, W, M_0)$, where P and T are disjoint sets of respectively *places* and *transitions*, $W : (P \times T) \cup (T \times P) \to \mathbb{N}$ is an arc weight function, and $M_0 : P \to \mathbb{N}$ is an *initial marking* (in general, any mapping $M : P \to \mathbb{N}$ is a marking). We will use the standard conventions concerning the graphical representation of PT-nets, as illustrated in Figure 3(a).

For every element $x \in P \cup T$, we denote

$${}^{\bullet}x = \{y \mid W(y, x) > 0\} \text{ (pre-set of } x\text{)},$$

$$x^{\bullet} = \{y \mid W(x, y) > 0\} \text{ (post-set of } x\text{)}.$$

If $x \in T$, we will call $p \in {}^{\bullet}x$ a pre-place of x and $p \in x^{\bullet}$ a post-place of x. The dot-notation extends in the usual way to sets of elements, for example, ${}^{\bullet}X = \bigcup_{x \in X} {}^{\bullet}x$.

Moreover, for every place $p \in P$ and step $\alpha \in \mathcal{P}(T)$, we denote:

$$W(p, \alpha) = \sum_{t \in T} \alpha(t) \cdot W(p, t) \text{ and } W(\alpha, p) = \sum_{t \in T} \alpha(t) \cdot W(t, p) .$$

In other words, $W(p, \alpha)$ gives the number of tokens that the firing of α removes from p, and $W(\alpha, p)$ is the total number of tokens inserted into p after the execution of α.

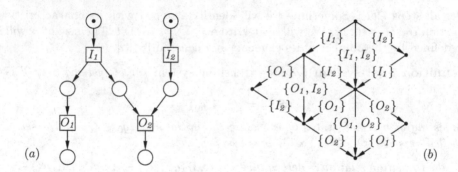

Fig. 3. Net N (a), and its concurrent reachability graph $CRG(N)$ (b)

Given a PT-net $\mathcal{N} = (P, T, W, M_0)$, a step $\alpha \in \mathcal{P}(T)$ is *enabled* and may be *fired* at a marking M if, for every place $p \in P$:

$$M(p) \geq W(p, \alpha) . \tag{1}$$

We denote this by $M[\alpha\rangle$. (For a singleton step $\alpha = \{t\}$, we will write $M[t\rangle$ rather than $M[\{t\}\rangle$.) Firing such a step leads to the marking M', for every place $p \in P$ defined by:

$$M'(p) = M(p) - W(p, \alpha) + W(\alpha, p) . \tag{2}$$

We denote this by $M[\alpha\rangle M'$.

The *concurrent reachability graph* $CRG(\mathcal{N})$ of $\mathcal{N}$ is the st-system $CRG(\mathcal{N}) = ([M_0\rangle, A, M_0)$ over T where:

$$[M_0\rangle = \{M_n \mid \exists \alpha_1, \ldots, \alpha_n \ \exists M_1, \ldots, M_{n-1} \ \forall 1 \leq i \leq n \ : \ M_{i-1}[\alpha_i\rangle M_i\} \tag{3}$$

is the set of reachable markings and $(M, \alpha, M') \in A$ iff $M[\alpha\rangle M'$. Figure 3(b) shows the concurrent reachability graph of the PT-net in Figure 3(a). Furthermore, we will call $\alpha_1 \ldots \alpha_n$, as in the formula (3), a *step sequence* and write $M_0[\alpha_1 \ldots \alpha_n\rangle M_n$.

Definition 2 (sequential conflict). *Two distinct transitions, $t, t' \in T$, are in* sequential conflict *at a marking M if $M[t\rangle$ and $M[t'\rangle$, but $M[tt'\rangle$ does not hold.*

Definition 3 (concurrent conflict). *Two distinct transitions, $t, t' \in T$, are in* concurrent conflict *at a marking M if $M[t\rangle$ and $M[t'\rangle$, but $M[\{t, t'\}\rangle$ does not hold.*

Note that sequential conflict implies concurrent conflict, but not necessarily vice versa.

Definition 4 (safe net). *A PT-net $\mathcal{N} = (P, T, W, M_0)$ is safe if*

$$\forall p \in P \ \forall M \in [M_0\rangle : \ M(p) \leq 1 .$$

In view of the above definition, the markings of safe nets can be treated as subsets of the set of places P, where a marking is a set of places for which $M(p) = 1$.

3 Step Persistence in Nets

Muller's speed independent theory provided a unique method for guaranteeing hazard-freeness of asynchronous circuits [14]. The *semimodularity* condition in this work required that an excitation of a circuit element must not be removed until absorbed by the system [18]. This condition was identified by Keller in [11] [2] to be the same as the property of persistence in his named transition system model for asynchronous parallel computation. Thus satisfying the property of persistence became one of the key requirements when designing hazard-free asynchronous circuits.

Later, the idea of persistence was investigated in many papers, for example, in [1,2,3,4,7,13,19]. However, with the exception of [7], it was only considered in the context of sequential executions of systems, and defined for transitions (rather than steps) as follows:

Definition 5 (persistent net, [13]). *A PT-net* $\mathcal{N} = (P, T, W, M_0)$ *is persistent if, for all distinct transitions* $t, t' \in T$ *and any reachable marking* M, $M[t\rangle$ *and* $M[t'\rangle$ *imply* $M[tt'\rangle$.

We can re-write this definition from the point of view of single transition as follows:

Definition 6 (persistent transition). *A transition* $t \in T$ *is persistent in a PT-net* $\mathcal{N} = (P, T, W, M_0)$ *if*

$$\forall M \in [M_0\rangle \ \ \forall t' \in T \setminus \{t\} : \ M[t\rangle \wedge M[t'\rangle \Longrightarrow M[t't\rangle.$$

The following definition gives three versions (A, B and C) of a definition of a persistent step. In each case, we try to capture the fact that a persistent step, which is enabled at some reachable marking M, cannot be disabled by another enabled step. The difference in the versions lies either in the different understanding of what 'not to be disabled' means or what we mean by a 'different' step. Notice that the introduced notions of a persistent step are defined globally and the required conditions must be satisfied at all the markings, where a candidate for a persistent step is enabled. A more detailed analysis of persistent steps, which considers also steps that are locally persistent, is presented in [12].

Definition 7 (persistent step in a net). *A step* $\alpha \in \mathcal{P}(T)$ *is* A-*persistent,* B-*persistent and* C-*persistent in a PT-net* $\mathcal{N} = (P, T, W, M_0)$ *if respectively the following hold:*

(A) $\forall M \in [M_0\rangle \ \forall \beta \in \mathcal{P}(T) : \ M[\alpha\rangle \wedge M[\beta\rangle \wedge \beta \neq \alpha \qquad \Longrightarrow M[\beta(\alpha \setminus \beta)\rangle$

(B) $\forall M \in [M_0\rangle \ \forall \beta \in \mathcal{P}(T) : \ M[\alpha\rangle \wedge M[\beta\rangle \wedge \beta \cap \alpha = \varnothing \Longrightarrow M[\beta\alpha\rangle$

(C) $\forall M \in [M_0\rangle \ \forall \beta \in \mathcal{P}(T) : \ M[\alpha\rangle \wedge M[\beta\rangle \wedge \beta \neq \alpha \qquad \Longrightarrow M[\beta\alpha\rangle \,.$

As can be easily seen, each of the three versions is a conservative extension of the standard definition of a persistent transition (see Definition 6). A-persistence

---

[2] Keller ([11]) was the first to consider persistence in the context of Petri nets.

requires that only unexecuted part of a step α should not be disabled, while B-persistence and C-persistence insist on continued enabledness of the persistent step α. In B-persistence, two steps are considered different if their intersection is empty, while for A-persistence and C-persistence it is enough if different steps do not coincide (but they can have non-empty intersection). It turns out that A-persistence and B-persistence are equivalent, as is shown in the following proposition.

Proposition 1. *Let* $\alpha \in \mathcal{P}(T)$ *be a step of a PT-net* $\mathcal{N} = (P, T, W, M_0)$. *Then* α *is* A*-persistent in* $\mathcal{N}$ *iff* α *is* B*-persistent in* $\mathcal{N}$.

Proof. Suppose that α and $\beta \in \mathcal{P}(T)$ are two different non-empty steps enabled at a marking M of $\mathcal{N}$ (notice that empty step is trivially persistent according to A, B or C-persistence defined in Definition 7).
$\implies$ Let α be A-persistent in $\mathcal{N}$, and so $M[\beta(\alpha \setminus \beta)\rangle$. Suppose $\beta \cap \alpha = \varnothing$. Then we have $\alpha \setminus \beta = \alpha$, implying $M[\beta\alpha\rangle$. Hence, α is B-persistent in $\mathcal{N}$.
$\impliedby$ Let α be B-persistent in $\mathcal{N}$, and so $M[(\beta \setminus \alpha)\alpha\rangle$ as $(\beta \setminus \alpha) \cap \alpha = \varnothing$. Hence, for every place $p \in P$:

$$M(p) - W(p, \beta \setminus \alpha) + W(\beta \setminus \alpha, p) \geq W(p, \alpha)$$

which implies:

$$M(p) - W(p, \beta) + W(p, \beta \cap \alpha) + W(\beta, p) - W(\beta \cap \alpha, p) \geq W(p, \alpha) .$$

Hence we have

$$\begin{aligned} M(p) - W(p, \beta) + W(\beta, p) &\geq W(p, \alpha) - W(p, \beta \cap \alpha) + W(\beta \cap \alpha, p) \\ &= W(p, \alpha \setminus \beta) + W(\beta \cap \alpha, p) \\ &\geq W(p, \alpha \setminus \beta) \end{aligned}$$

implying $M[\beta(\alpha \setminus \beta)\rangle$. Hence α is A-persistent. □

It is easy to see that C-persistence is stronger than the other two notions. Figure 4 (taken from [12]) shows an example of a step, $\{t, \bar{t}\}$, which is A-persistent, but not C-persistent. The step $\{t, \bar{t}\}$ there is only enabled at marking M_3, where also $\{\bar{t}\}$ and $\{t\}$ steps are enabled. After executing step $\{\bar{t}\}$, step $\{t, \bar{t}\}$ is still enabled, but after executing step $\{t\}$, only unexecuted part of $\{t, \bar{t}\}$ (that means $\{\bar{t}\}$) is enabled.

For a safe PT-net $\mathcal{N}$, C-persistent non-singleton steps are built out of transitions lying on self-loops. To show this, we first prove an auxiliary result.

Proposition 2. *Let* α *be a* C*-persistent step of a safe PT-net* $\mathcal{N} = (P, T, W, M_0)$ *enabled at a reachable marking* M *of* $\mathcal{N}$. *Then* ${}^\bullet(\alpha \cap \beta) = (\alpha \cap \beta)^\bullet$ *for every step* $\beta \neq \alpha$ *enabled at* M *in* $\mathcal{N}$.

Proof. Let $\alpha \cap \beta \neq \varnothing$ (for $\alpha \cap \beta = \varnothing$ the proposition holds). Suppose $p \in {}^\bullet(\alpha \cap \beta)$ for some step $\beta \neq \alpha$ enabled at M in $\mathcal{N}$. This and $M[\alpha\rangle$ (as α is enabled at M) imply $M(p) = 1$. Since α is C-persistent, there exists a marking M' such

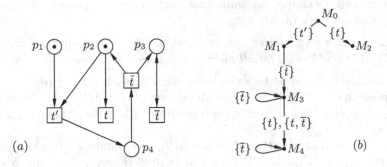

Fig. 4. A safe net $\mathcal{N}$ (a), and its concurrent reachability graph $CRG(\mathcal{N})$ (b). Step $\{t, \bar{t}\}$ is A-persistent, but not C-persistent

that $M[\beta\rangle M'[\alpha\rangle$. Again, as $M'[\alpha\rangle$ and $p \in {}^{\bullet}(\alpha \cap \beta)$, we have $M'(p) = 1$. From $M[\beta\rangle M'$, we have $M'(p) = M(p) - W(p, \beta) + W(\beta, p)$, and so $\sum_{t \in T} \beta(t) \cdot W(t, p) = \sum_{t \in T} \beta(t) \cdot W(p, t)$. Let $\beta = \{t_1, \ldots, t_n\}$. Then, $W(t_1, p) + \ldots + W(t_n, p) = W(p, t_1) + \ldots + W(p, t_n)$ and, by $\mathcal{N}$ being safe, all the arc weights in this formula are 0 or 1. We now consider two cases:

1. There is $i \leq n$ such that $W(t_i, p) = 1$ and $W(p, t_i) = 0$. Since $M[t_i\rangle$ (as $M[\beta\rangle$), $M(p) = 1$ and $\mathcal{N}$ is safe, we have a contradiction, because t_i, when fired, would deposit another token in p.
2. There is $j \leq n$ such that $W(t_j, p) = 0$ and $W(p, t_j) = 1$. This means that there is $i \leq n$, $W(t_i, p) = 1$ and $W(p, t_i) = 0$. But this was already ruled out by the first case. So, contradiction again.

As a result, for each transition $t_i \in \beta$, $W(p, t_i) = W(t_i, p)$. Hence $p \in (\alpha \cap \beta)^{\bullet}$. Consequently, ${}^{\bullet}(\alpha \cap \beta) \subseteq (\alpha \cap \beta)^{\bullet}$.

Suppose now that $p \in (\alpha \cap \beta)^{\bullet} \setminus {}^{\bullet}(\alpha \cap \beta)$. Then, by $M[\alpha \cap \beta\rangle$ and the safeness of $\mathcal{N}$, $M(p) = 0$. Hence, by $M[\alpha\rangle$ and $M[\beta\rangle$, we must have $p \notin {}^{\bullet}\alpha \cup {}^{\bullet}\beta$. Consequently, since there is M'' such that $M[\beta\alpha\rangle M''$, we obtain $M''(p) \geq 2$, a contradiction with $\mathcal{N}$ being safe. Hence ${}^{\bullet}(\alpha \cap \beta) = (\alpha \cap \beta)^{\bullet}$.   □

Theorem 1. *Let α be a non-singleton C-persistent step of a safe PT-net $\mathcal{N} = (P, T, W, M_0)$ which is enabled in at least one reachable marking. Then all the transitions of α lie on self-loops, i.e., ${}^{\bullet}t = t^{\bullet}$ for $t \in \alpha$.*

Proof. If $\alpha = \varnothing$ the theorem holds. Let $|\alpha| \geq 2$. Suppose that $t \in \alpha$ and M be a reachable marking enabling α. Since $\{t\} \neq \alpha$ and $M[t\rangle$ for any marking M such that $M[\alpha\rangle$, we have, from Proposition 2, ${}^{\bullet}(\alpha \cap \{t\}) = (\alpha \cap \{t\})^{\bullet}$. Hence ${}^{\bullet}t = t^{\bullet}$.   □

We now want to relate the persistence of a step with the persistence of its constituent transitions in safe nets. We first consider A-persistent steps, but as we already know the results will also hold for B-persistent steps.

First, we prove a simple, but important, fact concerning pre-sets and post-sets of transitions in steps of safe nets.

Fact 1. *If α is a step enabled in a reachable marking of a safe PT-net $\mathcal{N}$, then $({}^\bullet t \cup t^\bullet) \cap ({}^\bullet u \cup u^\bullet) = \varnothing$, for all distinct transitions $t, u \in \alpha$.*

Proof. Let $t, u \in \alpha$ and M be a reachable marking such that $M[\alpha\rangle$.

Suppose that $p \in {}^\bullet t \cap {}^\bullet u$. Since $M[\alpha\rangle$, we obtain $M[\{t, u\}\rangle$. That means $M(p) \geq W(p, t) + W(p, u) = 2$, a contradiction with $\mathcal{N}$ being safe. As a result, ${}^\bullet t \cap {}^\bullet u = \varnothing$.

Suppose now that $p \in t^\bullet \cap u^\bullet$. Since $M[\alpha\rangle$, we obtain $M[\{t, u\}\rangle M'$. Hence $M'(p) = M(p) - W(p, t) - W(p, u) + W(t, p) + W(u, p)$. As t and u cannot share a pre-place, $W(p, t)$ and $W(p, u)$ cannot both be 1. If one of them has p as its pre-place, then $M(p) = 1$, as otherwise one of the transitions would not be enabled at M, and both are enabled at M. So, the right hand side of the equation yields 2, but the left hand side cannot as the net is safe. We have a contradiction. As a result, $t^\bullet \cap u^\bullet = \varnothing$.

Suppose now that $p \in t^\bullet \cap {}^\bullet u$. By $M[u\rangle$, $M(p) = 1$. On the other hand, we know that $p \notin {}^\bullet t \cap {}^\bullet u$ and so $p \in t^\bullet \setminus {}^\bullet t$. Since $M[t\rangle M''$, for some marking M'', we obtain $M''(p) = 2$, a contradiction with $\mathcal{N}$ being safe. Hence, $t^\bullet \cap {}^\bullet u = \varnothing$. □

Theorem 2. *Let α be a step in a safe PT-net $\mathcal{N} = (P, T, W, M_0)$ which is enabled in at least one reachable marking. If all the transitions in α are persistent in $\mathcal{N}$, then α is A-persistent in $\mathcal{N}$.*

Proof. Let M be a reachable marking and $\beta \neq \alpha$ be a step in $\mathcal{N}$ such that $M[\alpha\rangle$ and $M[\beta\rangle$. We need to show that $M[\beta(\alpha \setminus \beta)\rangle$.

Assume that $\alpha \cap \beta = \{t_1, \ldots, t_m\}$, $\alpha \setminus \beta = \{w_1, \ldots, w_n\}$ and $\beta \setminus \alpha = \{u_1, \ldots, u_k\}$. Note that all the transitions in these three sets are different. From $M[\beta\rangle$ we have $M[t_1 \ldots t_m u_1 \ldots u_k\rangle$. Now, since each w_i is persistent and enabled at M, we have that $M[t_1 \ldots t_m u_1 \ldots u_k w_1 \ldots w_n\rangle$. Since α and β are steps in a safe net $\mathcal{N}$ enabled at some marking (M), we have, from Fact 1, that transitions in α and β have disjoint pre-sets and post-sets. Hence we have $M[\beta(\alpha \setminus \beta)\rangle$. □

We now consider C-persistent steps. In this case the antecedent in the implication is stronger.

Theorem 3. *Let α be a step in a safe PT-net $\mathcal{N} = (P, T, W, M_0)$ which is enabled in at least one reachable marking. If all the transitions in α are persistent and lie on self-loops in $\mathcal{N}$, then α is C-persistent in $\mathcal{N}$.*

Proof. Let M be a reachable marking and $\beta \neq \alpha$ be a step in $\mathcal{N}$ such that $M[\alpha\rangle$ and $M[\beta\rangle$. We need to show that $M[\beta\alpha\rangle$.

Proceeding similarly as in the previous proof we can show that

$$M[t_1 \ldots t_m u_1 \ldots u_k w_1 \ldots w_n\rangle \, .$$

Now, since all transitions in α lie on self-loops we further obtain

$$M[t_1 \ldots t_m u_1 \ldots u_k t_1 \ldots t_m w_1 \ldots w_n\rangle \, .$$

Hence we have $M[\beta\alpha\rangle$. □

In both Theorem 2 and Theorem 3, the implications in the opposite direction do not hold. A counterexample is shown in the Figure 5.

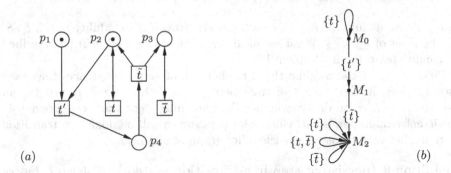

Fig. 5. A safe net $\mathcal{N}$ (a), and its concurrent reachability graph $CRG(\mathcal{N})$ (b). A persistent step $\alpha = \{t, \bar{t}\}$ of $\mathcal{N}$ contains a non-persistent transition t.

Observe that step $\alpha = \{t, \bar{t}\}$ in Figure 5 is both A-persistent and C-persistent, but $t \in \alpha$ is not persistent at $M_0 = \{p_1, p_2\}$, because there exists $t' \neq t$ such that $M_0[t\rangle$ and $M_0[t'\rangle$, but $M_0[t't\rangle$ does not hold.

4   Pruning Reachability Graphs

The main motivation for studying persistent steps in this paper is to discover which sets of transitions can be executed synchronously and therefore be treated as some kind of 'atomic actions', giving rise to new 'bigger' transitions, which would execute in a 'hazard-free' way. We will call them *bundles*. Looking at our application area of asynchronous circuits, bundling actions would reduce signal management by merging concurrent signals into one event. This merging must be done in a consistent fashion. The best candidates for bundles are persistent steps, but if we want to form 'bigger' transitions from them, we must make sure that one enabled persistent step does not include another enabled persistent step. All the transitions in a bundle must always appear together, in the same configurations. In the ideal situation (we say ideal, because it might be difficult to achieve), we do not want to allow, for example, three persistent steps $\{a, b\}$, $\{a\}$ and $\{b\}$ to be enabled in a given transition system. We need to choose: either to opt for $\{a, b\}$ and delete $\{a\}$ and $\{b\}$, or the other way round. Therefore, we need to develop an algorithm which, for a given net $\mathcal{N} = (P, T, W, M_0)$, would allow us to prune its reachability graph $CRG(\mathcal{N})$ in such a way that all persistent steps would satisfy an additional 'non-inclusion' condition. The 'pruned' transition system would represent the desired behaviour, which then we would like to implement in a form of a Petri net in a process of synthesis.

First we define a sub-st-system, which will be obtained as a result of pruning a given reachability graph.

Definition 8 (sub-st-system). *An st-system* $STS = (Q, A, q_0)$ *is a* sub-st-system *of an st-system* $STS' = (Q', A', q_0)$ *if* $Q \subseteq Q'$, $A \subseteq A'$ *and, for every* $q \in Q$, $active_{STS}(q) = active_{STS'}(q)$. *We denote this by* $STS \preccurlyeq STS'$.

In the above definition, En_{STS} of a 'properly pruned' reachability graph STS' will be a set of *bundles*. What we mean by 'properly pruned' will be described by conditions stated in Problem 1.

First, we need to re-define the three notions of step persistence, that were used for nets, in the context of transition systems. Once we start pruning an st-system, we need to decide whether the remaining steps that were previously persistent remain persistent. Checks for persistence will be done in a transition system (that might not be a reachability graph of any net).

Definition 9 (persistent step in a transition system). *A step* $\alpha \in En_{STS}$ *is* A-persistent, B-persistent *and* C-persistent *in an st-system* $STS = (Q, A, q_0)$ *if respectively the following hold for all states* $q \in Q$ *and steps* β *such that* $\xleftarrow{\alpha} q \xrightarrow{\beta}$:

$$(A) \qquad \beta \neq \alpha \implies q \xrightarrow{\beta(\alpha \setminus \beta)}$$
$$(B) \qquad \beta \cap \alpha = \varnothing \implies q \xrightarrow{\beta \alpha}$$
$$(C) \qquad \beta \neq \alpha \implies q \xrightarrow{\beta \alpha} .$$

Remark 1. A step α is A-persistent in a net $\mathcal{N}$ iff α is A-persistent in a transition system $CRG(\mathcal{N})$. The same can be said in the case of B- and C-persistence.

We have the following relationships between the three step persistence notions defined for transition systems.

Proposition 3. *Let* $STS = (Q, A, q_0)$ *be a st-system.*

1. *If* $\alpha \in En_{STS}$ *is* A-*persistent, then it is also* B-*persistent.*
2. *If* $\alpha \in En_{STS}$ *is* C-*persistent, then it is also* B-*persistent.*

Proof. (1) Let $q \in Q$ and $\xleftarrow{\alpha} q \xrightarrow{\beta}$ be such that $\beta \cap \alpha = \varnothing$. Since $\alpha \in En_{STS}$ is A-persistent, we have $q \xrightarrow{\beta(\alpha \setminus \beta)}$. Hence $q \xrightarrow{\beta \alpha}$ which means that $\alpha \in En_{STS}$ is B-persistent.

(2) Follows directly from Definition 9. □

Note that in the class of general step transition systems, B-persistence does not imply A-persistence of steps, as it was proved for nets. Indeed, let $\alpha \in En_{STS}$ be B-persistent step in STS, and $\beta \neq \alpha$ and $q \in Q$ be such that $\xleftarrow{\alpha} q \xrightarrow{\beta}$. We know that $\beta \cap (\alpha \setminus \beta) = \varnothing$. However, with such assumptions, we cannot in general guarantee that $q \xrightarrow{\alpha \setminus \beta}$. Though latter is true for concurrent reachability graphs of PT-nets, we must also consider step transition systems resulting from the pruning of such reachability graphs.

Problem 1. *Let $\mathcal{N}$ be a PT-net and $CRG(\mathcal{N})$ be its concurrent reachability graph. Find an st-system STS such that $STS \preccurlyeq CRG(\mathcal{N})$ and additionally satisfying $(D)\&(E)$ or $(D)\&(F)$, where the three conditions are defined as follows:*

(D) All steps in En_{STS} are B-persistent in STS.[3]
(E) $\alpha \not\subset \beta$ for all nonempty different steps $\alpha, \beta \in En_{STS}$.
(F) $\alpha \not\subset \beta$ for all states q and all nonempty different steps $\alpha, \beta \in En_{STS}(q)$.

We denote this respectively by

$$STS \preccurlyeq^{global}_{pers} CRG(\mathcal{N}) \quad and \quad STS \preccurlyeq^{local}_{pers} CRG(\mathcal{N}).$$

We also refer to the condition described in (E) as global non-inclusion, and to the condition described in (F) as local non-inclusion.

The difference between $\preccurlyeq^{global}_{pers}$ and $\preccurlyeq^{local}_{pers}$ is that the latter only requires non-inclusion of bundles locally for each state, whereas the former insists that non-inclusion holds globally. We therefore have

Proposition 4. *$STS \preccurlyeq^{global}_{pers} CRG(\mathcal{N})$ implies $STS \preccurlyeq^{local}_{pers} CRG(\mathcal{N})$.*

In our first attempt to solve Problem 1, we will concentrate on PT-nets that are persistent according to Definition 5. We then have the following result.

Theorem 4. *If $\mathcal{N}$ is persistent (according to Definition 5), then there is at least one STS satisfying $STS \preccurlyeq^{global}_{pers} CRG(\mathcal{N})$.*

Proof. It suffices to take $CRG(\mathcal{N})$ and delete all non-singleton (nonempty) steps. □

As the above proof produces completely sequential solution, we will now search for a more concurrent one. We will also require that the original PT-net is not only persistent, but as well safe.

Proposition 5. *If $\mathcal{N}$ is persistent (according to Definition 5) and safe, then every step $\alpha \in En_{CRG(\mathcal{N})}$ is B-persistent in $CRG(\mathcal{N})$.*

Proof. Let $\alpha \in En_{CRG(\mathcal{N})}$. If $\mathcal{N}$ is persistent (according to Definition 5), all transitions in α are persistent (according to Definition 6). Hence, from Theorem 2 and the fact that $\mathcal{N}$ is safe, we have that α is A-persistent in $\mathcal{N}$, and also B-persistent in $\mathcal{N}$ (see Proposition 1). Following Remark 1, we conclude that α is B-persistent in $CRG(\mathcal{N})$. □

The above proposition guarantees B-persistence of steps in $CRG(\mathcal{N})$ of a persistent and safe net $\mathcal{N}$, but the non-inclusion conditions $((E)$ or $(F))$ are, in general, not satisfied in $CRG(\mathcal{N})$, as for all its states q, $q \xrightarrow{\alpha}$ implies $q \xrightarrow{\beta}$ for any step $\beta \subset \alpha$. To satisfy the non-inclusion conditions, we need to prune $CRG(\mathcal{N})$ in such a way that B-persistence of steps is maintained. We will now

---

[3] Alternatively, we could require A-persistence or C-persistence. We opted here for B-persistence, because it is the weakest of the three notions.

explore what happens if we decide to prune all but the maximal steps at every reachable marking.

In what follows the st-system $CRG^{max}(\mathcal{N})$ is obtained from $CRG(\mathcal{N})$, the concurrent reachability graph of a PT-net $\mathcal{N}$, by deleting at every reachable marking M, all the arcs labelled by non-maximal non-empty[4] steps, and then removing the nodes that became unreachable from the initial marking by the removal of such steps.

Proposition 6. $CRG^{max}(\mathcal{N}) \preccurlyeq CRG(\mathcal{N})$.

Proof. Follows from definitions and the fact that for each enabled step there is a maximal step enabled at the same marking.   □

Proposition 7. $CRG^{max}(\mathcal{N})$ *satisfies condition* (F) *from Problem 1.*

Proof. Follows from the fact that maximal steps are in-comparable (see definition of maximal steps in Section 2.1).   □

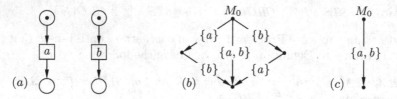

Fig. 6. A persistent and safe net $\mathcal{N}$ (a), its concurrent reachability graph $CRG(\mathcal{N})$ (b), and $CRG^{max}(\mathcal{N}) \preccurlyeq_{pers}^{local} CRG(\mathcal{N})$ obtained in the pruning procedure (c)

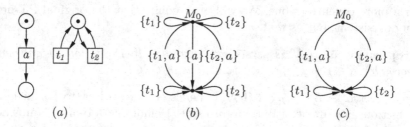

Fig. 7. A persistent and safe net $\mathcal{N}$ (a), its concurrent reachability graph $CRG(\mathcal{N})$ (b), and $CRG^{max}(\mathcal{N}) \preccurlyeq_{pers}^{local} CRG(\mathcal{N})$ obtained in the pruning procedure (c)

Figures 6, 7, 8 and 9 show the examples of persistent and safe nets for which the described pruning procedure works as their $CRG^{max}(\mathcal{N})$ graphs contain only B-persistent steps. In all the mentioned examples the pruned reachability graph satisfies $CRG^{max}(\mathcal{N}) \preccurlyeq_{pers}^{local} CRG(\mathcal{N})$, and in case of the example in Figure 6, we even have $CRG^{max}(\mathcal{N}) \preccurlyeq_{pers}^{global} CRG(\mathcal{N})$. So, the pruning procedure helped to

---

[4] For technical reasons we do not want to delete empty steps, as they might be important in future algorithms.

achieve local non-inclusion without jeopardising B-persistence of the remaining steps. Notice, however, that in Figures 7(c) and 9(c) this B-persistence in initial markings is achieved only, because the steps enabled there are not disjoint and therefore satisfy B-persistence condition trivially.

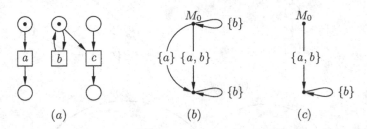

Fig. 8. A persistent and safe net $\mathcal{N}$ (a), its concurrent reachability graph $CRG(\mathcal{N})$ (b), and $CRG^{max}(\mathcal{N}) \preceq^{local}_{pers} CRG(\mathcal{N})$ obtained in the pruning procedure (c)

In general, pruning non-maximal steps may make some of the remaining steps non-B-persistent. Figure 10 shows that the initially enabled step $\{b\}$ is not B-persistent after the pruning procedure. After executing step $\{a\}$ it is not longer enabled. Instead step $\{b, c\}$ is enabled, because it was the maximal step in the marking M. Having said that, we propose a weaker version of condition (B) which holds for safe and persistent PT-nets.

Proposition 8. *If $\mathcal{N}$ is persistent (according to Definition 5) and safe, then, for every marking M in $CRG^{max}(\mathcal{N})$, $\xleftarrow{\alpha} M \xrightarrow{\beta}$ implies:*

$$(B') \ \beta \cap \alpha = \varnothing \implies \exists \gamma: \ \alpha \subseteq \gamma \wedge M \xrightarrow{\beta\gamma} .$$

Proof. From Proposition 5 we know that $M \xrightarrow{\beta\alpha}$ in $CRG(\mathcal{N})$. Moreover, there is a maximal step γ available (as it remained after pruning) after executing β from M such that $\alpha \subseteq \gamma$. Hence $M \xrightarrow{\beta\gamma}$ in $CRG^{max}(\mathcal{N})$. □

Hence, pruning non-maximal steps may result in the loss of persistence when $\alpha \subset \gamma$ in (B'). In such a case we may, however, 'repair' $\mathcal{N}$ by making the step γ non-enabled. The mechanism for achieving this is simple, namely we select one transition from α, one transition from $\gamma \setminus \alpha$, and make sure that they cannot be executed simultaneously.

Let $\mathcal{N}$ be a PT-net and $t \neq u$ be two transitions. Then $\mathcal{N}_{t \leftrightarrow u}$ is a PT-net obtained from $\mathcal{N}$ by adding a new place p marked initially with one token, and such that $W(p, t) = W(t, p) = W(p, u) = W(u, p) = 1$. This construction is illustrated in Figure 11, where we try to fix the problem of the net $\mathcal{N}$ in Figure 10. We added a new place p and chose b and c as our t and u (the only choice in this example) creating a new net $\mathcal{N}' = \mathcal{N}_{b \leftrightarrow c}$. The new place disables the concurrent step $\{b, c\}$ at M leaving only the singleton steps $\{b\}$ and $\{c\}$ enabled at M. They are now maximal steps at this marking. In fact, in this simple example,

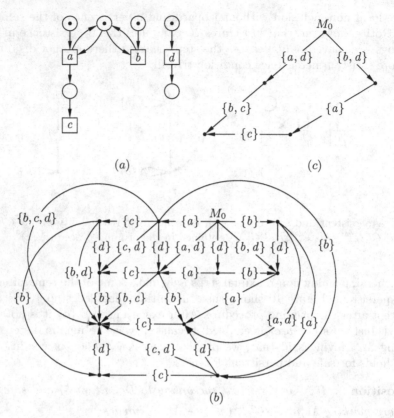

Fig. 9. A persistent and safe net $\mathcal{N}$ (a), its concurrent reachability graph $CRG(\mathcal{N})$ (b), and $CRG^{max}(\mathcal{N}) \preceq_{pers}^{local} CRG(\mathcal{N})$ obtained in the pruning procedure (c)

we have only singleton steps in the concurrent reachability graph, which makes it maximal without pruning.

In the following propositions we show that after the proposed modification the net generates a reachability graph, which is the sub-st-system of the reachability graph of the initial net. Also, the modified net is still persistent (according to Definition 5) and safe.

Proposition 9. *Let $\mathcal{N}$ be persistent (according to Definition 5) and safe net. The reachable markings of $CRG(\mathcal{N}_{t\leftrightarrow u})$ and $CRG(\mathcal{N})$ are the same, if we identify each reachable marking M of $\mathcal{N}$ with the reachable marking $M \cup \{p\}$ of $\mathcal{N}_{t\leftrightarrow u}$. Furthermore, $CRG(\mathcal{N}_{t\leftrightarrow u}) \preceq CRG(\mathcal{N})$.*

Proof. Follows from definitions.   □

Proposition 10. *If $\mathcal{N}$ is persistent (according to Definition 5) and safe, then $CRG(\mathcal{N}_{t\leftrightarrow u})$ is also persistent (according to Definition 5) and safe.*

Proof. Follows from definitions.   □

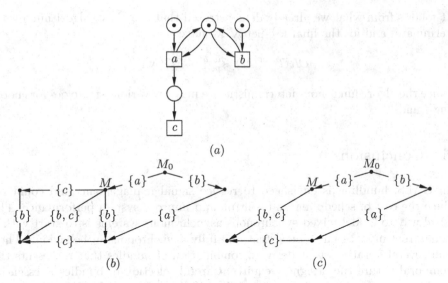

(a)

(b)   (c)

Fig. 10. A persistent and safe net $\mathcal{N}$ (a), its concurrent reachability graph $CRG(\mathcal{N})$ (b), and $CRG^{max}(\mathcal{N})$ obtained in the pruning procedure, which does not satisfy $CRG^{max}(\mathcal{N}) \preccurlyeq_{pers}^{local} CRG(\mathcal{N})$ (c)

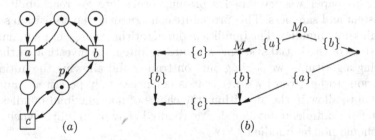

(a)   (b)

Fig. 11. Fixing the problem in the example in Figure 10: persistent and safe net $\mathcal{N}' = \mathcal{N}_{b \leftrightarrow c}$ (a), its concurrent reachability graph $CRG(\mathcal{N}') = CRG^{max}(\mathcal{N}')$, which trivially satisfies $CRG^{max}(\mathcal{N}') \preccurlyeq_{pers}^{local} CRG(\mathcal{N}')$ (and also $CRG^{max}(\mathcal{N}') \preccurlyeq_{pers}^{global} CRG(\mathcal{N}')$) (b)

We can now propose a dynamic way of pruning embodied by the following algorithm:

> **while** $\neg(CRG^{max}(\mathcal{N}) \preccurlyeq_{pers}^{local} CRG(\mathcal{N}))$ **do**
>
> > **choose** M, α, β, γ in $CRG^{max}(\mathcal{N})$ *satisfying* (B') *with* $\alpha \subset \gamma$
> >
> > **choose** $t \in \alpha$, $u \in \gamma \setminus \alpha$
> >
> > $\mathcal{N} := \mathcal{N}_{t \leftrightarrow u}$

It follows from what we already demonstrated that the above algorithm always terminates and for the final PT-net $\mathcal{N}$ we have:

$$CRG^{max}(\mathcal{N}) \preccurlyeq_{pers}^{local} CRG(\mathcal{N}) .$$

Since the algorithm is non-deterministic, we may try various strategies for choosing t and u.

5   Conclusions

In GALS, bundling is envisaged to reduce signal management, and could reduce the cost of scheduling and control, and improve system performance. The ideal way to model mixed synchronous-asynchronous systems is to start with a concurrent model that is persistent and fully asynchronous in behaviour. Then run several iterations that derive a combination of bundles that represents the temporal nature the designer requires. Careful selection of bundles is essential so that the pruned behaviour of the fully asynchronous model still exhibits some characteristics of its parent and is persistent. Step persistence is hence an important characteristic that will guarantee true persistent behaviour for mixed synchronous-asynchronous models.

In this paper we developed a pruning procedure for reachability graphs of persistent and safe nets. This procedure constructs a step transition system that contains only bundles. The bundles in our algorithm represent maximally concurrent steps of the initial system. In future we intend to investigate other possible pruning algorithms, weakening our constraints and allowing the initial system's behaviour to be given by a net that is not necessarily persistent. Furthermore, we plan to allow in the algorithms the choice of non-maximal bundles in certain cases. For example, input signals are usually behaving in fully asynchronous way and should not be bundled.

Acknowledgments. We would like to thank the anonymous referees for their comments and useful suggestions. This research was supported by the EPSRC GAELS and UNCOVER projects, the 973 Program Grant 2010CB328102, and NSFC Grant 61133001.

References

1. Barylska, K., Ochmański, E.: Levels of persistency in place/transition nets. Fundamenta Informaticae 93, 33–43 (2009)
2. Barylska, K., Mikulski, Ł., Ochmański, E.: On persistent reachability in Petri nets. Information and Computation 223, 67–77 (2013)
3. Best, E., Darondeau, P.: Decomposition theorems for bounded persistent Petri nets. In: van Hee, K.M., Valk, R. (eds.) PETRI NETS 2008. LNCS, vol. 5062, pp. 33–51. Springer, Heidelberg (2008)

4. Best, E., Darondeau, P.: Separability in persistent Petri nets. In: Lilius, J., Penczek, W. (eds.) PETRI NETS 2010. LNCS, vol. 6128, pp. 246–266. Springer, Heidelberg (2010)
5. Chapiro, D.M.: Globally-asynchronous locally-synchronous systems. PhD Thesis, Stanford University (1984)
6. Cortadella, J., Kishinevsky, M., Kondratyev, A., Lavagno, L., Yakovlev, A.: Logic synthesis for asynchronous controllers and interfaces. Springer Series in Advanced Microelectronics, vol. 8. Springer (2002)
7. Dasgupta, S., Yakovlev, A.: Desynchronisation technique using Petri nets. Electronic Notes in Theoretical Computer Science 245, 51–67 (2009)
8. Davis, A., Nowick, S.M.: An introduction to asynchronous circuit design. The Encyclopedia of Computer Science and Technology (1997)
9. Gurkaynak, F., Oetiker, S., Kaeslin, H., Felber, N., Fichtner, W.: GALS at ETH Zurich: Success or failure? In: Proceedings of the 12th IEEE International Symposium on Asynchronous Circuits and Systems (2006)
10. Iyer, A., Marculescu, D.: Power and performance evaluation of globally asynchronous locally synchronous processors. In: 29th International Symposium on Computer Architecture, pp. 158–168. IEEE Computer Society (2002)
11. Keller, R.: A fundamental theorem of asynchronous parallel computation. In: Tse-Yun, F. (ed.) Parallel Processing. LNCS, vol. 24, pp. 102–112. Springer, Heidelberg (1975)
12. Koutny, M., Mikulski, Ł., Pietkiewicz-Koutny, M.: A taxonomy of persistent and nonviolent steps. In: Colom, J.-M., Desel, J. (eds.) PETRI NETS 2013. LNCS, vol. 7927, pp. 210–229. Springer, Heidelberg (2013)
13. Landweber, L.H., Robertson, E.L.: Properties of conflict-free and persistent Petri nets. JACM 25, 352–364 (1978)
14. Muller, D.E., Bartky, W.S.: A theory of asynchronous circuits. In: Proceedings of an International Symposium on the Theory of Switching, pp. 204–243. Harvard University Press (1959)
15. Myers, C.: Asynchronous circuit design. Wiley (2004)
16. Sparso, J., Furber, S.: Principles of asynchronous circuit design: a systems perspective. Kluwer Academic Publishers (2001)
17. Stevens, K.S., Gebhardt, D., You, J., Xu, Y., Vij, V., Das, S., Desai, K.: The future of formal methods and GALS design. Electronic Notes in Theoretical Computer Science 245, 115–134 (2009)
18. Yakovlev, A.: Designing self-timed systems. VLSI System Design VI, 70–90 (1985)
19. Yakovlev, A., Koelmans, A., Semenov, A., Kinniment, D.: Modelling, analysis and synthesis of asynchronous control circuits using Petri nets. Integration, the VLSI Journal 21, 143–170 (1996)

A Taxonomy of Persistent and Nonviolent Steps

Maciej Koutny[1], Łukasz Mikulski[1,2], and Marta Pietkiewicz-Koutny[1]

[1] School of Computing Science
Newcastle University
Newcastle upon Tyne, NE1 7RU, U.K.
[2] Faculty of Mathematics and Computer Science
Nicolaus Copernicus University
Toruń, Chopina 12/18, Poland

Abstract. A concurrent system is persistent if throughout its operation no activity which became enabled can subsequently be prevented from being executed by any other activity. This is often a highly desirable (or even necessary) property; in particular, if the system is to be implemented in hardware. Over the past 40 years, persistence has been investigated and applied in practical implementations assuming that each activity is a single atomic action which can be represented, for example, by a single transition of a Petri net. Recently, it turned out that to deal with the synthesis of GALS systems one also needs to consider activities represented by steps, each step being a set of simultaneously executed transitions. Moving into the realm of step based execution, semantics creates a wealth of new fundamental problems and questions. In particular, there are different ways in which the standard notion of persistence could be lifted from the level of sequential semantics to the level of step semantics. Moreover, one may consider steps which are persistent and cannot be disabled by other steps, as well as steps which are nonviolent and cannot disable other steps. In this paper, we provide a classification of different types of persistence and nonviolence, both for steps and markings of PT-nets. We also investigate behavioural and structural properties of such notions.

Keywords: persistence, nonviolence, step semantics, Petri net, taxonomy, behaviour, structure.

1 Introduction

A concurrent system is persistent [2–4, 7] if throughout its operation no activity which became enabled can subsequently be prevented from being executed by any other activity. This is often a highly desirable (or even necessary) property; in particular, if the system is to be implemented in hardware [5, 8]. Over the past 40 years, persistence has been investigated and applied in practical implementations assuming that each activity is a single atomic action which can be represented, for example, by a single transition of a Petri net (used as a formal representation of a concurrent system). In other words, persistence was considered assuming the sequential execution semantics of concurrent systems.

J.-M. Colom and J. Desel (Eds.): PETRI NETS 2013, LNCS 7927, pp. 210–229, 2013.

Recently, in the paper [6] we argued that the notion of persistence is restricted and in dealing with the synthesis of GALS systems one also needs to consider activities represented by steps (sets of simultaneously executed transitions). Moving into the realm of step based execution semantics creates a wealth of new fundamental problems and intriguing questions, some of which have been addressed in [6]. In particular, there are different ways in which the standard notion of persistence could be lifted from the level of sequential semantics to the level of step semantics. For example, if part of an enabled has been executed by another step, should we insist on the whole delayed step to be still enabled, or just its remaining part? Moreover, one may consider steps which are persistent and cannot be disabled by other steps, as well as steps which are nonviolent [1, 2] and cannot disable other steps. In this paper, we aim at providing a classification of different types of persistent and nonviolent steps taking PT-nets to be the system model in which the discussion is carried out. Moreover, we introduce and investigate persistence and nonviolence at the level of markings of PT-nets. We then investigate behavioural and structural properties of notions pertaining to persistence and nonviolence both for the general PT-nets and safe PT-nets.

The paper is organised as follows. In the next section, we present basic notions and notations used throughout. Section 3 introduces various types of persistent and nonviolent steps of transitions in PT-nets, and Section 4 provides their taxonomy. The following section extends the discussion of persistence and nonviolence to markings of PT-nets. In Section 6, we investigate the basic properties of persistent and nonviolent steps of transitions in PT-nets, and then, in Section 7, we focus specifically on the class of safe PT-nets.

2   Preliminaries

A PT-*net* is a tuple $N = (P, T, W, M_0)$, where P and T are finite disjoint sets of respectively *places* and *transitions*, $W : (P \times T) \cup (T \times P) \to \mathbb{N}$ is an arc weight function, and $M_0 : P \to \mathbb{N}$ is the *initial marking*. In general, any mapping $M : P \to \mathbb{N}$ is a *marking* of N, and if M' is a marking such that $M(p) \geq M'(p)$, for all $p \in P$, then we denote $M \geq M'$. We also use the standard conventions concerning the graphical representation of nets.

A *step* α of N is a set of its transitions, $\alpha \subseteq T$. We will use $\alpha, \beta, \gamma, \dots$ to range over the set of steps. For every place $p \in P$, $W(p, \alpha) = \sum_{t \in \alpha} W(p, t)$ and $W(\alpha, p) = \sum_{t \in \alpha} W(t, p)$. Intuitively, $W(p, \alpha)$ gives the number of tokens that the firing of α removes from p, and $W(\alpha, p)$ is the total number of tokens inserted into p. The pre-places and post-places of a step α are respectively defined as $^\bullet\alpha = \{p \in P \mid W(p, \alpha) > 0\}$ and $\alpha^\bullet = \{p \in P \mid W(\alpha, p) > 0\}$. For technical reasons, we consider the empty step which has no pre-places nor post-places. A singleton step $\alpha = \{t\}$ is often denoted by t, and by a *non-singleton* step we mean any step that is not a singleton, including the empty step.

A step α is *enabled* and may be *fired* at a marking M if $M(p) \geq W(p, \alpha)$, for every place $p \in P$. We denote this by $M[\alpha\rangle$. Firing such an enabled step leads to the marking M' defined by $M'(p) = M(p) - W(p, \alpha) + W(\alpha, p)$, for every place $p \in P$. We denote this by $M[\alpha\rangle M'$.

A *step sequence from a marking* M is a (possibly empty) sequence of steps $\sigma = \alpha_1 \ldots \alpha_n$ such that there are markings $M_1, \ldots, M_{n+1}$ satisfying $M = M_1$ and $M_i[\alpha_i\rangle M_{i+1}$, for every $i \leq n$. We denote this by $M[\sigma\rangle$ and $M[\sigma\rangle M_{n+1}$. If $M = M_0$ then M_{n+1} belongs to the set $[M_0\rangle$ of *reachable* markings of N.

The *concurrent reachability graph* $CRG(N)$ of N is defined as a labelled directed graph $CRG(N) = ([M_0\rangle, A, M_0)$, where the reachable markings of N are vertices, the initial marking is the initial vertex, and the set of arcs is given by $A = \{(M, \alpha, M') \mid M \in [M_0\rangle \land M[\alpha\rangle M'\}$. In the diagrams, we omit arcs labelled by the empty step.

A PT-net N is *ordinary* if $W((P \times T) \cup (T \times P)) \subseteq \{0, 1\}$, and *safe* if $M(P) \subseteq \{0, 1\}$, for every $M \in [M_0\rangle$. It can be seen that a safe PT-net without non-active transitions (i.e., transitions that are not enabled at any reachable marking) is ordinary.

Note that being a safe PT-net does not depend on the chosen semantics, i.e., the sequential semantics where only singleton steps are executed, or the full step semantics. In what follows, a step α of a PT-net:

- is *active* if there is a reachable marking which enables it.
- is *positive* if $W(\alpha, p) \geq W(p, \alpha)$, for every $p \in P$.
- is *disconnected* if $({}^\bullet t \cup t^\bullet) \cap ({}^\bullet t' \cup t'^\bullet) = \varnothing$, for all distinct transitions $t, t' \in \alpha$.
- *lies on self-loops* if $W(p, t) = W(t, p)$, for all $t \in \alpha$ and $p \in P$.

Clearly, the empty step lies on self-loops, and if α lies on self-loops then it is also positive. We also have:

Fact 1. *If* $M[\alpha\rangle$ *and* $M' \geq M$, *then* $M'[\alpha\rangle$.

Fact 2. *If* $M[\alpha\rangle$ *and* $\beta \subseteq \alpha$, *then* $M[\beta(\alpha \setminus \beta)\rangle$.

Fact 3. *A step* α *is enabled at a reachable marking* M *of a safe PT-net iff* α *is disconnected and consists of transitions enabled at* M.

3   Persistence and Nonviolence

In its standard form, persistence is stated as a property of nets executed according to the sequential semantics.

Definition 1 (persistent net, [7]). *A* PT-*net* N *is* persistent *if, for all transitions* $t \neq t'$ *and any reachable marking* M *of* N, $M[t\rangle$ *and* $M[t'\rangle$ *imply* $M[tt'\rangle$.

The above definition captures a property of the entire system represented by the PT-net (see Figure 1). If one is interested in a fine-grained preservation of executability of actions, it is natural to re-phrase it in terms of individual transitions.

Definition 2 (nonviolent/persistent transition). *Let* t *be a transition enabled at a marking* M *of a* PT-*net* N. *Then:*

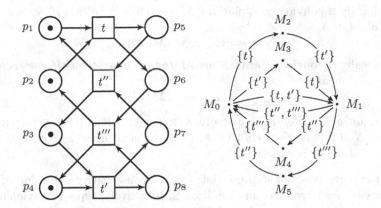

Fig. 1. A persistent safe PT-net and its concurrent reachability graph (arcs labelled by the empty step are omitted)

- t *is* locally nonviolent at M *if, for every transition t' enabled at M,*

$$t' \neq t \implies M[tt'\rangle .$$

- t *is* locally persistent at M *if, for every transition t' enabled at M,*

$$t' \neq t \implies M[t't\rangle .$$

Moreover, an active transition t is globally nonviolent *(or globally persistent) in N if it is locally nonviolent (resp. locally persistent) at every reachable marking of N at which it is enabled.*

The net-oriented and transition-oriented definitions are closely related as, due to the symmetric roles played by t and t' in Definition 1, we immediately obtain the following.

Proposition 1. *Let N be a* PT-net. *Then the following are equivalent:*

- *N is persistent.*
- *N contains only globally nonviolent transitions.*
- *N contains only globally persistent transitions.*

We will now introduce the central definitions of this paper, in which we lift the notions of persistence and nonviolence from the level of individual transitions to the level of steps.

Definition 3 (nonviolent step). *Let α be a step enabled at a marking M of a* PT-net *N. Then:*

- *α is* locally A-nonviolent at marking M *(or* LA-*nonviolent) if, for every step β enabled at M,*

$$\beta \neq \alpha \implies M[\alpha(\beta \setminus \alpha)\rangle$$

- α is locally B-nonviolent at marking M *(or* LB-*nonviolent) if, for every step* β *enabled at* M,
$$\beta \cap \alpha = \varnothing \implies M[\alpha\beta\rangle$$
- α is locally C-nonviolent at marking M *(or* LC-*nonviolent) if, for every step* β *enabled at* M,
$$\beta \neq \alpha \implies M[\alpha\beta\rangle$$

Moreover, an active step α *is globally* A/B/C-*nonviolent (or* GA/GB/GC-*nonviolent) in* N *if it is respectively* LA/LB/LC-*nonviolent at every reachable marking of* N *at which it is enabled.*

Each of the three types of step nonviolence is a conservative extension of transition nonviolence introduced in Definition 2. Intuitively, type-A nonviolence requires that only the unexecuted part of a delayed step is kept enabled, and so it is 'protected' by α. Type-B and type-C nonviolence, however, insist on maintaining the enabledness of the whole delayed step.

Definition 4 (persistent step). *Let* α *be a step enabled at a marking* M *of a* PT-*net* N. *Then:*

- α *is locally* A-*persistent at marking* M *(or* LA-*persistent) if, for every step* β *enabled at* M,
$$\beta \neq \alpha \implies M[\beta(\alpha \setminus \beta)\rangle$$
- α *is locally* B-*persistent at marking* M *(or* LB-*persistent) if, for every step* β *enabled at* M,
$$\beta \cap \alpha = \varnothing \implies M[\beta\alpha\rangle$$
- α *is locally* C-*persistent at marking* M *(or* LC-*persistent) if, for every step* β *enabled at* M,
$$\beta \neq \alpha \implies M[\beta\alpha\rangle$$

Moreover, an active step α *is globally* A/B/C-*persistent (or* GA/GB/GC-*persistent) in* N *if it is respectively* LA/LB/LC-*persistent at every reachable marking of* N *at which it is enabled.*

Again, each of the three types of step persistence is a conservative extension of transition persistence introduced in Definition 2. Type-A persistence requires that only unexecuted part of a delayed step is kept enabled, and in this case a persistent step can fail to fully 'survive'. Type-B and type-C persistence, however, insist on preserving the enabledness of a whole step. Note that in type-B of nonviolence and persistence, two steps are considered distinct if they are disjoint, whereas in the other two cases it is enough that they are different, and so they can have a nonempty intersection. Note that the empty step is trivially nonviolent and persistent in any possible sense.

Note: Since, as we prove later, type-A and type-B nonviolence (as well as persistence) are equivalent, in the examples we discuss only the type-A and type-C variants.

Moving from sequential to step semantics changes the way we perceive the persistence of PT-nets introduced by the standard Definition 1. In particular, in the sequential semantics, by Proposition 1, all transitions in a persistent net are both globally nonviolent and globally persistent. In the step semantics the situation is different. Consider, for example, the PT-net in Figure 1. It is persistent, and all of its active steps are both GA-persistent and GA-nonviolent. However, its nonempty steps fail to be LC-persistent or LC-nonviolent at some of the markings that enable them. More precisely, $\{t\}$, $\{t'\}$ and $\{t, t'\}$ are neither LC-persistent nor LC-nonviolent at M_0, while $\{t''\}$, $\{t'''\}$ and $\{t'', t'''\}$ are neither LC-persistent nor LC-nonviolent at M_1. This should not come as a surprise, as the type-C of persistence (or nonviolence) is a demanding property. Type-A of persistence and nonviolence, on the other hand, are close in spirit to their sequential counterparts.

A duality of the nonviolent and persistent steps is illustrated in Figure 2, where:

- $\{t\}$ is both a GA-nonviolent and GC-nonviolent step, but neither LA-persistent nor LC-persistent at M_0.
- $\{t'\}$ is both a GA-persistent and GC-persistent step, but neither LA-nonviolent nor LC-nonviolent at M_0.

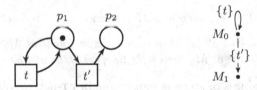

Fig. 2. A safe PT-net illustrating the duality of persistence and nonviolence

A step can be both nonviolent and persistent. For example, if we merge p_1 and p_2 in Figure 2, making both t and t' lie on self-loops, then $\{t\}$ and $\{t'\}$ become GA/GC-nonviolent/persistent.

4   Relating Persistent and Nonviolent Steps

In this section we investigate the expressiveness of different notions of persistence and nonviolence. Directly from Definitions 3 and 4 we have the following.

Proposition 2. *Let α be a step enabled at a reachable marking M of a PT-net N. Then, respectively:*

1. *If α is GA/GB/GC-nonviolent in N, then α is LA/LB/LC-nonviolent at M.*
2. *If α is GA/GB/GC-persistent in N, then α is LA/LB/LC-persistent at M.*

We then obtain a number of inclusions between different types of persistent and nonviolent steps which all hold for general PT-nets.

Proposition 3. *Let α be a step and M be a marking of a* PT-*net N. Then:*

1. *α is* LA-*nonviolent at M iff α is* LB-*nonviolent at M.*
2. *α is* LA-*persistent at M iff α is* LB-*persistent at M (cf. [6]).*

Proof. Assume that α is enabled at M, and β is another step enabled at M.

(1) Suppose that α is LA-nonviolent at M and $\beta \cap \alpha = \varnothing$. Then $M[\alpha(\beta \setminus \alpha)\rangle$ and $\beta \setminus \alpha = \beta$. Hence $M[\alpha\beta\rangle$, and so α is LB-nonviolent at M.
Conversely, suppose α is LB-nonviolent at M and $\beta \neq \alpha$. Then $M[\alpha(\beta \setminus \alpha)\rangle$ as $(\beta \setminus \alpha) \cap \alpha = \varnothing$ and $M[\beta \setminus \alpha\rangle$ (cf. Fact 2). Hence, α is LA-nonviolent at M.

(2) Suppose that α is LA-persistent at M and $\beta \cap \alpha = \varnothing$. Then $M[\beta(\alpha \setminus \beta)\rangle$ and $\alpha \setminus \beta = \alpha$. Hence $M[\beta\alpha\rangle$, and so α is LB-persistent at M.
Conversely, suppose that α is LB-persistent at M and $\beta \neq \alpha$. Then $M[(\beta \setminus \alpha)\alpha\rangle$ as $(\beta \setminus \alpha) \cap \alpha = \varnothing$ and $M[\beta \setminus \alpha\rangle$ (cf. Fact 2). Hence $M[(\beta \setminus \alpha)(\alpha \cap \beta)(\alpha \setminus \beta)\rangle$ (cf. Fact 2). Thus, by $M[\beta\rangle$, $M[\beta(\alpha \setminus \beta)\rangle$. Hence α is LA-persistent at M.   □

Corollary 1. *Let α be a step of a* PT-*net N. Then:*

1. *α is* GA-*nonviolent in N iff α is* GB-*nonviolent in N.*
2. *α is* GA-*persistent in N iff α is* GB-*persistent in N (cf. [6]).*

Proposition 4. *Let α be a step and M a marking of a* PT-*net N. Then:*

1. *If α is* LC-*nonviolent at M, then α is* LA-*nonviolent at M.*
2. *If α is* LC-*persistent at M, then α is* LA-*persistent at M.*

Proof. Since enabledness of steps is monotonic in PT-nets (see Fact 2), the two implications follow directly from Definitions 3 and 4, where the statements for LC-persistence and LC-nonviolence have stronger consequents.   □

Corollary 2. *Let α be a step of a* PT-*net N. Then:*

1. *If α is* GC-*nonviolent in N, then α is* GA-*nonviolent in N.*
2. *If α is* GC-*persistent in N, then α is* GA-*persistent in N.*

The implications in Propositions 2 and 4 (for type-A) cannot be reversed. A counterexample is provided in Figure 3, where $\{t\}$ is both LA-nonviolent and LA-persistent at M_3. However, it is neither LC-nonviolent nor LC-persistent at M_3 as well as it is neither GA-nonviolent nor GA-persistent (because of M_0).

The implications in Corollary 2 cannot be reversed. A counterexample is provided in Figure 3, where $\{t, t''\}$ is both GA-nonviolent and GA-persistent, but it is neither GC-nonviolent nor GC-persistent. As this step is only enabled at marking M_3, it fails to be LC-nonviolent or LC-persistent as well. Moreover, in Figure 3, $\{t'''\}$ is a step that is type-A and type-C globally nonviolent and persistent, because it is only enabled at one marking M_1, and no other nonempty step is enabled at M_1.

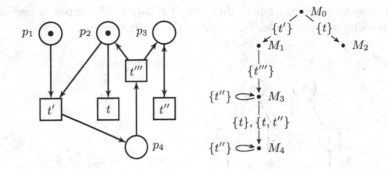

Fig. 3. A safe PT-net for (II,V,VI) in Figure 7 and (I,II) in Figure 9

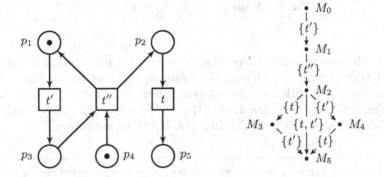

Fig. 4. A safe PT-net for (IV) in Figure 7

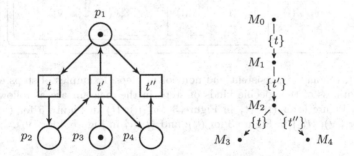

Fig. 5. A safe PT-net for (III) in Figure 7

Figure 4 shows that a step $\{t\}$ may be GA-persistent, but only LC-persistent (at M_4). Step $\{t\}$ is not GC-persistent, because it is not LC-persistent at M_2. The same example can be used when considering nonviolence.

Figure 5 shows an example of a step, $\{t\}$, that is LC-nonviolent, LA-nonviolent, LC-persistent and LA-persistent at M_0, but it is neither GC-nonviolent nor GA-nonviolent nor GC-persistent nor GA-persistent.

There may be steps in PT-nets that fail to satisfy all the types of persistency and nonviolence; for example, $\{t, t''\}$ and $\{t'\}$ in Figure 6.

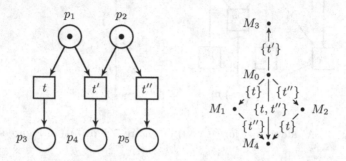

Fig. 6. A safe PT-net for (I) in Figure 7

Finally, there are PT-nets where all steps are neither persistent nor nonviolent whatever type (A or C) we choose. For example, take the net in Figure 6 and delete p_2, p_5 and t'' with all adjacent arcs. Then, the only non-empty steps in the concurrent reachability graph are $\{t\}$ and $\{t'\}$, and they prevent each other from being persistent. As a result, they also fail to be nonviolent.

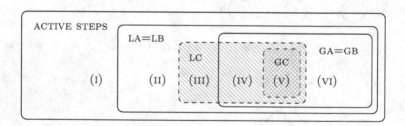

Fig. 7. A taxonomy of persistent and nonviolent steps. Examples of steps exhibiting the nonemptiness of the specific kinds of steps in the diagram are as follows: $\{t, t''\}$ and $\{t'\}$ in Figure 6 for (I); $\{t\}$ in Figure 3 for (II); $\{t\}$ in Figure 5 for (III); $\{t\}$ in Figure 4 for (IV); $\{t'''\}$ in Figure 3 for (V); and $\{t, t''\}$ in Figure 3 for (VI).

The relationships between different types of persistent and nonviolent steps are summarised in the diagram of Figure 7. As the relationships are the same for persistence or nonviolence, the diagram simply refers to different types of persistence or nonviolence.

5   Persistent and Nonviolent Markings

In this section, we focus on steps enabled at individual markings. A marking will be persistent (or nonviolent) according to a given type of persistence (or nonviolence) if all steps that it enables satisfy the corresponding definition of persistence (or nonviolence). Interestingly, in such markings, if all enabled steps are A (B or C) persistent they all are A (B or C) nonviolent, and vice versa. In a way, such markings create an environment where steps do not interfere with each other.

Definition 5 (nonviolent/persistent marking). *Let M be a reachable marking of a PT-net N. Then:*

- *M is A/B/C-nonviolent in N if every step enabled at M is respectively LA/LB/LC-nonviolent at M.*
- *M is A/B/C-persistent in N if every step enabled at M is respectively LA/LB/LC-persistent at M.*

Proposition 5. *A reachable marking of a PT-net is A/B/C-persistent iff it is A/B/C-nonviolent, respectively.*

Proof. By Definition 5, a reachable M is A-persistent in a PT-net N iff each step α enabled at M is LA-persistent at M. The latter in turn is equivalent to:

$$\forall \alpha : M[\alpha\rangle \implies (\forall \beta : \alpha \neq \beta \wedge M[\beta\rangle \implies M[\beta(\alpha \setminus \beta)\rangle))$$
$$\Leftrightarrow \forall \alpha, \beta : \alpha \neq \beta \wedge M[\alpha\rangle \wedge M[\beta\rangle \implies M[\beta(\alpha \setminus \beta)\rangle$$
$$\Leftrightarrow \forall \alpha, \beta : \alpha \neq \beta \wedge M[\alpha\rangle \wedge M[\beta\rangle \implies M[\alpha(\beta \setminus \alpha)\rangle$$
$$\Leftrightarrow \forall \alpha : M[\alpha\rangle \implies (\forall \beta : \alpha \neq \beta \wedge M[\beta\rangle \implies M[\alpha(\beta \setminus \alpha)\rangle)).$$

The last line is equivalent to stating that each step α enabled at M is LA-non-violent at M. Hence, by Definition 5, M is A-nonviolent in N.

The equivalences for types B and C can be shown in a similar way.   □

Proposition 6. *A reachable marking of a PT-net N is A-persistent (or A-non-violent) in N iff it is B-persistent (resp. B-nonviolent) in N.*

Proof. Follows directly from Definitions 4 and 5, and Propositions 3 and 5.   □

Proposition 7. *If a reachable marking of a PT-net N is C-persistent (or C-nonviolent) in N, then it is A-persistent (resp. A-nonviolent) in N.*

Proof. Follows directly from Definitions 4 and 5, and Propositions 4 and 5.   □

The implications in Proposition 7 cannot be reversed, and a suitable counterexample is provided in Figure 3, where M_3 is both A-persistent and A-nonviolent marking but it is neither C-persistent nor C-nonviolent. Notice that all the nonempty steps enabled at M_3 (i.e., $\{t\}$, $\{t''\}$ and $\{t, t''\}$) are LA-persistent at this marking, making it A-persistent (and A-nonviolent, see Proposition 5). However, $\{t\}$ and $\{t, t''\}$ are neither LC-persistent nor LC-nonviolent at M_3.

The relationships between different types of persistent and nonviolent markings are summarised in the diagram of Figure 9. As the relationships are the same for persistence or nonviolence, the diagram simply refers to different types of persistence or nonviolence.

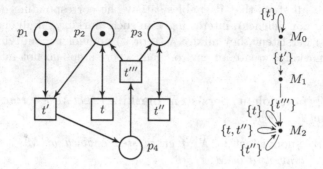

Fig. 8. A safe PT-net for (III) in Figure 9

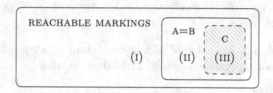

Fig. 9. A taxonomy of persistent and nonviolent markings. Examples of markings exhibiting the nonemptiness of the specific kinds of markings in the diagram are as follows: M_0 in Figure 3 for (I); M_3 in Figure 3 for (II); and M_2 in Figure 8 for (III).

6   Persistent and Nonviolent Steps in PT-nets

In this section, we investigate general properties of persistent and nonviolent steps. The first question we address is whether persistence and nonviolence of steps can be 'inherited' by their substeps. For general PT-nets, the answer turns out to be positive only for local persistence.

Proposition 8. *Let* $\gamma \subseteq \alpha$ *be two steps and* M *be a reachable marking of a* PT-*net. Then:*

1. *If* α *is* LA-*persistent at* M, *then* γ *is* LA-*persistent at* M.
2. *If* α *is* LC-*persistent at* M, *then* γ *is* LC-*persistent at* M.

Proof. From $\gamma \subseteq \alpha$, $M[\alpha\rangle$ and Fact 2, we have $M[\gamma\rangle$. Hence γ is active. Moreover, we assume $\varnothing \neq \gamma \neq \alpha$ as otherwise the results are obvious.

(1) Let $\beta \neq \gamma$ be a step enabled at M. If $\beta \neq \alpha$ then, as α is LA-persistent at M, we have $M[\beta(\alpha \setminus \beta)\rangle$. Hence, by $\gamma \setminus \beta \subseteq \alpha \setminus \beta$ and Fact 2, $M[\beta(\gamma \setminus \beta)\rangle$. If $\beta = \alpha$ then $M[\beta(\gamma \setminus \beta)\rangle$ as $M[\alpha\varnothing\rangle$ and $\gamma \setminus \beta = \varnothing$.

(2) Let $\beta \neq \gamma$ be a step enabled at M. If $\beta \neq \alpha$ then, as α is LC-persistent at M, we have $M[\beta\alpha\rangle$. Hence, by $\gamma \subseteq \alpha$ and Fact 2, $M[\beta\gamma\rangle$. If $\beta = \alpha$ and $\neg M[\alpha\gamma\rangle$, then we proceed as follows. By $M[\alpha\rangle$, there is $p \in P$ such that $M(p) - W(p, \alpha) +$

$W(\alpha, p) < W(p, \gamma)$. On the other hand, since α is LC-persistent at M and $\gamma \neq \alpha$ is enabled at M, we have $M[\gamma\alpha\rangle$. Thus $M(p) - W(p, \gamma) + W(\gamma, p) \geq W(p, \alpha)$. As a result, $W(p, \alpha) + W(p, \gamma) - W(\gamma, p) < W(p, \gamma) + W(p, \alpha) - W(\alpha, p)$, and so $W(\gamma, p) > W(\alpha, p)$, yielding a contradiction.   □

Proposition 8 does not hold for globally persistent steps and their substeps, whether we consider A-persistence or C-persistence. Figure 8 shows an example of a step, $\{t, t''\}$, which is both GA-persistent and GC-persistent, but its substep $\{t\}$ is neither GA-persistent nor GC-persistent, because of marking M_0. Furthermore, Proposition 8 extended to nonviolent steps does not hold, even for ordinary PT-nets. Figure 10 provides a counterexample, where $\{t, t''\}$ is both LA-nonviolent and LC-nonviolent at M_0 (in fact it is both GA-nonviolent and GC-nonviolent, as it is enabled nowhere else), but its substep $\{t\}$ is neither LA-nonviolent nor LC-nonviolent at M_0.

Type-C persistence and nonviolence are very demanding properties, and can only be satisfied by steps of a very particular kind. The presence of type-C persistent or nonviolent steps has therefore some structural implications for nets and their reachability graphs. The next result gives sufficient conditions for being a globally nonviolent step.

Theorem 1. *Each active positive step of a* PT*-net is both* GC*-nonviolent and* GA*-nonviolent.*

Proof. Let $M[\alpha\rangle M'$ and $\beta \neq \alpha$ be a step enabled at M. From $M' \geq M$ (as α is positive) and Fact 1 it follows that $M'[\beta\rangle$. Hence $M[\alpha\beta\rangle$, and so α is GC-nonviolent. Moreover, by Corollary 2, α is also GA-nonviolent.   □

The next result gives necessary conditions for being a GC-persistent step. Intuitively, the intersection of a GC-persistent step with any other step enabled at the same marking consumes at most the same resources (tokens) as it produces. This should not come as a surprise, because in C-persistence the intersection of two enabled steps at a given marking must be able to fire twice in a row.

Proposition 9. *Let α be a* GC*-persistent non-singleton step enabled at a reachable marking M of a* PT*-net. Then, for every step $\beta \neq \alpha$ enabled at M, $\alpha \cap \beta$ is a positive step.*

Proof. Suppose that the step $\gamma = \alpha \cap \beta$ is not positive (which implies that neither α nor β is empty), and so there is $p \in P$ such that $W(p, \gamma) > W(\gamma, p)$. We consider two cases.

Case 1: $\alpha \not\subseteq \beta$. Since $M[\alpha\rangle$ and $\gamma \subseteq \alpha$, we have $M[\gamma\rangle$. Also, since $\alpha \not\subseteq \beta$, $\gamma \neq \alpha$. As α is GC-persistent, there exists a marking M_1 such that $M[\gamma\rangle M_1[\alpha\rangle$. We can repeat this construction, replacing M with M_1, as α is globally C-persistent. In fact, we can repeat this construction $k = M(p) + 1$ times, obtaining $M[\gamma\rangle M_1[\gamma\rangle M_2[\gamma\rangle M_3 \ldots M_k[\alpha\rangle$.

We then observe that $M(p) - M_1(p) = W(p, \gamma) - W(\gamma, p) \geq 1$. Similarly, $M_i(p) - M_{i+1}(p) \geq 1$, for $i = 1, \ldots, k - 1$. Hence $M(p) - M_k(p) \geq k = M(p) + 1$ and so $M_k(p) < 0$ which is obviously impossible, yielding a contradiction.

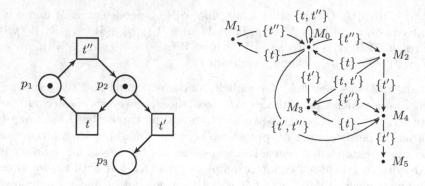

Fig. 10. An ordinary PT-net for the discussion of Proposition 8

Case 2: $\alpha \subset \beta$. Then $\gamma = \alpha \cap \beta = \alpha$. As α is a non-empty and non-singleton step, we can split it into two disjoint nonempty subsets: $\alpha = \gamma = \gamma' \uplus \gamma''$. Since $M[\alpha\rangle$ and $\gamma', \gamma'' \subseteq \alpha$, we have $M[\gamma'\rangle$ and $M[\gamma''\rangle$. Also, $\gamma' \neq \alpha$ and $\gamma'' \neq \alpha$. As α is GC-persistent, there exists a marking M' such that $M[\gamma'\rangle M'[\alpha\rangle$. Now, we can repeat this construction, for M' and step γ'', obtaining: $M[\gamma'\rangle M'[\gamma''\rangle M_1[\alpha\rangle$ or $M[\gamma'\gamma''\rangle M_1[\alpha\rangle$. We can repeat this construction, now starting at M_1, as α is globally C-persistent. In fact, we can repeat this construction $k = M(p) + 1$ times, obtaining $M[\gamma'\gamma''\rangle M_1[\gamma'\gamma''\rangle M_2[\gamma'\gamma''\rangle M_3 \ldots M_k[\alpha\rangle$.

We now observe that, by $\gamma = \gamma' \uplus \gamma''$, we have $W(p, \gamma) = W(p, \gamma') + W(p, \gamma'')$ and $W(\gamma, p) = W(\gamma', p) + W(\gamma'', p)$. The rest of the proof is then similar as in Case 1.   □

In Proposition 9, one cannot drop the assumption that α is a non-singleton step. Consider, for example, Figure 11 and take $\alpha = \{t'\}$ and $\beta = \{t, t'\}$, which are two different steps enabled at M_0. Although α is GC-persistent, the intersection $\alpha \cap \beta = \{t'\}$ is not a positive step, as $W(p_1, t') > W(t', p_1)$. Similarly, one cannot drop the assumption that α is GC-persistent. Consider, for example, Figure 12 and take $\alpha = \{t, t'\}$, and $\beta = \{t\}$, which are two different steps jointly enabled at several markings, such as M_0. Although α is a non-singleton step, the intersection $\alpha \cap \beta = \{t\}$ is not a positive step as $W(p_1, \alpha \cap \beta) > W(\alpha \cap \beta, p_1)$.

The implication in Proposition 9 cannot be reversed, and a counterexample is provided in Figure 13, where $\alpha = \{t, t'\}$ is a non-singleton step enabled (only) at M_0. There are three other nonempty steps enabled at M_0, viz. $\{t\}$, $\{t'\}$ and $\{t''\}$. Clearly, all $\alpha \cap \{t\} = \{t\}$ and $\alpha \cap \{t'\} = \{t'\}$ and $\alpha \cap \{t''\} = \varnothing$ are positive steps. However, α is not GC-persistent as it is not enabled after the execution of $\{t''\}$.

Finally, Proposition 9 cannot be re-stated for nonviolence, and a counterexample is provided in Figure 10, where $\alpha = \{t, t''\}$ is a GC-nonviolent non-singleton step, and $\beta = \{t\}$ is another step enabled together with α at M_0. However, $\alpha \cap \beta$ is not a positive step as $W(p_2, \alpha \cap \beta) > W(\alpha \cap \beta, p_2)$.

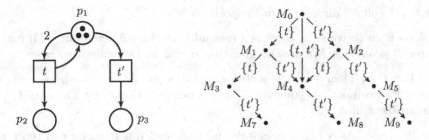

Fig. 11. A PT-net for the discussion of Proposition 9

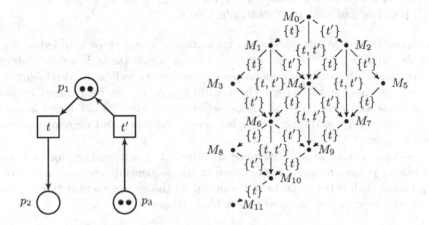

Fig. 12. A PT-net for the discussion of Proposition 9

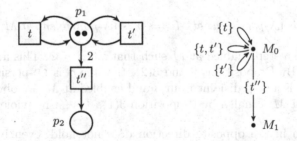

Fig. 13. A PT-net for the discussion of Proposition 9

Theorem 2. *Let α be a GC-persistent non-singleton step of a PT-net N, and γ be a subset of α. Then:*

1. *For every reachable marking M enabling α, γ is LC-persistent at M.*
2. *γ is GC-nonviolent in N.*

Proof. (1) Follows directly from Proposition 8.

(2) As α is an active step, there is a reachable marking M enabling α. If $\alpha = \varnothing$ the result is obvious. We now assume that $\alpha \neq \varnothing$ and consider two cases.

Case 1: $\gamma \subset \alpha$. Then $\gamma \neq \alpha$ is a step enabled at M. As α is GC-persistent, we can use Proposition 9 to conclude that γ is positive. Hence, by Theorem 1, γ is GC-nonviolent in N.

Case 2: $\gamma = \alpha$. As α is a non-empty and non-singleton step, we can represent it as a disjoint union of two nonempty subsets: $\alpha = \gamma' \uplus \gamma''$. From $M\langle\alpha\rangle$ and Fact 2, we have $M[\gamma'\rangle$ and $M[\gamma''\rangle$. Moreover, as α is GC-persistent, we can use Proposition 9 to conclude that both γ' and γ'' are positive steps. Therefore, γ is also positive, and so it is GC-nonviolent in N.   □

In Proposition 9, the intersection $\alpha \cap \beta$ of two different steps enabled at some reachable marking will be able to fire twice in a row (as α is GC-persistent). As a result, $\alpha \cap \beta$ can be seen as a persistent step as well as a nonviolent step at markings that enable α (cf. Theorem 2(2)). As α can be covered by such intersections, GC-persistence of a non-singleton α implies its GC-nonviolence. In a way, in type-C case, the boundary between persistence and nonviolence blurs to some extent.

Type-A persistence and nonviolence are different in nature. They follow closely the ideas of persistence and nonviolence in the sequential case and, intuitively, complement each other. The next result shows the complementarity of A-nonviolent and A-persistent steps at reachable markings.

Theorem 3. *Let M be a reachable marking of a PT-net N. If there are two disjoint steps α and β enabled at M such that every enabled step at M is a subset of their union, then the following holds:*

$$\alpha \text{ is LA-persistent at } M \implies \beta \text{ is LA-nonviolent at } M.$$

Proof. Let γ be a step enabled at M such that $\gamma \cap \beta = \varnothing$. This and $\gamma \subseteq \alpha \cup \beta$ implies $\gamma \subseteq \alpha$. By Propositions 8 and 3(2), the step γ is LB-persistent, and so $M[\beta\gamma\rangle$. Since γ is a step disjoint from β and enabled at M, we obtain that β is LB-nonviolent at M. Finally, by Proposition 3(1), β is LA-nonviolent at M.   □

The implication in the opposite direction does not hold, even in the case of ordinary nets. A counterexample is presented in Figure 10. Taking $\alpha = \{t'\}$ and $\beta = \{t, t''\}$, we see that β is a LA-nonviolent step at M_0, but α is not LA-persistent at M_0. We can, for example execute step $\{t\}$ at M_0 which leads us to marking M_1, where α is not enabled.

We end this section with a result that gives sufficient conditions for a step of an ordinary PT-net to be GC-nonviolent. It is a counterpart of Theorem 1 formulated for the general PT-nets. Although the conditions here are more restrictive than those in Theorem 1, they are linked to the structure of a net rather than its behaviour.

Theorem 4. *Let α be an active step of an ordinary* PT-*net N. If α lies on self-loops, then α is* GC-*nonviolent in N.*

Proof. Since all the transitions in α lie on self-loops, α is positive. Hence, by Theorem 1, α is GC-nonviolent in N.   □

7 Persistent and Nonviolent Steps in Safe PT-nets

In the case of safe PT-nets, we can identify more interesting properties of persistent or nonviolent steps and, in particular, link them to the structure of the nets. We start with results which are concerned with the structural properties related to C-persistence and C-nonviolence.

Proposition 10 (c.f. [6]). *Let α be a step which is* LC-*persistent or* LC-*nonviolent at a reachable marking M of a safe* PT-*net N. Then $^\bullet(\alpha \cap \beta) = (\alpha \cap \beta)^\bullet$, for every step $\beta \neq \alpha$ enabled at M.*

Proof. The result for LC-persistent α was proven in [6]. We therefore assume that $\alpha = \{t_1, \ldots, t_n\}$ is LC-nonviolent.

Suppose that $p \in {}^\bullet(\alpha \cap \beta)$, for some step $\beta \neq \alpha$ enabled at M. Clearly, $M(p) = 1$. Moreover, by Fact 3, α is disconnected.

Since α is LC-nonviolent, there is a marking M' such that $M[\alpha\rangle M'[\beta\rangle$. As $M'[\beta\rangle$ and $p \in {}^\bullet(\alpha \cap \beta)$, we have $M'(p) = 1$. Hence $M'(p) = M(p) - W(p, \alpha) + W(\alpha, p)$, and so $W(t_1, p) + \cdots + W(t_n, p) = W(p, t_1) + \cdots + W(p, t_n)$. By N being safe, all the arc weights in this formula are 0 or 1. Moreover, α is disconnected. It therefore follows that $W(p, t_i) = W(t_i, p)$, for each t_i. Hence $p \in (\alpha \cap \beta)^\bullet$, and so ${}^\bullet(\alpha \cap \beta) \subseteq (\alpha \cap \beta)^\bullet$.

Suppose now that $p \in (\alpha \cap \beta)^\bullet \setminus {}^\bullet(\alpha \cap \beta)$. Then, by $M[\alpha \cap \beta\rangle$ and the safeness of N, $M(p) = 0$. Hence, by $M[\alpha\rangle$ and $M[\beta\rangle$, we must have $p \notin {}^\bullet\alpha \cup {}^\bullet\beta$. Consequently, since there is M'' such that $M[\alpha\beta\rangle M''$, we obtain $M''(p) \geq 2$, a contradiction with N being safe. Hence ${}^\bullet(\alpha \cap \beta) \supseteq (\alpha \cap \beta)^\bullet$, and so the result holds.   □

As a result, we can link LC-persistence and LC-nonviolence with the structural property of lying on self-loops.

Theorem 5 (c.f. [6]). *Let α be a non-singleton step which is* LC-*persistent or* LC-*nonviolent at a reachable marking M of a safe* PT-*net N. Then α lies on self-loops.*

Proof. Suppose that $t \in \alpha$. Since $\{t\} \neq \alpha$ is a step enabled at M, by Proposition 10, ${}^\bullet(\alpha \cap \{t\}) = (\alpha \cap \{t\})^\bullet$. Hence ${}^\bullet t = t^\bullet$.   □

Corollary 3. *Let α be a non-singleton active step of a safe* PT-*net N. Then α is* GC-*nonviolent iff α lies on self-loops.*

Proof. Follows from Theorems 4 and 5.   □

Theorem 6 ([6]). *Let α be an active step of a safe* PT*-net N. If all the transitions in α are globally persistent and lie on self-loops, then α is* GC*-persistent in N.*

Theorem 6 can be seen as a counterpart of Theorem 4 which was proven for ordinary nets. The latter is in fact stronger as we only need to assume that α lies on self-loops. We note that the implication in Theorem 6 cannot be reversed, and a suitable counterexample is provided in Figure 8, where $\{t'\}$ is a GC-persistent step, but it does not lie on self-loops. Moreover, Theorem 6 cannot be lifted to the level of ordinary nets. Figure 14 provides a counterexample, where $\alpha = \{t, t'\}$ is neither locally nor globally C-persistent step even though both t and t' are globally persistent transitions lying on self-loops.

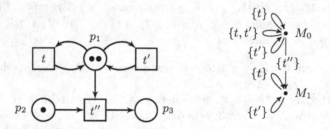

Fig. 14. An ordinary PT-net for the discussion of Theorem 6

It is interesting to see whether persistence or nonviolence are preserved by taking substeps. For general PT-nets, we only had results concerning the LA, LC and GC persistent steps. Here we can obtain similar results about nonviolence. Moreover, for the type-C of nonviolence the result holds globally.

Proposition 11. *Let $\gamma \subseteq \alpha$ be two steps and M be a reachable marking of a safe* PT*-net N. Then:*

1. *If α is* LA*-nonviolent at M then γ is* LA*-nonviolent at M.*
2. *If α is* LC*-nonviolent at M then γ is* LC*-nonviolent at M.*
3. *If α is* GC*-nonviolent in N then γ is* GC*-nonviolent in N.*

Proof. As the case $\alpha = \gamma$ is obvious, below we assume that $\gamma \subset \alpha$. Also, we assume that γ is nonempty, as the empty step trivially satisfies Definition 3.

(1) From $M[\alpha\rangle$ and $\gamma \subset \alpha$, we have $M[\gamma\rangle$. Let $\beta \neq \gamma$ be a step enabled at M. We need to prove that $M[\gamma\rangle M''[\beta \setminus \gamma\rangle$, for some marking M''. We consider two cases.

Case 1: $\beta \neq \alpha$. Since α is LA-nonviolent at M, we have $M[\alpha(\beta \setminus \alpha)\rangle$. Hence, for every place $p \in {}^\bullet(\beta \setminus \alpha)$, $p \in {}^\bullet\alpha$ implies $p \in \alpha^\bullet$. Furthermore, since α is disconnected (by Fact 3), we have that, for every place $p \in {}^\bullet(\beta \setminus \alpha)$ and transition $t \in \alpha$, $p \in {}^\bullet t$ implies $p \in t^\bullet$. As a result, for every place $p \in {}^\bullet(\beta \setminus \alpha)$, $p \in {}^\bullet\gamma$ implies $p \in \gamma^\bullet$. Hence, by $M[\beta\rangle$, we obtain that $M''[\beta \setminus \alpha\rangle$. We further observe that, by Fact 2, we get $M''[(\alpha \setminus \gamma) \cap \beta\rangle$. It therefore follows that all the transitions in $(\beta \setminus \alpha) \cup (\alpha \setminus \gamma) \cap \beta = \beta \setminus \gamma$ are enabled at M''. Moreover, as $\beta \setminus \gamma \subseteq \beta$ and $M[\beta\rangle$, we obtain from Fact 3 that the step $\beta \setminus \gamma$ is disconnected. Hence, again by Fact 3, $M''[\beta \setminus \gamma\rangle$.

Case 2: $\beta = \alpha$. Since $M[\alpha\rangle$ and $\gamma \subset \alpha$, by Fact 2, $M[\gamma\rangle M''[\alpha \setminus \gamma\rangle$.

(2) Since $\varnothing \neq \gamma \subset \alpha$, α is a non-singleton step. Thus, by Theorem 5, α lies on self-loops. Hence γ also lies on self-loops, and so we have $M[\gamma\rangle M[\beta\rangle$ as required.

(3) Since α is GC-nonviolent, it is LC-nonviolent at some reachable marking M. Proceeding similarly as in (2), we get that α lies on self-loops and, consequently, that γ lies on self-loops. Then, from Theorem 1, proved for general PT-nets, we obtain that γ is GC-nonviolent.   □

Proposition 11 does not hold for ordinary PT-nets, and Figure 10 shows a counterexample. The step $\{t, t''\}$ there is both GA and GC-nonviolent as well as LA and LC-nonviolent at M_0 (the only marking which enables it), but its substep $\{t\}$ is neither LA-nonviolent nor LC-nonviolent at M_0 (as once it is executed, the previously enabled step $\{t'\}$ becomes disabled). Also, Proposition 11(1) cannot be generalised to GA-nonviolent steps, and a suitable counterexample is provided in Figure 3, where $\{t, t''\}$ is a GA-nonviolent step, but its substep $\{t\}$ is not GA-nonviolent (as after executing $\{t\}$ at M_0, an enabled step $\{t'\}$ becomes disabled).

Theorem 7. *Let α be a GC-nonviolent step of a safe PT-net N. Then all the transitions in α are globally nonviolent in N.*

Proof. Let $t \in \alpha$. Consider a marking M enabling α, and so $M[t\rangle$. Then, from $\{t\} \subseteq \alpha$, the fact that α is GC-nonviolent and Proposition 11(3), we obtain that $\{t\}$ is GC-nonviolent in N. This means in particular that, for any reachable marking M of N enabling $\{t\}$, if $\{t'\} \neq \{t\}$ is a step enabled at M, we have $M[\{t\}\{t'\}\rangle$. We can therefore conclude that t, as a transition (rather than a step), is globally nonviolent (see Definition 2).   □

The above result does not hold for ordinary PT-nets, and a suitable counterexample is provided in Figure 10 which we used to demonstrate that Proposition 11 does not hold for ordinary PT-nets. In the latter case, we took a singleton substep of a GC-nonviolent step of an ordinary PT-net and showed that it disables another singleton step. The two singleton steps can be treated as transitions here.

The next two results give sufficient conditions for a step to be GA-persistent or GA-nonviolent in terms of the transitions it contains.

Theorem 8 (c.f. [6]). *Let α be an active step of a safe PT-net N. If all the transitions in α are globally persistent (nonviolent) in N, then α is GA-persistent (resp. GA-nonviolent) in N.*

Proof. In the case of persistence, the result was proven in [6]. We therefore assume that all the transitions in α are globally nonviolent. Let M be a reachable marking and $\beta \neq \alpha$ be a step in N such that $M[\alpha\rangle$ and $M[\beta\rangle$. Note that, by Fact 3, β is disconnected. We need to show that $M[\alpha(\beta \setminus \alpha)\rangle$.

Assume that $\alpha = \{t_1, \ldots, t_m\}$ and $\beta \setminus \alpha = \{u_1, \ldots, u_k\}$. From $M[\alpha\rangle$ and Fact 2, we have $M[t_1 \ldots t_m\rangle M'$. Now, for each transition u_i, since $M[u_i\rangle$ and every t_i is globally nonviolent, we have that $M'[u_i\rangle$. Thus, by Fact 3, $M'[\beta \setminus \alpha\rangle$, and so $M[\alpha(\beta \setminus \alpha)\rangle$.   □

The two implications in Theorem 8 cannot be reversed, and a suitable counterexample is provided in Figure 3, where a GA-nonviolent and GA-persistent step $\{t, t''\}$ contains a transition t that is neither globally nonviolent nor globally persistent (because of the marking M_0).

The last result concerning persistence and nonviolence in safe PT-nets shows that they can complement each other. It is a counterpart of Theorem 3, but here the result holds in both directions due to Proposition 11, which was not available for the general nor ordinary PT-nets.

Theorem 9. *Let M be a reachable marking of a safe PT-net N. If there are two disjoint steps α and β enabled at M such that every enabled step at M is a subset of their union, then the following holds:*

$$\alpha \text{ is LA-persistent at } M \iff \beta \text{ is LA-nonviolent at } M.$$

Proof. The ($\Longrightarrow$) implication follows from Theorem 3. To show the ($\Longleftarrow$) implication, let γ be a step enabled at M such that $\gamma \cap \alpha = \varnothing$. Since $\gamma \subseteq \alpha \cup \beta$, we have $\gamma \subseteq \beta$. As β is LA-nonviolent at M we have, from Proposition 11(1), that γ is LA-nonviolent at M. By Proposition 3(1), we have that γ is also LB-nonviolent at M, and so $M[\gamma\alpha\rangle$. Since γ is a step, disjoint from from α and enabled at M, we get that α is LB-persistent at M. Using Proposition 3(2), we can then conclude that α is LA-persistent at M.   □

8   Conclusions

In this paper, we initiated a detailed investigation of different notions of persistence and nonviolence in the step based semantics of concurrent systems. Among the problems and issues we plan to investigate in future are the phenomenon of confusion formulated for steps rather than transitions, and less restrictive notions of persistence and nonviolence, such as that of k-persistence. As the theory developed so far does not allow auto-concurrency, we plan to consider what is the impact of allowing steps to be multisets rather than sets of transitions.

Acknowledgement. We would like to thank the anonymous referees for their comments and useful suggestions. This research was supported by a fellowship funded by the "Enhancing Educational Potential of Nicolaus Copernicus University in the Disciplines of Mathematical and Natural Sciences" Project POKL.04.01.01-00-081/10, the EPSRC GAELS and UNCOVER projects, the 973 Program Grant 2010CB328102, and MSFC Grant 61133001.

References

1. Barylska, K., Ochmański, E.: Levels of persistency in Place/Transition nets. Fundamenta Informaticae 93, 33–43 (2009)
2. Barylska, K., Mikulski, Ł., Ochmański, E.: On persistent reachability in Petri nets. Information and Computation 223, 67–77 (2013)
3. Best, E., Darondeau, P.: Decomposition theorems for bounded persistent petri nets. In: van Hee, K.M., Valk, R. (eds.) PETRI NETS 2008. LNCS, vol. 5062, pp. 33–51. Springer, Heidelberg (2008)
4. Best, E., Darondeau, P.: Separability in persistent Petri nets. Fundamenta Informaticae 113, 179–203 (2011)
5. Dasgupta, S., Potop-Butucaru, D., Caillaud, B., Yakovlev, A.: Moving from weakly endochronous systems to delay-insensitive circuits. Electronic Notes in Theoretical Computer Science 146, 81–103 (2006)
6. Fernandes, J., Koutny, M., Pietkiewicz-Koutny, M., Sokolov, D., Yakovlev, A.: Step persistence in the design of GALS systems. Technical Report 1349, School of Computing Science, Newcastle University (2012)
7. Landweber, L.H., Robertson, E.L.: Properties of conflict-free and persistent Petri nets. Journal of the ACM 25, 352–364 (1978)
8. Yakovlev, A., Koelmans, A., Semenov, A., Kinniment, D.: Modelling, analysis and synthesis of asynchronous control circuits using Petri nets. INTEGRATION, the VLSI Journal 21, 143–170 (1996)

Colouring Space - A Coloured Framework for Spatial Modelling in Systems Biology

David Gilbert[1], Monika Heiner[2], Fei Liu[3], and Nigel Saunders[1]

[1] School of Information Systems, Computing and Mathematics
Brunel University, Uxbridge, Middlesex UB8 3PH, UK
{david.gilbert,nigel.saunders}@brunel.ac.uk
[2] Computer Science Institute, Brandenburg University of Technology
Postbox 10 13 44, 03013 Cottbus, Germany
monika.heiner@informatik.tu-cottbus.de
[3] Harbin Institute of Technology
West Dazhi Street 92, 150001 Harbin, China
liufei@hit.edu.cn

Abstract. In this paper we introduce a technique to encode spatial attributes of dynamic systems using coloured Petri nets and show how it can be applied to biological systems within the spirit of BioModel Engineering. Our approach can be equally applied to qualitative, stochastic, continuous or hybrid models of the same physical system, and can be used as the basis for multiscale modelling. We illustrate our approach with two case studies, one from the continuous and one from the stochastic paradigm. In this paper we only discuss the case of finite colours, and by unfolding our method can take advantage of all the analytical machinery and simulation techniques that have been developed for the uncoloured family of Petri net classes.

Keywords: Coloured Petri nets, qualitative, stochastic, continuous, hybrid Petri nets, spatial modelling, biomolecular networks, Systems Biology, BioModel Engineering.

1 The Coloured Framework

In this paper we build on [16,20], where we have introduced our methodology for the use of a structured family of Petri net classes which enables the investigation of biological systems using various complementary modelling abstractions comprising the qualitative and quantitative paradigms. In the following we focus on the use of the coloured subset of the previously introduced framework [20] – coloured qualitative Petri nets ($\mathcal{QPN}^{\mathcal{C}}$), coloured stochastic Petri nets ($\mathcal{SPN}^{\mathcal{C}}$), coloured continuous Petri nets ($\mathcal{CPN}^{\mathcal{C}}$), and coloured hybrid Petri nets ($\mathcal{HPN}^{\mathcal{C}}$); Fig. 1 recalls our coloured framework.

We extend our approach by considering biochemical processes evolving in space, which we illustrate with two case studies. In our spatial modelling approach we discretise space using coloured Petri nets, and in this paper we investigate the use of finite discrete colour sets. This ensures the following three

J.-M. Colom and J. Desel (Eds.): PETRI NETS 2013, LNCS 7927, pp. 230–249, 2013.
© Springer-Verlag Berlin Heidelberg 2013

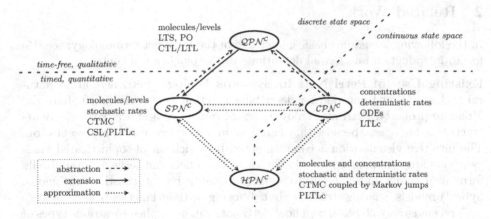

Fig. 1. The coloured unifying framework integrating four degrees of abstraction

features which are crucial for our BioModel Engineering principles (uniformity, reuse and conciseness) [17].

First and most importantly, the spatial modelling principle can be equally applied to all paradigms (qualitative, stochastic, continuous, and hybrid), i.e., once a Petri net model has been enriched with colour-encoded space, it can be easily transformed into any other net class while preserving all spatial attributes.

Second, all space-related information is encoded in colour and corresponding net annotations, such as colour sets, functions, and guards, which can be effortlessly reused in many models. Moreover, changing the notion of space or just some spatial attributes only requires the adaptation of those colour-related definitions, and the net structure itself needs not to be touched.

Third, the use of *a priori* finitely discretised space preserves the analysibility of the models, in particular we retain the discrete state space in both the qualitative and stochastic settings. All analysis and simulation techniques, which have been developed for uncoloured Petri nets over the last two decades, can be immediately reused by automatic unfolding.

The main contributions of our paper are

- a framework to encode space by coloured Petri nets, which can be equally applied in a qualitative, stochastic, continuous, or hybrid setting,
- a set of basic colour-related definitions which can be easily applied to a wide range of spatial scenarios,
- two substantial biological case studies illustrating the framework.

This paper is organised as follows. In the next section we recall some related work to set the background of our contribution. Afterwards we introduce our modelling approach of colour-encoded space by means of a popular case study in the continuous paradigm (Section 3), before applying it to a second case study in the stochastic paradigm (Section 4). We conclude our paper with a brief overview of the tools used and the summary.

2   Related Work

In the following we assume basic knowledge of the Petri net terminology; see [19] for an introduction and formal definitions in the biochemical context.

Existing Uses of Petri Nets in Systems Biology. Petri nets are a natural and established notation for describing reaction networks – both share the bipartite property. Petri nets enjoy a formal semantics and are particularly attractive to biologists, because they can 'buy in' to the executable representation. The intuitive visualisation is complemented by a rich set of sophisticated analysis techniques, supported by reliable tools. Petri nets can serve as an umbrella formalism comprising a family of related (qualitative, stochastic, continuous, hybrid) models, sharing structure, but differing in their kinetics [16].

A recent survey [2] has shown how Petri nets can be applied to various types of biological processes at different abstraction levels, illustrating this with a rich set of case studies. Most of these focus on the molecular level; however examples at the multi-cellular level include the signal-response behaviour of an organism [28], and developmental processes in multi-cellular pattern formation [7, 9, 23].

Current Challenges to Systems Biology Due to Complexity and Multiscale Issues. A drawback of current modelling approaches, including Petri nets, are their limitation to relatively small networks. Biological systems can be represented as networks which themselves typically contain regular (network) structures, and/or repeated occurrences of network patterns. This organisation occurs in a hierarchical manner, reflecting the physical and spatial organisation of the organism. Thus a further challenge is to represent the structure inherent in biological systems.

Coloured Petri Nets are high-level nets and a well-established modelling formalism. They have been used for over 20 years for the specification and analysis of communication protocols, distributed systems, automated production systems, work flows, and VLSI chip design [22]. They allow the description of similar network structures in a concise and well-founded way, providing a flexible template mechanism for network designers, and their combination with hierarchical structuring mechanisms is extremely powerful [21].

In coloured Petri nets, tokens can be distinguished via their colours. This allows for the discrimination of species (molecules, metabolites, proteins, secondary substances, genes, etc.). In addition, colours can be used to distinguish between sub-populations of a species in different locations (cytosol, nucleus, etc.).

Each place is assigned a colour set, specifying the kind of tokens which can reside on the place. A guard is associated with each transition, specifying which coloured tokens are required for firing, and each arc is allocated an inscription specifying the kind of tokens flowing through it. Coloured Petri nets with finite colour sets can be automatically unfolded into uncoloured Petri nets, which then permits the application of all of the existing powerful Petri net analysis and simulation techniques. Vice versa, uncoloured Petri nets can be folded into coloured Petri nets, if partitions of the place and transition sets are given. These partitions of the uncoloured net define the colour sets of the coloured net. As with hierarchical Petri nets, the conversion between uncoloured and coloured Petri nets changes the style of representation, but does not change the actual net structure of the underlying reaction network.

An attractive advantage of coloured Petri nets is their possibility to easily increase the size of a model consisting of many similar subnets by just adding colours, compare Fig. 2. This permits, e.g., concise representations of the uncoloured multi-cellular models of C. Elegans discussed in [7, 9]. These models consist of six (almost) identical network patterns, one for each cell. In a coloured version, the network pattern can be represented only once and the different cells are reflected in the coloured annotations of the net [23]. Another scenario for deploying colour to simulate a bacterial infection can be found in [8].

Colouring Space. In this paper we deploy colour to specify (biochemical) processes evolving in space. We develop a spatial specification style which can be equally applied in all modelling paradigms. This facilitates smooth movement between the modelling paradigms and the qualitative, stochastic, continuous or hybrid interpretation of the same Petri net. See [23] for more details and formal definitions of the structured family of coloured Petri net classes used in this paper, and [26] for all tool-related features.

In the continuous paradigm, our approach using discretised space corresponds to discretising partial differential equations. An alternative approach to model and solve partial differential equations using (discrete) Petri nets, based on the probably simplest time concept possible for this purpose (maximal steps, maximal auto-concurrency) is discussed in [3, 4]. A more elaborated comparison with other approaches to treat spatial properties is beyond the scope of this paper.

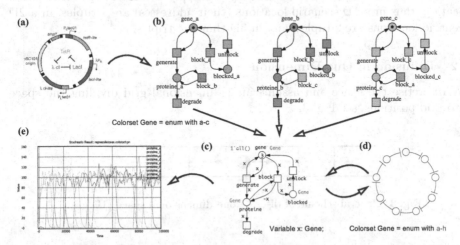

Fig. 2. The repressilator - a genes regulatory cycle [6]. (a) Schematic diagram for three genes. (b) Uncoloured Petri net model for three genes using logical transitions. (c) Folding of similar subnets into a coloured Petri net. (d) Schematic diagram for the generalised repressilator with nine genes. Modelling is accomplished by adjusting the colour set.(e) Stochastic simulation plot of the underlying uncoloured stochastic Petri net. See [20] for the explanation of annotations.

3   Continuous Paradigm

In this section we focus on the continuous part of the framework, illustrating it by means of a case study elaborated over two spatial dimensions.

3.1   Case Study 1: Diffusion

Background. We focus here on diffusion, which is a basic process occurring in biochemical systems with parameters over time and space. It can be regarded as the simplest form of passive mobility. Diffusion goes from regions of higher concentration to regions of lower concentration (Ficks laws) [10] where the diffusion flux is proportional to the minus gradient of concentrations.

Example 1. *One molecular species (here cyclic adenosine monophosphate – cAMP) diffuses continuously in space; i.e., it evolves simultaneously over time and space. The state-dependent diffusion rate follows mass/action kinetics, i.e., the rate is defined by the product of the species involved times some constant, summing up all dependencies on pressure, temperature, etc. The observation shall start with a high concentration (e.g., 100) in the middle of the space, with all other space positions initially set to 0.*

We are going to discuss this example in different scenarios, specifically 1- and 2-dimensional space (1D, 2D), using coloured Petri nets. We use the concept of colour to efficiently represent repeated structures in a continuous Petri net - i.e. to encode repeated elements of a set of ODEs. Each repeated element is associated with a colour, represented by a positive integer; sets of colours are thus discrete and finite. More specifically, we apply colour to represent spatial location; thus in a 1D scenario locations (their addresses) are 1-tuples, in a 2D scenario locations are 2-tuples, and in 3D they are triples.

3.2   Diffusion in One Dimension

We discretise the space and assume an 1-dimensional grid dividing the space into grid positions; see Fig. 3.

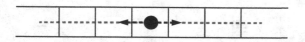

Fig. 3. General scheme of discrete one-dimensional space (1D grid)

A corresponding continuous Petri net is given in Fig. 4 modelling a discrete, 1-dimensional space comprising five grid positions - the five Petri net places $cAMP_i$, while the Petri net transitions model diffusion between neighbouring grid positions. The two outer places stand for the equivalence classes of the boundary space positions and beyond.

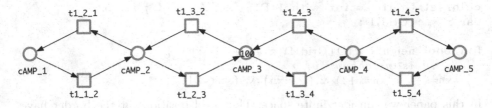

Fig. 4. Continuous Petri net for diffusion in one dimension. The space is discretised into five positions. The value of the middle position is initially set to 100, all other positions to zero, which is the default value, usually not given in graphics.

A continuous Petri net uniquely defines a system of Ordinary Differential Equations (ODEs) [14,31], with one equation for each place (variable). The rates of pre-transitions increase its value, thus defining plus terms, while the rates of post-transitions decrease its value, thus defining minus terms. Denoting the rate of a transition t_j by $v(t_j)$, and the set of pre-transitions (post-transitions) of a place c by ${}^\bullet c$ ($c^\bullet$), we get the generating Equation (1).

$$\frac{dc_i}{dt} = \sum_{t_j \in {}^\bullet c_i} v(t_j) - \sum_{t_j \in c_i {}^\bullet} v(t_j) \tag{1}$$

Assuming the diffusion rates $v(t_j)$ to follow mass/action kinetics with the common rate parameter k, we get the Equations (2)–(6) for the continuous Petri net in Fig. 4; for sake of readability we abbreviate $cAMP_i$ by c_i.

$$\frac{dc_1}{dt} = k \cdot c_2 - k \cdot c_1 \tag{2}$$

$$\frac{dc_2}{dt} = k \cdot c_1 + k \cdot c_3 - 2 \cdot k \cdot c_2 \tag{3}$$

$$\frac{dc_3}{dt} = k \cdot c_2 + k \cdot c_4 - 2 \cdot k \cdot c_3 \tag{4}$$

$$\frac{dc_4}{dt} = k \cdot c_3 + k \cdot c_5 - 2 \cdot k \cdot c_4 \tag{5}$$

$$\frac{dc_5}{dt} = k \cdot c_4 - k \cdot c_5 \tag{6}$$

We obtain a general pattern for an arbitrary, but static size of the discrete, 1-dimensional space by folding the (continuous) Petri net in Fig. 4 into a coloured (continuous) Petri net. For this purpose we introduce the following definitions.

const D1 = **int with** 5;   // grid size
const MIDDLE = **int with** D1/2+1;

```
colorset Grid1D = int with 1–D1;        // grid positions
var x,y : Grid1D;
```

```
fun bool neighbour1D(Grid1D x, Grid1D xn) {
    // xn is neighbour of x
    (xn=x−1 | xn=x+1) & (1<=xn) & (xn<=D1) };
```

In this paper we consider finite space. Thus grid positions at the border have fewer neighbours than inner grid positions. We obtain the coloured continuous Petri net given in Fig. 5, where colours serve as addresses in the spatial grid. Changing the grid position of a token now just means recolouring the token.

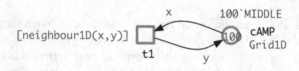

Fig. 5. Coloured continuous Petri net for diffusion in one dimension. The initial marking assigns the value 100 to the MIDDLE grid position, $v(t1) = MassAction(k)$.

Unfolding the coloured Petri net in Fig. 5 with $D1 = 5$ yields exactly the continuous Petri net given in Fig. 4, and thus in turn the ODEs (2)–(6). Changing the constant $D1$ adapts the model pattern to a specific grid size, which permits convenient model scaling, e.g., to increase the spatial resolution.

3.3   Diffusion in Two Dimensions

The generalisation to the 2-dimensional case using a Cartesian grid, see Fig. 6, is rather straightforward. We basically need to extend the definitions required for annotating the coloured Petri net while keeping the Petri net structure as it is.

We start off with a neighbourhood relation where inner grid positions have four neighbours, see Fig. 6(a), which is encoded in the function $neighbour2D4$. The corresponding coloured Petri net is given in Fig. 7(a), and its unfolding for $D1 = D2 = 5$ in Fig. 8. All transitions follow the same kinetic rate pattern.

```
const D1 = int with 5;          // grid size first dimension
const D2 = D1;                   // grid size second dimension
const MIDDLE = int with D1/2+1;
```

```
colorset CD1 = int with 1–D1;          // row index
colorset CD2 = int with 1–D2;          // column index
colorset Grid2D = product with CD1 x CD2;        // 2D grid
```

```
var x,a : CD1;
var y,b : CD2;

fun bool neighbour2D4 (CD1 x, CD2 y, CD1 xn, CD2 yn) {
  // (xn,yn) is one of the up to four neighbours of (x,y)
      (xn=x & yn=y-1) | (xn=x & yn=y+1)
  | (yn=y & xn=x-1) | (yn=y & xn=x+1)
  & (1<=xn & xn<=D1) & (1<=yn & yn<=D2) };
```

Next we consider a variation of the neighbourhood relation where each inner grid position has eight neighbours; see Fig. 6(b). We introduce three functions.

```
fun bool neighbour2D8 (CD1 x,CD2 y,CD1 xn,CD2 yn) {
  // (xn,yn) is one of the up to eight neighbours of (x,y)
      (xn=x-1 | xn=x | xn=x+1 ) & (yn=y-1 | yn=y | yn=y+1)
  & (!(xn=x & yn=y))
  & (1<=xn & xn<=D1) & (1<=yn & yn<=D2) };

fun bool lateral (CD1 x,CD2 y,CD1 xn,CD2 yn) {
      (xn=x & yn=y-1) | (xn=x & yn=y+1)
  | (yn=y & xn=x-1) | (yn=y & xn=x+1) };

fun bool diagonal (CD1 x,CD2 y,CD1 xn,CD2 yn) {
      (xn=x-1 & yn=y-1) | (xn=x+1 & yn=y-1)
  | (xn=x-1 & yn=y+1) | (xn=x+1 & yn=y+1) };
```

The latter two functions are used to appropriately set the rate functions, assuming that it takes longer to reach a diagonal neighbour than a lateral one:

$$v(t1) = \begin{cases} lateral(x,y,a,b) & : & MassAction(k) \\ diagonal(x,y,a,b) & : & MassAction(k/DIAGONAL), \end{cases} \tag{7}$$

with $\text{DIAGONAL} = \sqrt{2}$. The corresponding coloured Petri net is given in Fig. 7(b), and its unfolding for D1=D2=5 in Fig. 8.

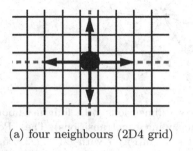   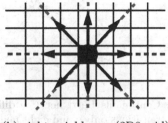

(a) four neighbours (2D4 grid)   (b) eight neighbours (2D8 grid)

Fig. 6. General scheme of discrete two-dimensional space with two different neighbourhood relations

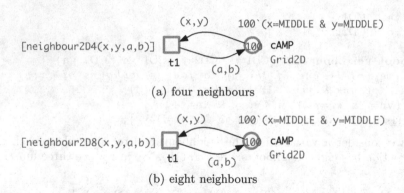

(a) four neighbours

(b) eight neighbours

Fig. 7. $\mathcal{CPN}^C$ for diffusion in two dimensions with two different neighbourhood relations. The difference consists of the neighbour function used as transition guard and the rate functions; (a) $v(t1) = MassAction(k)$, (b) see Equation (7).

Remarks: The coloured Petri nets in Fig. 5 and 7 all share the same structure, they differ in their colour-related annotations. It is obvious how to adjust the definitions to other neighbourhood relations.

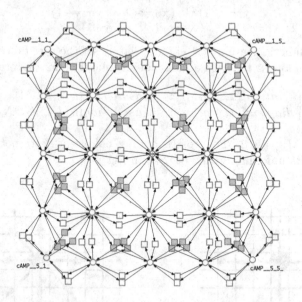

Fig. 8. Continuous Petri nets for diffusion in two dimensions with four neighbours (white transitions only), and eight neighbours (including grey transitions). These Petri nets have been generated by unfolding the two $\mathcal{CPN}^C$ in Fig. 7 with $D1 = D2 = 5$.

3.4   Computational Experiments

For the time being, all computational experiments are undertaken by unfolding coloured Petri nets which is automatically performed in the background, and numerically solving the underlying ODEs which again are generated automatically. Both transformation steps and the continuous simulation itself are features of Snoopy, the tool used in this paper, see also Section 5. In other words, the $\mathcal{CPN}^{\mathcal{C}}$ serve as a kind of very high-level description of ODEs. Note that the mapping from $\mathcal{CPN}^{\mathcal{C}}$ to ODEs is unique but not vice-versa [31].

The key challenge when unfolding coloured Petri nets is to compute all transition instances, which suffers from combinatorial explosion. However, when the number of transition instances is only determined by guards (logical expressions), which is the case in our scenario, a constraint satisfaction approach [32] can be employed. As each coloured transition can be considered separately, the unfolding can be easily parallelised by multiple threads to take advantage of state-of-the-art multi-core computer architectures. We have used the efficient search strategies of *Gecode* [12] to substantially improve the unfolding efficiency of coloured Petri nets; for more details see [23, 27].

The unfolding of any $\mathcal{CPN}^{\mathcal{C}}$ version of our gradient example yields extremely large continuous Petri nets. It is easy to see that in the 2-dimensional case the number of places always equals $D1D2$, while the number of transitions amounts to $4D1D2 - 2(D1 + D2)$ for a 2D4 grid, and $8D1D2 - 6(D1 + D2) + 4$ for a 2D8 grid, respectively. To give an example, the unfolding of an 120×120 2D4 grid (used in Fig. 10, last row) generates 14,400 places and 57,120 transitions, with an unfolding time of about 25 seconds (on a standard laptop computer). The generated continuous Petri net in turn is transformed into ODEs according to formula (1), i.e., the number of places determines the number of ordinary differential equations, and the number of transitions the total number of terms in these equations to be simulated.

The actual simulation, i.e., the numerical integration of the generated ODEs takes a couple of seconds and yields time traces for each unfolded place, see Fig. 9. These traces are converted into heat maps, one for each time step, i.e., a sequence of heat maps eventually visualises the evolution in time and space, see Fig. 10.

We performed a couple of experiments to test the scalability of our model. Model scaling also usually requires adjustments of the initial marking and rate parameters. To maximise the flexibility of our model we slightly changed the specification style of the initial marking. We introduced a couple of constants (including LB – lower bound, UB – upper bound to specify a rectangle) which eventually permit the specification of the range of grid positions set to 100 in the initial marking in a better adjustable manner:

$$100\,{}^{\backprime}((\text{LB} <=\text{x} \ \& \ \text{x} <= \text{UB}) \ \& \ (\text{LB} <=\text{y} \ \& \ \text{y} <= \text{UB})).$$

To reach equivalent states in the same simulation time, we need to scale the parameters by the square of the resolution factor; see Fig. 10 for some results.

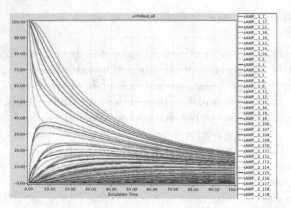

Fig. 9. Simulation plot of the ODEs generated from a $\mathcal{CPN}^{\mathcal{C}}$, illustrating approaching to the future steady state where all concentrations will be equal

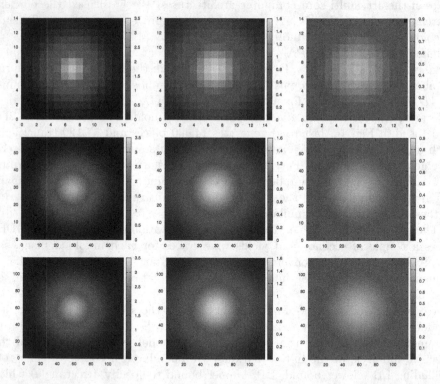

Fig. 10. Continuous simulation results for diffusion in two dimensions with four neighbours in space resolutions 15×15, 60×60, and 120×120. The three snapshots given for each resolution are taken at simulation time 25 (left), 50 (middle), and 100 (right column).

4   Stochastic Paradigm

In this section we focus on the stochastic part of the framework, reusing the colour definitions which we introduced in the previous section. Diffusion can be treated stochastically using the laws of Brownian motion, for example embodied in the Gilllespie algorithm [15]. However, the mapping is straightforward and instead we present a more sophisticated and challenging biological example.

4.1   Case Study 2: Bacterial Colony Growth – Phase Variation

We study phase variation in bacterial cell colonies which grow in space. We developed a Coloured Stochastic Petri Net which allows us to substantially extend the method applied in [30] to computationally predict the sector-like patterning characteristic of such colonies, see e.g., Figure 1 in [1].

Background. A common microbial stochastic mechanism is phase variation, in which gene expression is controlled by a reversible genetic mutation, rearrangement, or modification. Phase variation has traditionally been considered in the context of 'contingency genes' in which a sub-population is continuously generated which is pre-adapted to repeated environmental transitions, often to immune selective changes. However recent re-consideration, in the light of stochastic processes in genes under other forms of regulation, suggests an important potential role in bacterial specialization and differentiation, and the generation of structured bacterial populations.

Example 2. *We consider a colony of bacteria with two phenotypes A and B, which develop over time by cell division. Cell division may involve cell mutation, and back-mutation alternates phenotypes; see Fig. 11. The observation should start with one bacterium of phenotype A. We are interested in the proportion of phenotypes in the cell generations, and how their spatial distribution evolves over time.*

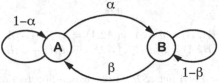

Fig. 11. Phase variation within bacterial colonies - basic scheme. Mutation from A to B happens at rate α, and backward mutation at rate β.

4.2   Step-Wise Modelling

In the following we describe the step-wise approach which we have employed to construct our $\mathcal{SPN}^C$ model. We start off with a basic model of phase variation between two states in bacterial colonies as discussed in [30] which did not model spatial aspects, and encode this as a stochastic Petri net. Next we enrich the basic model with a 2D8 grid, where the parent remains in-situ, and the child is displaced by one grid position. Finally, we refine our model by controlling colony spreading and thickness. Our stochastic spatial model permits us to describe the development of sector-like pattering typical of phase variation in bacterial colonies.

Step 1 – Basic Model of Phase Variation. We start with the equations taken from the previous *deterministic* model of phase variation [30], which describe *synchronous* growth in cell colonies with two phenotypes A and B, modelled here by two corresponding variables indexed by the discrete time steps. These equations include the assumption that "if phase variation occurs, the progeny consists of one A and one B."

$$A_{n+1} = 2d_A(1 - \alpha)A_n + d_A\alpha A_n + d_B\beta B_n \tag{8}$$

$$B_{n+1} = 2d_B(1 - \beta)B_n + d_B\beta B_n + d_A\alpha A_n \tag{9}$$

Here, d_A and d_B specify the fitness, i.e., the proportions of A or B, respectively, that survive to division.

Previously [30], behaviour was explored by iterating the equations on a spreadsheet. We develop a Petri net model that is directly executable by playing the token game which facilitates its comprehension, and permits the exploration of the behaviour by standard analysis and simulation techniques. Our initial stochastic Petri net, see Fig. 12, corresponds to Equations (8)–(9), but adopts an *asynchronous* modelling approach so that cells divide individually.

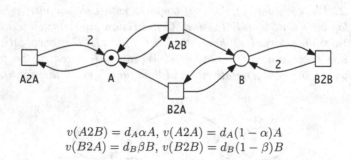

$$v(A2B) = d_A\alpha A, \; v(A2A) = d_A(1 - \alpha)A$$
$$v(B2A) = d_B\beta B, \; v(B2B) = d_B(1 - \beta)B$$

Fig. 12. Stochastic Petri net ($\mathcal{SPN}$) corresponding to Equations (8)–(9)

Model parameters (taken from [30])

- *mutation rates* α (forward), and β (backward): in the range of $10^{-2} - 10^{-5}$; e.g. *high:* $\alpha = \beta = 0.0025$, *medium:* $\alpha = \beta = 0.0005$, *low:* $\alpha = \beta = 0.00005$; α and β could also take different values;
- *relative fitness* f – ratio of phenotype survival probability: $f = d_A/d_B$; typical values: f = 1.0 (no fitness difference), 0.99, 0.9, 0.5.

Derived Measures of Interest

– *Total number of bacteria.* The n-th generation in a synchronous model yields 2^n bacteria. Vice versa, if we know the total number *total* of bacteria generated by asynchronous cell division, then we can obtain the corresponding synchronous generation counter n by

$$n = \log_2 total \tag{10}$$

For example, 26 synchronous generations (which may develop in about 24 hours) end up with a total population size of approximately $67 \cdot 10^6$.

– *Proportion of A and B.*

$$propA = \frac{A}{A+B}; \quad propB = \frac{B}{A+B} \tag{11}$$

Simulating the stochastic model allows us to observe asynchronous population growth such that cells divide individually. Each event (firing of a transition) corresponds to the division of one cell. Consequently, the size of the population will grow in steps by 1, see Fig. 13, in contrast with the synchronous model. Depending on the setting for the output steps of the simulator we may not be able to observe all events in the simulation trace.

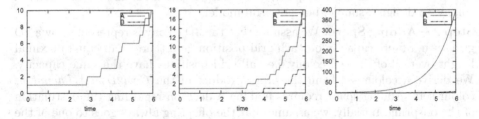

Fig. 13. Two single stochastic simulation runs, and one continuous run; $\alpha = \beta = 0.0025$, $d_A = d_B = 1$, i.e., no fitness advantage

Observations

– Keeping a relative fitness of 1 while extending the simulation time allows us to observe that the variables A and B will finally be almost identical, meaning their proportions will finally approach 50%.

– Likewise, keeping the mutation rates equal and giving one mutant a fitness advantage over the other, e.g. using a fitness ratio of 0.9, then the mutants with the greater fitness will finally outnumber the mutants with lower fitness and the proportion of the latter ones in the total population approaches zero over time.

Starting from simulation traces like the ones given in Fig. 13, all diagrams presented and discussed in [30] can be derived by some post-processing.

To prepare for the modelling of cell colonies in space we fold our first (uncoloured) Petri net. For this purpose we introduce two colour sets, $Phenotype = \{a, b\}$, and $DivisionType = \{replicate, mutate\}$. These definitions allow us to fold the two places A and B into one coloured place *cell* with the colour set *Phenotype*, and to fold the four transitions into the coloured transition *division*. We obtain the basic model given in Fig. 14.

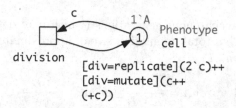

Fig. 14. $\mathcal{SPN}^C$ as $\mathcal{SPN}$ short-hand notation; unfolding this $\mathcal{SPN}^C$ generates the $\mathcal{SPN}$ in Fig. 12. See listing in Fig 15 for the related definitions.

The derivation of our final model, see Fig. 15, from the basic model, see Fig. 14, requires three further steps: adding space, controlling colony spreading, and controlling thickness. We deliberately ignore some complexities, e.g. nutrition and oxygen which are responsible for the vertical structure of the bacterial colony, to design a simple, but powerful model.

Step 2 – Adding Space. We assume that the 3D colony is represented by a 2D grid with a finite capacity on each grid position, and there is an equal maximal height over all of the cell colony, i.e., all grid positions have the same capacity. We derive a colour set from the cross product of the *Grid2D* and *Phenotype* colour sets. Adding space requires making a decision regarding the destination of the offspring. Initially, we assume that the offspring always goes to one of the neighbouring positions which is chosen stochastically.

In this case study we are concerned with mutation rates and their influence on the system behaviour. So their total values have to be kept constant. Introducing space means technically to multiply transitions (basically one for each direction per grid position). To counterbalance this effect, we scale the transition rates by dividing them by the number of grid positions and by N, with N being the number of neighbours. With other words, all transitions (which we get by unfolding) make four equivalence classes, and the sum of all rates in one equivalence class is kept constant, independent of the grid size and the neighbourhood relation. Thus, the total rates in the phase variation model with space are the same as in the phase variation model without space.

Step 3 – Controlling Colony Spreading. Cells do not actively move; as a result of cell division they can either pile up on the parent's grid position or be displaced to a neighbouring position. To model this phenomenon, we add an alternative transition which allows an offspring to stay with its parent. Thus, the rate functions need now to be scaled by N+1.

To control the tendency between staying with the parent (*division*1) or going to a neighbouring position (*division*2), we introduce a preference factor γ, which may vary between 0 and $\gamma_{\max}$ without changing the total division rate (sum of rates of *division*1 and *division*2). For this purpose, we define $\gamma_H = \gamma/\gamma_{\max}$, and $\gamma_N = (\gamma_{\max} - \gamma)/\gamma_{\max}$ to further scale the rate functions correspondingly.

Increasing γ increases the preference to stay with the parent, while decreasing γ increases the preference to displace. Setting γ to $\gamma_{\max}$ precludes the ability to go to a neighbour, thus the size of the colony is restricted by the capacity of one grid position. Setting γ to zero precludes staying with the parent. Cells then have the tendency to first occupy all grid positions, before the thickness increases simultaneously over the whole colony patch.

Step 4 – Controlling Thickness. The bacteria generated by cell division can pile up on top of each other and thus increase the colony thickness at that grid position. This thickness is limited because of the cells' requirements for access to oxygen and nutrients. In order to control the thickness we introduce a constant POOLSIZE, which limits the maximum number of cells at a certain grid position. We set POOLSIZE to give room for 26 generations. See the listing in Fig. 15 for a summary of all required colour-related definitions.

4.3   Computational Experiments

All computational experiments are done on the automatically unfolded Petri nets. Unfolding our coloured Petri net for a 101×101 grid yields an uncoloured Petri net with 30,603 places and 362,404 transitions with an unfolding time of 630 seconds. The unfolded Petri net is simulated using the Gillespie algorithm [15]. One stochastic simulation run takes about 40 minutes. The output comprises a pair of simulation traces for each grid position, corresponding to the two phenotypes A and B, similar to Fig. 9, with each run behaving differently.

The analysis considers the development over time of the proportion of the given genotype in the total population, and the patterning into characteristic segments. This requires converting the stochastic simulations into 2D representations, see Fig. 16, and analysing the development of the 2D sector-like patterns over time. We expect that the model will finally permit the prediction of mutation rates and fitness by counting and measuring pattern segments, which in the future could give new insights into the population dynamics of mutation. Currently, our model predicts behavior which has not been measured so far in the wet lab — the model generates a time series description of the evolution of the patterns in cell colony (indicated in Fig. 16), while wet lab data just give a snapshot of the final state.

5   Tools

All Petri nets in this paper were constructed with Snoopy [29], recently extended to support coloured Petri nets [20, 23]. Simulations were done with Snoopy's built-in stochastic simulator and Marcie [18]. Simulation traces have been further

```
const  D1 = int  with  101;
const  D2 = D1;
const  MIDDLE = int  with  D1/2+1;
const  POOLSIZE = int  with  7000;
const  POOLSIZE_1 = int  with  POOLSIZE-1;

colorset  Phenotype = enum  with  A, B;
colorset  DivisionType = enum  with  replicate, mutate;
colorset  CD1 = int  with  1-D1;
colorset  CD2 = int  with  1-D2;

colorset  Grid2D = product  with  CD1 x CD2;
colorset  Grid = product  with  Grid2D x Phenotype;

var  c : Phenotype;
var  div : DivisionType;
var  x,xn : CD1;
var  y,yn : CD2;

fun  bool  neighbour2D8(CD1 x,CD2 y,CD1 xn,CD2 yn) { . . . };
fun  bool  lateral(CD1 x,CD2 y,CD1 xn,CD2 yn) { . . . };
fun  bool  diagonal(CD1 x,CD2 y,CD1 xn,CD2 yn) { . . . };
```

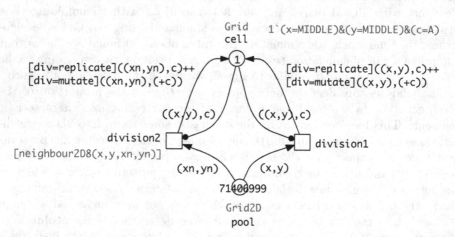

Fig. 15. $\mathcal{SPN}^{\mathcal{C}}$ for the final spatial model of phase variation. The integers on the places give the total number of tokens of any colour.

Fig. 16. 2D representation of a single stochastic run showing the development of binary phase variation in a cell colony over time and space, and illustrating the development of sector-like patterns. Density of the two phenotypes is represented by yellow and dark blue, respectively. Each run looks differently due to the built-in stochasticity.

processed by customised Java (Python) programs, and finally visualised with Gnuplot (matplotlib). Snoopy and the models in Snoopy format can be obtained from http://www-dssz.informatik.tu-cottbus.de. Thus, all our results can be easily reproduced by the interested reader.

6   Summary

In this paper we have deployed colour to specify biochemical processes evolving in time and space. The spatial modelling style presented can be applied to a wide range of biological and also technical application scenarios. The framework we have introduced covers qualitative, stochastic, continuous and hybrid modelling paradigms It exploits existing simulation techniques and analytical machinery by unfolding to uncoloured nets.

Due to page restrictions we have only presented continuous and stochastic case studies, but it is obvious how to apply our spatial modelling approach to qualitative and hybrid examples. It is also straightforward how to extend the colouring principle to a 3-dimensional space, or how to adapt it to different notions of space; e.g., using polar coordinates.

The coloured Petri nets which we have presented might give an impression of simplicity, which just underlines the power of abstraction by folding into coloured models. Crucially, this technique enables a new approach to multiscale modelling, and we have elsewhere illustrated this by using coloured stochastic and continuous Petri nets to model planar cell polarity in Drosophila fly wing [11, 13,17]. A 2-dimensional space is organised as a regular honeycomb lattice of cells which is interpreted over a regular grid by tuning the neighbourhood functions. Each position in the grid contains a subgrid describing the intracellular level. Further case studies deploying coloured Petri nets for spatial modelling problems can be found in [5, 13, 24, 25].

Our modelling style supports BioModel engineering by the established *separation of concerns* principle. Changing the notion of space just requires the appropriate adaptation of the definition of the colour sets, the functions specifying the neighbourhood relation, and the transition rate functions. The net structure itself needs not to be altered. All colour-related definitions can be reused via Snoopy's export/import functionality.

Our current ongoing work includes the development of visualisation and model checking over spatial patterns in multiple dimensions and scales, as well as non-rectangular geometries. Future work will address computational challenges due to the fact that currently simulations must be performed at the unfolded level rather than at the coloured level.

Acknowledgments. This research has been partially funded by the British EPSRC Research Grant EP I036168/1 and the German BMBF Research Grant 0315449H. The computational experiments undertaken in this project would not have been possible without the constant support by Mostafa Herajy, Martin Schwarick, and Christian Rohr. We acknowledge as well the support by Mary Ann Blätke and Jan Wegener in producing some of the data plots.

References

1. Appelmelka, B.J., Monteirob, M.A., Martinc, S.L., Morand, A.P., Vandenbroucke-Graulsa, C.M.: Why Helicobacter pylori has Lewis antigens. Trends in Microbiology 8(12), 565–570 (2000)
2. Baldan, P., Cocco, N., Marin, A., Simeoni, M.: Petri Nets for Modelling Metabolic Pathways: a Survey. J. Natural Computing (9), 955–989 (2010)
3. Bertens, L.: Computerised modelling for developmental biology: an exploration with case studies. Ph.D. thesis, Leiden University (September 2012)
4. Bertens, L., Kleijn, J., Hille, S., Koutny, M., Heiner, M., Verbeek, F.: Modeling biological gradient formation: combining partial differential equations and Petri nets. Tech. Rep. CS-TR-1379, Univ. of Newcastle upon Tyne, School of CS (2013)
5. Blätke, M., Dittrich, A., Rohr, C., Heiner, M., Schaper, F., Marwan, W.: JAK/-STAT signalling - an executable model assembled from molecule-centred modules demonstrating a module-oriented database concept for systems and synthetic biology. Molecular BioSystem (2013)
6. Blossey, R., Cardelli, L., Phillips, A.: Compositionality, Stochasticity and Cooperativity in Dynamic Models of Gene Regulation. HFSP Journal 1(2), 17–28 (2008)
7. Bonzanni, N., Krepska, E., Feenstra, K., Fokkink, W., Kielmann, T., Bal, H., Heringa, J.: Executing multicellular differentiation: quantitative predictive modelling of c. elegans vulval development. Bioinformatics 25, 2049–2056 (2009)
8. Carvalho, R., Kleijn, J., Meijer, A., Verbeek, F.: Modeling Innate Immune Response to Early Mycobacterium Infection. Computational and Mathematical Methods in Medicine 2012, Article ID 790482 (2012)
9. Chen, L., Masao, N., Kazuko, U., Satoru, M.: Simulation-based model checking approach to cell fate specification during C. Elegans vulval development by hybrid functional Petri net with extension. BMC Systems Biology 42, 3 (2009)
10. Fick, A.: Über Diffusion (in German). Annalen der Physik 170(1), 59–86 (1855)
11. Gao, Q., Gilbert, D., Heiner, M., Liu, F., Maccagnola, D., Tree, D.: Multiscale Modelling and Analysis of Planar Cell Polarity in the Drosophila Wing. IEEE/ACM Transactions on Computational Biology and Bioinformatics 99(PrePrints) (2012)
12. Gecode: Gecode: An open constraint solving library (2011), http://www.gecode.org

13. Gilbert, D., Heiner, M.: Petri nets for multiscale Systems Biology. Brunel University, Uxbridge (2011), http://multiscalepn.brunel.ac.uk/
14. Gilbert, D., Heiner, M.: From Petri nets to differential equations - an integrative approach for biochemical network analysis. In: Donatelli, S., Thiagarajan, P.S. (eds.) ICATPN 2006. LNCS, vol. 4024, pp. 181–200. Springer, Heidelberg (2006)
15. Gillespie, D.: Exact stochastic simulation of coupled chemical reactions. The Journal of Physical Chemistry 81(25), 2340–2361 (1977)
16. Heiner, M., Gilbert, D.: How Might Petri Nets Enhance Your Systems Biology Toolkit. In: Kristensen, L.M., Petrucci, L. (eds.) PETRI NETS 2011. LNCS, vol. 6709, pp. 17–37. Springer, Heidelberg (2011)
17. Heiner, M., Gilbert, D.: Biomodel engineering for multiscale systems biology. Progress in Biophysics and Molecular Biology (2012)
18. Heiner, M., Rohr, C., Schwarick, M.: MARCIE – Model Checking and Reachability Analysis Done Efficiently. In: Colom, J.-M., Desel, J. (eds.) PETRI NETS 2013. LNCS, vol. 7927, pp. 389–399. Springer, Heidelberg (2013)
19. Heiner, M., Gilbert, D., Donaldson, R.: Petri nets for systems and synthetic biology. In: Bernardo, M., Degano, P., Zavattaro, G. (eds.) SFM 2008. LNCS, vol. 5016, pp. 215–264. Springer, Heidelberg (2008)
20. Heiner, M., Herajy, M., Liu, F., Rohr, C., Schwarick, M.: Snoopy - A Unifying Petri Net Tool. In: Haddad, S., Pomello, L. (eds.) PETRI NETS 2012. LNCS, vol. 7347, pp. 398–407. Springer, Heidelberg (2012)
21. Huber, P., Jensen, K., Shapiro, R.: Hierarchies in coloured Petri nets. In: Rozenberg, G. (ed.) APN 1990. LNCS, vol. 483, pp. 313–341. Springer, Heidelberg (1991)
22. Jensen, K., Kristensen, L.: Coloured Petri nets: modelling and validation of concurrent systems. Springer (2009)
23. Liu, F.: Colored Petri Nets for Systems Biology. Ph.D. thesis, BTU Cottbus, Dep. of CS (January 2012)
24. Liu, F., Heiner, M.: Modeling membrane systems using colored stochastic Petri nets. Nat. Computing (2013)
25. Liu, F., Heiner, M.: Multiscale modelling of coupled Ca^{2+} channels using coloured stochastic Petri nets. IET Systems Biology (2013)
26. Liu, F., Heiner, M., Rohr, C.: Manual for Colored Petri Nets in Snoopy. Tech. Rep. 02-12, Brandenburg University of Technology Cottbus, Dep. of CS (March 2012)
27. Liu, F., Heiner, M., Yang, M.: An efficient method for unfolding colored Petri nets. In: Proc. 2012 Winter Simulation Conference (WSC 2012). IEEE, Berlin (2012)
28. Marwan, W., Sujathab, A., Starostzik, C.: Reconstructing the regulatory network controlling commitment and sporulation in *Physarum polycephalum* based on hierarchical Petri net modeling and simulation. J. of Theoretical Biology 236(4), 349–365 (2005)
29. Rohr, C., Marwan, W., Heiner, M.: Snoopy - a unifying Petri net framework to investigate biomolecular networks. Bioinformatics 26(7), 974–975 (2010)
30. Saunders, N., Moxon, E., Gravenor, M.: Mutation rates: estimating phase variation rates when fitness differences are present and their impact on population structure. Microbiology 149(2), 485–495 (2003)
31. Soliman, S., Heiner, M.: A unique transformation from ordinary differential equations to reaction networks. PlosONE 5(12), e14284 2010)
32. Tsang, E.P.K.: Foundations of Constraint Satisfaction. Academic Press, London (1993)

The Vehicle Relocation Problem in Car Sharing Systems: Modeling and Simulation in a Petri Net Framework

Monica Clemente[1], Maria Pia Fanti[2], Agostino M. Mangini[2], and Walter Ukovich[1]

[1] Department of Engineering and Architecture, University of Trieste, Italy
[2] Department of Electrical and Information Engineering, Polytechnic of Bari, Italy
monica.clemente@phd.units.it,
{fanti,mangini}@deemail.poliba.it, ukovich@units.it

Abstract. The paper proposes an user-based solution for the vehicle relocation problem in car sharing systems. In particular, the impact of a mechanism of economic incentives based on a real time monitoring of vehicles distribution among the parking areas is assessed. We consider three different operative conditions that are described by the Unified Modeling Language and modeled in a Timed Petri Net (TPN) framework. In order to show and compare the effectiveness of the adopted management strategies, the real case study of an electric-car sharing system is evaluated and simulated in the TPN environment. The results underline how the proposed solution leads to an improvement of the overall system performances, by highlighting at the same time the limits of such a strategy.

Keywords: Car Sharing, Timed Petri Net, Simulation, Sustainable Mobility.

1   Introduction

Over the past few decades the environmental and socio-economic problems linked to the mobility in urban areas have underlined the need of reducing the massive use of private vehicles. In this context, systems in which a common fleet of vehicles is shared among multiple users (the so-called shared-use vehicle systems) have reached great popularity [1], mainly in the forms of car sharing.

Specifically, in a car sharing system every user can autonomously rent a car according to her/his needs and for a period that can be very short, unlike the traditional car rental. The benefits of such an offer are clear and can be summarized as follows:

1. general improvement of transport efficiency, thanks to the decrease of the total number of vehicles required to meet the travel demand and to a more rational use of mobility opportunities;
2. sharing among multiple users of fixed costs usually related to the possession of a private vehicle (purchase, insurance, maintenance);
3. reduction of carbon emissions;
4. increased use of public transport.

J.-M. Colom and J. Desel (Eds.): PETRI NETS 2013, LNCS 7927, pp. 250–269, 2013.

Simultaneously, the energy dependence of the mobility on fossil fuels, characterized by great instability both in prices and in supplies, has increased the interest in Electric Vehicles (EVs). However, high prices and limited range are big obstacles to their complete diffusion: fleets of shared-use vehicle systems can represent a concrete opportunity to overcome these initial drawbacks and to support the achievement of electric mobility.

In order to obtain an efficient implementation of such systems and thus limit the undeniable competitive advantage of private traditional vehicles, a detailed planning is required. In particular, there are many issues that affect the performances and the success of an electric-car sharing organization.

First, the determination of the optimal fleet size, based on the actual mobility demand, is fundamental and different approaches can be found in the related literature: e.g., George and Xia [2] propose a closed queuing network model for solving the fleet-sizing problem in a general vehicle rental system, while Nakayama et al. [3] attempt to optimize the design and the operation of an EV-sharing system using genetic algorithms.

At the same time, depot location is a parameter which greatly influences the service utilization rate [4], as well as pricing policies [5], [6] and vehicle reservation systems [7]. Another important feature that determines the popularity of a generic car sharing system is its flexibility: conventionally users are required to pick-up and return vehicles at the same station, but this strictly binds the conformation of users' travels; on the other hand, in the so-called *one-way rental systems* an user can return the rented vehicle at any station, making the service more useful, but at the same time introducing some management complexities and thus increasing management costs. In particular, in this type of service the balance in the distribution of vehicles among stations during the day must be guaranteed through a relocation mechanism, and two main categories of vehicle redistribution strategies can be found in literature: *operator-based* strategies and *user-based* strategies [8]. In order to support the management in dealing with these key issues and make the service more user-friendly, the application of the modern Information and Communication Technology (ICT) seems to be essential [9], [10].

This paper is part of this framework and addresses the application of ICT for the management of a one-way rental-electric-car sharing service. In particular, the effectiveness of a system of economic incentives settled to ensure a *user-based* rebalancing of the number of vehicles parked in each station throughout the day is assessed. Incentives are intended to influence the travel behavior of the users according to the system conditions, monitored in real time. To show the impact of the proposed solution, a model of an electric-car sharing system has been developed in a Timed Petri Net (TPN) framework. Indeed, TPN allows concisely representing in an unified structure both static and dynamic aspects of the considered system, thanks to its twofold representation, graphical and mathematical. In particular, the graphical aspect enables a concise way to design and verify the model, while the mathematical description allows simulating the considered system in software environments, by considering different dynamic conditions. Three different operative conditions are evaluated and modeled: the first one describes an easy management of the considered service; in the second one users are always encouraged to return the vehicle as soon as possible; in the third operative condition incentives to the users, which depend on the real time monitoring of the system balance conditions, are introduced. In order to show and compare the effectiveness of the adopted management strategies, the proposed vehicle relocation strategies are applied to the real case of the electric-car sharing system of Pordenone, a town of the North of Italy. The simulation studies and results in the TPN environment show how the proposed solution leads to an improvement of the overall system performances, by highlighting at the same time the limits of such a strategy.

The remainder of the paper is structured as follows. Section 2 presents a brief overview on the TPN used in this paper, while Section 3 describes the considered problem and the proposed solution. In Section 4 a case study is described and then the related TPN model is detailed; Section 5 presents system behavior simulations under the three different operative conditions and discusses the results. Finally, in Section 6 the conclusions are summarized.

2   Basics of Timed Petri Nets

This section recalls some basic definitions on the TPN formalism used in the paper.

2.1   Net Structure

A Petri net (PN) [11] is a bipartite graph described by the four-tuple $PN=(P, T, Pre, Post)$, where P is a set of places with cardinality m, T is a set of transitions with cardinality n, Pre: $P{\times}T{\rightarrow}\mathbb{N}^{m{\times}n}$ and $Post$: $P{\times}T{\rightarrow}\mathbb{N}^{m{\times}n}$ are the *pre-* and the *post-incidence matrices*, respectively, which specify the arcs connecting places and transitions. More precisely, for each $p{\in}P$ and $t{\in}T$ element $Pre(p,t)$ ($Post(p,t)$) is a natural number indicating the arc multiplicity if an arc going from p to t (from t to p) exists, and it equals 0 otherwise. Note that $\mathbb{N}$ is the set of non-negative integers. Matrix $C=Post-Pre$ is the $m{\times}n$ incidence matrix of the PN.

The state of a PN is given by its current marking, which is a mapping $M: P{\rightarrow}\mathbb{N}^m$, assigning to each place of the net a non-negative number of tokens. A PN system $<PN, M_0>$ is a net PN with an initial marking M_0.

Classical PN do not convey any notion of time, but in order to represent systems with temporal constraints TPN have been introduced: TPN are obtained from PN by associating a firing time to each transition of the net [12]. In particular, there are three types of timed transitions: immediate transitions (represented with bars), stochastic transitions (represented with boxes) and deterministic transitions (represented with black boxes). More formally, a TPN is a six-tuple $TPN=(P, T, Pre, Post, F, RS)$, where $P, T, Pre, Post$ have the same meaning as described above. Moreover, function $F: T \rightarrow \mathbb{R}_0^+$ specifies the timing associated to each transition, where $\mathbb{R}_0^+$ is the set of non-negative real numbers. In particular, $F(t_j)=\delta_j$ specifies the timing associated to the timed deterministic transitions, and $F(t_j)= 1/\lambda_j$ is the average firing delay each stochastic transition, where λ_j is the average transition firing rate. Finally, RS: $T \rightarrow \mathbb{R}^+$ is a function that associates a probability value called *random switch* to conflicting transitions.

2.2   Net Dynamics

A transition $t_j \in T$ is enabled at M if and only if (iff) it holds: $M \geq Pre(\cdot, t_j)$ and we write $M[t_j$ to denote that $t_j \in T$ is enabled at marking M. When fired, t_j produces a new marking M', denoted by $M[t_j > M'$, that is computed by the PN state equation $M' = M + C\vec{t_j}$, where $\vec{t_j}$ is the n-dimensional firing vector corresponding to the j-th canonical basis vector. Moreover, the enabling degree of transition $t_j \in T$ at M is equal to $enab(M, t_j) = max\{k \in \mathbb{N} \mid M \geq k \cdot Pre(\cdot, t_j)\}$.

If t_j is infinite-server semantics, we associate to it a number of clocks that is equal to $enab(M; t_j)$ [12]. Each clock is initialized to a value that is equal to the time delay of t, if t_j is deterministic, or to a random value depending on the distribution function of t_j, if t_j is stochastic. On the contrary, if a discrete transition is k-server semantics, then the number of clocks that are associated to t_j is equal to $min\{k, enab(M; t_j)\}$. The values of clocks associated to t_j decrease linearly with time, and t_j fires when the value of one of its clocks is null (if $\bar{k}$ clocks reach simultaneously a null value, then t_j fires $\bar{k}$ times). Note that in the paper we consider transitions using infinite server semantics and enabling memory policy. This means that if a transition enabling degree is reduced by the firing of a different transition, then the disabled clocks have no memory of this in future enabling [12]- [13].

3   The Vehicle Relocation Problem

In traditional car sharing organizations, users are required to return the rented vehicle to the same parking area where it has been picked-up and, so, only round trips are possible: this configuration of the system, usually called two-way rental, simplifies certainly the management of the service, but at the same time it seems to be suitable only for leisure or sporadic trips, limiting the number of potential users [14].

In the so called one-way rental systems, instead, customers can rent a vehicle in one station and return it to a different one. However, since demand for vehicles is not evenly distributed among the parking areas and not all destinations have the same popularity, this strategy of management can lead quickly to an imbalance in the number of cars available in each station. For this reason, an accurate planning of the service which prevents or minimizes vehicle imbalance is necessary, as well as vehicle relocation activities. This issue becomes more critical when the system fleet is composed by electric vehicles, since customers have the pressing need to find an available parking stall at the destination station in order to recharge, when necessary, the rented vehicle.

In literature, two main relocation strategies are identified: the operator-based strategy and the user-based strategy [7], [8], [15]. In the operator-based relocation technique system staff relocates vehicles among the parking areas [16], [17]. On the other hand, when an user-based strategy is adopted, users themselves ensure the rebalancing of the system with their travel behavior, influenced through different types of incentives and a strategic planning of the service rules [7]. This technique is particularly advantageous both from an economic and an environmental point of view, since no additional trips without customers are conducted. An example of this approach is described by Uesugi *et al.* in [18], where a method for assigning the optimum number of vehicles to users, according to distribution of parked vehicles, is presented.

3.1   The Proposed User-Based Strategies

This paper presents and assesses two particular user-based relocation techniques. The proposed solutions consist of an incentive mechanism that provides to the users, in real time, suggestions of behavior based on the actual vehicles allocation: more precisely, through ICT tools installed on board of each vehicle, the system suggests to the users in which infrastructure or within what time to give back the rented vehicle. If a customer follows the provided directions, he has a right to a discount on the total cost of the rental.

In order to study the outcomes derived from the introduction of such a mechanism, we compare three different operative conditions. In the first one, no incentive system is taken into account and vehicles are relocated only at the end of the working day. In the second case, users are always encouraged to return the rented vehicle as soon as possible, regardless of the system balance conditions. In the third operative condition the vehicles distribution among the parking areas is monitored at regular time intervals and, whenever the system is unbalanced, suitable travel conformations are suggested in real time to the users.

3.2   The UML Description

With the aim of obtaining a clear and synthetic representation of the three different service management strategies just described, the Unified Modeling Language (UML) [19] has been used: in particular, activity diagrams have been developed. This kind of diagrams is useful in fact to model dynamic aspects of a system through the description of the actions that, chained together, represent a process occurring in the system itself. Every activity can involve different actors: in activity diagrams the so-called swim lanes are used to show which actors are responsible for which actions and so diagrams are divided into columns. The main elements of this type of diagrams are: the initial activity (represented with a solid circle); the final activity (represented with a bull's eye symbol); general activities (represented with a rectangle with rounded edges); arcs connecting activities and representing flows; forks and joins, depicted by horizontal split and modeling concurrent activities and actions respectively beginning and ending at the same time; decisions, represented with diamonds and modeling alternative flows; signals representing activities sending or receiving a message. In particular, there are two types of messages: input signals, shown by a concave polygon, and output signals, shown by a convex polygon.

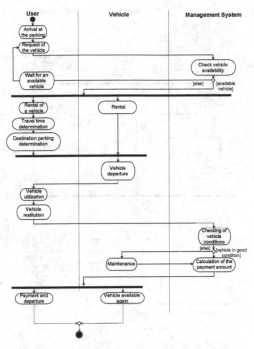

Fig. 1. First operative condition: relocations at the end of the working day

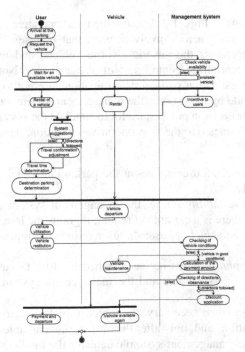

Fig. 2. Second operative condition: users encouraged to return the vehicle as soon as possible

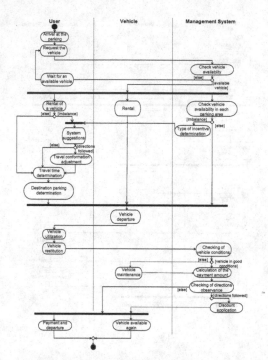

Fig. 3. Third operative condition: real time monitoring of the system and incentive mechanism

The activity diagrams of Figs. 1-3 describe the three operative conditions previously introduced. There are three actors involved: user, that represents the generic service user; vehicle, that represents the generic electric vehicle; and management system, which represents the centralized control system of the service. Note that "Travel time determination" and "Destination parking determination" are not to be intended as preventive statements made by the users to the management system, but rather as a simple schematization of their decision process, useful to simulate the system behavior.

Six main phases characterize the first operative condition, detailed in *activity diagram of Fig. 1*:

1. *vehicle request phase:* an user arrives at the parking area and makes a request for an available vehicle;
2. *checking vehicle availability phase*: the management system checks the vehicles availability and, if there is a car not yet rented, it grants the hire;
3. *rental and use of the vehicle phase*: the user refines the rental of the vehicle and makes his trip;
4. *vehicle restitution phase*: when he has finished his trip, the user brings back the vehicle in a service parking area and the management system checks the vehicle conditions;
5. *maintenance phase*: when necessary, the rented vehicle is carried to a car park for maintenance operations and, only after this, it is again available in a parking area;
6. *payment phase*: the management system calculates the total cost of the rental, then the user pays and leaves the system.

In addition to these phases, in second operative condition (Fig. 2) also the *incentive to users* phase and the *checking of directions observance* phase must be taken into account.

Finally, the third operative condition (Fig. 3) is characterized also by the *system balance condition* phase, when the management system checks the number of vehicles available in each parking area and, on the basis of that, determines the type of incentive that must be suggested to users.

4 A Case Study

4.1 The System Description

In order to show the modeling approach and to investigate the impacts of the proposed management strategies, the paper analyzes the real case study of the electric-car sharing service of Pordenone, a town of the North of Italy. Although it is a system of limited size, the results achieved in this context can be easily adapted and extended to any generic car sharing company: in fact, the TPN model here developed is modular and each parking area is represented by the same structure, as detailed below.

The considered car sharing service includes 10 electric vehicles, made available to the users in two parking areas (named *P3* and *P5* on the map of Fig. 4); three more parking spaces are actually planned by the management (*P1*, *P2*, *P4*). Obviously, since it is an electric-car sharing system, these areas are not ordinary parking stalls, but they have also to be equipped with infrastructures for the vehicles recharging. The service is regulated by the following rules:

- electric cars can move within Limited Traffic Zone (ZTL) every day, unless there are special events;
- parking within the ZTL is permitted in specified areas (blue parking areas) and no payment is due;
- it is not allowed to leave the municipality territory;
- rented vehicles must be given back by 8:15 pm;
- rented vehicles can be given back in any parking area (such system is therefore a one-way rental system).

Fig. 4. Parking areas location of the electric-car sharing system of Pordenone

It is clear that the described one-way rental system requires a relocating mechanism that ensures the balance of the number of vehicles available in each station throughout the day, according to the mobility demand. In particular, since in the current management policy relocations of vehicle are carried out only at the end of the working day, hereafter we refer to the first operative condition depicted in Fig. 1 as *case as is*. Management strategy of Fig. 2 is named *case to be-offline* instead, since there is not a real time monitoring of the system, while the introduction of the economic incentives based on the actual balance conditions of the service characterizes the so-called *case to be-online*. More in detail, the following strategy has been defined, with four types of incentives depending on the system conditions:

7. *both in P3 and in P5 there are more than 2 available vehicles*: the system is still balanced and no suggestion has to be provided to the customers (*incentive type*: 0);
8. *both in P3 and in P5 there are2 or less available vehicles*: customers are encouraged to give back the vehicle as soon as possible (*incentive type*: 1);
9. *in P3 there are 2 or less available vehicles*: users are encouraged to give back the vehicle in *P3* (*incentive type*: 2);
10. *in P5 there are 2 or less available vehicles*: users are encouraged to give back the vehicle in *P5* (*incentive type*: 3).

When a customer follows the received suggestions, he obtains a discount of 20% on the total rental cost.

4.2   The Timed Petri Net Model

This section describes the TPN modeling technique used to represent an *electric-car sharing* service. In particular, the model developed in this paper is referred to the *car sharing* service of Pordenone, but it can be easily applied, with few modifications, to any generic *car sharing* system.

To model the system behavior, the following assumptions have been made:

- *Maximum waiting time*: it is reasonable to assume that each user is willing to wait up to 10 minutes for an available vehicle, then he leaves the parking without having been served.
- *Maintenance*: it is assumed that only one car park is available for the vehicles maintenance, with the capability to operate on a single vehicle at a time. Furthermore, the probability that a vehicle needs service after the rental period is considered not negligible (10 %) and, among the vehicles in maintenance, 99% is available again in one of the two parking areas after one hour, while 1% requires an eight-hours service. Note that maintenance time also includes recharging operations needed by EVs.
- *Number of available vehicles*: 10 vehicles are assumed to be available in the service, initially equally distributed between parking *P3* and parking *P5*.

In particular, these assumptions are based on the real operative conditions of the service of Pordenone, collected through interviews with the system operators.

The TPN represented in Fig. 5 models the two cases *as is* and *to be-offline* described by the UML diagrams of Fig. 1 and Fig. 2, respectively. On the other hand, the TPN shown in Fig. 6 models the case *to be-online* described by the UML diagram of Fig. 3. Tables 1-5 show the meaning of places and transitions of the two TPNs. Moreover, activity diagrams have been translated into TPN models by a resource oriented approach, using the same guidelines described in [20].

In this context we consider the transitions using infinite server semantics and enabling memory policy to model the waiting time of each actor (user or vehicle) modeled by the TPN. Indeed, a number of clocks, initialized to the waiting time, is associated by the transitions to the actor operations.

In particular, in Fig. 5 the case *as is* is modeled: the firing probability (called *random switch*, RS) of conflicting transitions is represented with red color, while the firing delay δ [h] of deterministic transitions and the exponential distribution parameter λ [1/h] of stochastic transitions are labeled in blue.

As mentioned above, a modular TPN model has been developed: each parking area is represented by the same structure (a submodel that represents users' arrival and waiting for a vehicle, one for the travel time determination, one that models the destination determination and, finally, a submodel that represents the evaluation of the need of maintenance), while the maintenance operations are modeled by a structure that can be identically repeated on the basis of the available number of car parks in the service (Fig. 5).

More in detail, the users arrive to the parking P3 (transition t_1) or to the parking P5 (transition t_2). They wait for an available vehicle (places p_1 and p_2). If the waiting time is greater than 10 minutes, the user decides not to use the car sharing service (transition t_{25} or t_{26}). Hence, the conflict between transition t_3 (t_4) and transition t_{25} (t_{26}) is resolved only by the timing. On the contrary, the user rents the vehicle in parking P3 (transition t_3) or in parking P5 (transition t_4). Now, the user communicates the travel time: it may be of 20 minutes (transition t_5 if the user is in parking P3 or transition t_7 if the user is in parking P5) or superior and about 1 hour (transition t_6 if the user is in parking P3 or transition t_8 if the user is in parking P5). Note that the conflict between the transition t_5 (t_7) and t_6 (t_8) is now solved by the random switch RS. The return of the vehicle can be in parking P3 or parking P5, though the user rents the vehicle in a different parking. Transitions t_9, t_{10}, t_{11} and t_{12} model the behavior of the rented vehicle. For example, transition t_9 (t_{11}) represents the utilization for 20 minutes of a vehicle that will be returned in parking P3 (P5), while transition t_{10} (t_{12}) represents the utilization for about 1 hour of a vehicle that will be returned in parking P3 (P5). The conflicts between transition t_9 and t_{11}, and between t_{10} and t_{12} are solved by the function RS.

When the user returns the vehicle, two cases are possible. If the vehicle is in good condition, it is deposited in the parking P3 or P5 (transitions t_{14} and t_{15}, respectively); then the user pays and leaves the parking (transitions t_{17} for parking P3 and transition t_{18} for parking P5). If the vehicle is not in good condition, then maintenance is required (transitions t_{13} and t_{16}). Maintenance can be short (transition t_{19}) or long (transition t_{20}). After the maintenance, the vehicle is reported to the parking P3 or P5 (transitions t_{21}, t_{22}, t_{23} and t_{24}). All the conflicts are solved by function RS, as illustrated in Fig. 5.

Moreover, note in Fig. 5 the different modeling approach for the two possible travel times: while travels that last 20 minutes are represented by deterministic transitions, longer trips are modeled by an exponential distribution, since users pay less attention to respect the limit of an hour and more delays or accidents may occur during the rental period. At the same time, note that TPN of Fig. 5 represents also the case *to be-offline*, since no real time monitoring of the system is considered: the effects of the incentive to return the rented vehicle as soon as possible are modeled with the variation of RS of transitions t_9 and t_{10} from 0.60 to 0.70, which means that the probability that a customer returns the vehicle after 20 minutes varies from 0.60 to 0.70 (and, so, a customer gives back the car after almost an hour with a probability of 0.30). Fig. 3 shows the *case to be-online*: real time monitoring of the system is modeled by the introduction of the places *Alarm 1* and *Alarm 2*.

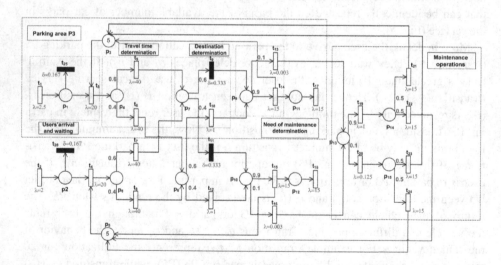

Fig. 5. The TPN system modeling the case *as is*

Table 1. Meaning and initial marking of places in the TPN of Fig. 5

Name	Meaning	$M_o(p_i)$
p_1	user waiting for an available vehicle in P3	0
p_2	user waiting for an available vehicle in P5	0
p_3	parking P3 capacity	5
p_4	rented vehicle in P3	0
p_5	parking P5 capacity	5
p_6	rented vehicle in P5	0
p_7	selected travel time is 20 min	0
p_8	selected travel time is 60 min	0
p_9	selected destination parking is P3	0
p_{10}	selected destination parking is P5	0
p_{11}	vehicle again available in P3	0
p_{12}	vehicle again available in P5	0
p_{13}	vehicle maintenance required	0
p_{14}	short maintenance (1 hour)	0
p_{15}	long maintenance (8 hours)	0

Table 2. Meaning and average transition firing rates λ $[h^{-1}]$ of the stochastic transitions in the TPN of Fig. 5

Name	Meaning	λ_s
t_1	user arrival at parking P3	2.5
t_2	user arrival at parking P5	2
t_3	vehicle rental at parking P3	20
t_4	vehicle rental at parking P5	20
t_5	user decision to travel 20 min	40
t_6	user decision to travel 60 min	40
t_7	user decision to travel 20 min	40
t_8	user decision to travel 60 min	40
t_{10}	destination parking determination (P3) during a travel of 60 min	1
t_{12}	destination parking determination (P5) during a travel of 60 min	1
t_{13}	vehicle picked up for maintenance	0.003
t_{14}	vehicle restitution at parking P3	15
t_{15}	vehicle restitution at parking P5	15
t_{16}	vehicle picked up for maintenance	0.003
t_{17}	user departure from parking P3	15
t_{18}	user departure from parking P5	15
t_{19}	short maintenance operations	1
t_{20}	long maintenance operations	0.125
t_{21}	vehicle return to P3 after maintenance	3
t_{22}	vehicle return to P5 after maintenance	3
t_{23}	vehicle return to P3 after maintenance	3
t_{24}	vehicle return to P5 after maintenance	3

Table 3. Meaning and firing delays δ_d [h] of deterministic transitions in the TPN of Fig. 5

Name	Meaning	δ_d
t_9	destination parking determination (P3) during a travel of 20 min	0.333
t_{11}	destination parking determination (P5) during a travel of 20 min	0.333
t_{25}	user departure from parking P3 without being served	0.167
t_{26}	user departure from parking P5 without being served	0.167

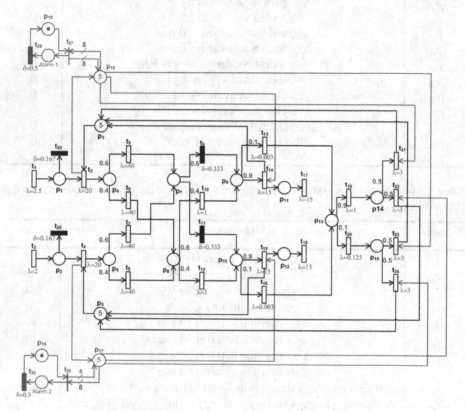

Fig. 6. The TPN system modeling case *to be-online*

Table 4. Meaning and initial marking of new places in the TPN of Fig. 6

Name	Meaning	$M_o(p_i)$
p_{16}	monitored number of available vehicles in P3	5
p_{17}	monitored number of available vehicles in P5	5
Alarm 1	incentive mechanism active (P3)	0
Alarm 2	incentive mechanism active (P5)	0
p_{18}	Alarm 1 capacity	1
p_{19}	Alarm 2 capacity	1

Table 5. Meaning and firing delays δ_d [h] of new deterministic transitions in the TPN of Fig. 6

Name	Meaning	δ_d
t_{27}	incentive mechanism activation (P3)	0
t_{28}	incentive communication to users	0.5
t_{29}	incentive mechanism activation (P5)	0
t_{30}	incentive communication to users	0.5

Table 6. Effects of the different types of incentive on the *random switches* of conflicting transitions

Type of incentive	Condition	Effect
0	$M(p_{16}) < 8$ and $M(p_{17}) < 8$	$RS(t_5)=RS(t_7)=0.60$ $RS(t_6)=RS(t_8)=0.40$ $RS(t_9)=RS(t_{10})=0.60$ $RS(t_{11})=RS(t_{12})=0.40$
1	$M(p_{16}) >= 8$ and $M(p_{17}) >= 8$	$RS(t_5)=RS(t_7)=0.70$ $RS(t_6)=RS(t_8)=0.30$
2	$M(p_{16}) >= 8$	$RS(t_9)=RS(t_{10})=0.70$ $RS(t_{11})=RS(t_{12})=0.30$
3	$M(p_{17}) >= 8$	$RS(t_9)=RS(t_{10})=0.45$ $RS(t_{11})=RS(t_{12})=0.55$

If *Alarm 1* (*Alarm 2*) is marked, then the vehicle availability in parking area P3 (P5) is low and so the apt incentive is communicated to users: the incentive lasts for half an hour and then the system is monitored again. In Tables 4-5 the meaning of new places and transitions is shown.

In particular, the impact of the real time suggestions on the customers' behavior has been modeled with the variation of the *RS* of conflicting transitions, as reported in Table 6. Note that the effects of incentives on the determination of the destination parking are not the same for P3 and P5: the second one is, in fact, less attractive for users and so customers are more unwilling to choose it as travel destination even with the promise of a discount.

5   The Case Study Simulation and Results

5.1   Simulation Specification

In order to analyze the performances of the electric-car sharing service and to evaluate the impact of the incentive mechanism previously described, numerical simulations have been performed. In particular, three scenarios (*A, B* and *C*), characterized by different levels of system congestion, are considered: users' inter-arrival times in *Scenario A*, in *Scenario B* and in *Scenario C* are reported in Table 7. (Note that the exponential distribution parameters λ of transitions t_1 and t_2 depicted in Figs. 5-6 and in Table 2 are referred to *Scenario A*).

Table 7. Users' inter-arrival times [min] in *Scenarios A, B, C* in both parking areas

	P3	P5
Scenario A	24	30
Scenario B	12	20
Scenario C	10	15

In all the considered scenarios, the three different operative conditions introduced in Section 4 are taken into account (*case as is, case to be-offline, case to be-online*). The following three performance indices are defined:

- *Level of Service (LOS)*: this performance index is expressed in terms of average fraction of served users and can be determined as follows:

$$LOS = \frac{|t_3| + |t_4|}{|t_1| + |t_2|}, \tag{1}$$

where the operator $|t|$ indicates the number of firing of transition t during a simulation. We recall that transition t_3 (t_4) fires if an user utilizes the electric car of the parking *P3* (*P5*). On the other hand, the firing number of transitions t_1 and t_2 represents the total number of users arriving to the parking *P3* and *P5*, respectively.

- *Company revenue R*: such an index is defined as the sum of the total travel cost supported by each user, considering the possibility of monetary incentives.

  In a car sharing service, each user has to pay both a *distance charge* (0.20 €/km) and an *hourly charge* (5 €/hour). In the presented context only the *hourly charge* is taken into account, since it is the cost that is influenced by the incentive mechanism. The value of R (in euros) can be determined as follows:

$$R = [hourly\ charge] \cdot \begin{bmatrix} (1 - discount) \cdot \delta_9 \cdot |t_9| + (1 - discount) \cdot \delta_{11} \cdot |t_{11}| \\ + \frac{1}{\lambda_{10}} \cdot |t_{10}| + \frac{1}{\lambda_{12}} \cdot |t_{12}| \end{bmatrix}. \tag{2}$$

The transitions t_9, t_{10}, t_{11} and t_{12} represent the decision of the user to utilize the electric car for 20 minutes or for about 1 hour. Hence, $\delta_9 \cdot |t_9|$, $\delta_{11} \cdot |t_{11}|$, $\frac{1}{\lambda_{10}} \cdot |t_{10}|$ and $\frac{1}{\lambda_{12}} \cdot |t_{12}|$ represent the total time of utilization of an electric car. Moreover, in case an incentive strategy is applied, the term (1-*discount*) has to be necessarily added to (2). In the case study we assume *discount*=0.2.

- *The average car sharing company gain G*: it is calculated as the difference between the company revenue R and the monetary penalty P defined in Section 4:

$$P = [monetary\ penalty] \cdot (|t_{25}| + |t_{26}|) \text{ €}. \tag{3};$$

$$G = R - P \text{ €} \tag{4}.$$

In the case study we assume *monetary penalty*=5€ and it has been introduced in order to quantify the damage that each user that leaves the system without having been served represents for the company; the inability to satisfy the demand, in fact, does not represent only a loss of earning for the company, but it has also repercussions on the image and the attractiveness of the service itself. In particular, the firing of t_{25} or t_{26} represents the disappointed user who goes away without using the car sharing service, because he waited more than 10 minutes.

The TPN model of Figs. 5 and 6 are simulated in the MATLAB software environment [21]. Such a matrix-based engineering software appears particularly appropriate for simulating the dynamics of TPN based on the matrix formulation of the marking update, as well as to describe and simulate PN systems with a large number of places and transitions. The performance indices have been evaluated by long simulation run of 21600 minutes (car sharing service is operative for 12 hours per day, 30 days per month), with a transient period of 30 minutes. In particular, the estimates of the performance indices are deduced by 50 independent replications, with a 95% confidence interval. Besides, the half width of the confidence interval is about 2.2% in the worst case, confirming the sufficient accuracy of the performance indices estimation. Finally, considering that the average CPU time for a simulation run is about 408 seconds on a PC equipped with a 1.73 GHz processor and 1 GB RAM, the presented modeling and simulation approach can be applied to large and complex systems.

5.2   Simulation Results

Simulation results are depicted in Figs. 7-9. In Fig. 7 the average fraction of served users is reported by comparing case *as is* and case *to be-offline*. It is interesting to note that an incentive mechanism that does not consider the instantaneous vehicle balance conditions of the service does not improve significantly the *LOS*.

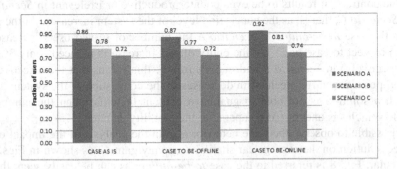

Fig. 7. The average fraction of served users for each case in all the considered scenarios

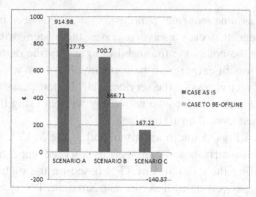

Fig. 8. Average company gain comparison between case *as is* and case *to be-offline* in all the considered scenarios

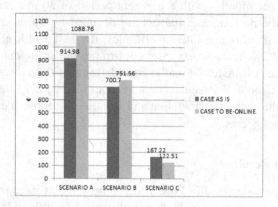

Fig. 9. Average company gain comparison between case *as is* and case *to be-online* in all the considered scenarios

On the contrary, it results to be even counterproductive or irrelevant in *Scenario B* and in *Scenario C*, that is, as the congestion level of the system increases. At the same time, in the case *to be-online* in *Scenario A* the number of served users increases of 6% with respect to case *as is*, but the entity of this increase is reduced (only 3% and 2%, respectively) in *Scenario B* and in *Scenario C*: this means that the effectiveness of the proposed incentive mechanism decreases as the congestion of the system grows and such a solution is not able to guarantee evident benefits when the number of available vehicles is undersized compared to the mobility demand.

It is possible to observe the same behavior also in the analysis of the impact of the proposed solution on the average car sharing company gain *G*, as shown in Figs. 8-9. In particular, Fig. 8 is referred to the case *to be-offline*: as can be easily seen, the introduction of the incentive mechanism leads to a reduction of the company gain (with a peak of - 49% in *Scenario B*) and to a real economic loss in *Scenario C*. On the other hand, when the typology of information suggested to the users is based on the current system conditions (Fig. 9), the company gain is higher in the first two scenarios, but the effect in *Scenario B* is less pronounced (19% vs. 7%); in *Scenario C*, on the contrary, there is a decrease of the company gain, so not even a real time monitoring of the number of vehicles is sufficient to ensure an enhancement of the system performances when it turns out to be too congested.

Summing up, the introduction of an incentive mechanism based on a continuous monitoring of the vehicles distribution among the parking areas improves the *LOS* of a car sharing system with positive economical outcomes for the company itself, but it cannot disregard a prior and coherent sizing of the system; on the other hand, a mechanism which does not consider the actual balance conditions of the system does not turn out to be a concrete solution for the user-based vehicle relocation problem.

Compared to the techniques described in other works, the considered approach does not take into account the possibility of *ridesharing* (users share a ride in a single vehicle) or *trip splitting* (multiple users that have to travel between the same origin and destination drive separate vehicles), and this represents an advantage for customers who can be unwilling to share the same vehicle with strangers or to travel separately from their acquaintances. However, the effectiveness of the proposed solution highly depends on users' participation and so the percentage of discount on the total travel cost has to be strategically determined.

6 Conclusion

The possibility for a car sharing customer to pick-up a vehicle in a parking area and return it to a different one after the rental period is a feature that deeply influences the popularity of a car sharing organization, since it guarantees flexibility in users' trip conformation. However, since the vehicle demand is not uniformly distributed, such a system can lead quickly to an imbalance in the number of available cars in each station. For this reason, a vehicle relocation system is fundamental in order to ensure an acceptable level of service.

This paper presents two particular user-based vehicle relocation techniques. In particular, according to the instantaneous number of cars in each parking area, a real time strategy suggests to the customers, through ICT tools installed on board each vehicle, the best travel behavior useful to support the rebalancing of the system. In particular, in order to assess the impact of the proposed solution, the electric-car sharing system of Pordenone (Italy) has been considered and three operative conditions have been taken into account. The considered case study and the management strategies are modeled in a *Timed Petri Net* (TPN) framework and a simulation study compares and assesses the system performances in different scenarios.

The obtained results underline that a system which does not consider the instantaneous balance conditions of the service and which suggests always to customers to return the rented vehicle as soon as possible is not a solution for the imbalance problem and it leads to economic losses for the car sharing organization. On the other hand, a simple ICT application and the real time monitoring of the system can increase the number of served users and, therefore, improve the overall service performance with economic benefits for the company. However, the effectiveness of this solution decreases as the congestion level of the system grows and this fact underlines that such an action is not able to overcome problems linked to an undersizing of the service in terms of number of vehicles initially made available in each station.

Future research will address a more in-depth behavioral analysis of the users' level of acceptance of real time suggestions and, on the basis of this, the optimization of the percentage of discount on the total travel cost that has to be applied when a customer follows the received suggestions.

Acknowledgments. This paper is supported by the project "Un Electric Car Club per il FVG", financed by the region Friuli Venezia Giulia, Italy, within the framework "contributi a sostegno della ricerca, dello sviluppo, dell''innovazione e del trasferimento tecnologico per lo sviluppo di sistemi per la mobilità individuale finalizzati alla riduzione di consumi e di emissioni, ai sensi dell'articolo 16 della legge regionale 11 agosto 2010 n. 14".

References

1. Barth, M., Shaheen, S.A.: Shared-Use Vehicle Systems: Framework for Classifying Carsharing, Station Cars, and Combined Approach. Transportation Research Record 1791, Paper No. 02-3854, 105–112 (2002)
2. George, D.K., Xia, C.H.: Fleet-sizing and Service Availability for a Vehicle Rental System via Closed Queueing Networks. European Journal of Operational Research 211(1), 198–207 (2011)
3. Nakayama, S., Yamamoto, T., Kitamura, R.: Simulation Analysis for the Management of an Electric Vehicle-Sharing System: Case of the Kyoto Public-Car System. Transportation Research Record 1791, Paper No. 02-2653, 99–104 (2002)
4. Correia, G.H.A., Antunes, A.P.: Optimization Approach to Depot Location and Trip Selection in One-Way Carsharing Systems. Transportation Research Part E (48), 233–274 (2012)
5. Duncan, M.: The Cost Saving Potential of Carsharing in a US Context. Transportation 38(2), 363–382 (2011)
6. Shuster, T.D., Byrne, J., Corbett, J., Schreuder, Y.: Assessing the Potential Extent of Car Sharing: A new Method and its Implications. Transportation Research Record: Journal of the Transportation Research Board, No. 1927, 174–181 (2005)
7. Barth, M., Todd, M., Xue, L.: User-Based Vehicle Relocation Techniques for Multiple-Station Shared-Use Vehicle Systems. In: Proceedings of the 2004 Transportation Research Board Annual Meeting, Washington D.C (January 2004)
8. Cepolina, E.M., Farina, A.: Urban Car Sharing: An Overview of Relocation Strategies. WIT Transactions on the Built Environment 128, 419–431 (2012)
9. Masuch, N., Lutzenberger, M., Ahrndt, S., Hessler, A., Albayrak, S.: A Context-Aware Mobile Accesible Electric Vehicle Management System. In: Proceedings of the Federated Conference on Computer Science and Information Systems, pp. 305–312 (2011)
10. Manzini, R., Pareschi, A.: A Decision Support System for the Car Pooling Problem. Journal of Transportation Technologies (2), 85–101 (2012)
11. Peterson, J.L.: Petri Net Theory and the Modeling of Systems. Prentice Hall, Englewood Cliffs (1981)

12. Ajmone Marsan, M., Balbo, G., Conte, G., Donatelli, S., Franceschinis, G.: Modelling with Generalized Stochastic Petri Nets. J. Wiley & Sons (1995)
13. David, R., Alla, H.: Discrete, Continuous and Hybrid Petri Nets. Springer (2004)
14. Jorge, D., Correia, G., Barnhart, C.: Testing the Validity of the MIP Approach for Locating Car Sharing Stations in One-Way Systems. In: Euro Working Group on Transportation, 15th edn., Paris, September 10-13 (2012)
15. Weikl, S., Bogenberger, K.: Relocation Strategies and Algorithms for free-floating Car Sharing Systems. In: 15th International IEEE Conference on Intelligent Transportation Systems, Anchorage, Alaska, USA, September 16-19 (2012)
16. Barth, M., Todd, M.: Simulation Model Performance Analysis of a Multiple Station Shared Vehicle System. Transportation Research, Part C (7), 237–259 (1999)
17. Kek, A.G.H., Cheu, R.L., Meng, Q., Fung, C.H.: A Decision Support System for Vehicle Relocation Operations in Carsharing Systems. Transportation Research Part E (45), 149–158 (2009)
18. Uesugi, K., Mukai, N., Watanabe, T.: Optimization of Vehicle Assignment for Car Sharing Systems. In: Apolloni, B., Howlett, R.J., Jain, L. (eds.) KES 2007, Part II. LNCS (LNAI), vol. 4693, pp. 1105–1111. Springer, Heidelberg (2007)
19. Booch, G.J., Rumbaugh, J., Jacobson, I.: The Unified Modeling Language User Guide. Addison-Wesley, Reading (1998)
20. Fanti, M.P., Mangini, A.M., Dotoli, M., Ukovich, W.: A Three Level Strategy for the Design and Performance Evaluation of Hospital Departments. IEEE Transactions on Systems, Man and Cybernetics: Systems (99), 1–15 (2013)
21. MathWorks, Inc, The Mathworks, MATLAB Release Notes for Release 14, Natick, MA (2006)

Net-Based Analysis of Event Processing Networks – The Fast Flower Delivery Case

Matthias Weidlich[1], Jan Mendling[2], and Avigdor Gal[1]

[1] Technion – Israel Institute of Technology, Haifa, Israel
weidlich@tx.technion.ac.il, avigal@ie.technion.ac.il
[2] Wirtschaftsuniversität Wien, Vienna, Austria
mendling@wu.ac.at

Abstract. Event processing networks emerged as a paradigm to implement applications that interact with distributed, loosely coupled components. Such a network consists of event producers, event consumers, and event processing agents that implement the application logic. Event processing networks are typically intended to process an extensive amount of events. Hence, there is a need for performance and scalability evaluation at design time. In this paper, we take up the challenge of modelling event processing networks using coloured Petri nets. We outline how this type of system is modelled and illustrate the formalisation with the widely used showcase of the Fast Flower Delivery Application (FFDA). Further, we report on the validation of the obtained coloured Petri net with an implementation of the FFDA in the ETALIS framework. Finally, we show how the net of the FFDA is employed for analysis with CPN-Tools.

1   Introduction

Complex event processing is a paradigm that builds on concepts from database technology enhanced with dynamic processing capabilities. So-called event processing networks (EPNs) [1] are at the centre of complex event processing systems. The overall system behaviour of such a network is decomposed into a set of *event producers* that generate events. Those are processed by *event processing agents* who create further events that are relevant to *event consumers*. Often, event producers can also be event consumers, such that EPNs are not simply a complex kind of event-condition-action pipeline, but rather a cybernetic system that observes events in the real-world and, based thereon, coordinates action.

There exists a plethora of approaches for implementing event processing networks and dealing with its intrinsic challenges [2]. A general problem in this context, though, is to analyse the overall behaviour of an EPN. Yet, there is currently no generally accepted formal model for complex event processing. The potential of utilising coloured Petri nets to this end stems from their capability of specifying concurrency in an explicit manner with support for typing of events. Indeed, this merit has been recognized already for active database systems [3], which promote rule-based processing in a non-distributed environment.

J.-M. Colom and J. Desel (Eds.): PETRI NETS 2013, LNCS 7927, pp. 270–290, 2013.
© Springer-Verlag Berlin Heidelberg 2013

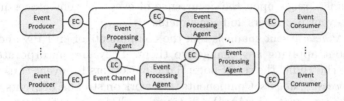

Fig. 1. Schematic representation of an event processing network

In this paper, we investigate the application of coloured Petri nets for specifying and analysing EPNs. We turn to the case of the Fast Flower Delivery Application (FFDA) [1]. This real-world case is promoted by the Event Processing Technical Society[1] and has become a de-facto benchmark for demonstrating the capabilities of event processing systems. Our contribution is the mapping of concepts from EPNs to coloured Petri nets with a discussion of design choices. Further, we report on the validation of the coloured Petri net obtained for the FFDA with its implementation in ETALIS, an open-source event processing engine. Finally, we demonstrate the merits of analysis and simulation capabilities for this domain, thereby contributing to the formal foundations of EPNs and opening this emerging field for Petri net analysis.

The paper is structured as follows. Section 2 introduces the main concepts of event processing networks. Section 3 sketches the Fast Flower Delivery Application. Section 4 defines the coloured Petri net model for this application case. Section 5 is devoted to the validation of this model with the implementation of the application in ETALIS. Section 6 summarises findings from analysing the Petri net. Section 7 discusses related work and Section 8 concludes the paper.

2   Background: Complex Event Processing

Following [1], Section 2.1 presents the essentials of event processing networks (EPNs) and Section 2.2 outlines event pattern detection.

2.1   Event Processing Networks

Event types and events. An event is a happening of interest, an *'occurrence within a particular system or domain'* [1]. Events are typed and an event type is a specification for a set of events with related semantics and structure. A common model for events is attribute-based, i.e., each event has a set of (required or optional) attributes organised as key-value pairs. For instance, an event of type 'delivery request' may be characterised by a number of attributes and values, such as 'pickup = 24.09.12' and 'time = 3 days'.

Event producers and consumers. An EPN, as illustrated in Fig. 1, consists of event producers and event consumers. Event producers emit events, eventually

---

[1] http://www.ep-ts.com

event consumers react upon the occurrence of events. Event processing agents act as both event producers and consumers.

Event channels. Event channels link the components of an EPN and forward events without applying any changes to them. They may incorporate routing mechanisms that limit the set of potential input events for event consumers.

Event processing agents. Components that work on streams of events are called event processing agents (EPA). We distinguish EPAs that (1) filter events, (2) transform events, and (3) detect event patterns. A filter EPA performs a selection of events, typically based on the event attributes. A transformation EPA takes events as input, processes them, and produces a set of derived events as output using a stateful or stateless data transformation operation. A pattern detection EPA defines a complex detection logic and outputs derived events.

Event contexts. Event processing agents work on events that are considered to be *relevant*. This relevance is determined by the event context, which is defined along different dimensions [1]. Most prominently, the temporal dimension partitions events based on their occurrence time, e.g., using a sliding window. Then, event detection concerns only events that occurred within the same window. Events may also be partitioned based on space, external state, or segments of attribute values. For the aforementioned example, one may require joint processing of events of type 'delivery request' and 'delivery bid'. Still, only relevant events of both types shall be considered, e.g., events that occur within a window of two hours and match in their attribute values (e.g., the bid refers to a request).

2.2   Detecting Event Patterns

Pattern detection EPAs use a set of standard patterns to compose complex event patterns (aka complex events). Based on [1,4], we first discuss these patterns and then elaborate on event processing policies, which disambiguate semantics of event patterns. For illustration, circles, stars, and squares in Fig. 2 depict events of types A, B, and C, arranged by their time of occurrence.

Common Event Patterns. The *all* pattern defines a conjunction of event types. In Fig. 2, for instance, the pattern *all*(A, B, C) is detected for the events $\{a1, b1, c1\}$. The *any* requires one event of one of the given event types to occur (precise semantics depend

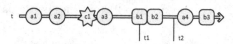

Fig. 2. An event stream example

on the event processing policies, see below). In Fig. 2, *any*(A, B, C) detects the event $a1$. The *absence* pattern refers to the absence of events of a certain type, e.g., we detect *absence*(D) for the example.

Besides these basic patterns, there are threshold and dimensional patterns. A threshold pattern defines an aggregation operation and a threshold. An example is the *count* pattern. Instantiated as *count*$(A, 3)$ it detects the events $a1$, $a2$, and $a3$. Threshold patterns also refer to an assertion over the *min/max/average/n-highest/n-lowest* values of event attributes. For instance, the pattern *max*$(A, att, >, 0.5)$ is detected if the average of the attributes *att*

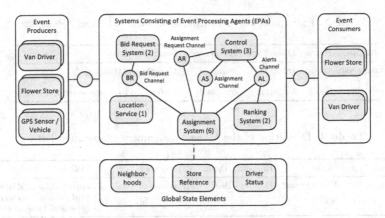

Fig. 3. The event processing network of the Fast Flower Delivery Application. For each system, the number of contained EPAs is given in brackets.

of events of type A is larger than 0.5. Dimensional patterns refer to time and space. The *sequence* pattern defines a list of event types and detects events that occur in the respective order. Pattern $sequence(A, A, B)$, for instance, detects the events $a1$, $a2$, and $b1$.

Event Processing Policies. The *evaluation policy* determines when the pattern is evaluated, either 'immediate' for each event, or 'deferred' until the temporal event partition closes. For Fig. 2, pattern $and(A, B)$ may be detected at time $t1$ upon the occurrence of $b1$, or postponed (time $t2$) until the time window closes. The *cardinality policy* defines whether a pattern is detected a 'single' time, a 'bounded' number of times, or 'unrestrictedly' often. With 'unrestricted', the pattern $and(A, B)$ is detected multiple times for Fig. 2, e.g., for $\{a1, b1\}$ and $\{a4, b3\}$. The *repeated type policy* considers events of the same type with the values 'first', 'last', 'override', and 'every'. For the example, pattern $and(A, B)$ detects events $a1$ and $b1$ (value 'first') or $a3$ and $b1$ (value 'last'). The *consumption policy* defines whether events that are detected by a pattern are 'consumed' or available for 'reuse'. This policy controls whether a pattern with event $a3$ is detected with both, event $b1$ and $b2$, or whether $b2$ is detected together with $a2$.

3 The Fast Flower Delivery Application

This section introduces the Fast Flower Delivery Application (FFDA), which is specified in detail in [1]. It has been utilized by vendors of event processing systems to demonstrate their capabilities.[2]

The FFDA allows a consortium of flower stores to rely on a network of independent van drivers to process their flower deliveries. The event processing

---

[2] Implementations of the FFDA are listed at
http://www.ep-ts.com/content/view/79/112/

Table 1. Excerpt of the definition of the event Delivery Request

Attribute Name	Data Type	Occurrence	Semantic Role
requestId	Integer	required	Common attribute
...	...	...	...
neighbourhood	Location	optional	Reference to neighbourhood entity

Table 2. Definition of the Manual assignment preparation EPA

Pattern Type	Context	Relevant Types	Parameters	Policies
n-highest	Bid interval	Delivery Bid	Attribute = ranking Count = 5	cardinality = single evaluation = deferred repeated type = every

network of the FFDA is illustrated in Fig. 3. Here, flower stores, van drivers, and GPS sensors in their vehicles act as event producers. Flower stores and van drivers also consume events. The FFDA functionality is provided by five systems, each consisting of a set of event processing agents. These are connected by event channels and may access global state elements that capture the information on neighbourhoods, the flower stores, and the current status of a driver.

The functionality of the application is best described in five phases. First, in the bid phase, a flower store places a Delivery Request in the FFDA. The Bid Request System enriches the request with the minimal ranking required for drivers to process the delivery. The same system then sends Bid Request events to drivers in nearby neighbourhoods. The position of drivers is tracked in the global state element Driver Status. It is regularly updated by GPS Location events processing by the Location Service using the Neighbourhoods global state element.

In the assignment phase, drivers respond to the requests with a Delivery Bid. The Assignment System correlates requests with bids and conducts an automatic or manual assignment. The former is implemented by taking the first matching bid and emitting an Assignment event. It fixes the pickup time and delivery time and is consumed by a driver. A manual assignment starts two minutes after recording a request. The Assignment System sends the five highest-ranked bids to the flower store using an Assignment Request event. The store chooses one, leading to an Assignment event. If there are no bids for a delivery, an event of type No Bidders Alert is emitted by the Assignment System. If an Assignment Request is not handled within one minute, the Control System creates an Assignment Not Done alert.

Once a driver picks up a delivery from a flower store, the store emits a Pickup Confirmation. The successful delivery is acknowledged by a Delivery Confirmation event, created by the driver's mobile device upon signature of the recipient. The Control System monitors the process and creates a Pickup Alert, if a Pickup Confirmation is not recorded within five minutes after the committed time. A Delivery Alert is created if a Delivery Confirmation is late by more than 10 minutes.

Table 3. Overview of the formalisation of EPN concepts with CPN concepts

EPN Concept	CPN Concept
Event Type	Product colour set or list colour set with singleton tokens
Event	Token of colour of event type colour set or entry of list of singleton token of list colour set
Global State Element	Place of product colour set or place of list colour set with singleton token
Type of Event Producer	Transition per type of created event
Type of Event Consumer	Transition per type of consumed event
Event Channel	Places and arcs
Event Context	Timestamp of token, colour of a token, or distinguished place
Event Processing Agent	Transition or subnets

Each time a driver processed 20 deliveries, the Ranking System evaluates the performance. It creates a Ranking Increase event, if no Delivery Alert events have been recorded within that period. It creates a Ranking Decrease event, if a driver caused more than five Delivery Alert events. Both events trigger changes of the Driver Status global state element, which captures the rankings of drivers.

The complete specification of the FFDA can be found in [1]. Table 1 and 2 show excerpts to illustrate the definition of an event and an EPA.

4   A CPN Model of an EPN

This section describes how an EPN in general and the FFDA in particular are modelled as a Coloured Petri Net (CPN). We introduce the formalisation of EPN elements following the structure used in Section 2 to introduce EPNs. An overview of the formalisation is given in Table 3. The mapping of all EPN elements is illustrated with the FFDA. Our CPN model of the FFDA comprises all systems that directly influence its execution (i.e., report generation is neglected). The resulting CPN comprises 40 transitions and 76 unique places. Due to space limitations, we show only excerpts of the model.[3] We close this section with a reflection on limitations and lessons learnt from modelling the FFDA.

We assume the reader to be familiar with basic CPN notions and notations, see also [5,6]. Further, we rely on the CPN-Tools framework [7] and, thus, assume a basic understanding of the notations used by this tool.

4.1   Event Types, Events, and Global State Elements

We consider two alternatives to represent an event type, using a product colour set or list colour set with a singleton token. In the first case, the type of the colour set is defined as the product of colour sets needed to represent the event

---
[3] The model is available at http://matthiasweidlich.com/projects/ffda.cpn

attributes. To enable performance analysis of an EPN, colour sets representing event types have to be timed. Following this line, an event is then represented as a coloured token of a place with the according colour set.

An example in the FFDA is the event type Delivery Request, which has six attributes of type integer or string, requestID, store, addresseeLocation, required-PickupTime, requiredDeliveryTime, neighborhood. Neglecting the addressee location which is not used during processing, the colour set is defined as:

```
colset DelReq = product ID * STR  * PICK * DEL * NBH timed;
```

An alternative representation of event types utilises a list colour set for which the base type is derived as discussed above. Places of this colour set are marked with a single token in all markings (henceforth referred to as a singleton token). Then, an event is represented as an entry of the list carried by such a token.

For instance, the FFDA defines an event type Pickup Confirmation with two attributes, requestID and store. We defined the respective colour sets as follows:

```
colset PickupConf = product ID*STR; colset lPickupConf = list PickupConf;
```

The choice of which formalisation to apply for event types (and, thus events) depends on how the events are processed by the components of the EPN. If processing is concerned mainly with the existence or absence of an event, instead of its separate processing, a formalisation based on a singleton token carrying a list of event entries is more appropriate. For instance, colour set PickupConf is not timed since only the existence or absence of events of that type influences the FFDA. The modelled list, in turn, can easily be checked for an element satisfying a certain criterion, which allows for implementing inhibitor arcs.

Finally, global state elements are presented by a single place for which the colour set is defined as discussed for event types. Again, one may either chose a place of a product colour set or rely on an according list colour set.

4.2   Event Producers and Consumers

Creation and consumption of events is realised by transitions accessing places representing the event type to be produced or consumed, respectively. An event processing component may produce or consume events of different types. A flower store in the FFDA, for instance, creates events of type Delivery Request and Pickup Confirmation. Events of different types are often created (or consumed) independent of each other. Hence, we decided to model creation (consumption) of events of different types by separate transitions, even if the events are emitted by the same instance or type of event producer (consumer). The concrete instance of an event producer (consumer) is captured as part of the event payload, such that creation (consumption) of events of a certain type is modelled with a single transition for all event producers (consumers) of the same type.

Events may be created based on assumptions on the EPN environment, such that the trigger mechanisms are part of the model configuration. Further, creation of an event may be done in reference to other events consumed or created earlier. Then, the event producer relies on the state of the EPN besides the

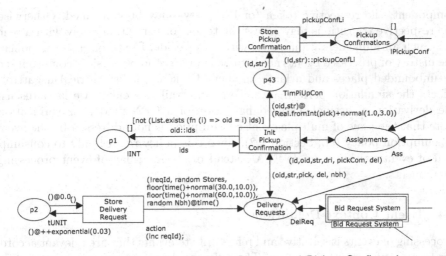

Fig. 4. Excerpt for the creation of Delivery Request and Pickup Confirmation events

configuration of the model. Both trigger mechanisms are illustrated in the CPN excerpt of the FFDA in Fig. 4. Here, the creation of Delivery Request events is modelled by transition **Store Delivery Request**, which is enabled only at certain times. The depicted configuration allows for firing of the transition every 30 time units (say minutes) on average, following an exponential distribution. Upon firing, the transition creates a Delivery Request event with according payload.

Creation of Pickup Confirmation events is done in reference to Delivery Request and Assignment events. Transition **Init Pickup Confirmation** schedules the occurrence of a Pickup Confirmation event. To ensure that at most one Pickup Confirmation event is created per Delivery Request, the guard condition of the transition accesses a place that keeps a list of all processed request identifiers. Upon firing, transition **Init Pickup Confirmation** further creates a token carrying the id of the original request with a normally distributed deviation from the scheduled pickup time. At that time, transition **Store Pickup Confirmation** may fire to create an actual Pickup Confirmation event.

4.3 Event Channels

Events are routed between event processing components through event channels. In a CPN, we model them with places of a colour set representing an event type and arcs that connect these places with transitions representing event processing components. Depending on whether such a component produces or consumes events to or from a channel, the respective transition has the respective place in its postset or preset. Thus, routing decisions made by event channels are directly reflected in the structure of the CPN model.

Modelling event consumption is intricate though. In an EPN, events are like signals that are broadcasted. Hence, only if an event is consumed by a single component, the respective token (or list entry) may be consumed. Otherwise, the respective transitions may access the token (or list entry) solely with a read arc. Such an implementation comes with a downside. First, for each component the history of processed events needs to be kept. Second, analysis is compromised by unbounded places and a large number of tokens (or list entries) negatively affects the simulation performance. To countervail this effect, we later discuss the derivation of a single-case configuration of a CPN model for verification. Note that the issue of multiple event consumption is not addressed by the event consumption policy (cf., Section 2.2), since this policy relates only to consumption of events within a single EPA instead of the interplay of event processing components.

4.4   Event Context Definitions

Processing of events is selective and refers only to events that are relevant according to a event context definition. The different dimensions of an event context are represented in the CPN model as follows.

Temporal. The temporal dimension is reflected by timestamps of tokens in the CPN. The correlation of events within a certain time window is realized as follows: at the start of the window, a token is created in a distinguished place with its timestamp set to the end of the window.

Segmentation-Oriented. Segmentation defined over attributes of event types is realised by matching attribute values of token colours, either as part of arc inscriptions or transitions inscriptions.

Spatial. The spatial dimension is encoded using coordinates of the underlying spatial model in the payload of events. As such, the realisation resembles the one for segmentation, just using operations over spatial coordinates.

State-Oriented. A global state is represented by a distinguished place, as introduced for global state elements. A transition of an EPA that relies on this state, thus, has a read arc to this place to access the current state value.

As an example in the FFDA, consider the event context Pickup Interval. It is a composite context, defined by a segmentation context Request and a temporal context Temporal Pickup Interval. The former is induced by the identifier of a delivery request and used, for instance, in the aforementioned definition of colour set DelReq (Section 4.1). The context Temporal Pickup Interval is initiated by an event of type Assignment and terminated by the occurrence of a Pickup Confirmation. Further, it defines an expiration offset as five time units later than the pickup time taken from the Assignment event. Combining the formalisations for temporal and segmentation-oriented event contexts, the composite context Pickup Interval is modelled as a distinguished place. The colour set of this place is based on the type of event attributes used for segmentation. It is also timed, so that tokens carry the pickup time plus five time units as a timestamp. We depict the respective excerpt of the CPN, when discussing the EPA that relies on this context.

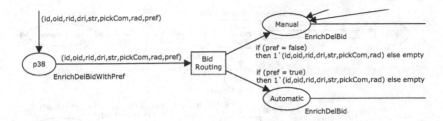

Fig. 5. Excerpt for the Bid Routing EPA (type 'filter')

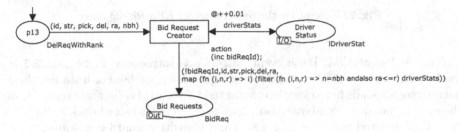

Fig. 6. Excerpt for the Bid Request Creator EPA (type 'enrich')

4.5 Event Processing Agents

EPAs either filter or transform events, or they detect event patterns.

Filter EPA. A pure filter EPA is stateless. Depending on a filter expression, events are classified as satisfying or not satisfying the filter. A filter EPA is modelled as a transition that consumes (or reads as discussed above) a token (or list entry) and creates a token (or list entry) on one of two places of equal colour set, depending on the applied filter expression. This way, filtering is explicitly represented and details on the filter EPA can be collected during simulation.

The FFDA contains a filter EPA in the Bid Request System. The Bid Routing EPA, which filters delivery bids, is shown in Fig. 5. The transition creates tokens based on the value of the token colour (value of **pref**) that represents the preference of the flower store requesting delivery for doing a manual (**pref** = 'false') or automatic assignment (**pref** = 'true') of drivers.

Transformation EPA. A transformation EPA changes events by translation, composition, aggregation, enrichment, splitting, or projection. It is represented by a transition that applies the respective operation on tokens or the colour values of tokens, or list entries of singleton tokens or values of these list entries. As an example, consider the Bid Request Creator EPA that takes a Delivery Request event and creates a Bid Request event. As part of that, the Driver Status element is accessed to select those drivers that meet the requirements in terms of a minimal ranking and have last been seen in the same neighbourhood. In the respective CPN excerpt, shown in Fig. 6, the latter is ensured by the arc inscription filtering the list of drivers carried by the singleton token in **Driver Status** (function **filter**). From the resulting list, only the identifiers of drivers are selected (function **map**).

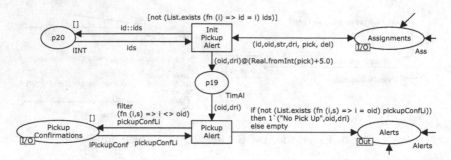

Fig. 7. Excerpt for the pattern detect EPA Pickup Alert

Pattern Detect EPA. Basic event patterns as introduced in Section 2.2 are modelled using formalisations of Boolean expressions, see [8]. For instance, the *all* pattern corresponds to a transition connected to places of colour sets representing the event types of the pattern. Also, absence of events is traced back to inhibitor arcs, implemented using list colour sets and transition guard conditions, cf., [9]. The challenges in the formalisation of pattern detect EPAs, however, stem from the interplay of the event patterns with event contexts and evaluation policies.

We first illustrate the interplay of event detection and event contexts with the Pickup Alert EPA of the FFDA. It uses the Pickup Interval event context (see above) to detect the absence of Pick Confirmation events. The excerpt for the EPA is shown in Fig. 7. As discussed above, the Pickup Interval event context is modelled by a token with an according value and timestamp that is created in the place with colour set TimAl. The timestamp represents the closing of the event context, which enables transition Pickup Alert. Upon firing, this transition accesses a list of a singleton token, in which each entry represents a Pickup Confirmation. The actual alert, an event of type Pickup Alert, is emitted by the creation of a token in the place of which the colour set represents all types of alerts. This token is created only if the list of delivery confirmations does not contain an according entry. Since Delivery Confirmation events are consumed only by the Pickup Alert EPA and only one alert may be raised in the event context of a delivery, the entry representing the processed Delivery Confirmation event is removed from the list of the singleton token of place Pickup Confirmations.

To illustrate the influence of evaluation policies on the formalisation, Fig. 8 depicts the CPN for the Evaluation System comprising the Ranking Increase and Ranking Decrease EPAs. Both EPAs refer to the Driver Evaluation event context. This composite context relates to segmentation per driver and non-overlapping groups of 20 Delivery Confirmation events, represented as follows: transition Init Context accesses the list carried by a singleton token in place Delivery Confirmations. The transition can fire, if there is an entry in the list that has not been processed, which is implemented by the transition guard and a separate place to keep track of processed entries. Upon firing, the transition selects one entry of this list that has not been processed before (see the action of the code segment of the transition). For this entry, tokens of colour set INT are created in three places. Here,

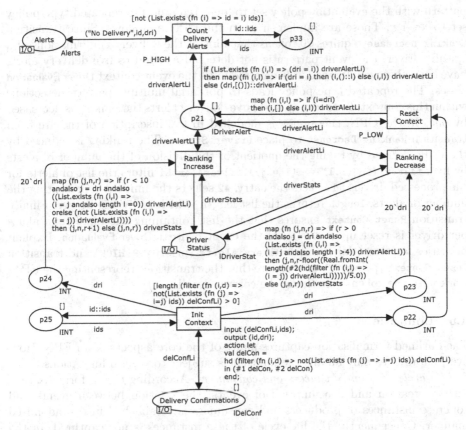

Fig. 8. CPN for the pattern detect EPAs Ranking Increase and Ranking Decrease

the token value represents the identifier of a driver. 20 tokens of the same value in these places, therefore, represent 20 confirmed deliveries for the respective driver. Independent thereof, transition Count Delivery Alerts consumes tokens that represent Deliver Alert events and updates the list of the token of the place with colour set lDriverAlert. An entry of the list associates a driver identifier with a list of type UNIT to count the delivery alerts per driver.

Once 20 delivery confirmations have materialised as tokens, transition Ranking Increase is enabled. It represents the EPA of the same name and upon firing, may change the list of the token in place Driver Status. If no delivery alerts have been recorded (no entry in driverAlertLi for the driver or the entry contains zero alerts), the ranking of the driver is increased by one by changing the list carried by the token in place Driver Status.

Further, transition Ranking Decrease is enabled once 20 delivery confirmations have been observed for a driver. The transition accesses a token with the list of the number of alerts per driver (variable driverAlertLi). If at least five alerts have been recorded, the driver's ranking is decreased. The EPA relies on a count pattern with the evaluation policy set to 'deferred' and the repeated type policy set to 'every'. Those are implemented in the CPN as follows. Since transition `Ranking Decrease` requires 20 tokens that hint at the delivery confirmations of a specific driver, the event pattern is not detected as soon as five delivery alerts have been observed. Instead, it is deferred until the event context Driver Evaluation closes. The repeated type policy leads to potential multiple pattern detections within this context. For every group of five delivery alerts the ranking is decreased by one. In the CPN, this policy is modelled in the inscription of the arc from transition `Ranking Decrease` to place `Driver Status`. The ranking is reduced by the value derived by taking the quotient of the division of the number of alerts by five: `filter (fn (i,l) => (i = j))` `driverAlertLi` filters the list of alerts for the processed driver, `hd` selects the entry, `#2` selects the unit list representing the number of alerts, `length` returns the list length, which is divided by five. Finally, transition `Reset Context` ensures that the list containing the number of alerts per driver is reset upon closing of the event context Driver Evaluation. Explicit priorities for transition `Count Delivery Alerts` (priority `P_HIGH`) and transition `Reset Context` (priority `P_LOW`) ensure that the transitions representing the EPAs work with the tokens containing the correct details.

4.6   Limitations

The outlined formalisation captures many of the core aspects of an EPN. However, it provides only an abstraction that is subject to several limitations.

Life-cycle of event producers and consumers. According to the formalisation, during creation and consumption of events, the relation between events and concrete instances of producers and consumers is established in a randomised manner. Consequently, the life-cycle of these instances is not captured. In the model obtained for the FFDA, a van driver, for instance, may be assigned to deliveries with equal pickup times at different stores. We argue, however, that a realistic model of the life-cycle of event producer or consumer instances may even allow for such cases. It may well be that a driver commits to equal pickup times at different stores and plans to pick up one delivery shortly before the committed time, and a second one with a small delay.

Model of space. An EPN is typically distributed. The outlined formalisation considers the spatial dimension as part of event attributes. This requires an approximation of location changes of event producers and event consumers, which is a limitation. For the FFDA, the representation of the spatial dimension is limited to a finite set of GPS coordinates emitted by the GPS Sensor event producer. The impact of different locations on the creation or consumption of events is not taken into account though. For instance, Delivery Confirmation events may be created for a driver at different locations within a time frame that would not allow to change location accordingly.

Model of time. Our approach assumes that event creation and consumption is homogeneous. The configuration of the model allows for defining probability distributions, upon which the event creation or consumption is based. Still, this limits the expressiveness of the model. More fine-grained control, e.g., different distributions at different days or hours of a day, is not part of the model. Although one may imagine to include those aspects in the model (e.g., a calendar may be modelled as a CPN), the time approximation will be an inherent limitation of the model. An example for the FFDA are Delivery Request events, for which creation is independent of days or hours of a day.

4.7 Lessons Learnt from the FFDA

Finally, we highlight the lessons learnt from modelling the FFDA.

Usage of hierarchies. The complexity of an EPN such as the FFDA calls for hierarchical modelling. In particular, it turned out to be reasonable to refine transitions representing a specific system of the FFDA with a subnet that includes the mapping of the according event processing agents. Also, the utilisation of fusion sets, i.e., multiple representations of a single place, for places that represent global state elements proved valuable for reducing model complexity.

Representation of event consumption. We discussed earlier that consumption of events in an EPN does not necessarily correspond to consumption of a token (or deleting an entry from a list), since the respective event representation may be relevant for multiple transitions that model event processing components. We outlined workarounds that rely on non-modifying access. However, this requires to keep a state (the knowledge on events processed already) even for EPAs that are actually stateless. As such, these workarounds complicate the model and may lead to an increased state space considered during simulation.

Modelling event processing policies. Even though the basic event patterns are easy to capture using CPN concepts, the interplay with the event context definitions and event processing policies imposes serious challenges for any EPN formalisation. For complex cases, it proved useful to separate the event pattern detection step into several steps in the CPN model. That is, the activation of an EPA as imposed by the event context definitions is decoupled from the detection logic using separate transitions. An example for this approach is the presented formalisation of the Evaluation System.

Overlapping event contexts. Different EPAs may represent alternative processing options (e.g., manual or automatic assignment in the FFDA). Their formalisation is particularly challenging once the event context definitions overlap partially in terms of considered events (e.g., automatic assignment relates only to the Request context; manual assignment relates to the composite Bid Interval context, which includes the Request context). Then, the activation of alternative EPAs may not be separated from the actual pattern detection. Events that are not part of the overlap of event contexts (e.g., delivery bids under manual processing that arrive after the bid interval has passed) have to be treated separately.

5 Model Validation for the Implementation in ETALIS

We illustrated our approach to formalising event processing networks with the FFDA as it is specified in [1]. In order to validate the model, we considered the concrete implementation of the FFDA in the Event-driven Transaction Logic Inference System (ETALIS).[4] Below, we first describe the FFDA implementation in ETALIS before turning to the model configuration and validation.

Implementation. ETALIS is an open-source system for Complex Event Processing published under the GNU GPL. It is implemented in Prolog and provides a declarative, rule-based language for the description of complex events. The system provides means for event pattern filtering, enrichment, and detection, and supports reasoning over events. ETALIS is shipped with an implementation of the FFDA that comprises 36 Prolog clauses, which are grounded in the ETALIS Language for Events (ELE). Events are expressed as ground atoms, i.e., predicates that contain only constants as arguments (the event's payload). The actual application logic of the FFDA is implemented in ELE using event patterns and event rules. Event patterns are defined based on predicates, conditions, and their temporal relations; event rules define implications between event patterns and predicates representing complex events.

Following this model, events produced outside of the FFDA are given as a sequence of facts. The ETALIS implementation of the FFDA provides an interface that accepts such a stream of facts. Here, the delay between the event occurrences is explicitly modelled using sleep statements that suspend execution.

Model Configuration and Setup. Configuration of the CPN model of the FFDA requires selecting the number and types of instances of event producers and event consumers. Further, global state elements have to be initialised and the creation of events by event producers needs to be controlled by defining probability distributions for their occurrence. To this end, we selected configuration values that seem to be reasonable against the setting in which the FFDA shall be operating. In most use cases, these values are either predefined (e.g., the number of participating drivers) or need to be estimated based on expectation and past experience (e.g., the frequency of delivery requests). We also introduced latency for event processing and consumption.

As for the validation, we used CPN-Tools to simulate the FFDA and collected all processed events. Then, we extracted all events created by event producers that are not part of the implementation, i.e., the events that are the input to the FFDA, such as Delivery Request events. These events were converted into an ELE event stream, which was fed into ETALIS. Again, we collected all produced events during processing, which allowed us to compare the events produced by the FFDA as implemented in ETALIS with those obtained by simulation of the CPN model. Based on this comparison, we refined the configuration of the model to resolve minor deviations observed.

Validation Results. For the final validation, we consider the *presence* and the *time* of alert events produced by the FFDA. The FFDA implementation in ETALIS produced 56 of such events, whereas the simulation created 58 events. Investigation of the respective event streams reveals that in two cases, alerts have been created only in the simulation since the time window for raising the

---

[4] http://code.google.com/p/etalis/

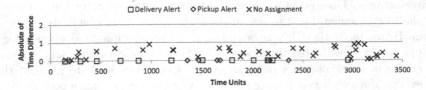

Fig. 9. Delay of correlated events in the CPN based simulation and ETALIS

alert was met by less than one time unit. Since the CPN simulation relies on timestamps defined as real numbers as opposed to the integer based timestamps in ETALIS, the alert events were present only in the simulation. Turning to the occurrence time, Fig. 9 illustrates the observed delay for correlated alert events. For Delivery Alert and Pickup Alert events, we observe no difference in both event streams. For No Assignment alerts, there is a delay of up to 0.93 time units. This delay also stems from the different time resolution in the simulation (real numbers) and ETALIS (integers). Since the first two types of alerts are raised at predefined timestamps, the delay is visible only for No Assignment alerts that are raised based on the internal event context definition. However, the observed delay stays below one time unit, meaning that the simulation is kept in sync with the ETALIS implementation when processing advances. Hence, the CPN model is indeed suited for evaluating the performance of the FFDA.

6   Analysis of the EPN

Once the CPN model for an event processing network has been created and validated, it may be used for analysis. Below, we first discuss the verification of basic properties before we turn the focus on the simulation of the EPN.

6.1   Properties

To investigate basic properties, it is reasonable to obtain a *single-case configuration* of the CPN model. Since the creation of some events is time based, the execution of the EPN may be non-terminating. Also, modelling of event channels as discussed in Section 4.3 leads to unbounded places if the number of events created by event producers is not limited. Therefore, we create a single-case configuration of the CPN model by restricting the enabling of transitions representing event producers, so that they may fire only a certain number of times.

For the FFDA, a reasonable notion of a case is the processing of one Delivery Request, created by a single firing of transition `Store Delivery Request` representing the event producer. Also, the second event producer that creates events only based on time, i.e., the GPS Sensor of a vehicle, also needs to be restricted.

Given a single-case configuration of a CPN, we investigate several properties.

Boundedness. Once event producers have been restricted accordingly, the CPN shall be bounded, i.e., the size of the state space should be finite.

Place Bounds. If the CPN is bounded, the concrete maximal bounds for all places are investigated. In most cases, there is a clear dependency between events of different types (e.g., an event of one type leads to at most three events of another type), which shall be reflected in the place bounds.

Deadlocks. Deadlocking markings shall reflect valid end states of the EPN. That is, the marking shall represent a state in which the processing of the single case is (successfully or not) finished.

Dead Transitions. Dead transitions represent event processing components that are never executed for the respective single case. These components shall indeed be not involved according to the chosen configuration.

We derived different single-case configurations for the CPN representing the FFDA and used the state space tool of the CPN-Tools framework to analyse the aforementioned properties. We obtained the following results. For all single-case configurations, the state space of the system was bounded, with the size varying between 80 - 350 states. An analysis of the place bounds revealed that only a few places have a bound larger than one. Since we merged the event types for alert events in the course of formalisation, a single colour set and a single place represent alerts of different types. A single delivery request may cause multiple alerts of different types, which leads to a bound above one for the respective place. All other places with a bound above one represent global state elements, for which each token models one data entry. The number of tokens for these places is constant in all markings. Further, all deadlocking markings for all single-case configurations turned out to be valid final states of the EPN. Finally, for each configuration, we observed several dead transitions. However, those transitions had to be dead according to the configuration. For example, the transition modelling automatic assignment of delivery bids is dead, if the configuration includes a delivery request of a store preferring manual assignment.

6.2   Simulation

Simulation of a configured model of the EPN allows for conclusions on its performance. Bottlenecks can be investigated and the influence of different assumptions on the environment is made explicit.

To demonstrate the merits of simulation of EPNs, we conducted a simulation experiment. Here, the configuration of the FFDA comprises a consortium of six flower stores in a city with 10 neighbourhoods. There are 20 van drivers working for the consortium (initial rankings and current locations for them are randomised). Delivery requests are created every 30 time units (we assume minutes) on average, following an exponential distribution. Further configuration of event creation includes the GPS sensor (normal distribution with mean 5 and variance 2), the placement of bids for the delivery (exponential with intensity 5), the creation of manual assignments (exponential, 1.5), and the creation of pickup and delivery confirmations (offset to the committed time by a normal distribution with mean 1 (4) and variance 3 (6)).

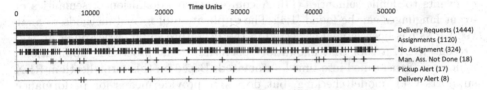

Fig. 10. Events observed during a run of 50,000 transition firings (20 drivers)

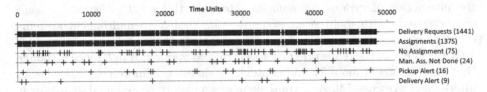

Fig. 11. Events observed during a run (40 drivers) corresponding to the same simulation time as in Fig. 10

We used CPN-Tools to run a simulation of 50,000 transition firings. The run covers a simulation time of 48,516 time units. Interpreting a time unit as a minute, we simulated more than 800 hours of execution. The events observed during execution are listed for the most important types in Fig. 10. More than 1,400 delivery requests have been created, which led to nearly 2,300 delivery bids and around 1,100 deliveries. We also observe only a few alerts that relate to manual assignments that have not been completed in time (once the assignment system has forwarded delivery bids to a flower store) or delayed pickups and deliveries. In contrast, there is a rather high number of assignment alerts. Those relate to bid requests that are not answered by any driver. The observed alert events hint at an insufficient number of drivers that participate in the network.

To investigate this issue, we conducted a second simulation with 40 van drivers (other configuration values remain unchanged). To achieve the same simulated time, we executed 62,650 transition firings. The results are visualised in Fig. 11. In comparison, we observe significant differences between both experiments. The second run shows more delivery bids from drivers per request, i.e., more than three bids per delivery, whereas the ratio was less than 1:2 in the first run. As a consequence, the number of assignment alerts drops and only every 19th delivery request does not lead to an assignment, whereas this was the case for every 4th to 5th request in the first run.

7 Related Work

Event languages are at the very core event processing networks. Different formalisations of event languages have been presented. The TESLA language [10] is grounded in first order logic and translates rules for complex events into logical expressions. Other work formulates algebraic expressions over temporal streams of events to define semantics [11]. A comprehensive definition of semantics of event languages can be found [1,4]. The application of formal methods to analyse EPNs has been largely neglected in the literature so far. Ericsson et al. [12] present REX, a tool that allows for specifying event processing applications, which are transformed into timed automated. Based on this formalisation, REX supports CTL model checking, but does not provide means for performance evaluation.

Event processing networks capture complex behaviour, which results from the interaction of various autonomous actors. In this regard, they share some characteristics with multi-agent systems [13]. Similar to multi-agent systems, event producers, consumers and processing agents show autonomy in working on a local view of the real-world in a decentralised manner. A difference is though that the notions of mobility and context dependence is more emphasized with multi-agent systems [14]. As a consequence, agents are represented as tokens in CPN formalisations [15]. In this context, object-oriented Petri nets have been extensively used for modelling [16] and analysis [17]. While agent coordination and knowledge representation is an issue for multi-agent systems [18], event processing networks are more concerned with an explicit representation of the decision rules. In practice, these decision rules are encoded with an event query language [2]. As we aimed to demonstrate, the usage of CPN bears the potential of fine-tuning these decision rules to best fit the environmental factors.

Petri nets have further been used for simulation and analysis in diverse fields of application, such as IT-service processes [19], network protocols [20], and embedded systems [21]. For workflow modelling, the peculiarities of modelling reactive systems – EPNs are reactive by definition – with Petri nets have been worked out in [22]. The reactive semantics proposed in this work may allow for modelling a subset of EPN concepts even without relying on the CPN formalism.

8   Conclusion

In this paper, we took up the challenge of modelling event processing networks with coloured Petri nets for analysis and simulation. We show how essential parts of EPNs are represented in a CPN and turned to the case of the Fast Flower Delivery Application in particular. Since the FFDA is a de-facto benchmark for demonstrating the capabilities of event processing systems, our work covers a broad range of concepts and aspects of EPNs. We reported on the experiences on modelling the FFDA and presented a validation of our model with an implementation in ETALIS. Finally, the potential of a CPN formalisation of event processing networks for analysis of system properties and simulation was outlined.

Applications realised as EPNs process large amounts of events, which calls for appropriate tools to analyse the behaviour and performance of an EPN. Our work addresses this demand by leveraging CPNs as a well-established formalism. Still, automation of a net-based formalisation of EPNs is hindered by several factors. First, despite the advancements on the formal definition of event languages, there is a lack of a generally accepted specification language for EPNs. Second, expressiveness of the languages used to define event contexts, evaluation policies, and the event detection logic is varying. We showed that these aspects have a large influence on the formalisation. Further work is needed to consolidate the different approaches to EPN modelling, so that boundaries of net-based formalisations can be made explicit.

Another direction for future work is the derivation of model configurations for simulation once a CPN model has been created and validated for an EPN. That relates in particular to the application of data mining techniques to infer event distributions and dependencies between event producing components.

Acknowledgements. We thank Guy Hareuveni for his support on the model validation and the CPN-Tools team for creating this magnificent piece of software.

References

1. Etzion, O., Niblett, P.: Event Processing in Action. Manning Publications (2010)
2. Voisard, A., Ziekow, H.: Architect: A layered framework for classifying technologies of event-based systems. Inf. Syst. 36(6), 937–957 (2011)
3. Gatziu, S., Dittrich, K.R.: Detecting composite events in active database systems using Petri nets. In: RIDE-ADS, pp. 2–9 (1994)
4. Adi, A., Etzion, O.: Amit - the situation manager. VLDB J. 13(2), 177–203 (2004)
5. Jensen, K., Kristensen, L.M.: Coloured Petri Nets - Modelling and Validation of Concurrent Systems. Springer (2009)
6. van der Aalst, W., Stahl, C.: Modeling Business Processes: A Petri Net-Oriented Approach. MIT Press (2011)
7. Vinter Ratzer, A., et al.: CPN Tools for editing, simulating, and analysing Coloured Petri Nets. In: [23], pp. 450–462
8. Steggles, L.J., Banks, R., Wipat, A.: Modelling and analysing genetic networks: From boolean networks to Petri nets. In: Priami, C. (ed.) CMSB 2006. LNCS (LNBI), vol. 4210, pp. 127–141. Springer, Heidelberg (2006)
9. Mulyar, N., van der Aalst, W.: Patterns in Colored Petri Nets. BETA Working Paper Series WP 139. Eindhoven University of Technology (2005)
10. Cugola, G., Margara, A.: Tesla: a formally defined event specification language. In: Bacon, J., Pietzuch, P.R., Sventek, J., Çetintemel, U. (eds.) DEBS, pp. 50–61. ACM (2010)
11. Hinze, A., Voisard, A.: EVA: An event algebra supporting adaptivity and collaboration in event-based systems. ICSI Technical Report TR-09-006 (2009)
12. Ericsson, A., Pettersson, P., Berndtsson, M., Seiriö, M.: Seamless formal verification of complex event processing applications. In: Jacobsen, H.A., Mühl, G., Jaeger, M.A. (eds.) DEBS, pp. 50–61. ACM (2007)
13. Ferber, J.: Multi-Agent Systems: An Introduction to Distributed Artificial Intelligence. Addison-Wesley (1999)
14. Kwon, O.B.: Modeling and generating context-aware agent-based applications with amended Colored Petri Nets. Expert Syst. Appl. 27(4), 609–621 (2004)

15. Marzougui, B., Hassine, K., Barkaoui, K.: A new formalism for modeling a multi agent systems: Agent Petri nets. J. of Softw. Eng. and Appl. 3(12), 1118–1124 (2010)
16. Moldt, D., Wienberg, F.: Multi-agent-systems based on Coloured Petri Nets. In: Azéma, P., Balbo, G. (eds.) ICATPN 1997. LNCS, vol. 1248, pp. 82–101. Springer, Heidelberg (1997)
17. Köhler, M., Moldt, D.: Analysis of mobile agents using invariants of object nets. ECEASST 12 (2008)
18. Weyns, D., Holvoet, T.: A colored Petri net for a multi-agent application. In: Proceedings of Modeling Objects, Components and Agents (MOCA 2002), Aarhus, Denmark, pp. 121–140 (2002)
19. Bartsch, C., Mevius, M., Oberweis, A.: Simulation of it service processes with Petri-nets. In: Feuerlicht, G., Lamersdorf, W. (eds.) ICSOC 2008. LNCS, vol. 5472, pp. 53–65. Springer, Heidelberg (2009)
20. Gordon, S., Billington, J.: Analysing the WAP Class 2 wireless transaction protocol using coloured Petri nets. In: Nielsen, M., Simpson, D. (eds.) ICATPN 2000. LNCS, vol. 1825, pp. 207–226. Springer, Heidelberg (2000)
21. Junior, M.N.O., Neto, S., Maciel, P., Lima, R., Ribeiro, A., Barreto, R., Tavares, E., Braga, F.: Analyzing software performance and energy consumption of embedded systems by probabilistic modeling: An approach based on Coloured Petri Nets. In: Donatelli, S., Thiagarajan, P.S. (eds.) ICATPN 2006. LNCS, vol. 4024, pp. 261–281. Springer, Heidelberg (2006)
22. Eshuis, R., Dehnert, J.: Reactive Petri nets for workflow modeling. In: [23], pp. 296–315
23. van der Aalst, W.M.P., Best, E. (eds.): ICATPN 2003. LNCS, vol. 2679, pp. 337–356. Springer, Heidelberg (2003)

Hierarchical Conformance Checking
of Process Models Based on Event Logs

Jorge Munoz-Gama[1], Josep Carmona[1], and Wil M.P. van der Aalst[2]

[1] Universitat Politecnica de Catalunya, Barcelona, Spain
[2] Eindhoven University of Technology, Eindhoven, The Netherlands
{jmunoz,jcarmona}@lsi.upc.edu, w.m.p.v.d.aalst@tue.nl

Abstract. Process mining techniques aim to extract knowledge from event logs. *Conformance checking* is one of the hard problems in process mining: it aims to diagnose and quantify the mismatch between observed and modeled behavior. Precise conformance checking implies solving complex optimization problems and is therefore computationally challenging for real-life event logs. In this paper a technique to apply hierarchical conformance checking is presented, based on a state-of-the-art algorithm for deriving the subprocesses structure underlying a process model. Hierarchical conformance checking allows us to decompose problems that would otherwise be intractable. Moreover, users can navigate through conformance results and zoom into parts of the model that have a poor conformance. The technique has been implemented as a ProM plugin and an experimental evaluation showing the significance of the approach is provided.

Keywords: Process Mining, Conformance Checking, Process Diagnosis.

1 Introduction

Process mining emerged as a crucial discipline for addressing challenges related to Business Process Management (BPM) and "Big Data"[1]. Information systems record an overwhelming amount of data representing the footprints left by process executions. For example, Boeing jet engines can produce 10 terabytes of operational information every 30 minutes, and Walmart logs may store one million customer transactions per hour [2].

Process mining tackles three challenges relating event data (i.e., log files) and process models: *discovery* of a process model from an event log, *conformance checking* given a process model and a log, and *enhancement* of a process model with the information obtained from a log. Process mining research has produced a multitude of algorithms that demonstrated to be of great value for undertaking small or medium-sized problem instances. However, it is well-accepted that most of the existing algorithms are unable to handle problems of industrial size. In our opinion, as has happened in other areas (VLSI, manufacturing, among others) where the size of the input prevents from solving the problem straight away, the next generation of algorithms for process mining must incorporate high-level

J.-M. Colom and J. Desel (Eds.): PETRI NETS 2013, LNCS 7927, pp. 291–310, 2013.
© Springer-Verlag Berlin Heidelberg 2013

techniques that enable the problem decomposition in a divide-and-conquer style. This paper proposes decomposition techniques for conformance checking.

In spite of its significance, few conformance checking algorithms exist in the literature. The seminal work by Rozinat *et al.* [3] was the first in formalizing the problem and enumerating the four dimensions to consider for determining the adequacy of a model in describing a log: *fitness*, *precision*, *generalization* and *simplicity*. In this paper we focus on the first two: fitness, that quantifies the capability of the model in reproducing the traces of the log, and precision, that quantifies how precise is the model in representing the log. In this paper, we will decompose state-of-the-art approaches for conformance checking. On the one hand, we will use *alignment techniques*. These provide a crucial step into relating model and log traces for unfitting models or models with duplicate/invisible transitions [4,5]. On the other hand, a novel precision technique based on superposing log and model behavior and detecting accordingly model's escaping points will allow us to quantify precision [6]. Both approaches have been combined in order to derive a robust estimation of precision [5]. The high-level technique presented in this paper is grounded on the aforementioned work.

To decompose process models, we use a technique to create the so-called *Refined Process Structure Tree* (RPST) originally proposed in [7]. This decomposition computes fragments of the net that have a single-entry and a single-exit node, thus intuitively resembling isolated subprocesses within the general model. Remarkably, it is a structural decomposition that can be computed in linear time on the underlying graph structure. Additionally, a tree-like structure may be derived representing the hierarchy between the computed components. This tree will be used in the strategy presented in this paper for hierarchical conformance checking: each component in the tree is processed in order to satisfy the requirements for conformance checking of [5] (e.g., find an initial and final marking, determine whereas it belongs to a cyclic part of the net), and finally a conformance checking problem instance is solved on the component and the log projected into the participating activities. Unlike the decomposition approaches proposed in [8,9], our approach uses a hierarchy of semantically meaningful process fragments. The RPST structure recursively splits the process into smaller fragments that are understandable for the analyst. Moreover, the hierarchy can be used to navigate to problematic parts of the process.

The approach has been implemented as a ProM plugin, and experiments have been performed on different dimensions, ranging from a comparison between the manual decomposition made by a human and the one presented in this paper, to the application of the technique to a set of benchmarks. Also, experiments illustrating the differences with the passage-based approach [8] are reported.

The paper is organized as follows: Sect. 2 presents the theoretical background of the paper. Sect. 3 presents the main stages of the methodology for hierarchical conformance checking, while Sect. 4 describes how to apply the approach to sound and safe workflow nets (a important class of Petri nets tailored towards business processes). Sect. 5 describes related work. Sect. 6 shows several experiments illustrating the usefulness of the suggested approach. Finally, conclusions and future work are reported in Sect. 7.

2   Preliminaries

The starting point for conformance checking are an *event log* and a *model*. An event log records the execution of all cases (i.e. process instances). Each case is described by a trace, i.e., an activity sequence. Different cases may have exactly the same trace, i.e., an event log is a multiset of traces. A model represents the possible flows of the process. In this approach we use Petri net as a formal model of processes. A Petri Net PN is a tuple (P, T, A, m_i, m_o) where P and T represent finite sets of places and transitions, respectively, with $P \cap T = \emptyset$. Function $A : (P \times T) \cup (T \times P) \to \mathbb{N}$ represents the weighted flow relation. The markings over a Petri net are defined as *multisets* and we use multiset notation accordingly, e.g., $m = [p^2, q]$ or $m = [p, p, q]$ for a marking m with $m(p) = 2$, $m(q) = 1$, and $m(x) = 0$ for all $x \notin \{p, q\}$. The initial and final markings of PN are denoted as m_i and m_o respectively.[1] Given a marking m and a set n, $m_{\downarrow n}$ denotes the projection of m on n, e.g., $[a, a, b, c]_{\downarrow [a,c]} = [a, a, c]$.

3   Methodology for Hierarchical Conformance Checking

In this section we introduce the approach for decomposing a model in a hierarchical manner in order to perform conformance checking. Algorithm 1 presents the approach, which has three stages: the *Decomposition Stage* (Sec. 3.1) where the model is decomposed into components, the *Post-processing Stage* (Sec. 3.2) where the components are processed to enable the analysis, and finally the *Conformance Stage* (Sec. 3.3) where conformance is analyzed for every component.

3.1   Decomposition Stage

The first stage corresponds with the *decomposition* phase: the model is decomposed into hierarchical components, i.e., the decomposition is not a partitioning (which is the case in [8]) but instead for each component different of the net itself there is always another component containing it. This enables the analysis at different degrees of granularity and independence, similar to zooming in and out using online maps, to get a better understanding of the underlying process. In order to be able to navigate through different layers of the model properly, three ingredients are needed: 1) a subprocesses decomposition, 2) a hierarchy relating these subprocesses, and 3) a mechanism to enhance and propagate additional information about the components within the hierarchy.

---

[1] In some approaches such as [3,6] only an initial marking is considered. However, in [5] (cornerstone of the conformance checking of this paper) the authors introduce analogously the need of the final marking.

Algorithm 1. Decomposed Conformance algorithm

procedure DECONF(net, log, m_i, m_o)
    $\{c_1 \ldots c_n\} \leftarrow$ DECOMPOSE(net)   ▷ Decomposition Stage
    $h \leftarrow$ BUILD_HIERARCHY($\{c_1 \ldots c_n\}$)
    $h^* \leftarrow$ ENRICH_HIERARCHY(h, net)

    for all subprocess $c \in h^*$ **do**
        $pn_c \leftarrow$ COMPUTE_PETRI_NET(c, net)   ▷ Post-processing Stage
        $m_{ic} \leftarrow$ COMPUTE_INITIAL_MARKING(c, net, m_i)
        $m_{oc} \leftarrow$ COMPUTE_FINAL_MARKING(c, net, m_o)
        $log_c \leftarrow$ COMPUTE_LOG(log, pn_c)

        COMPUTE_CONFORMANCE($log_c, pn_c, m_{ic}, m_{oc}$)   ▷ Conformance Stage
    end for
end procedure

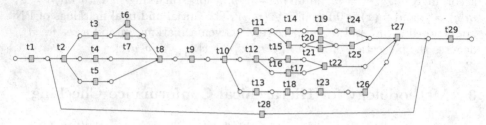

Fig. 1. Illustrative example of process modeled as a Petri net

Subprocesses Decomposition. Given that the final goal of this approach is the conformance analysis and the diagnosis of potential deviations, we propose a decomposition based on the identification of subprocesses within the main process. In particular, these subprocesses that match the *Single-Entry-Single-Exit* pattern, also known as *SESEs* [10], are detected. SESE represents a well-defined part of a general process. Note that, this decomposition refers to the structure of the model (not its behavior). Therefore, the SESE detection is not performed directly over the model, but over the underlying graph of the model, called *workflow graph*.

In our case, given a Petri net $PN = (P, T, A)$ we define its workflow graph simply as graph $G = (V, E)$ with no distinction between places and transitions, i.e., $V = P \cup T$ and $E = \{(x, y) \in V \times V | A(x, y) > 0\}$. For example, considering the example process of Fig. 1 modeled as a Petri net, its corresponding workflow graph is shown in Fig. 2. Similar representations can be obtained for other modeling notations such as BPMN, EPCs, or UML, and therefore, the hierarchical conformance approach presented in this paper is fully generic and can be applied to other notations. However, in this work, we focus on Petri nets.

In the remaining definitions, the following context is assumed: Let G be the workflow graph of a given Petri net, and let $G_F = (V_F, F)$ be a connected subgraph of G formed by a set of edges F and the vertexes V_F induced by F.

Definition 1 (Interior and Boundary nodes [7]). *A node $x \in V_F$ is interior with respect to G_F iff it is connected only to nodes in V_F; otherwise x is a boundary node of G_F.*

For instance, in the graph of Fig. 2, given the subgraph composed by the edges in D and its induced vertexes, $t2$ and $t8$ are boundary nodes, while the rest are interior nodes. A boundary node is an entry or exit node if additional requirements are satisfied.

Definition 2 (Entry and Exit nodes [7]). *A boundary node x of G_F is an entry of G_F iff no incoming edge of x belongs to F or if all outgoing edges of x belong to F. A boundary node x of G_F is an exit of G_F iff no outgoing edge of x belongs to F or if all incoming edges of x belong to F.*

Following with the example above, $t2$ would be an entry, and $t8$ an exit. Now we can define a SESE as follows:

Definition 3 (SESE, Trivial SESE and Canonical SESE [7]). *$F \subseteq E$ is a SESE (Single-Exit-Single-Entry) of graph $G = (V, E)$ iff G_F has exactly two boundary nodes: one entry and one exit. A SESE is trivial if it is composed by a single edge. F is a canonical SESE of G if it does not overlap with any other SESE of G, i.e., given any other SESE F' of G, they are nested ($F \subseteq F'$ or $F' \subseteq F$), or they are disjoint ($F \cap F' = \emptyset$). By definition, the source of a workflow graph is an entry to every fragment it belongs to and the sink of the net is an exit from every fragment it belongs to.*

For the sake of clarity and unless otherwise is stated, in the rest of the paper we will refer to canonical SESEs simply as SESEs. Note also that the SESEs are defined as a set of edges (not as subgraphs). However, for simplicity we will refer also to the subgraph as SESE when the context is clear. Given that a single edge is a SESE, in the remainder of the paper we will only consider SESEs above a given threshold size[2]. Figure 2 shows the canonical SESEs of the example in Fig. 1 having size greater than 4 nodes (i.e., A,...R).

Component Hierarchical Structure. Next we construct the Refined Process Structure Tree (RPST) between canonical SESEs. This tree-like structure will group all non-overlapping siblings at the same level that give rise, in an upper level, to the canonical SESE that includes all of them. This will allow us to navigate through the different levels of the tree thus providing views at different level of granularity.

---

[2] On Sect. 6.1 we provide an experiment to estimate this threshold size based on manual decompositions.

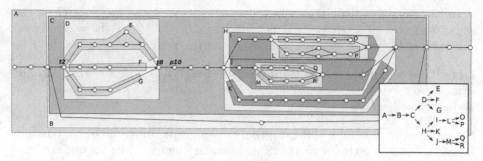

Fig. 2. Workflow graph of the example in Fig. 1, and its canonical SESEs with size greater than 4 nodes. On the bottom-right corner we show the corresponding RPST.

Definition 4 (Refined Process Structure Tree (RPST) [7]). *The* Refined Process Structure Tree (RPST) *of G is the tree composed by the set of all its canonical SESEs, such that, the parent of a canonical SESE F is the smallest canonical SESE that strictly contains F. The root of the tree is the entire graph, the leaves are the trivial SESEs.*

Figure 2 shows the RPST for the SESEs found in example of the same figure. Note that, due to the definition of canonical SESE, all siblings of a tree node will never have overlapping edges. However, it is not required that the union of all the siblings results in the entire parent canonical fragment (i.e., the parent of a SESE may have arcs not included in any of its children). For example, the edge (*t8,p10*) belonging to *C* is neither included on *D* nor *H*. The computation of the canonical SESEs and it's corresponding RPST of a workflow graph is a well studied problem in the literature, and can be computed in linear time. In [11], the authors proposed the original algorithm for constructing an RPST. In [7,10], the computation of the RPST is considerably simplified by introducing a pre-processing step that reduces the implementation effort considerably. The existence of RPST is guaranteed, i.e., the trivial RPST with the net as root and the trivial SESEs as leaves is always possible. Besides providing an explicit hierarchy among SESEs, the RPST structure satisfies additional properties useful for our approach: the RPST is *unique* (i.e., same graph will always result in the same RPST) and *modular* (i.e., a local change in the workflow graph only results in a local change of the tree).

Structure Enhancement. Next, we enrich the hierarchical structure obtained (the RPST) with additional information that may be used to improve the conformance checking result. In this step we detect when a subprocess can be repeated more than once in a process execution. The idea behind this is that in order to perform the conformance checking analysis correctly, one must determine whether the subprocess reflected in a component can appear more than once within the same trace, i.e., within the same instance of the whole process, this subprocess can be executed several times. Such knowledge is highly relevant for the conformance evaluation.

In order to incorporate this information, first we must determine the cyclic behavior of the model. We will use this information to determine if a component may be iterated to reproduce a trace. Finally, this information is going to be transmitted to the RPST, and propagated through the tree. We will use the structural theory of Petri nets to determine potential iterative behaviors. In particular, we will use *T-invariants* [12] to determine potential repetitive behavior (see Sect. 4 for further details).

3.2   Component Post-processing Stage

To perform a multilayer fine-grained conformance checking as the one presented in this paper, we must first derive the log and the net for each one of the subprocesses considered. For the log, the complete log refers to actions of the whole model. Therefore, we must project it only over the actions involved in the component we are analyzing, i.e., to analyze a component x that models only the set of tasks T_x, we must remove from the log all events not in T_x. Besides the net itself, we must determine also the initial and final markings of the Petri net for analyzing conformance. This step, and how it is performed, depends on the type of Petri net considered. In Sect. 4 we will show how to perform this processing for a well-known class of Petri nets used to model business processes.

3.3   Conformance Stage

Finally, the last stage is the actual conformance checking, i.e., ultimate goal of the approach presented in this paper. In process mining, checking the conformance refers to the procedure of analyzing if the model considered is an appropriate and faithful representation of the reality reflected in the log. Literature clearly shows that conformance of a model with respect to a log is a multidimensional property. There is consensus on four orthogonal dimensions: fitness, precision, generalization and simplicity [1,3,13]. In this paper, we will focus on the first two: fitness and precision. However, we would like to stress that the whole approach presented in this paper is not restricted to any particular conformance algorithm, i.e., it can be used in combination with any other conformance technique for the four dimensions.

Fitness measures if the model is able to represent all the behavior in the log. For example, consider the models and event log shown in Fig 3. Model a) fits perfectly the log d) because it is able to reproduce both traces. However, model b) fails to reproduce the second trace. Precision, on the other hand, measures if the model is precise modeling the log. For instance, a) models the log precisely (i.e., no more behavior is allowed than the one reflected), while c) is completely imprecise (it allows much more behavior than the one in the log, and therefore, it underfits the process it is meant to describe).

The difference between the hierarchical conformance checking (as presented in this paper) and existing approaches is that the conformance analysis is

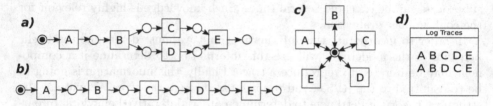

Fig. 3. Illustrative examples of fitness and precision with respect to a log d). Model a) fits better than b), and is more precise than c).

computed at different levels (not only in the complete model and log). By combining the conformance results of the different layers, we will be able to navigate and perform a complete diagnosis analysis of our system in both a top-down and bottom-up way. The fitness and precision checking proposed in this paper is based on cost-optimal alignment between traces in the log and possible runs of the model. Such alignments are then used as reference to compute fitness and precision (cf. Fig 4). The technique presented in [5] only requires a log and a Petri Net with an initial and final marking.

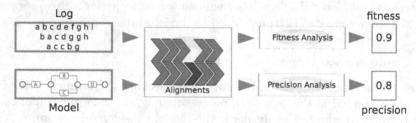

Fig. 4. Conformance checking based on Cost-Optimal Alignment

Cost-Optimal Alignment. In order to check conformance, given a trace in the log, we need to find the model execution (trace) that best represents such log trace. This is done by aligning the trace with all possible executions of the model, assigning a cost to each alignment, and choosing the optimal one. An alignment is defined as sequence of pairs (called *moves*), where each pair can be: a) a 'move' in the model b) a 'move' in the log, and c) a 'move' both in the model and the log. For example, here are two possible alignments between the only two traces of the model in Fig.3a (ABCDE and ABDCE) and the log trace ABCDE:

$$\gamma_1 = \frac{\begin{vmatrix} A & B & C & D & E \end{vmatrix}}{\begin{vmatrix} A & B & C & D & E \end{vmatrix}} \quad \gamma_2 = \frac{\begin{vmatrix} A & B & C & D & \perp & E \end{vmatrix}}{\begin{vmatrix} A & B & \perp & D & C & E \end{vmatrix}}$$

For each alignment γ_1 and γ_2, the upper row represents moves in the log, and the lower row represents moves in the model. The alignment between the trace and the execution ABCDE (shown as γ_1) is composed of synchronized moves (i.e., moves in both model and log), whereas the other alignment (γ_2) contains both a move on the log $(C, \bot)$ and a move on the model $(\bot, C)$ to be able to align the model execution and the trace. To measure the cost of an alignment, we define a cost function. For instance, we define a function that, for each pair, assigns a cost of '0' when it is a synchronized pair, and '1' otherwise (model and log disagree). The total cost of the alignment is the sum of the costs of individual moves[3]. Given such function, the costs of the alignments in the example above are 0 for γ_1 and 2 for γ_2. In [4,14], the authors propose various approaches to efficiently compute an optimal alignment for a given log trace.

Precision. In order to compute precision from the optimal alignments, the bottom rows of all optimal alignments are used to build a prefix automaton describing the modeled behavior observed in reality. Note that the alignments help to squeeze observed behavior into the model even in case of deviations. Then, this prefix automaton describing the actual observed behavior is compared with the modeled behavior, leading to the detection of the so called *imprecisions*, i.e., points in the process where the model allows for more behavior that actually observed in the event log. Due to space constraints, we refer to [5] for further details. The main difference between the approach in the literature, and the one proposed here, is that precision checking is performed for every subprocess (i.e., SESE), not only for the complete system.

The differences are illustrated using the following example: imagine the initial model of Fig.1, and consider a log composed of two traces: $t_1 t_2 t_4 t_5 t_3 t_6 t_8[\ldots]$ and $t_1 t_2 t_5 t_4 t_3 t_7 t_8[\ldots]$ and the model of Fig.1 (for sake of simplicity in this example we focus only on the part of the model between t_2 and t_8). The corresponding prefix automaton shown Fig. 5 reflects two main imprecisions: those nodes marked in dark gray represent the possibility in the model of executing concurrently the three branches (the branch starting at t_4, the one at t_5 and the one at t_3). The other imprecision (in light gray) is derived from the choice between t_6 and t_7. The process has three concurrent branches (notice that the log only reflects the concurrency among the $t4$ and $t5$ ones), and all the behavior modeled in each individual branch is reflected in the log (including the the choice between t_6 and t_7 where both options appear in the log). And this is precisely what we can see in a precision analysis using the hierarchical approach: the conformance analysis of SESE D (cf. Fig. 2) is similar to the previous approach, i.e., will derive a similar prefix automaton. However, in the hierarchical approach the analysis will also be done for the interior SESEs E, F and G, which will reflect a perfect precision, e.g., E precisely represents the projected log for (traces $t_2 t_3 t_6 t_8[\ldots]$ and $t_2 t_3 t_7 t_8[\ldots]$). A process-analyst can see that the conformance problems in D are not within any of its interior subprocesses but instead is a problem related to the sequencing of the subprocesses.

Fitness. Once the optimal alignment for each trace is found, the non-sync moves are used to detect the fitness anomalies, i.e., points where the model does not

---

[3] For the sake of simplicity, we use the default unit cost function. However, arbitrary complex user-defined cost functions can be used [4].

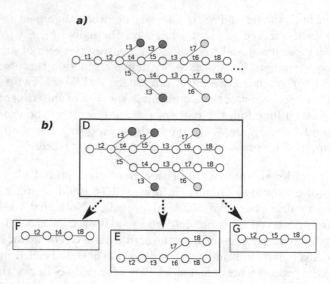

Fig. 5. Imprecisions for the example: a) Prefix automaton for the model of Fig. 1, b) Prefix automata for SESEs F, E and G

reflect the log, or points where the log does not reflect the model. Various fitness metrics have been proposed, penalizing such anomalies. Due to space constraints, we refer to [4] for further details. Similar to the precision case, our hierarchical helps to diagnose of the process: one can navigate through the hierarchy, discarding subprocesses that are perfectly fitting, and focusing the analysis only on those that have fitness problems. For example, given long and repetitive traces such as $t_1 t_2 t_3 t_4 t_5 t_6 t_7 t_8$ [...] $t_{28} t_2 t_3 t_4 t_5 t_6 t_7 t_8$ [...] $t_{28} t_2 t_3 t_4 t_5 t_6 t_7 t_8$ [...] t_{29}, note that t_6 and t_7 are sequential in this trace. Looking at component E, one may clearly diagnose the fitness problem i.e., t_6 and t_7 are in conflict in the model.

4   The Case of Business Processes

In this section we apply the approach on a particular subclass of Petri nets tailored towards business processes: sound and safe workflow nets [15].

Workflow nets are a well studied model in the Business Process Management literature, and therefore, they are the target of various conformance approaches, (e.g., [3]). This section instantiates the methodology presented in the previous section to this class of nets. In the remainder we consider Petri nets satisfying the following seven conditions. The process models used for conformance checking must be a *workflow* net: 1) there is a single source place *in*, i.e., $\{p \in P | {}^\bullet p = \emptyset\} = \{in\}$, 2) there is a single sink place *out* , i.e., $\{p \in P | p^\bullet = \emptyset\} = \{out\}$, 3) every node is on a path from *in* to *out*. The workflow net must be *sound*: 4)have the option to complete, i.e., starting from the initial marking (just a token in place *in*), it is always possible to reach the marking with one token in place *out* (marking [*out*]), 5) have proper completion, i.e., at the moment a token is put in place *out*, all the other places should be empty, 6) there are no dead transitions starting from the initial state [*in*]. And finally, the sound workflow net must be *safe*: 7) the number of tokens of any place at any time must be at most one.

We now present the details of the post-processing stage for safe and sound workflow nets. In particular, given $G_F = (V_F, F)$, a SESE obtained during the decomposition phase, we describe how to build a SESE-Net SN' from it, and how to determine its initial and final markings. In other words, we aim to build a Petri net $SN' = (P', T', A', i', o', p'_i, p'_o, m'_i, m'_o)$, where P', T', A' define the net, p'_i, p'_o are the source and sink places of the net, m'_i, m'_o are its initial and final markings, and o', i' define the nodes of the Petri net (transition or place) representing the single-entry and the single-exit of the SESE.

In the remainder of this section the following context is assumed: Let $WF = (P, T, A, p_i, p_o, m_i, m_o)$ be the workflow net to be analyzed, and $G = (V, E)$ its corresponding workflow graph. Let $G_F = (V_F, F)$ define a canonical SESE of G, and let $i, o \in V_F$ be the entry and exit nodes of G_F, respectively.

Given that the SESE decomposition is performed over the workflow graph, the first step is to determine which nodes of the original net WF are included in SN'. In other words, for all $x \in V_F : x \in P \implies x \in P'$ and $x \in T \implies x \in T'$. Similar, the projection over the arcs is done, i.e., if $x \in F$ then $x \in A'$. For example, given the SESE D of the running example, shown in Fig. 6a, $t_2 \ldots t_8$ and $p_3 \ldots p_9$ are the transitions and places of the original net included in this SESE-net, respectively.

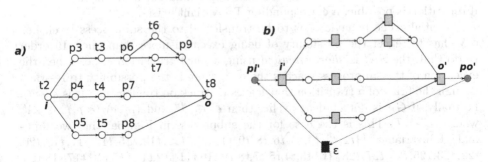

Fig. 6. a) SESE and b) its corresponding SESE-Net

Note that, by definition, G_F contains a single start and end node: i and o. Therefore, we define i', o' as $i' = i$ and $o' = o$. The existence of such single-entry and single-exit is an appropriate property in order to determine the source and sink places of the SN'. Since the SESE decomposition is performed over the workflow graph (where there is no distinction between places and transitions), it is not guaranteed that i' and o' are places (they may be transitions). This is the case of the component D of Fig. 6. In such cases where the entry (or the exit) node is a transition, a pre-processing of the SN' is required, i.e., an artificial place is created and linked with an artificial arc to such transition. The

Fig. 7. Pre-processing of the SESE-net to guarantee a place-bordered net

transformation is informally illustrated in Fig. 7. An example is shown in Fig. 6a, where both the entry and the exit nodes are transitions. Therefore, two artificial places are created (and its corresponding arcs) to represent the p'_i and p'_o of the SESE-net, as it is shown in Fig. 6b.

As discussed in the previous section, a net can contain cyclic behavior, i.e., subprocesses that are repeated several times within the main process execution. In the decomposition phase, the cyclic behavior is detected by determining those SESEs that are covered by T-invariants (i.e., if all transitions in a SESE belong to some T-invariant, we assume that the component can be repeated) and the hierarchical structure is enhanced with this information. Formally, a T-invariant is a vector $\mathbf{x} : T \to \mathbb{Z}$ such that $\mathbf{C} \cdot \mathbf{x} = 0$, where $\mathbf{C}$ is the *incidence matrix* of the net [12]. Intuitively, a T-invariant is a set of transitions such that the marking reached after executing them is the same as the one before its execution. Clearly, T-invariants provide only an approximation of the real repetitive behavior, but we have seen in practice that this heuristic approximation works fine. Moreover, if repetition is possible, a corresponding T-invariant exists.

Potential cyclic behavior has to be transferred to the subprocess level, i.e., SN' has to reflect the possibility of being executed more than once. In order to do that, the SN' is *short-circuited* using a silent transition ϵ, i.e., when the execution of the subprocess reaches the end, there is the possibility to re-start, though the firing of a transition that leaves no track on the log (i.e., it is *silent*). Formally, if G_F is detected as cyclic, then $\epsilon \in T'$ and $(p'_o, \epsilon), (\epsilon, p'_i) \in A'$, where $\epsilon \notin T$. This is the case for the subprocess in Fig. 6a: the two minimal T-invariants $\{\{t2, t3, t4, t5, t6, t8, t9, t10, t11, t12, t13, t15, t16, t17, t18, t20, t22, t23, t25, t26, t27, t28\}, \{t2, t3, t4, t5, t7, t8, t9, t10, t11, t12, t13, t15, t16, t17, t18, t20, t22, t23, t25, t26, t27, t28\}\}$ of the initial net (Fig.1) include the transitions of the SESE and therefore it is tagged as cyclic. Hence, a silent transition short-circuit is created as it is shown in Fig. 6b.

Next, we need to determine the initial and final markings (m'_i and m'_o) of the subprocess. Given the workflow net nature of the subprocess and the original workflow net, we define the initial and final markings as $[p'_i]$ and $[p'_i]$ respectively (cf. Fig. 6b).

After post-processing we obtain a new Petri net SN' modeling the subprocess behavior. When iteration is detected, due to the type of nets we are restricting in this section (safe and sound workflow nets), the modification done to the original SESE-net enables the iterative behavior in the modified SESE-net. The final SESE-net (with or without iteration), is used to do conformance analysis.

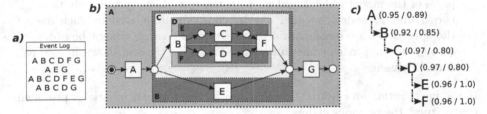

Fig. 8. The hierarchical approach is non-monotonic with respect to subprocess inclusion. Each node in the hierarchy c) of the SESEs detected in b) is annotated with a pair of real numbers, representing the fitness and precision of the corresponding SESE-net for the log obtained after projecting the event log of a) for the participating activities. For instance, the fitness (first number in the nodes) for nodes A (0.95), B (0.92) and C (0.97) is neither increasing nor decreasing.

Likewise it is done in [9] for fitness, it should be possible to establish formal guarantees relating both fitness and precision metrics evaluated in the subprocesses with respect to the ones of the original net. However, since neither fitness nor precision are monotonic with respect to subprocess inclusion, the weakest conditions under which a formal guarantee relating model and subprocesses fitness/precision can be given will be investigated in the future. The example in Fig. 8 illustrates the non-monotonic nature of the hierarchical approach.

5   Related Work

As it was mentioned in Sec.1, few conformance checking algorithms exists in the literature, e.g., [16,3,4,6,5,17]. However, given the complexity of the conformance analysis, recently a technique for decomposed conformance checking has appeared: in [8] the authors propose a non-hierarchical decomposition based on passages, also meant to be applied to the same class of Petri nets as the one considered in this paper. A passage is a pair of two non-empty sets of transitions (X, Y) such that the set of direct successors of X is Y, whilst the set of direct predecessors of Y is X. Any graph can be decomposed into minimal passages representing a partitioning of its transition set, and in [8] authors demonstrate how both discovery and conformance problems can be decomposed using passages.

There are significant differences between the passage decomposition approach and the one presented in this paper. We enumerate here the most important ones:

1. The passages approach computes a 1-level partitioning, whereas our approach derives a hierarchy of components. Note that once 1-level passages are obtained, they may be united to form higher-level passages, since the union of passages is a passage. However, algorithms for this high-level post-process to form a hierarchy have not been proposed nor implemented.
2. The components in the passages approach represent causality fragments between two sets of transitions, whilst components in our approach denote

parts of the model which interface with the rest of the system only through two boundary nodes. In other words, components derived by each one of the approaches are incomparable, i.e., a typical passage cannot be obtained through our approach whilst a typical SESE (like the one shown in Fig. 6a) is not a passage.

In the next section an empirical comparison for a set of benchmarks is provided that witness the previous claims.

6   Experimental Results

In this section we present the experimental results supporting the claims made in earlier sections. Most of the models have been generated with the PLG tool [18]. Two main type of experiments are provided in two subsections: Sect. 6.1 is structured in two parts: in part I we compare the hierarchical approach presented in this paper with manual decompositions performed by some persons, and in part II we test the approach through a set of experiments, comparing it with the approaches in the literature and especially with the one presented in [8]. Sect. 6.2 describes an experiment illustrating how the approach can be applied for a large example, where it is not possible to handle large subprocesses and therefore only nodes of the RPST that have a size less than a certain value can be analyzed in practice. An implementation of the approach presented in this paper can be found as a plugin within the ProM Framework[4].

6.1   Empirical Evaluation

Part I: Similarities with Human-Made Model Decomposition
The first set of experiments is designed to study one of the strong points of the approach proposed: the process-like decomposition (based on Single Entry Single Exit components) and its intuitive relation with the mental schema of process analyst. A set of process models ($man01$ to $man06$) has been prepared in order to be manually decomposed by possible actors of the approach, and the resulting components have been compared with the ones obtained by the approach of this paper. The set of benchmarks is composed by six Petri nets modeling plausible processes. Each model contains the most common patterns found on process modeling: choice, concurrency, sequencing, invisible tasks, and loops. The manual decomposition has been performed by 7 persons, all them with complete liberty to decide their own decomposition. The results of the experiments are shown in Table 1[5]. The table shows, the number of transitions and places of the initial Petri net, whether all persons used hierarchy in the decomposition (hierarchy ?), the (average of the 7 persons) maximum number of levels in the hierarchy (h- max), and the corresponding maximum number of

---

[4] Under the package *JorgeMunozGama*.
[5] In the link http://www.lsi.upc.edu/~jmunoz/files/PN2013benchmarks.zip the reader can inspect the manual decompositions.

Table 1. Manual decomposition compared with the hierarchical approach

	$|T|$	$|P|$	hierarchy ?	h-max	h-SESE	$|C|$	$|Size(C)| < 5$	% SESE-like
man01	44	45	Yes	3.4	8	63	14%	91%
man02	16	16	Yes	2.4	5	28	7%	97%
man03	31	34	Yes	3.3	6	50	16%	89%
man04	31	31	Yes	3.3	7	51	16%	96%
man05	48	50	Yes	3.7	7	79	14%	92%
man06	45	51	Yes	3.6	8	71	3%	94%

levels in the hierarchy computed by our approach (with a 4-nodes threshold). Also, the number of components computed by all the persons is shown ($|C|$), and the percentage of components having less than 5 nodes reported ($|Size(C)| < 5$). Finally, we provide the percentage of the components that satisfy the SESE property (column % SESE-like).

The first conclusion raised from the experiment is related with the intuitive use of hierarchy for decomposition. This is one of the main differences of the approach proposed in this paper compared with other approaches such as [8]. The results show that in all benchmarks, all persons use hierarchy. Second, the average number of levels in the hierarchy used by the 7 persons is smaller than the one computed by our approach. The main reasons for that is that the persons have not considered components that differ from each other of very few nodes which is the case for a parent and sibling having a high degree of overlapping. For instance, in Fig. 2, the only difference between SESE A and SESE B is the four edges in A not in B. This suggests to refine the SESE detection algorithm to collapse in the RPST nodes that differ on very few edges, in order to remove this type of redundancy, e.g., collapse nodes A and B as one single SESE in the example.

Note that persons prefer larger and process-like components instead of small and oversimple ones. For example, from column $|C|$, the percentage of small components ranges from 3% to 16% (and in most of the cases they correspond with components of size four: a choice of two transitions). This motivates the use of a threshold in the approach proposed in this paper. And finally, there is a direct relation between the components suggested manually and the ones provided by the approach, i.e., the vast majority of the components correspond with SESEs, as it is shown in *% SESE-like* column.

Part II: Evaluation through a Set of Benchmarks
This second set of experiments focuses on the actual hierarchical conformance checking. The experiment is composed by six models of different sizes, and their corresponding logs. For each of the benchmark model and log combinations, the decomposition is obtained and the conformance analysis is performed per component. The results are shown in Table 2 (comparison with respect to the passage approach), and Table 3, where the results for approaches in the literature [4,5,6] are provided. Notice that the conformance checking performed in this paper

Table 2. Conformance Results: passages vs. hierarchical

	$|T|$	$|P|$	$|C|$	comp	node	max	min	$|C|$	comp	node	max	min	comp	node	max	min
				Passages					*Hierarchical*							
				Fitness					Fitness				Precision			
lu01	81	74	50	1	1	1	1	48	1	1	1	1	.958	.761	1	.293
lu02	45	51	29	.897	.898	.942	.84	34	.926	.912	1	.883	.978	.899	1	.649
lu03	86	80	50	1	1	1	1	55	1	1	1	1	.918	.719	1	.327
lu04	43	38	23	1	1	1	1	27	1	1	1	1	.837	.710	1	.343
lu05	91	82	50	.976	.977	1	.939	59	.914	.915	1	.480	.955	.789	1	.362
lu06	59	54	36	.988	.988	1	.955	28	.992	.989	1	.98	.92	.689	1	.272

Table 3. Conformance Results: results for non-decomposition approaches in the literature

	Fitness [4]	Precision [5]	[6]
lu01	1	.301	.301
lu02	.884	.65	.72
lu03	1	.335	.335
lu04	1	.35	.35
lu05	.965	.22	.22
lu06	.988	.283	.295

provides metrics for SESE components. Therefore in Table 2 we present some intuitive metrics that combine the individual metrics into one single metric, to have an intuitive estimation of the conformance of the whole model. In Table 2 the number of places and transitions per benchmark are reported, and for each one of the decomposition approaches (passages or hierarchical), the number of components obtained (column $|C|$) is reported. The table also provides the average fitness/precision with respect to the number of components (*comp*), and with respect to the size of the components (*node*), i.e., in *node* larger components have larger weight on the average fitness/precision computation. The table provides the maximum and minimum fitness/precision value obtained for each benchmark (max and min)[6]. Finally, Table 3 reports the conformance analysis (fitness and precision) by other approaches in the literature.

Regarding the comparison between the passages approach and the hierarchical, it should be noted that there is a tendency for the passages approach to derive less components, which in fact are considerably small. On the fitness reported, both approaches report perfect fitness for the fitting models *lu01*, *lu03* and *lu04*, and some differences for the rest. Interestingly, in *lu05* a component with a fitness value of 0.480 is detected, representing the diagnosis of a fitness problem in a particular subprocess of the model.

---

[6] Only the fitness calculation is implemented in the approach [8].

By comparing the values in Table 3 of the non-decomposition approaches in the literature with the ones in Table 2, a clear tendency of the decomposition techniques to provide higher average values (both in the comp and the node columns) is detected. This is especially manifested in the precision dimension, where although the precision problems are still detected and equally evaluated as in the non-decomposition approaches (see column min for the hierarchical precision), they are reported at the level of the subprocess, thus identifying the real portion of the model that represents the conformance problem.

Hence, the results for the hierarchical approach, divided into components, provide an useful tool for diagnosis. This is complemented by the tree visualization provided by the implementation, where the conformance results of each component are displayed over its corresponding node in the hierarchy using a intuitive code of colors, aiming at localizing problems within the process (see Fig. 9). As an example, in some processes such as *lu03* (shown in Fig. 9), the differences in precision between comp and node is considerable, reflecting that while the vast majority of components have a high precision, the larger components have a low one. This is confirmed in the conformance results obtained for each particular component: while the two largest components have a precision of 0.33 and 0.32 respectively (shown in the figure as nodes in red in the RPST), the rest of components (of significant smaller size) have precisions close to or exactly 1.0. An introspective view of this conformance problem shows that a big component allows the concurrent execution of three subprocesses, each one having no conformance problems. However, in the log there is a real sequencing on these three components, and therefore a precision problem has been identified between these three subprocesses.

6.2 Handling a Large Conformance Problem

The hierarchical approach presented in this paper can be used to limit the complexity of the conformance analysis by bounding the size of the conformance instances to solve. This may be done by selecting a partitioning in the RPST selecting those SESEs whose size do not exceed a given threshold size. This section illustrates how this strategy can be used to perform conformance analysis for problems of industrial size.

With the PLG tool, we have created a model of 693 nodes, depicted in Fig. 10, and its corresponding log[7]. Then we have used the approach in [4] to estimate the fitness of the model with respect to the log, establishing a time limit of 1 hour. Within this limit, the aforementioned approach did not completed the fitness computation. If instead the hierarchical approach is applied together with a restriction on the size of subprocesses to consider (we have set 200 nodes as maximum SESE size), we have been able to analyze conformance for the subprocesses forming the partitioning shown in Fig. 10 in less than 2 minutes.

---

[7] To ease subprocess identification, we have used colors in Fig. 10 which will only be visible in the electronic version or in a colored printed version of the paper.

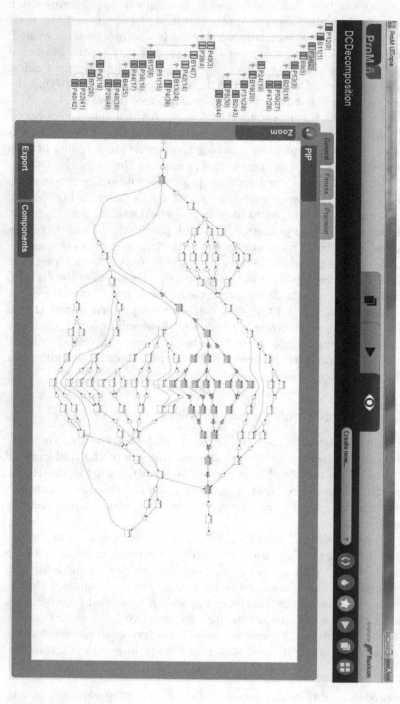

Fig. 9. The hierarchical view of conformance checking: the tree-view (left panel) provides the RPST together with the conformance results. Fitness and precision metrics for each SESE (node in the RPST) are provided as two colored squares, following a semaphore code: green (100-90%), yellow (89-75%), orange (74-50%) and red (49-0 %). By selecting a particular node in the RPST, the corresponding SESE is highlighted in the right panel.

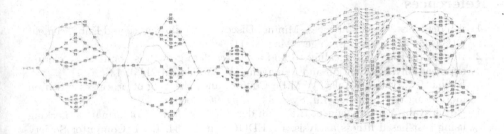

Fig. 10. Partitioning based on SESEs for conformance analysis: we have highlighted in colors the SESEs for which a conformance analysis has been done

This conformance analysis revealed, for instance, that the subprocess in green (leftmost) has a poor precision (0.318) and therefore it may be inspected in isolation to determine the causes of the conformance problem, while the process in pink has a perfect precision of 1.0.

7   Conclusions and Future Work

In this paper we presented a hierarchical approach for conformance checking that enables process analysts to investigate deviations between modeled and observed behavior. The technique combines previous work on conformance checking and model decomposition techniques in order to identify subprocesses within a given model for which an isolated conformance checking can be done, offering a hierarchical structure that can be used to navigate through the conformance results. The experiments show both the usefulness of the approach and the difference with related techniques.

The future work aims to extend the result in various directions. We plan to investigate the theoretical guarantees and extend the proposed technique to a larger class of Petri nets. Regarding algorithms, we plan to study new algorithms to improve both the quality and the performance of the proposed methodology, e.g., proposing a reduction of the RPST to avoid too much overlapping between SESEs and finding ways to propagate in a bottom-up manner the conformance results. The latter is very important to be able for real-life conformance checking where performance is an issue. Finally, although the approach as been proved to handle models of industrial size, further experiments considering real-life models and logs will reveal the real diagnosis value of the contribution presented in this paper.

Acknowledgements. This work has been partially supported by the Spanish Ministerio de Educación (AP2009-4959) and by the projects TIN-2011-22484 and TIN-2007-66523.

References

1. van der Aalst, W.M.P.: Process Mining: Discovery, Conformance and Enhancement of Business Processes. Springer (May 2011)
2. Rogers, S.: Big data is scaling BI and analytics-data growth is about to accelerate exponentially. Information and Management - Brookfield 21(5), 14 (2011)
3. Rozinat, A., van der Aalst, W.M.P.: Conformance checking of processes based on monitoring real behavior. Inf. Syst. 33(1), 64–95 (2008)
4. Adriansyah, A., van Dongen, B.F., van der Aalst, W.M.P.: Conformance checking using cost-based fitness analysis. In: EDOC, pp. 55–64. IEEE Computer Society (2011)
5. Adriansyah, A., Munoz-Gama, J., Carmona, J., van Dongen, B.F., van der Aalst, W.M.P.: Alignment based precision checking. In: La Rosa, M., Soffer, P. (eds.) BPM Workshops 2012. LNBIP, vol. 132, pp. 137–149. Springer, Heidelberg (2013)
6. Munoz-Gama, J., Carmona, J.: A General Framework for Precision Checking. International Journal of Innovative Computing, Information and Control (IJI-CIC) 8(7B), 5317–5339 (2012)
7. Polyvyanyy, A., Vanhatalo, J., Völzer, H.: Simplified computation and generalization of the refined process structure tree. In: Bravetti, M., Bultan, T. (eds.) WS-FM 2010. LNCS, vol. 6551, pp. 25–41. Springer, Heidelberg (2011)
8. van der Aalst, W.M.P.: Decomposing process mining problems using passages. In: Haddad, S., Pomello, L. (eds.) PETRI NETS 2012. LNCS, vol. 7347, pp. 72–91. Springer, Heidelberg (2012)
9. van der Aalst, W.M.P.: Decomposing Petri nets for process mining: A generic approach. Technical Report BPM-12-20, BPM Center (September 2012)
10. Polyvyanyy, A.: Structuring process models. PhD thesis, University of Potsdam (2012)
11. Vanhatalo, J., Völzer, H., Koehler, J.: The refined process structure tree. In: Dumas, M., Reichert, M., Shan, M.-C. (eds.) BPM 2008. LNCS, vol. 5240, pp. 100–115. Springer, Heidelberg (2008)
12. Silva, M., Teruel, E., Colom, J.M.: Linear algebraic and linear programming techniques for the analysis of place or transition net systems. In: Reisig, W., Rozenberg, G. (eds.) APN 1998. LNCS, vol. 1491, pp. 309–373. Springer, Heidelberg (1998)
13. Rozinat, A., de Medeiros, A.K.A., Günther, C.W., Weijters, A.J.M.M.T., van der Aalst, W.M.P.: The need for a process mining evaluation framework in research and practice. In: ter Hofstede, A.H.M., Benatallah, B., Paik, H.-Y. (eds.) BPM Workshops 2007. LNCS, vol. 4928, pp. 84–89. Springer, Heidelberg (2008)
14. Adriansyah, A., Sidorova, N., van Dongen, B.F.: Cost-based fitness in conformance checking. In: Caillaud, B., Carmona, J., Hiraishi, K. (eds.) ACSD, pp. 57–66. IEEE (2011)
15. van der Aalst, W.M.P., van Hee, K.M., ter Hofstede, A.H.M., Sidorova, N., Verbeek, H.M.W., Voorhoeve, M., Wynn, M.T.: Soundness of workflow nets: classification, decidability, and analysis. Formal Asp. Comput. 23(3), 333–363 (2011)
16. Cook, J., Wolf, A.: Software Process Validation: Quantitatively Measuring the Correspondence of a Process to a Model. ACM Transactions on Software Engineering and Methodology 8(2), 147–176 (1999)
17. Weerdt, J.D., Backer, M.D., Vanthienen, J., Baesens, B.: A Multi-Dimensional Quality Assessment of State-of-the-Art Process Discovery Algorithms Using Real-Life Event Logs. Information Systems 37(7), 654–676 (2012)
18. Burattin, A., Sperduti, A.: PLG: A framework for the generation of business process models and their execution logs. In: zur Muehlen, M., Su, J. (eds.) BPM 2010 Workshops. LNBIP, vol. 66, pp. 214–219. Springer, Heidelberg (2011)

Discovering Block-Structured Process Models from Event Logs - A Constructive Approach

Sander J.J. Leemans, Dirk Fahland, and Wil M.P. van der Aalst

Department of Mathematics and Computer Science, Eindhoven University of Technology,
P.O. Box 513, NL-5600 MB, Eindhoven, The Netherlands

Abstract. Process discovery is the problem of, given a log of observed behaviour, finding a process model that 'best' describes this behaviour. A large variety of process discovery algorithms has been proposed. However, no existing algorithm guarantees to return a fitting model (i.e., able to reproduce all observed behaviour) that is sound (free of deadlocks and other anomalies) in finite time. We present an extensible framework to discover from any given log a set of block-structured process models that are sound and fit the observed behaviour. In addition we characterise the minimal information required in the log to rediscover a particular process model. We then provide a polynomial-time algorithm for discovering a sound, fitting, block-structured model from any given log; we give sufficient conditions on the log for which our algorithm returns a model that is language-equivalent to the process model underlying the log, including unseen behaviour. The technique is implemented in a prototypical tool.

Keywords: process discovery, block-structured process models, soundness, fitness.

1 Introduction

Process mining techniques aim to extract information from event logs. For example, the audit trails of a workflow management system or the transaction logs of an enterprise resource planning system can be used to discover models describing processes, organisations and products. The most challenging process mining problem is to learn a process model (e.g., a Petri net) from example traces in some event log. Many process discovery techniques have been proposed. For an overview of process discovery algorithms, we refer to [13]. Unfortunately, existing techniques may produce models that are unable to replay the log, may produce erroneous models and may have excessive run times.

Which process model is 'best' is typically defined with respect to several quality criteria. An important quality criterion is soundness. A process model is *sound* if and only if all process steps can be executed and some satisfactory end state is always reachable. In most use cases, an unsound process model can be discarded without considering the log that it should represent. Another model quality criterion is fitness. A model has perfect *fitness* with respect to a log if it can reproduce all traces in the log. The quality criterion *precision* expresses whether the model does not allow for too much behaviour, *generalisation* expresses that the model will allow future behaviour that is currently

J.-M. Colom and J. Desel (Eds.): PETRI NETS 2013, LNCS 7927, pp. 311–329, 2013.
© Springer-Verlag Berlin Heidelberg 2013

absent in the log.[10] Other model quality criteria exist, for which we refer to [23]. In this paper, we focus on soundness and fitness, as so far no existing discovery algorithm guarantees to return a sound fitting model in finite time.

In addition to finite run time, there are other desirable properties of process discovery algorithms. In reality, the log was produced by some real-life process. The original process is *rediscoverable* by a process discovery algorithm if, given a log that contains enough information, the algorithm returns a model that is equivalent to the original process using some equivalence notion. For instance, *language-rediscoverability* holds for an algorithm that returns a model that is language-equivalent to the original model - even if the log obtained from the original model contains less behaviour. *Isomorphic-rediscoverability* holds for an algorithm that returns a model that is isomorphic to (a representation of) the original model. The amount of information that is required to be in the log is referred to as *log completeness*, of which the most extreme case is total log completeness, meaning that all possible behaviour of the original process must be present in the log. A process discovery technique is only useful if it assumes a much weaker notion of completeness. In reality one will rarely see all possible behaviour.

Many process discovery algorithms [5,25,26,24,7,11,27,17,4,19,9,3] using different approaches have been proposed in the past. Some techniques guarantee fitness, e.g., [27], some guarantee soundness, e.g. [9], and others guarantee rediscoverability under some conditions, e.g., [5]. Yet, there is essentially no discovery algorithm guaranteeing to find a sound, fitting model in finite time for all given logs.

In this paper, we use the block-structured process models of [9,3] to introduce a framework that guarantees to return sound and fitting process models. This framework enables us to reason about a variety of quality criteria. The framework uses any flavour of block-structured process models: new blocks/operators can be added without changing the framework and with few proof obligations. The framework uses a divide and conquer approach to decompose the problem of discovering a process model for a log L into discovering n subprocesses of n sublogs obtained by splitting L. We explore the quality standards and hard theoretically founded limits of the framework by characterising the requirements on the log under which the original model can be rediscovered.

For illustrative purposes, we give an algorithm that uses the framework and runs in polynomial time for any log and any number of activities. The framework guarantees that the algorithm returns a sound fitting model. The algorithm works by dividing the activities of the log over a number of branches, such that the log can be split according to this division. We characterise the conditions under which the algorithm returns a model that is language-equivalent to the original process. The algorithm has been prototypically implemented using the ProM framework [12].

In the following, we first discuss related work. Section 3 explains logs, languages, Petri nets, workflow nets and process trees. In Section 4 the framework is described. The class of models that this framework can rediscover is described in Section 5. In Section 6 we give an algorithm that uses the framework and we report on experimental results. We conclude the paper in Section 7.

2 Related Work

A multitude of process discovery algorithms has been proposed in the past. We review typical representatives with respect to guarantees such as soundness, fitness, rediscoverability and termination. Techniques that discover process models from ordering relations of activities, such as the α algorithm [5] and its derivatives [25,26], guarantee isomorphic-rediscoverability for rather small classes of models [6] and do not guarantee fitness or soundness. Semantics-based techniques such as the language-based region miner [7,8], the state-based region miner [11], or the ILP miner [27] guarantee fitness but neither soundness nor rediscoverability. Frequency-based techniques such as the heuristics miner [24] guarantee neither soundness nor fitness. Abstraction-based techniques such as the Fuzzy miner [17] produce models that do not have executable semantics and hence guarantee neither soundness nor fitness nor any kind of rediscoverability.

Genetic process discovery algorithms [4,19] may reach certain quality criteria if they are allowed to run forever, but usually cannot guarantee any quality criterion given finite run time. A notable exception is a recent approach [9,3] that guarantees soundness. This approach restricts the search space to block-structured process models, which are sound by construction; however, finding a fitting model cannot be guaranteed in finite run time.

The Refined Process Structure Tree [21] is a parsing technique to find block structures in process models by which soundness can be checked [15], or an arbitrary model can be turned into a block-structured one (if possible) [20]. However, these techniques only analyse or transform a given model, but do not allow to construct a sound or fitting model. The language-based mining technique of [8] uses regular expressions to pre-structure the input language (the log) into smaller blocks; this block-structuring of the log is then used during discovery for constructing a fitting, though possibly unsound, process model.

Unsound models can be repaired to become sound by simulated annealing [16], though fitness to a given log is not preserved. Non-fitting models can be repaired to become fitting by adding subprocesses [14], though soundness is not guaranteed. Hence, a more integrated approach is needed to ensure soundness and fitness. In the following we will propose such an integrated approach building on the ideas of a restriction to block-structured models[3,9], and of decomposing the given log into block-structured parts prior to model construction.

3 Preliminaries

Logs. We assume the set of all process activities Σ to be given. An *event* e is the occurrence of an activity: $e \in \Sigma$. A *trace* t is a possibly empty sequence of events: $t \in \Sigma^*$. We denote the empty trace with ϵ. A *log* L is a finite non-empty set of traces: $L \subseteq \Sigma^*$. For example, $\{\langle a, b, c \rangle, \langle a, c, b \rangle\}$ denotes a log consisting of two traces abc and acb, where for instance abc denotes that first a occurred, then b and finally c. The *size* of a log is the number of events in it: $||L|| = \sum_{t \in L} |t|$.

Petri Nets, Workflow Nets and Block-structured Workflow Nets. A *Petri net* is a bipartite graph containing places and transitions, interconnected by directed arcs. A transition models a process activity, places and arcs model the ordering of process activities. We assume the standard semantics of Petri nets here, see [22]. A *workflow net* is a Petri net having a single start place and a single end place, modeling the start and end state of a process. Moreover, all nodes are on a path from start to end[5]. A *block-structured workflow net* is a hierarchical workflow net that can be divided recursively into parts having single entry and exit points. Figure 1 shows a block-structured workflow net.

Process Trees. A *process tree* is a compact abstract representation of a block-structured workflow net: a rooted tree in which leaves are labeled with activities and all other nodes are labeled with operators. A process tree describes a language, an operator describes how the languages of its subtrees are to be combined.

We formally define process trees recursively. We assume a finite alphabet Σ of activities and a set $\bigoplus$ of operators to be given. Symbol $\tau \notin \Sigma$ denotes the silent activity.

– a with $a \in \Sigma \cup \{\tau\}$ is a process tree;
– Let $M_1, \ldots, M_n$ with $n > 0$ be process trees and let $\oplus$ be a process tree operator, then $\oplus(M_1, \ldots, M_n)$ is a process tree.

There are a few standard operators that we consider in the following: operator $\times$ means the exclusive choice between one of the subtrees, $\rightarrow$ means the sequential execution of all subtrees, $\circlearrowleft$ means the structured loop of loop body M_1 and alternative loop back paths $M_2, \ldots, M_n$, and $\wedge$ means a parallel (interleaved) execution as defined below. Please note that for $\circlearrowleft$, n must be ≥ 2.

To describe the semantics of process trees, we define the language of a process tree M as a recursive monotonic function $\mathcal{L}(M)$, using for each operator $\oplus$ a language join function $\oplus_l$:

$$\mathcal{L}(a) = \{\langle a \rangle\} \text{ for } a \in \Sigma$$
$$\mathcal{L}(\tau) = \{\epsilon\}$$
$$\mathcal{L}(\oplus(M_1, \ldots, M_n)) = \oplus_l(\mathcal{L}(M_1), \ldots, \mathcal{L}(M_n))$$

Each operator $\oplus$ has its own language join function $\oplus_l$. Each function takes several logs and produces a new log: $\oplus_l : 2^{\Sigma^*} \times \ldots \times 2^{\Sigma^*} \rightarrow 2^{\Sigma^*}$.

$$\times_l(L_1, \ldots, L_n) = \bigcup_{1 \leq i \leq n} L_i$$
$$\rightarrow_l(L_1, \ldots, L_n) = \{t_1 \cdot t_2 \cdots t_n | \forall i \in 1 \ldots n : t_i \in L_i\}$$
$$\circlearrowleft_l(L_1, \ldots, L_n) = \{t_1 \cdot t_1' \cdot t_2 \cdot t_2' \cdots t_m | \forall i : t_i \in L_1 \wedge t_i' \in \bigcup_{2 \leq j \leq n} L_j\}$$

To characterise $\wedge$, we introduce a set notation $\{t_1, \ldots, t_n\}_{\simeq}$ that interleaves the traces $t_1 \ldots t_n$. We need a more complex notion than a standard projection function due to overlap of activities over traces.

$$t \in \{t_1, \ldots, t_n\}_{\simeq} \Leftrightarrow \exists (f : \{1 \ldots |t|\} \rightarrow \{(j, k) | j \leq n \wedge k \leq |t_j|\}) :$$
$$\forall i_1 < i_2 \wedge f(i_1) = (j, k_1) \wedge f(i_2) = (j, k_2) : k_1 < k_2 \wedge$$
$$\forall i \leq n \wedge f(i) = (j, k) : t(i) = t_j(k)$$

where f is a bijective function mapping each event of t to an event in one of the t_i and $t(i)$ is the ith element of t. For instance, $\langle a, c, d, b \rangle \in \{\langle a, b \rangle, \langle c, d \rangle\}_{\simeq}$. Using this notation, we define $\wedge_1$:

$$\wedge_1(L_1, \ldots, L_n) = \{t | t \in \{t_1, \ldots, t_n\}_{\simeq} \wedge \forall i : t_i \in L_i\}$$

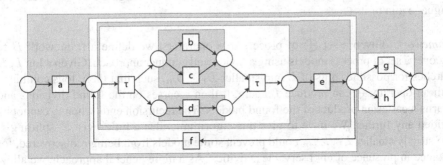

Fig. 1. A Petri net, modified from [2, page 196]. The rectangle regions denote the process tree nodes in $\rightarrow(a, \circlearrowleft(\rightarrow(\wedge(\times(b, c), d), e), f), \times(g, h))$.

Each of the process tree operators has a straightforward formal translation to a sound, block-structured workflow Petri net [9,3]. For instance, the Petri net shown in Figure 1 corresponds to the process tree $\rightarrow(a, \circlearrowleft(\rightarrow(\wedge(\times(b, c), d), e), f), \times(g, h))$. If one would come up with another process tree operator, soundness of the translation follows if the translation of the new process tree operator is sound in isolation. The four operators presented here translate to well-structured, free-choice Petri nets [1]; other operators might not.

The *size* of a model M is the number of nodes in M and is denoted as $|M|$: $|\tau| = 1$, $|a| = 1$ and $|\oplus(M_1, \ldots, M_n)| = 1 + \sum_i |M_i|$. Two process trees $M = \oplus(M_1, \ldots, M_n)$ and $M' = \oplus'(M_1', \ldots, M_n')$ are *isomorphic* if and only if they are syntactically equivalent up to reordering of children in the case of $\times$, $\wedge$ and the non-first children of $\circlearrowleft$.

If M is a process tree and L is a log, then L *fits* M if and only if every trace in L is in the language of M: $L \subseteq \mathcal{L}(M)$. A *flower model* is a process tree that can produce any sequence of Σ. An example of a flower model is the model $\circlearrowleft(\tau, a_1, \ldots, a_m)$ where $a_1, \ldots, a_m = \Sigma$.

As additional notation, we write $\Sigma(L)$ and $\Sigma(M)$ for the activities occurring in log L or model M respectively, not including τ. Furthermore, $Start(L)$, $Start(M)$ and $End(L)$, $End(M)$ denote the sets of activities with which log L and model M start or end.

4   Framework

In this section, we introduce a highly generic process discovery framework. This framework allows for the derivation of various process discovery algorithms with predefined guarantees. Then we prove that each model returned by the framework fits the log and that the framework describes a finite computation, both for any set of process tree operators having a corresponding monotonic language join function.

Requirement on the Process Tree Operators. The framework works independently of the chosen process tree operators. The only requirement is that each operator $\oplus$ must have a sensible language join function $\oplus_l$, such that the language of $\oplus$ reflects the language join of its $\oplus_l$.

Framework. Given a set $\oplus$ of process tree operators, we define a framework B to discover a set of process models using a divide and conquer approach. Given a log L, B searches for possible splits of L into smaller $L_1 \ldots L_n$, such that these logs combined with an operator $\oplus$ can produce L again. It then recurses on the found divisions and returns a cartesian product of the found models. The recursion ends when L cannot be divided any further. We have to give this algorithmic idea a little twist as splitting L into strictly smaller $L_1 \ldots L_n$ could prevent some models from being rediscovered, for instance in presence of unobservable activities. As a more general approach, we allow L to be split into sublogs having the same size as L. However, such splits that do not decrease the size of L may only happen finitely often. For this, we introduce a counter parameter ϕ, which has to decrease if a non-decreasing log split is made. Parameter ϕ essentially bounds the number of invisible branches that the discovered model can have.

The actual split is left to a function $select(L)$ that takes a log and returns a set of tuples $(\oplus, ((L_1, \phi_1), \ldots, (L_n, \phi_n)))$, where $\oplus$ is the operator identified to split L, and $L_1 \ldots L_n$ are the logs obtained by splitting L w.r.t. $\oplus$. Each log L_i has a corresponding counter parameter ϕ which bounds recursion on L_i. Various $select(L)$ functions can be defined, so we parameterise the framework B with this $select$ function.

function $B_{select}(L, \phi)$
    if $L = \{\epsilon\}$ **then**
        $base \leftarrow \{\tau\}$
    else if $\exists a \in \Sigma : L = \{\langle a \rangle\}$ **then**
        $base \leftarrow \{a\}$
    else
        $base \leftarrow \emptyset$
    end if
    $P \leftarrow select(L)$
    if $|P| = 0$ **then**
        if $base = \emptyset$ **then**
            return $\{\circlearrowright(\tau, a_1, \ldots, a_m) \text{ where } \{a_1, \ldots, a_m\} = \Sigma(L)\}$
        else
            return $base$
        end if
    end if
    return $\{\oplus(M_1, \ldots, M_n) | (\oplus, ((L_1, \phi_1), \ldots, (L_n, \phi_n))) \in P \wedge \forall i : M_i \in B(L_i, \phi_i)\} \cup base$
end function

Any $select$ function can be used, as long as the tuples it returns adhere to the following definition:

Definition 1. *For each tuple* $(\oplus, ((L_1, \phi_1), \ldots, (L_n, \phi_n)))$ *that select(L) returns, it must hold that*

$$L \subseteq \oplus_l(L_1, \ldots, L_n) \wedge$$
$$\forall i : ||L_i|| + \phi_i < ||L|| + \phi \wedge$$
$$\forall i : ||L_i|| \leq ||L|| \wedge$$
$$\forall i : \phi_i \leq \phi \wedge$$
$$\forall i : \Sigma(L_i) \subseteq \Sigma(L) \wedge$$
$$\oplus \in \bigoplus \wedge$$
$$n \leq ||L|| + \phi$$

In the remainder of this section, we will prove some properties that do not depend on a specific preference function *select*.

Theorem 2. *Assuming select terminates, B terminates.*

Proof. Termination follows from the fact that in each recursion, $||L|| + \phi$ gets strictly smaller and that there are finitely many recursions from a recursion step. By construction of *select*, $\Sigma(L_i) \subseteq \Sigma(L)$, and therefore Σ is finite. By construction of P, $n \leq ||L|| + \phi$, so there are finitely many sublogs L_i. Hence, *select* creates finitely many log divisions. Therefore, the number of recursions is finite and hence B terminates. □

Theorem 3. *Let $\bigoplus$ be a set of operators and let L be a log. Then $B(L)$ returns at least one process tree and all process trees returned by $B(L)$ fit L.*

Proof. Proof by induction on value of $||L|| + \phi$. Base cases: $||L|| + \phi = 1$ or $||L|| + \phi = 2$. Then, L is either $\{\epsilon\}$ or $\{\langle a \rangle\}$. By code inspection, for these L B returns at least one process tree and all process trees fit L.
Induction hypothesis: for all logs $||L'|| + \phi'$ smaller than $||L|| + \phi$, $B(L', \phi')$ returns at least one process tree and all process trees that B returns fit L' : $\forall ||L'|| + \phi' < ||L|| + \phi$: $|B(L')| \geq 1 \wedge \forall M' \in B(L') : L' \subseteq \mathcal{L}(M')$.
Induction step: assume $||L|| + \phi > 2$ and the induction hypothesis. Four cases apply:

- Case $L = \{\epsilon\}$, see base case;
- Case $L = \{\langle a \rangle\}$, see base case;
- Case P is empty, $L \neq \{\epsilon\}$ and $L \neq \{a\}$. Then B returns the flower model $\{\circlearrowright(\tau, a_1, \ldots, a_m)$ where $a_1, \ldots, a_m = \Sigma(L)\}$ and that fits any log.
- Case P is nonempty, $L \neq \{\epsilon\}$ and $L \neq \{a\}$. Let $M_1, \ldots, M_n$ be models returned by $B(L_1, \phi_1), \ldots, B(L_n, \phi_n)$ for some logs $L_1, \ldots, L_n$ and some counters $\phi_1, \ldots, \phi_n$. By construction of the P-selection step, $\forall i : ||L_i|| + \phi_i < ||L|| + \phi$. By the induction hypothesis, these models exist. As B combines these models in a cartesian product, $|B(L)| \geq 1$. By the induction hypothesis, $\forall i : L_i \subseteq \mathcal{L}(M_i)$. Using the fact that $\oplus_l$ is monotonic and the construction of M, we obtain $\oplus_l(L_1, \ldots, L_n) \subseteq \mathcal{L}(\oplus(M_1, \ldots, M_n)) = \mathcal{L}(M)$. By construction of P, $L \subseteq \oplus_l(L_1, \ldots, L_n)$, and by $\oplus_l(L_1, \ldots, L_n) \subseteq L(M)$, we conclude that $L \subseteq \mathcal{L}(M)$. We did not pose any restrictions on $M_1, \ldots, M_n$, so this holds for all combinations of $M_1, \ldots, M_n$ from the sets returned by B. □

5   Rediscoverability of Process Trees

An interesting property of a discovery algorithm is whether and under which assumptions an original process can be rediscovered by the algorithm. Assume the original process is expressible as a model M, which is unknown to us. Given is a log L of M: $L \subseteq \mathcal{L}(M)$. M is isomorphic-rediscoverable from L by algorithm B if and only if $M \in B(L)$. It is desirable that L can be as small as possible to rediscover M. In this section, we explore the boundaries of the framework B of Section 4 in terms of rediscoverability.

We first informally give the class of original processes that can be rediscovered by B, and assumptions on the log under which this is guaranteed. After that, we give an idea why these suffice to rediscover the model. In this section, the preference function *select* as used in B is assumed to be the function returning all log divisions satisfying Definition 1. Otherwise, the original model might be removed from the result set.

Class of Rediscoverable Models.   Any algorithm has a representational bias; B can only rediscover processes that can be described by process trees. There are no further limitations: B can rediscover every process tree. An intuitive argument for this claim is that as long as the log can be split into the parts of which the log was constructed, the algorithm will also make this split and recurse. A necessity for this is that the log contains 'enough' behaviour.

Log Requirements.   All process trees can be rediscovered given 'enough' traces in the log, where enough means that the given log can be split according to the respective process tree operator. Intuitively, it suffices to execute each occurrence of each activity in M at least once in L. Given a large enough ϕ, B can then always split the log correctly.

This yields the notion of *activity-completeness*. Log L is activity-complete w.r.t. model M, denote $L \diamond_a M$, if and only if each leaf of M appears in L at least once. Formally, we have to distinguish two cases. For a model M' where each activity occurs at most once and a log L',

$$L' \diamond_a M' \Leftrightarrow \Sigma(M') \subseteq \Sigma(L')$$

In the general case, where some activity $a \in \Sigma$ occurs more than once in M, we have to distinguish the different occurrences. For a given alphabet Σ consider a refined alphabet Σ' and a surjective function $f : \Sigma' \to \Sigma$, e.g., $\Sigma' = \{a_1, a_2, \ldots, b_1, b_2, \ldots\}$ and $a = f(a_1) = f(a_2) = \cdots, b = f(b_1) = f(b_2) = \cdots$, etc. For a log L' and model M' over Σ', let $f(L')$ and $f(M')$ denote the log and the model obtained by replacing each $a \in \Sigma'$ by $f(a) \in \Sigma$. Using this notation, we define for arbitrary log L and model M,

$$L \diamond_a M \Leftrightarrow \exists \Sigma', (f' : \Sigma' \to \Sigma), M', L' : f(L') = L \wedge f(M') = M \wedge L' \diamond_a M',$$

where each activity $a \in \Sigma'$ occurs at most once in M'.

Rediscoverability of Models. In order to prove isomorphic rediscoverability, we need to show that any log $L \diamond_a M$ can be split by B such that M can be constructed, given a large enough ϕ.

Theorem 4. *Given a large enough ϕ, for each log L and model M such that $L \subseteq \mathcal{L}(M)$ and $L \diamond_a M$ it holds that $M \in B(L)$.*

Proof. Proof by induction on model sizes. Base case: $|M| = 1$. A model of size 1 consists of a single leaf l. By $L \subseteq \mathcal{L}(M) \wedge L \diamond_a M$, L is $\{l\}$. These are handled by the $L = \{\epsilon\}$ or $L = \{\langle a \rangle\}$ clauses and hence can be rediscovered.

Induction hypothesis: all models smaller than M can be rediscovered: $\forall |M'| < |M| \wedge L' \subseteq \mathcal{L}(M) \wedge L' \diamond_a M' : M' \in B(L', \phi')$, for some number ϕ'.

Induction step: assume $|M| > 1$ and the induction hypothesis. As $|M| > 1$, $M = \oplus(M_1, \ldots, M_n)$ for certain $\oplus$, n and $M_1 \ldots M_n$. By $L \subseteq \mathcal{L}(M)$ and definition of $\mathcal{L}(M)$, there exist $L_1 \ldots L_n$ such that $\forall i : L_i \subseteq \mathcal{L}(M_i)$, $\forall i : L_i \diamond_a M_i$ and $L \subseteq \oplus_l(L_1, \ldots, L_n)$. By the induction hypothesis there exist $\phi_1 \ldots \phi_n$ such that $\forall i : M_i \in B(L_i, \phi_i)$. We choose ϕ to be large enough by taking $\phi = \max\{n, \phi_1 + 1, \ldots, \phi_n + 1\}$. By this choice of ϕ,

$$\forall i : ||L_i|| + \phi_i < ||L|| + \phi \wedge \phi_i \leq \phi$$

and

$$n \leq ||L|| + \phi$$

hold. By construction of $\oplus_l$,

$$\forall i : ||L_i|| \leq ||L||$$

By $|M| > 1$ and our definitions of $\times_l$, $\rightarrow_l$, $\wedge_l$ and $\circlearrowleft_l$, L does not introduce new activities:

$$\forall i : \Sigma(L_i) \subseteq \Sigma(L)$$

Hence, $(\oplus, ((L_1, \phi_1), \ldots, (L_n, \phi_n))) \in P$. By the induction hypothesis, $\forall M_i : M_i \in B(L_i)$. The models returned by $B(L_i)$ will be combined using a cartesian product, and as $M = \oplus(M_1, \ldots, M_n)$, it holds that $M \in B(L)$. $\qquad\qquad\square$

This proof shows that it suffices to pick ϕ to be the sum of the width and depth of the original model M in order to rediscover M from an activity-complete log L.

6 Discovering Process Trees Efficiently

The framework of Section 4 has a practical limitation: for most real-life logs, it is infeasible to construct the full set P. In this section, we introduce an algorithm B' that is a refinement of the framework B. B' avoids constructing the complete set P. The central idea of B' is to compute a log split directly based on the ordering of activities in the log. We first introduce the algorithmic idea and provide formal definitions afterwards. We conclude this section with a description of the classes of process trees that B' is able to language-rediscover and a description of our prototype implementation.

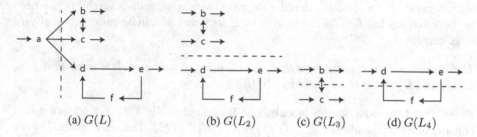

| (a) $G(L)$ | (b) $G(L_2)$ | (c) $G(L_3)$ | (d) $G(L_4)$ |

Fig. 2. Several directly-follows graphs. Dashed lines denote cuts.

6.1 Algorithmic Idea

The directly-follows relation, also used by the α-algorithm [5], describes when two activities directly follow each other in a process. This relation can be expressed in the *directly-follows graph* of a log L, written $G(L)$. It is a directed graph containing as nodes the activities of L. An edge (a, b) is present in $G(L)$ if and only if some trace $\langle \cdots, a, b, \cdots \rangle$ exists in L. A node of $G(L)$ is a *start node* if its activity is in $Start(L)$. We define $Start(G(L)) = Start(L)$. Similarly for end nodes in $End(L)$, and $End(G(L))$. The definition for $G(M)$ is similar. For instance, Figure 2a shows the directly-follows graph of log $L = \{\langle a, b, c\rangle, \langle a, c, b\rangle, \langle a, d, e\rangle, \langle a, d, e, f, d, e\rangle\}$.

The idea for our algorithm is to find in $G(L)$ structures that indicate the 'dominant' operator that orders the behaviour. For example, $G(L)$ of Fig 2a can be partitioned into two sets of activities as indicated by the dashed line such that edges cross the line only from left to right. This pattern corresponds to a sequence where the activities left of the line precede the activities right of the line. This is the decisive hint on how to split a given log when using the framework of Section 4. Each of the four operators $\times$, $\rightarrow$, $\circlearrowright$, $\wedge$ has a characteristic pattern in $G(L)$ that can be identified by finding a partitioning of the nodes of $G(L)$ into n sets of nodes with characteristic edges in between. The log L can then be split according to the identified operator, and the framework recurses on each of the split logs. The formal definitions are provided next.

6.2 Cuts and Components

Let $G(L)$ be the directly-follows graph of a log L. An *n-ary cut* c of $G(L)$ is a partition of the nodes of the graph into disjoint sets $\Sigma_1 \ldots \Sigma_n$. We characterise a different cut for each operator $\times$, $\rightarrow$, $\circlearrowright$, $\wedge$ based on edges between the nodes.

In a *exclusive choice cut*, each Σ_i has a start node and an end node, and there is no edge between two different $\Sigma_i \neq \Sigma_j$, as illustrated by Figure 3(left). In a *sequence cut*, the sets $\Sigma_1 \ldots \Sigma_n$ are ordered such that for any two nodes $a \in \Sigma_i, b \in \Sigma_j, i < j$, there is a path from a to b along the edges of $G(L)$, but not vice versa; see Figure 3(top). In a *parallel cut*, each Σ_i has a start node and an end node, and any two nodes $a \in \Sigma_i, b \in \Sigma_j, i \neq j$ are connected by edges (a, b) and (b, a); see Figure 3(bottom). In a *loop cut*, Σ_1 has all start and all end nodes of $G(L)$, there is no edge between nodes of different $\Sigma_i \neq \Sigma_j, i, j > 1$, and any edge between Σ_1 and $\Sigma_i, i > 1$ either leaves an end node of

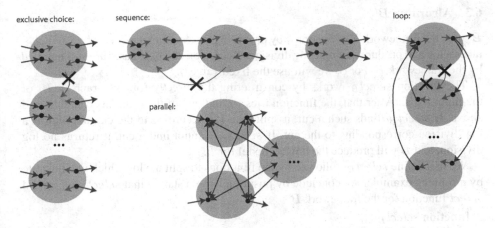

Fig. 3. Cuts of the directly-follows graph for operators $\times$, $\rightarrow$, $\wedge$ and $\circlearrowleft$

Σ_1 or reaches a start node of Σ_1; see Figure 3(right). An n-ary cut is *maximal* if there exists no cut of G of which n is bigger. A cut c is *nontrivial* if $n > 1$.

Let $a \rightsquigarrow b \in G$ denote that there exists a directed edge chain (path) from a to b in G. Definitions 5, 6, 7 and 8 show the formal cut definitions.

Definition 5. *An* exclusive choice cut *is a cut* $\Sigma_1, \ldots, \Sigma_n$ *of a directly-follows graph* G, *such that*

1. $\forall i \neq j \wedge a_i \in \Sigma_i \wedge a_j \in \Sigma_j : (a_i, a_j) \notin G$

Definition 6. *A* sequence cut *is an ordered cut* $\Sigma_1, \ldots, \Sigma_n$ *of a directly-follows graph* G *such that*

1. $\forall 1 \leq i < j \leq n \wedge a_i \in \Sigma_i \wedge a_j \in \Sigma_j : a_j \rightsquigarrow a_i \notin G$
2. $\forall 1 \leq i < j \leq n \wedge a_i \in \Sigma_i \wedge a_j \in \Sigma_j : a_i \rightsquigarrow a_j \in G$

Definition 7. *A* parallel cut *is a cut* $\Sigma_1, \ldots, \Sigma_n$ *of a directly-follows graph* G *such that*

1. $\forall i : \Sigma_i \cap Start(G) \neq \emptyset \wedge \Sigma_i \cap End(G) \neq \emptyset$
2. $\forall i \neq j \wedge a_i \in \Sigma_i \wedge a_j \in \Sigma_j : (a_i, a_j) \in G \wedge (a_j, a_i) \in G$

Definition 8. *A* loop cut *is a partially ordered cut* $\Sigma_1, \ldots, \Sigma_n$ *of a directly-follows graph* G *such that*

1. $Start(G) \cup End(G) \subseteq \Sigma_1$
2. $\forall i \neq 1 \wedge a_i \in \Sigma_i \wedge a_1 \in \Sigma_1 : (a_1, a_i) \in G \Rightarrow a_1 \in End(G)$
3. $\forall i \neq 1 \wedge a_i \in \Sigma_i \wedge a_1 \in \Sigma_1 : (a_i, a_1) \in G \Rightarrow a_1 \in Start(G)$
4. $\forall 1 \neq i \neq j \neq 1 \wedge a_i \in \Sigma_i \wedge a_j \in \Sigma_j : (a_i, a_j) \notin G$
5. $\forall i \neq 1 \wedge a_i \in \Sigma_i \wedge a_1 \in Start(G) : (\exists a_1' \in \Sigma_1 : (a_i, a_1') \in G) \Leftrightarrow (a_i, a_1) \in G$
6. $\forall i \neq 1 \wedge a_i \in \Sigma_i \wedge a_1 \in End(G) : (\exists a_1' \in \Sigma_1 : (a_1', a_i) \in G) \Leftrightarrow (a_1, a_i) \in G$

6.3   Algorithm B'

B' uses the framework B of Section 4 by providing a *select* function $select_{B'}$. $select_{B'}$ takes a log and produces a single log division. Recursion and base cases are still handled by the framework B. For ϕ, we will use the fixed value 0.

The function $select_{B'}$ works by constructing the directly follows graph $G(L)$ of the input log L. After that, the function tries to find one of the four cuts characterised above. If $select_{B'}$ finds such a cut, it splits the log according to the cut and returns a log division corresponding to the cut. If $select_{B'}$ cannot find a cut, it returns no log division, and B will produce the flower model for L.

We first define $select_{B'}$ followed by the functions to split the log which we illustrate by a running example. We conclude by posing a lemma stating that $select_{B'}$ is a valid *select* function for the framework B.

> **function** $select_{B'}(L)$
>   **if** $\epsilon \in L \vee \exists a \in \Sigma(L) : L = \{\langle a \rangle\}$ **then**
>   **return** $\emptyset$
>   **else if** $c \leftarrow$ a nontrivial maximal exclusive choice cut c of $G(L)$ **then**
> $\Sigma_1, \ldots, \Sigma_n \leftarrow c$
> $L_1, \ldots, L_n \leftarrow \text{EXCLUSIVECHOICESPLIT}(L, (\Sigma_1, \ldots, \Sigma_n))$
>   **return** $\{(\times, ((L_1, 0), \ldots, (L_n, 0)))\}$
>   **else if** $c \leftarrow$ a nontrivial maximal sequence cut c of $G(L)$ **then**
> $\Sigma_1, \ldots, \Sigma_n \leftarrow c$
> $L_1, \ldots, L_n \leftarrow \text{SEQUENCESPLIT}(L, (\Sigma_1, \ldots, \Sigma_n))$
>   **return** $\{(\rightarrow, ((L_1, 0), \ldots, (L_n, 0)))\}$
>   **else if** $c \leftarrow$ a nontrivial maximal parallel cut c of $G(L)$ **then**
> $\Sigma_1, \ldots, \Sigma_n \leftarrow c$
> $L_1, \ldots, L_n \leftarrow \text{PARALLELSPLIT}(L, (\Sigma_1, \ldots, \Sigma_n))$
>   **return** $\{(\wedge, ((L_1, 0), \ldots, (L_n, 0)))\}$
>   **else if** $c \leftarrow$ a nontrivial maximal loop cut c of $G(L)$ **then**
> $\Sigma_1, \ldots, \Sigma_n \leftarrow c$
> $L_1, \ldots, L_n \leftarrow \text{LOOPSPLIT}(L, (\Sigma_1, \ldots, \Sigma_n))$
>   **return** $\{(\circlearrowright, ((L_1, 0), \ldots, (L_n, 0)))\}$
>   **end if**
>   **return** $\emptyset$
> **end function**

Using the cut definitions, $select_{B'}$ divides the activities into sets $\Sigma_1 \ldots \Sigma_n$. After that, $select_{B'}$ splits the log.

The cuts can be computed efficiently using graph techniques. We will give an intuition: the exclusive choice cut corresponds to the notion of connected components. If we collapse both strongly connected components and pairwise unreachable nodes into single nodes, the collapsed nodes that are left are the Σs of the sequence cut. If both of these cuts are not present, then we "invert" the graph by removing every double edge, and adding double edges where there was no or a single edge present. In the resulting graph each connected component is a Σ_i of the parallel cut. If these cuts are not present, temporarily removing the start and end activities and computing the connected components in the resulting graph roughly gives us the loop cut. As shown in Lemma 16 in [18], the order in which the cuts are searched for is arbitrary, but for ease of proof and computation we assume it to be fixed as described in $select_{B'}$.

We define the log split functions together with a running example.

Consider the log $L = \{\langle a, b, c\rangle, \langle a, c, b\rangle, \langle a, d, e\rangle, \langle a, d, e, f, d, e\rangle\}$. $G(L)$ is shown in Figure 2a which has the sequence cut $\{a\}, \{b, c, d, e, f\}$. The log is then split by projecting each trace of L onto the different activity sets of the cut.

function SEQUENCESPLIT$(L, (\Sigma_1, \ldots, \Sigma_n))$
$\qquad \forall j : L_j \leftarrow \{t_j | t_1 \cdot t_2 \cdots t_n \in L \wedge \forall i \leq n \wedge e \in t_i : e \in \Sigma_i\}$
$\qquad$ **return** $L_1, \ldots, L_n$
end function

In the example, SEQUENCESPLIT$(L, (\{a\}, \{b, c, d, e, f\})) =$ $\{\langle a\rangle\}, \{\langle b, c\rangle, \langle c, b\rangle, \langle d, e\rangle, \langle d, e, f, d, e\rangle\}$. Call the second log L_2. $G(L_2)$ is shown in Figure 2b and has the exclusive choice cut $\{b, c\}, \{d, e, f\}$. The log is then split by moving each trace of L into the log of the corresponding activity set.

function EXCLUSIVECHOICESPLIT$(L, (\Sigma_1, \ldots, \Sigma_n))$
$\qquad \forall i : L_i \leftarrow \{t | t \in L \wedge \forall e \in t : e \in \Sigma_i\}$
$\qquad$ **return** $L_1, \ldots, L_n$
end function

In the example, EXCLUSIVECHOICESPLIT$(L_2, (\{b, c\}, \{d, e, f\})) = \{\langle b, c\rangle, \langle c, b\rangle\}$, $\{\langle d, e\rangle, \langle d, e, f, d, e\rangle\}$. Call the first log L_3 and the second log L_4. $G(L_3)$ is shown in Figure 2c and has the the parallel cut $\{b\}, \{c\}$. The log is split by projecting each trace for each activity set in the cut.

function PARALLELSPLIT$(L, (\Sigma_1, \ldots, \Sigma_n))$
$\qquad \forall i : L_i \leftarrow \{t|_{\Sigma_j} | t \subset L\}$
$\qquad$ **return** $L_1, \ldots, L_n$
end function

where $t|_X$ is a function that projects trace t onto set of activities X, such that all events remaining in $t|_X$ are in X. In our example, PARALLELSPLIT$(L_3, (\{b\}, \{c\})) =$ $\{\langle b\rangle\}, \{\langle c\rangle\}$. The directly-follows graph of the log $L_4 = \{\langle d, e\rangle, \langle d, e, f, d, e\rangle\}$ is shown in Figure 2d and has the loop cut $\{d, e\}, \{f\}$. The log is split by splitting each trace into subtraces of the loop body and of the loopback condition which are then added to the respective sublogs.

function LOOPSPLIT$(L, (\Sigma_1, \ldots, \Sigma_n))$
$\qquad \forall i : L_i \leftarrow \{t_2 | t_1 \cdot t_2 \cdot t_3 \in L \wedge$
$\qquad\qquad\qquad \Sigma(\{t_2\}) \subseteq \Sigma_i \wedge$
$\qquad\qquad\qquad (t_1 = \epsilon \vee (t_1 = \langle \cdots, a_1\rangle \wedge a_1 \notin \Sigma_i)) \wedge$
$\qquad\qquad\qquad (t_3 = \epsilon \vee (t_3 = \langle a_3, \cdots\rangle \wedge a_3 \notin \Sigma_i))\}$
$\qquad$ **return** $L_1, \ldots, L_n$
end function

In our example, LOOPSPLIT$(L_4, (\{d, e\}, \{f\})) = \{\langle d, e\rangle\}, \{\langle f\rangle\}$.

Framework B and $select_{B'}$ together discover a process model from the log $L =$ $\{\langle a, b, c\rangle, \langle a, c, b\rangle, \langle a, d, e\rangle, \langle a, d, e, f, d, e\rangle\}$ as follows. The only exclusive choice cut for $G(L)$ is $\{a, b, c, d, e, f\}$, which is a trivial cut. As we have shown before, a sequence cut for $G(L)$ is $\{a\}, \{b, c, d, e, f\}$. Then, $select_{B'}$ calls SEQUENCESPLIT, which returns two sublogs: $L_1 = \{\langle a \rangle\}$ and $L_2 = \{\langle b, c \rangle, \langle c, b \rangle, \langle d, e \rangle, \langle d, e, f, d, e \rangle\}$. Then, $select_{B'}$ returns $\{\rightarrow, (L_1, L_2)\}$. After that, B constructs the partial model $M = \rightarrow(B(L_1), B(L_2))$ and recurses.

Let us first process the log L_1. $B(L_1)$ sets $base$ to $\{a\}$, and $select_{B'}(L_1)$ returns $\emptyset$. Then, B returns the process tree a, with which the partially discovered model becomes $M = \rightarrow(a, B(L_2))$.

For $B(L_2)$, EXCLUSIVECHOICESPLIT splits the log in $L_3 = \{\langle b, c \rangle, \langle c, b \rangle\}$ and $L_4 = \{\langle d, e \rangle, \langle d, e, f, d, e \rangle\}$. The partially discovered model then becomes $M = \rightarrow(a, \times(B(L_3), B(L_4)))$.

For $B(L_3)$, there is no nontrivial exclusive choice cut and neither a nontrivial sequence cut. As we have shown before, PARALLELSPLIT splits L_3 into $L_5 = \{\langle b \rangle\}$ and $L_6 = \{\langle c \rangle\}$. M becomes $\rightarrow(a, \times(\wedge(B(L_5), B(L_6)), B(L_4)))$.

For $B(L_4)$, LOOPSPLIT splits L_4 into $L_7 = \{\langle d, e \rangle\}$ and $L_8 = \{\langle f \rangle\}$, such that M becomes $\rightarrow(a, \times(\wedge(B(L_5), B(L_6)), \circlearrowleft(L_7, L_8)))$.

After one more sequence cut ($B(L_7)$) and a few base cases ($B(L_5), B(L_6), B(L_7)$), B' discovers the model $\rightarrow(a, \times(\wedge(b, c), \circlearrowleft(\rightarrow(d, e), f)))$.

As B' uses a $select$ function to use the framework, we need to prove that $select_{B'}$ only produces log divisions that satisfy Definition 1.

Lemma 9. *The log divisions of L that $select_{B'}$ returns adhere to Definition 1.*

The proof idea of this lemma is to show that each of the clauses of Definition 1 holds for the log division $L_1 \ldots L_n$ that $select_{B'}$ chooses, using a fixed ϕ of 0: $L \subseteq \oplus_l(L_1, \ldots, L_n), \forall i : \|L_i\| < \|L\|, \forall i : \Sigma(L_i) \subseteq \Sigma(L), \oplus \in \bigoplus$ and $n \leq \|L\|$. For the detailed proof of this lemma, please refer to [18].

6.4   Language-Rediscoverability

An interesting property of a discovery algorithm is whether and under which assumptions a model can be discovered that is language-equivalent to the original process. It can easily be inductively proven that B' returns a single process tree for any log L. B' language-rediscovers a process model if and only if the mined process model is language-equivalent to the original process model that produced the log: $\mathcal{L}(M) = \mathcal{L}(B'(L))$ (we abuse notation a bit here), under the assumption that L is complete w.r.t. M for some completeness notion. Our proof strategy for language-rediscoverability will be to reduce each process tree to a normal form and then prove that B' isomorphically rediscovers this normal form. We first define the log completeness notion, after which we describe the class of models that B' can language-rediscover. We conclude with a definition of the normal form and the proof.

Log Completeness. Earlier, we introduced the notion of a directly-follows graph. This yields the notion of *directly-follows completeness* of a log L with respect to a model M, written as $L \diamond_{df} M$: $L \diamond_{df} M \equiv \langle \cdots, a, b, \cdots \rangle \in \mathcal{L}(M) \Rightarrow \langle \cdots, a, b, \cdots \rangle \in L \wedge Start(M) \subseteq Start(L) \wedge End(M) \subseteq End(L) \wedge \Sigma(M) \subseteq \Sigma(L)$. Intuitively, the directly-follows graphs M must be mappable on the directly-follows graph of L.

Please note that the framework does not require the log to be directly-follows complete in order to guarantee soundness and fitness.

Class of Language-Rediscoverable Models. Given a model M and a generated complete log L, we prove language-rediscoverability assuming the following model restrictions, where $\oplus(M_1, \ldots, M_n)$ is a node at any position in M:

1. Duplicate activities are not allowed: $\forall i \neq j : \Sigma(M_i) \cap \Sigma(M_j) = \emptyset$.
2. If $\oplus = \circlearrowleft$, the sets of start and end activities of the first branch must be disjoint: $\oplus = \circlearrowleft \Rightarrow Start(M_1) \cap End(M_1) = \emptyset$.
3. No τ's are allowed: $\forall i \leq n : M_i \neq \tau$.

A reader familiar with the matter will have recognised the restrictions as similar to the rediscoverability restrictions of the α algorithm [5].

Normal Form. We first introduce reduction rules on process trees that transform an arbitrary process tree into a normal form. The intuitive idea of these rules is to combine multiple nested subtrees with the same operator into one node with that operator.

Property 10.

$$\oplus(M) = M$$
$$\times(\cdots_1, \times(\cdots_2), \cdots_3) = \times(\cdots_1, \cdots_2, \cdots_3)$$
$$\rightarrow(\cdots_1, \rightarrow(\cdots_2), \cdots_3) = \rightarrow(\cdots_1, \cdots_2, \cdots_3)$$
$$\wedge(\cdots_1, \wedge(\cdots_2), \cdots_3) = \wedge(\cdots_1, \cdots_2, \cdots_3)$$
$$\circlearrowleft(\circlearrowleft(M, \cdots_1), \cdots_2) = \circlearrowleft(M, \cdots_1, \cdots_2)$$
$$\circlearrowleft(M, \cdots_1, \times(\cdots_2), \cdots_3) = \circlearrowleft(M, \cdots_1, \cdots_2, \cdots_3)$$

It is not hard to reason that these rules preserve language. A process tree on which these rules have been applied exhaustively is a *reduced* process tree. For a reduced process tree it holds that a) for all nodes $\oplus(M_1, \ldots, M_n)$, $n > 1$; b) $\times$, $\rightarrow$ and $\wedge$ do not have a direct child of the same operator; and c) the first child of a $\circlearrowleft$ is not a $\circlearrowleft$ and any non-first child is not an $\times$.

Language-Rediscoverability. Our proof strategy is to first exhaustively reduce the given model M to some language-equivalent model M'. After that, we prove that B' discovers M' isomorphically. We use two lemmas to prove that a directly-follows complete log in each step always only allows to 1) pick one specific process tree operator, and 2) split the log in one particular way so that M' is inevitably rediscovered.

Lemma 11. *Let $M = \oplus(M_1, \ldots, M_n)$ be a reduced model that adheres to the model restrictions and let L be a log such that $L \diamond_{df} M$. Then $select_{B'}$ selects $\oplus$.*

The proof strategy is to prove for all operators that given a directly-follows complete log L of some process tree $\oplus(\ldots)$, $\oplus$ will be the first operator for which $G(L)$ satisfies all cut criteria according to the order $\times, \rightarrow, \wedge, \circlearrowleft$, which is the order in which the operators are checked in $select_{B'}$. For instance, for $\times$, $G(L)$ cannot be connected and therefore will B will select $\times$. For more details, see [18].

Lemma 12. *For each reduced process tree $M = a$ (with a in Σ) or $M = \tau$, and a log L that fits and is directly-follows complete to M, it holds that $M = B'(L)$.*

This lemma is proven by a case distinction on M being either τ or $a \in \Sigma$, and code inspection of $select_{B'}$. For more details, see [18].

Lemma 13. *Let $M = \oplus(M_1, \ldots, M_n)$ be a reduced process tree adhering to the model restrictions, and let L be a log such that $L \subseteq \mathcal{L}(M) \wedge L \diamond_{\mathrm{df}} M$. Let $\{(\oplus, ((L_1, 0), \ldots, (L_n, 0)))\}$ be the result of $select_{B'}$. Then $\forall i : L_i \subseteq \mathcal{L}(M_i) \wedge L_i \diamond_{\mathrm{df}} M_i$.*

The proof strategy is to show for each operator $\oplus$, that the cut $\Sigma_1 \ldots \Sigma_n$ that $select_{B'}$ chooses is the correct activity division ($\forall i : \Sigma_i = \Sigma(M_i) = \Sigma(L_i)$). Using that division, we prove that the SPLIT function returns sublogs $L_1 \ldots L_n$ that are valid for their submodels ($\forall i : L_i \subseteq \mathcal{L}(M_i)$). We then show that each sublog produced by SPLIT produces a log that is directly-follows complete w.r.t. its submodel ($\forall i : L_i \diamond_{\mathrm{df}} M_i$). See [18] for details.

Using these lemmas, we prove language-rediscoverability.

Theorem 14. *If the model restrictions hold for a process tree M, then B' language-rediscovers M: $\mathcal{L}(M) = \mathcal{L}(B'(L))$ for any log L such that $L \subseteq \mathcal{L}(M) \wedge L \diamond_{\mathrm{df}} M$.*

We prove this theorem by showing that a reduced version M' of M is isomorphic to the model returned by B', which we prove by induction on model sizes. Lemma 12 proves isomorphism of the base cases. In the induction step, Lemma 11 ensures that $B'(L)$ has the same root operator as M, and Lemma 13 ensures that the subtrees of M' are isomorphically rediscovered as subtrees of $B'(L)$. For a detailed proof see [18].

Corollary 15. *The process tree reduction rules given in Property 10 yield a language-unique normal form.*

Take a model M that adheres to the model restrictions. Let $L \subseteq \mathcal{L}(M) \wedge L \diamond_{\mathrm{df}} M$ and $M' = B'(L)$. Let M'' be another model adhering to the model restrictions and fitting L. As proven in Lemma 16 in [18], the cuts the algorithm took are mutually exclusive. That means that at each position in the tree, only two options exist that lead to fitness: either the operator $\oplus \in \{\times, \rightarrow, \wedge, \circlearrowleft\}$ chosen by B', or a flower model. By Theorem 14, $B'(L)$ never chose the flower model. Therefore, $B'(L)$ returns the most-precise fitting process tree adhering to the model restrictions. According to the definitions in [10], M' is a model of perfect simplicity and generalisation: M' contains no duplicate activities (simplicity) and any trace that can be produced by M in the future can also be produced by M' (generalisation). By Corollary 15 and construction of Property 10, it is the smallest process tree model having the same language as well.

6.5   Tool Support

We implemented a prototype of the B' algorithm as the InductiveMiner plugin of the ProM framework [12], see http://www.promtools.org/prom6/. Here, we sketch its run time complexity and illustrate it with a mined log.

Run Time Complexity. We sketch how we implemented B' as a polynomial algorithm. Given a log L, B' returns a tree in which each activity occurs once, each call of B' returns one tree, and B' recurses on each node once, so the number of recursions is $O(|\Sigma(L)|)$. In each recursion, B' traverses the log and searches for a graph cut. In Section 6.2, we sketched how directly-follows graph cuts can be found using standard (strongly) connected components computations. The exclusive choice, parallel and loop cuts were translated to finding connected components, the sequence cut to finding strongly connected components. For these common graph problems, polynomial algorithms exist. B' is implemented as a polynomial algorithm.

Illustrative Result. To illustrate our prototype, we fed it a log, obtained from [2, page 195]: $L = \{\langle a, c, d, e, h\rangle, \langle a, b, d, e, g\rangle, \langle a, d, c, e, h\rangle, \langle a, b, d, e, h\rangle, \langle a, c, d, e, g\rangle,$ $\langle a, d, c, e, g\rangle, \langle a, b, d, e, h\rangle, \langle a, c, d, e, f, d, b, e, h\rangle, \langle a, d, b, e, g\rangle, \langle a, c, d, e, f, b, d, e, h\rangle,$ $\langle a, c, d, e, f, b, d, e, g\rangle, \langle a, c, d, e, f, d, b, e, g\rangle, \langle a, d, c, e, f, c, d, e, h\rangle,$ $\langle a, d, c, e, f, d, b, e, h\rangle, \langle a, d, c, e, f, b, d, e, g\rangle, \langle a, c, d, e, f, b, d, e, f, d, b, e, g\rangle,$ $\langle a, d, c, e, f, d, b, e, g\rangle, \langle a, d, c, e, f, b, d, e, f, b, d, e, g\rangle, \langle a, d, c, e, f, d, b, e, f, b, d, e, h\rangle,$ $\langle a, d, b, e, f, b, d, e, f, d, b, e, g\rangle, \langle a, d, c, e, f, d, b, e, f, c, d, e, f, d, b, e, g\rangle\}$. The result of our implementation is $M' = \rightarrow(a, \circlearrowleft(\rightarrow(\wedge(\times(b, c), d), e), f), \times(h, g))$. A manual inspection reveals that this model indeed fits the log.

Take an arbitrary model M that could have produced L such that L is directly-follows complete w.r.t. M. Then by Theorem 14, $\mathcal{L}(M) = \mathcal{L}(M')$.

7   Conclusion

Existing process discovery techniques cannot guarantee soundness, fitness, rediscoverability and finite run time at the same time. We presented a process discovery framework B and proved that B produces a set of sound, fitting models in finite time. We described the conditions on the process tree operators under which the framework achieves this. The process tree operators $\times$, $\rightarrow$, $\wedge$ and $\circlearrowleft$ satisfy these conditions. However, the framework is extensible and could be applied to other operators, provided these satisfy the conditions. Another way of positioning our work is that our approach is able to discover some τ transitions in models for which the α-algorithm fails.

To make the framework even more extensible, it uses a to-be-given preference function *select* that selects preferred log divisions. Soundness, fitness and framework termination are guaranteed for any *select* adhering to B. We showed that if the model underlying the log is a process tree, then B can isomorphically-rediscover the model.

To illustrate B, we introduced an algorithm B' that uses B and returns a single process tree. B' works by dividing the activities in the log into sets, after which it splits the log over those sets. We proved that $select_{B'}$ adheres to B, which guarantees us soundness, fitness and framework termination for any input log. We proved that if the model underlying the log is representable as a process tree that has no duplicate activities, contains no silent activities and where all loop bodies contain no activity that is both a start and end activity of that loop body, then B' language-rediscovers this model. The only requirement on the log is that it is directly-follows complete w.r.t. the model underlying it. We argued that B' returns the smallest, most-precise, most-general model adhering to the model restrictions, and runs in a time polynomial to the number of activities and the size of the log.

Future Work. It might be possible to drop the model restriction 2 of Section 6.4, which requires that the the the sets of start and end activities of the leftmost branch of a loop operator must be disjoint, when length-two-loops are taken into account and a stronger completeness requirement is put on the log. Moreover, using another strengthened completeness assumption on the log, the no-τ restriction might be unnecessary. We plan on performing an empirical study to compare our B' algorithm to existing techniques. Noise, behaviour in the log that is not in the underlying model, could be handled by filtering the directly-follows relation, in a way comparable to the Heuristics miner [24], before constructing the directly-follows graph.

References

1. van der Aalst, W.: Workflow verification: Finding control-flow errors using petri-net-based techniques. In: van der Aalst, W.M.P., Desel, J., Oberweis, A. (eds.) Business Process Management. LNCS, vol. 1806, pp. 161–183. Springer, Heidelberg (2000)
2. van der Aalst, W.: Process Mining: Discovery, Conformance and Enhancement of Business Processes. Springer (2011)
3. van der Aalst, W., Buijs, J., van Dongen, B.: Towards improving the representational bias of process mining. In: Aberer, K., Damiani, E., Dillon, T. (eds.) SIMPDA 2011. Lecture Notes in Business Information Processing, vol. 116, pp. 39–54. Springer, Heidelberg (2012)
4. van der Aalst, W.M.P., de Medeiros, A.K.A., Weijters, A.J.M.M.: Genetic process mining. In: Ciardo, G., Darondeau, P. (eds.) ICATPN 2005. LNCS, vol. 3536, pp. 48–69. Springer, Heidelberg (2005)
5. van der Aalst, W., Weijters, T., Maruster, L.: Workflow mining: Discovering process models from event logs. IEEE Transactions on Knowledge and Data Engineering 16(9), 1128–1142 (2004)
6. Badouel, E.: On the α-reconstructibility of workflow nets. In: Haddad, S., Pomello, L. (eds.) PETRI NETS 2012. LNCS, vol. 7347, pp. 128–147. Springer, Heidelberg (2012)
7. Bergenthum, R., Desel, J., Lorenz, R., Mauser, S.: Process mining based on regions of languages. In: Alonso, G., Dadam, P., Rosemann, M. (eds.) BPM 2007. LNCS, vol. 4714, pp. 375–383. Springer, Heidelberg (2007)
8. Bergenthum, R., Desel, J., Mauser, S., Lorenz, R.: Synthesis of Petri nets from term based representations of infinite partial languages. Fundam. Inform. 95(1), 187–217 (2009)
9. Buijs, J., van Dongen, B., van der Aalst, W.: A genetic algorithm for discovering process trees. In: 2012 IEEE Congress on Evolutionary Computation (CEC), pp. 1–8. IEEE (2012)
10. Buijs, J.C.A.M., van Dongen, B.F., van der Aalst, W.M.P., et al.: On the role of fitness, precision, generalization and simplicity in process discovery. In: Meersman, R., et al. (eds.) OTM 2012, Part I. LNCS, vol. 7565, pp. 305–322. Springer, Heidelberg (2012)
11. Cortadella, J., Kishinevsky, M., Lavagno, L., Yakovlev, A.: Deriving Petri nets from finite transition systems. IEEE Transactions on Computers 47(8), 859–882 (1998)

12. van Dongen, B.F., de Medeiros, A.K.A., Verbeek, H.M.W., Weijters, A.J.M.M., van der Aalst, W.M.P.: The ProM framework: A new era in process mining tool support. In: Ciardo, G., Darondeau, P. (eds.) ICATPN 2005. LNCS, vol. 3536, pp. 444–454. Springer, Heidelberg (2005)

13. van Dongen, B.F., de Medeiros, A.K.A., Wen, L.: Process mining: Overview and outlook of Petri net discovery algorithms. In: Jensen, K., van der Aalst, W.M.P. (eds.) ToPNoC II, LNCS, vol. 5460, pp. 225–242. Springer, Heidelberg (2009)

14. Fahland, D., van der Aalst, W.M.P.: Repairing process models to reflect reality. In: Barros, A., Gal, A., Kindler, E. (eds.) BPM 2012. LNCS, vol. 7481, pp. 229–245. Springer, Heidelberg (2012)

15. Fahland, D., Favre, C., Koehler, J., Lohmann, N., Völzer, H., Wolf, K.: Analysis on demand: Instantaneous soundness checking of industrial business process models. Data Knowl. Eng. 70(5), 448–466 (2011)

16. Gambini, M., La Rosa, M., Migliorini, S., Ter Hofstede, A.H.M.: Automated error correction of business process models. In: Rinderle-Ma, S., Toumani, F., Wolf, K. (eds.) BPM 2011. LNCS, vol. 6896, pp. 148–165. Springer, Heidelberg (2011)

17. Günther, C.W., van der Aalst, W.M.P.: Fuzzy mining–adaptive process simplification based on multi-perspective metrics. In: Alonso, G., Dadam, P., Rosemann, M. (eds.) BPM 2007. LNCS, vol. 4714, pp. 328–343. Springer, Heidelberg (2007)

18. Leemans, S., Fahland, D., van der Aalst, W.: Discovering block-structured process models from event logs - a constructive approach. Tech. Rep. BPM-13-06, Eindhoven University of Technology (April 2013)

19. de Medeiros, A., Weijters, A., van der Aalst, W.: Genetic process mining: an experimental evaluation. Data Mining and Knowledge Discovery 14(2), 245–304 (2007)

20. Polyvyanyy, A., Garcia-Banuelos, L., Fahland, D., Weske, M.: Maximal structuring of acyclic process models. The Computer Journal (2012), http://comjnl.oxfordjournals.org/content/early/2012/09/19/comjnl.bxs126.abstract

21. Polyvyanyy, A., Vanhatalo, J., Völzer, H.: Simplified computation and generalization of the refined process structure tree. In: Bravetti, M. (ed.) WS-FM 2010. LNCS, vol. 6551, pp. 25–41. Springer, Heidelberg (2011)

22. Reisig, W., Schnupp, P., Muchnick, S.: Primer in Petri Net Design. Springer-Verlag New York, Inc. (1992)

23. Rozinat, A., de Medeiros, A.K.A., Günther, C.W., Weijters, A.J.M.M., van der Aalst, W.M.P.: The need for a process mining evaluation framework in research and practice. In: ter Hofstede, A.H.M., Benatallah, B., Paik, H.-Y. (eds.) BPM Workshops 2007. LNCS, vol. 4928, pp. 84–89. Springer, Heidelberg (2008)

24. Weijters, A., van der Aalst, W., de Medeiros, A.: Process mining with the heuristics miner-algorithm. Technische Universiteit Eindhoven, Tech. Rep. WP 166 (2006)

25. Wen, L., van der Aalst, W., Wang, J., Sun, J.: Mining process models with non-free-choice constructs. Data Mining and Knowledge Discovery 15(2), 145–180 (2007)

26. Wen, L., Wang, J., Sun, J.: Mining invisible tasks from event logs. In: Dong, G., Lin, X., Wang, W., Yang, Y., Yu, J.X. (eds.) APWeb/WAIM 2007. LNCS, vol. 4505, pp. 358–365. Springer, Heidelberg (2007)

27. van der Werf, J.M.E.M., van Dongen, B.F., Hurkens, C.A.J., Serebrenik, A.: Process discovery using integer linear programming. In: van Hee, K.M., Valk, R. (eds.) PETRI NETS 2008. LNCS, vol. 5062, pp. 368–387. Springer, Heidelberg (2008)

Faster Verification of Partially Ordered Runs in Petri Nets Using Compact Tokenflows

Robin Bergenthum

Department of Software Engineering and Theory of Programming,
FernUniversität in Hagen
robin.bergenthum@fernuni-hagen.de

Abstract. In this paper we tackle the problem of verifying whether a labeled partial order (LPO) is executable in a Petri net. In contrast to sequentially ordered runs an LPO includes both, information about dependencies and independencies of events. Consequently an LPO allows a precise and intuitive specification of the behavior of a concurrent or distributed system. In this paper we consider Petri nets with arc weights, namely marked place/transition-nets (p/t-nets). Accordingly the question is whether a given LPO is an execution of a given p/t-net.

Different approaches exist to define the partial language (i.e. the set of executions) of a p/t-net. Each definition yields a different verification algorithm, but in terms of runtime all these algorithms perform quite poorly for most examples. In this paper a new compact characterization of the partial language of a p/t-net will be introduced, optimized with respect to the verification problem. The goal is to develop an algorithm to efficiently decide the verification problem.

1 Introduction

Specifications of concurrent or distributed systems are often formulated in terms of scenarios [31, 10, 35, 20]. A scenario can be represented by a labeled partial order (LPO), i.e. a partially ordered set of events. In many cases it is part of a system's specification that given LPOs should or should not be executable by the system. P/t-nets [30, 32], i.e. Petri nets with arc weights, are well suited for modelling concurrent or distributed systems and have a huge range of theoretical and practical applications [29, 2, 13, 11]. Thus, it is a natural question whether an LPO is an executions of a p/t-net. We refer to this question as the verification problem. Deciding the verification problem can help to check conformance of a system, uncover system faults or requirements, validate the system and evaluate design alternatives.

The set of all executable LPOs of a p/t-net is called its partial language. To define this language there exist several equivalent characterisations [18, 17, 19, 21, 28]. Each characterisation yields a different algorithm deciding the verification problem. The characterisations of the partial language are as follows:

- The set of runs. A run is an LPO which includes the transitive order relation between events of an occurrence net of the given p/t-net [18, 17].
- The set of executions. An LPO is an execution if each cut (i.e. a maximal set of unordered events) of the LPO is enabled after the occurrence of its prefix in the given p/t-net [19].

J.-M. Colom and J. Desel (Eds.): PETRI NETS 2013, LNCS 7927, pp. 330–348, 2013.

- The set of LPOs for which valid tokenflows describing the flow of tokens between events exist [21].
- The set of LPOs for which valid interlaced tokenflows describing a quadruple of flows of tokens between events on the skeleton arcs of the LPO exist [28].

The set of runs coincides with the set of executions of a p/t-net [23, 34]. The set of runs can be described using the definition of valid tokenflows [21]. A tokenflow is a function assigning non-negative integers to the set of arcs of an LPO. Each tokenflow describes a distribution of tokens with respect to a place of the p/t-net. Given two events a and b and a fixed place p, the value of a tokenflow function x on the arc (a, b) describes that the occurrence of a produces $x(a, b)$ tokens in place p that will be consumed by the occurrence of b. A tokenflow is valid if it respects the firing rule of the p/t-net in a sense that each event receives enough tokens to occur and no event has to pass more tokens then its occurrence produces. An LPO is an execution if there is a valid tokenflow for each place of a p/t-net. The notion of interlaced tokenflows and its equivalence to normal tokenflows is presented in [28].

The different characterisations of the partial language of the p/t-net lead to different algorithms solving the verification problem. Unfortunately, each known verification algorithm has big drawbacks dependent on the structure of the given LPO or the given p/t-net. The main ideas of the corresponding verification algorithms are as follows:

Algorithms using the notion of runs decide the verification problem by considering the set of occurrence nets of the given p/t-net. They check if the given LPO includes the partial order of events for one of the occurrence nets. The set of all occurrence nets can be calculated using so called unfolding algorithms [27, 15, 8]. Even if the number of occurrence nets is infinite these algorithms are able to calculate a set of occurrence nets such that the number of events of each net is equal to the number of events of the given LPO. The main problem is that the number of occurrence nets may be exponential in the number of events such that the runtime complexity of such an algorithm is in exponential time.

Algorithms using the notion of executions decide the verification problem by checking if each cut of the given LPO is enabled after the occurrence of its prefix. This time the number of cuts can be exponential in the number of events such that the runtime complexity of such an algorithm is in exponential time. However, the number of cuts is small if the given LPO is dense. In such a case the LPO describes a lot of dependencies between events and the verification algorithm is applicable. Given a thin LPO or an LPO describing reasonable concurrency between events this verification algorithm is insufficient.

Algorithms using the notion of valid tokenflows are by now the most elegant way to decide the verification problem. As stated above a tokenflow is a function describing the distribution of tokens between events with respect to a place. This distribution respects the transitive ordering of event in the LPO, such that a valid tokenflow can be constructed if and only if the LPO is a run of the p/t-net. The first implementation of a verification algorithm using the notion of valid tokenflows is given in [3, 4]. Each construction of a valid tokenflow is done by constructing maximal flows [16, 1] in so called associated flow networks. The runtime of this algorithm is in polynomial time. The main disadvantage of this algorithms is that its runtime highly depends on the size of the order relation of the LPO. If the LPO describes a lot of dependencies the verification algorithm is hardly applicable. A second approach to solve the verification problem using valid tokenflows is given in [24]. This approach is able to reduce the runtime by reducing the number of maximal flow problems to be solved. Still, its runtime is poor if the LPO is dense.

The most recent characterisation of the partial language of a given p/t-net, so called valid interlaced tokenflows, was given in [28]. An interlaced tokenflow is a quadruple of special tokenflows defined on every skeleton arc of the given LPO. The interplay between the components of a valid interlaced tokenflow ensures that they correspond to a valid tokenflow defined on every arc of the given LPO. The main advantage is that the number of skeleton arcs is the smallest representation of the number of all dependencies specified by the LPO. By now it is not known if the resulting verification algorithm is faster than the algorithm described in [24] since both algorithms are not yet implemented.

In this paper we introduce compact tokenflows as a new characterisation of the partial language of a p/t-net. Compact tokenflows are optimized to efficiently solve the verification problem. The main ideas leading to the concept of compact tokenflows are as follows:

- Compact tokenflows are based on normal tokenflows. A Tokenflow is defined on the arcs of an LPO and the number of arcs does not increase exponentially with the number of events. This is a necessary condition to receive an algorithm having polynomial worst case runtime complexity.
- LPOs include the complete transitive relation between events. In contrast to token-flows a compact tokenflow is defined on the skeleton of an LPO. The skeleton is the smallest representation of the transitive relation.
- In contrast to tokenflows a compact tokenflow abstracts from the history of tokens. Since in a compact tokenflow events are able to pass received tokens to later events, a compact tokenflow describes the sum of tokens produced by sets of events and not the number of tokens produced by each event.
- An interlaced tokenflow respects the history of each tokens by considering a four component flow of tokens. Using compact tokenflows only a single tokenflow is needed.

We will show that compact tokenflows can be constructed adopting the ideas presented in [24]. We will be able to introduce a new verification algorithm having an efficient runtime independent from the structure of the given LPO and p/t-net. This algorithm will be much faster than any known verification algorithm.

The paper is organized as follows. In Section 2 we describe LPOs, p/t-nets and their partial languages. In Section 3 we recapitulate the concepts of tokenflows and introduce the new concept of compact tokenflows. We prove that compact tokenflows are equivalent to normal tokenflows. In section 4 we describe the resulting verification algorithm. In Section 5 we present runtime experiments of the new verification algorithm comparing it to all alternative verification algorithms. Finally, Section 6 concludes the paper.

2   Labeled Partial Orders, p/t-Nets and Partial Languages

In this paper the following notations will be used. $\mathbb{N}$ denotes the non-negative integers. Given a finite set A, 2^A denotes the power set of A. $\mathbb{N}^A$ denotes the set of multisets over A. For $m \in \mathbb{N}^A$ we write $m = \sum_{a \in A} m(a) \cdot a$.

Definition 1. *A labeled partial order (LPO) is a triple $lpo = (V, <, l)$, where V is a finite number of events, $< \subseteq V \times V$ is a transitive and irreflexive relation over V and $l : V \to T$ is a labeling function assigning a label $t \in T$ to each event.*

Given an LPO $lpo = (V, <, l)$ and an event $e \in V$, the preset of e is denoted by $\bullet e = \{v \in V | v < e\}$. Given a set of events E, the preset of E is denote by $\bullet E = \bigcup_{v \in E} \bullet v$. A prefix is a set of events E such that $E = (E \cup \bullet E)$ holds. The maximal set of events having an empty preset is denoted by $min(V)$. A set of events C is called a co-set if $(v, v' \in C \Rightarrow v \not< v')$ holds. A cut is a co-set which is not included in any other co-set.

The transitive closure of a relation $\to$ is denoted by $\to^*$, the transitive reduction of a relation $\to$ is denoted by $\to^\circ$. Given a finite relation the transitive reduction is the smallest relation such that its transitive closure equals the primary LPO. Given an LPO $lpo = (V, <, l)$ the transitive reduction $<^\circ$ is called the skeleton of lpo. The graph $(V, <^\circ)$ forms a Hasse diagram and we call the labeled Hasse diagram $(V, <^\circ, l)$ the compact LPO of lpo.

In this paper concurrent or distributed system will be given by p/t-nets, i.e. Petri nets with arc weights.

Definition 2. *A marked place/transition-net (p/t-net) is a tuple $N = (P, T, W, m_0)$, where P and T are finite sets of places and transitions fulfilling $P \cap T = \emptyset$, $W : (P \times T) \cup (T \times P) \to \mathbb{N}$ is a multiset of edges and $m_0 : P \to \mathbb{N}$ is an initial marking.*

Given a p/t-net $N = (P, T, W, m_0)$, a transition $t \in T$ is enabled in N iff $\forall\, p \in P : W(p, t) \le m_0(p)$ holds. If a transition is enabled it may fire and change the given marking m_0 to a new marking m which is for each $p \in P$ given by $m(p) = m_0(p) + W(t, p) - W(p, t)$. A multiset of transitions $\tau \in \mathbb{N}^T$ (called a step of N) is enable in N iff $\forall\, p \in P : \sum_{t \in \tau} \tau(t) \cdot W(p, t) \le m_0(p)$ holds. If a step is enabled it may fire and change the given marking m_0 to a new marking m which is for each $p \in P$ given by $m(p) = m_0(p) + \sum_{t \in \tau} \tau(t) \cdot (W(t, p) - W(p, t))$.

Given a p/t-net a sequential run is a sequence of consecutively enabled and fired transitions or transition steps. Instead of the set of sequential runs we consider the partial language of a p/t-net. In contrast to sequentially ordered runs an LPO includes arbitrary dependencies and independencies between events.

Definition 3. *Given a p/t-net $N = (P, T, W, m_0)$ and an LPO $lpo = (V, <, l)$ with $l(V) \subseteq T$. The lpo is an execution of N iff for each $p \in P$ and each cut C of lpo : $m_0(p) + \sum_{e \in \bullet C}(W(l(e), p) - W(p, l(e))) \ge \sum_{e \in C} W(p, l(e))$ holds. We denote the set of all executions of N by $L(N)$, the so called partial language of N.*

Following Definition 3, we say that an LPO is executable with respect to a place p, if $m_0(p) + \sum_{e \in \bullet C}(W(l(e), p) - W(p, l(e))) \ge \sum_{e \in C} W(p, l(e))$ holds.

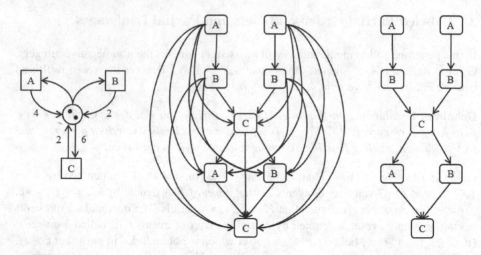

Fig. 1. A p/t-net, an executable LPO and its compact LPO

As stated in the introduction there are other equivalent characterizations of the partial language of a p/t-net. Any characterization leads to the same verification problem: Given a p/t-net N and an LPO lpo, decide if $lpo \in L(N)$ holds.

Figure 1 shows a p/t-net having only one place which is marked by two tokens in the initial marking, three transitions A, B, C as well as several weighted arcs. Arcs without attached numbers have the arc weight one. The LPO in the middle consists of eight events labeled with the transitions of the p/t-net. This LPO is an execution of the p/t-net according to Definition 3. The right side of Figure 1 depicts the corresponding compact LPO.

3   Tokenflows

The definition of valid tokenflows (see [21]) was originally invented to more easily prove the equivalence between runs and executions of a p/t-net. The original proof is quite complex and was introduced in [23, 34]. In [21] additionally an algorithm deciding the verification problem in polynomial time was obtained which was first implemented in [4] yielding a fast verification algorithm in case of a thin LPOs. [24] describes a second algorithm having $|V|$-times faster runtime (V denotes the number of events of the given LPO). This was the first step in the direction of an efficient verification algorithm, but still this algorithm performs poorly in case of a dense LPOs. After this tokenflows have been used for the synthesis of p/t-nets from LPOs [5, 6], synthesis from infinite sets of LPOs [7, 33, 28], for the definition of the partial language of more general classes of Petri nets [25], as well as for the unfolding of the partial language of p/t-nets [9, 8]. All in all the verification problem took a back seat. The idea of this paper is to find a characterisation of the partial language based on tokenflows which is optimized for the verification problem.

First, we will recall the concept of tokenflows invented in [21]. Given an LPO $lpo = (V, <, l)$ and a p/t-net $N = (P, T, W, m_0)$ such that $T \subseteq l(V)$ holds, a tokenflow x

is a function assigning a non-negative integer to each event and each arc of the LPO. Given an arc (v, v') of lpo the value $x(v, v')$ describes the number of tokens produced by the occurrence of v which are consumed by the occurrence of v'. Given an event v, the value $x(v)$ describes the number of tokens consumed by v from the initial marking. A tokenflow is valid with respect to a place $p \in P$, if it respects the occurrence rule for each event v of lpo in a sense that: the sum of ingoing tokens of v equals $W(p, l(v))$, the sum of outgoing tokens of v is less than $W(l(v), p)$ and the sum of tokens consumed from the initial marking by all events does not exceed the initial marking $m_0(p)$. Since such a valid distribution of tokens along the arcs of an LPO respects the described dependencies, a valid tokenflow coincides with a run of a p/t-nets considering p. If there exists a valid tokenflow for each place p of N a corresponding occurrence net of N can be found and vice versa.

Definition 4. *Given a p/t-net $N = (P, T, W, m_0)$ and an LPO $lpo = (V, <, l)$, such that $l(V) \subseteq T$ holds. A tokenflow is a function $x : (< \cup V) \to \mathbb{N}$. Given an event $v \in V$, by $in(v) = x(v) + \sum_{v' < v} x(v', v)$ we denote the inflow of v and by $out(v) = \sum_{v < v'} x(v, v')$ we denote the outflow of v.*

Given a place $p \in P$ a tokenflow is valid with respect to p iff the following conditions hold.

(I) $\forall v \in V : in(v) = W(p, l(v))$,
(II) $\forall v \in V : out(v) \leq W(l(v), p)$,
(III) $\sum_{v \in E} x(v) \leq m_0(p)$.

lpo is called a valid LPO of N iff for every $p \in P$ there is a valid tokenflow.

The main result of [21] is that the partial language of a p/t-net can be characterised using valid tokenflows.

Theorem 1. *([21]) Given a p/t-net N. The partial language $L(N)$ coincides with the set of valid LPOs of N.*

Theorem 1 leads to the verification algorithms given in [4] and [24]. Given a fixed place both algorithms construct a valid tokenflow by constructing maximal flows in associated flow networks derived from the LPO and the p/t-net. The big disadvantage of both algorithms is that the size of the corresponding flow networks is directly related to the number of arcs of the LPO. Although the runtime complexity of both algorithms is in polynomial time, both algorithms perform worse than an algorithm checking if each cut of an LPO is enabled after the occurrence of its prefix if the given LPO is dense.

In the following we want to introduce a new characterisation of the partial language of a p/t-net yielding an efficient verification algorithm. A compact tokenflow is defined on the skeleton of an LPO, i.e. its corresponding compact LPO. A (normal) tokenflow describes the complete distribution of tokens between events using the given transitive order relation of the LPO. In a compact tokenflow each event v can receive more tokens than its transition $l(v)$ consumes and v is allowed to pass those additional tokens on to later events. The main idea is that losing the set of transitive arcs is fine since every transitive arc can be represented by a path of skeleton arcs.

Definition 5. *Given a p/t-net $N = (P, T, W, m_0)$, an LPO $lpo = (V, <, l)$ such that $l(V) \subseteq T$ holds and the skeleton $\lhd$ of lpo. A compact tokenflow is a function $x :$ $(\lhd \cup V) \to \mathbb{N}$. Given an event $v \in V$, by $in^{\lhd}(v) = x(v) + \sum_{v' \lhd v} x(v', v)$ we denote the inflow of v and by $out^{\lhd}(v) = \sum_{v \lhd v'} x(v, v')$ we denote the outflow of v.*

Given a place $p \in P$ a compact tokenflow is valid with respect to p iff the following conditions hold.

(i) $\forall v \in V : in^{\lhd}(v) \geq W(p, l(v)),$
(ii) $\forall v \in V : out^{\lhd}(v) \leq in^{\lhd}(v) - W(p, l(v)) + W(l(v), p),$
(iii) $\sum_{v \in E} x(v) \leq m_0(p).$

lpo is called a valid compact LPO of N iff for every $p \in P$ there is a valid compact tokenflow.

Please note that the definition of a compact tokenflow can be generalised to using not only the skeleton arcs of an LPO, but using any subset of the partial order relation including the skeleton. Only the skeleton arcs are needed, but if an LPO is given by some acyclic graph all definitions and algorithms shown in this paper may be directly applied to this graph instead of its transitive reduction.

The left side of Figure 2 shows a valid tokenflow for the place and the p/t-net shown in Figure 1. Transitive arcs are not depicted if their value is equal to zero. The right side shows a valid compact tokenflow with respect to the same place.

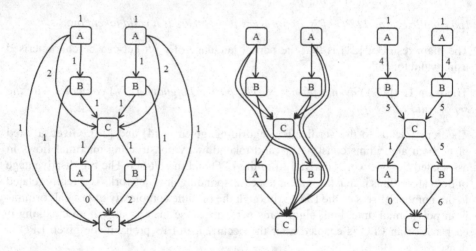

Fig. 2. A valid tokenflow and a valid compact tokenflow of the p/t-net shown in Figure 1

In the next step we will prove that the set of valid compact LPOs of a p/t-net N coincides with its partial language $L(N)$. Since the compact representation abstracts from the concrete distribution of tokens, in the sense that the history of a token is lost while it is passed on by events, this proof cannot be done similarly to the proof concerning valid LPOs given in [21]. We will show that for every valid tokenflow in an LPO there is a valid compact tokenflow. Given a place p and a valid tokenflow we will fold this tokenflow into a compact tokenflow. We will show that every possible folding yields a valid compact tokenflow. After that, given an LPO and a valid compact tokenflow for every place of N, we will proof that this LPO is an execution of N (in the sense of Definition 3).

Definition 6. *Given an LPO $lpo = (V, <, l)$, the skeleton $\lhd$ of lpo and $x : (< \cup V) \to \mathbb{N}$ a tokenflow. For each arc $(v, v') \in\, <$ there exists a path of skeleton arcs from v to v'. Consequently a function $\rho :\, < \to 2^{\lhd}$ can be found which maps every arc (v, v') of lpo to a path from v to v' in $\lhd$. Given a fixed function ρ for every skeleton arc (e, e') we define $(e, e')^* = \{(v, v') \in\, < \,|(e, e') \in \rho(v, v')\}$. Depending on ρ we are now able to define the associated compact tokenflow $x^{\lhd} : (\lhd \cup V) \to \mathbb{N}$.*

$$x^{\lhd}(a) = \begin{cases} \sum_{(v,v') \in (e,e')^*} x(v, v'), & \text{if } a = (e, e') \in \lhd, \\ x(e), & \text{if } a = e \in V. \end{cases}$$

The middle part of Figure 2 depicts a possible function ρ mapping every transitive arcs to a path of skeleton arcs. There may be different choices for ρ since often for some transitive arcs it is possible to choose between different paths of skeleton arcs using concurrent parts of the LPO. Given a skeleton arc, its compact tokenflow is the sum of tokenflow of all attached transitive arcs.

Theorem 2. *Given a p/t-net $N = (P, T, W, m_0)$, an LPO $lpo = (V, <, l)$, such that $l(V) \subseteq T$ holds, a place $p \in P$ and a tokenflow x which is valid with respect to p. Every possible associated compact tokenflow $x^{\lhd}$ is valid with respect to p.*

Proof. Given a fixed event $v \in V$, the image set $\rho(<)$ contains four different kinds of paths. Paths leading to v, paths beginning at v, paths including v and paths not including v. The inflow of v with respect to $x^{\lhd}$ is given by the sum of values of paths leading to v, paths including v and the value $x(v)$. By $\delta(v)$ we denote the sum of tokenflow reaching v via paths including v whereby $in^{\lhd}(v) = in(v) + \delta(v)$ holds. The outflow of v with respect to $x^{\lhd}$ is given by the sum of values of paths starting at v and paths including v whereby $out^{\lhd}(v) = out(v) + \delta(v)$ holds. The conditions (i), (ii) and (iii) of Definition 5 hold as follows:

$$in^{\lhd}(v) = in(v) + \delta(v)$$

$$\geq in(v)$$

$$\overset{(I)}{=} W(p, l(v)),$$

$$out^{\lhd}(v) = out(v) + \delta(v)$$

$$\overset{(I)}{=} out(v) + \delta(v) + in(v) - W(p, l(v))$$

$$= out(v) + in^{\lhd}(v) - W(p, l(v))$$

$$\overset{(II)}{\leq} W(l(v), p) + in^{\lhd}(v) - W(p, l(v)),$$

$$\sum_{v \in V} x^{\triangleleft}(v) = \sum_{v \in V} x(v)$$

$$\overset{(III)}{\leq} m_0(p).$$

Theorem 2 states that each valid tokenflow can be folded into a valid compact tokenflow. In a next step we need to prove the opposite direction. Since it is not clear in which way a compact tokenflow needs to be unfolded to get a valid tokenflow, we prove that if there exists a compact tokenflow which is valid with respect to a place p the LPO is an execution with respect to p.

Definition 7. *Given an LPO* $lpo = (V, <, l)$, *the skeleton* $\triangleleft$ *of lpo and a compact tokenflow* $x^{\triangleleft}$. *Given a set of events* $E \subseteq V$, *by* $E^{\ll} = \triangleleft \cap ((V \backslash E) \times E)$ *we denote the set of skeleton arcs leading to* E *and by* $E^{\gg} = \triangleleft \cap (E \times (V \backslash E))$ *we denote the set of skeleton arcs leaving* E. *By* $IN(E) = \sum_{(e,e') \in E^{\ll}} x^{\triangleleft}(e, e') + \sum_{e \in E} x^{\triangleleft}(e)$ *we denote the inflow of* E *and by* $OUT(E) = \sum_{(e,e') \in E^{\gg}} x^{\triangleleft}(e, e') + \sum_{e \in V \backslash E} x^{\triangleleft}(e)$ *we denote the outflow of* E.

Lemma 1. *Given a p/t-net* $N = (P, T, W, m_0)$, *an LPO* $lpo = (V, <, l)$ *with* $l(V) \subseteq T$, *a place* $p \in P$, *a compact tokenflow* $x^{\triangleleft}$ *which is valid with respect to* p *and a prefix* $E \subseteq V$ *of lpo. The following inequation holds:*

$$OUT(E) \leq m_0(p) + \sum_{v \in E} (W(l(v), p) - W(p, l(v))).$$

Proof. We will prove this by induction. Let $E = \emptyset$ be the empty prefix.

$$OUT(\emptyset) = \sum_{v \in V} x^{\triangleleft}(v)$$

$$\overset{(iii)}{\leq} m_0(p)$$

$$= m_0(p) + \sum_{v \in \emptyset} (W(l(v), p) - W(p, l(v))).$$

Let E' be a prefix of lpo and let $OUT(E') \leq m_0(p) + \sum_{v \in E'} (W(l(v), p) - W(p, l(v)))$ hold. Given a event $e \in V \backslash E'$ which is minimal in $V \backslash E'$. We define a new prefix $E = (E' \cup e)$.

$$OUT(E) = OUT(E') - in^{\triangleleft}(e) + out^{\triangleleft}(e)$$

$$\overset{(ii)}{\leq} OUT(E') - in^{\triangleleft}(e) + in^{\triangleleft}(e) - W(p, l(e)) + W(l(e), p)$$

$$\leq m_0(p) + \sum_{v \in E'} (W(l(v), p) - W(p, l(v))) - W(p, l(e)) + W(l(e), p)$$

$$= m_0(p) + \sum_{v \in E} (W(l(v), p) - W(p, l(v)))$$

Every prefix E of lpo can be constructed by consecutively appending events to the empty prefix (each event minimal in the set of not added events).

The next lemma states a relation between the number of tokens flowing into a cut C of events of an LPO and the number of tokens leaving the prefix given by $^{\bullet}C$.

Lemma 2. *Given an LPO lpo* $= (V, <, l)$*, the skeleton* $\lhd$ *of lpo and a compact token-flow* $x^{\lhd}$*. For each cut* C *of lpo the following inequation holds:* $IN(C) \leq OUT(^{\bullet}C)$.

Proof. Every skeleton arc of *lpo* leading to C starts at an event in $^{\bullet}C$. Since C is a cut every skeleton arc leaving $^{\bullet}C$ leads to C. Thereby intuitively both values should be equal. We get an inequation, because of the values of the tokenflow defined on events.

$$IN(C) = \sum_{(v,v')\in C\ll} x^{\lhd}(v, v') + \sum_{v\in C} x^{\lhd}(v)$$

$$= \sum_{(v,v')\in {}^{\bullet}C\gg} x^{\lhd}(v, v') + \sum_{v\in C} x^{\lhd}(v)$$

$$\leq \sum_{(v,v')\in {}^{\bullet}C\gg} x^{\lhd}(v, v') + \sum_{v\in\{V\backslash{}^{\bullet}C\}} x^{\lhd}(v)$$

$$= OUT(^{\bullet}C).$$

At this point, with the help of Lemma 1 and 2, we are able to prove the opposite direction to Theorem 2.

Theorem 3. *Given a p/t-net* $N = (P, T, W, m_0)$*, an LPO lpo* $= (V, <, l)$*, such that* $l(V) \subseteq T$ *holds, a place* $p \in P$ *and a compact tokenflow* $x^{\lhd}$ *valid with respect to* p*. lpo is executable with respect to* p.

Proof. For each cut C of *lpo* the following inequation hold:

$$\sum_{v\in C} W(p, l(v)) \overset{(i)}{\leq} \sum_{v\in C} in^{\lhd}(v)$$

$$\overset{(\text{Lemma 1})}{\leq} OUT(^{\bullet}C)$$

$$\overset{(\text{Lemma 2})}{\leq} m_0(p) + \sum_{v\in {}^{\bullet}C}(W(l(v), p) - W(p, l(v))).$$

We conclude this section with its main theorem. It states that the set of valid compact LPOs coincide with the partial language of a p/t-net. This theorem follows directly from Theorem 2 and Theorem 3.

Theorem 4. *Given a p/t-net* N*, the partial language* $L(N)$ *coincides with the set of valid compact LPOs of* N.

We have shown that compact tokenflows are equivalent to all other characterisations of the partial language of a p/t-net. In the next section we will present an efficient way to solve the verification problem using compact tokenflows. This algorithm will construct a valid compact tokenflow for each place of the p/t-net if such a tokenflow exists. This algorithm will only need to regard skeleton arcs, hence it is much faster then the algorithms presented in [4] and [24]. In contrast to normal tokenflows the new definition abstracts from the concrete distribution of tokens between events. Compared to interlaced tokenflows [28] only a single tokenflow is needed for every arc of the LPO.

4 Verification Algorithm

In this section we provide an algorithm solving the verification problem using compact tokenflows. Given a p/t-net $N = (P, T, W, m_0)$ and an LPO $lpo = (V, <, l)$ we will construct a flow network for every place of N such that constructing a maximal flow in each network will decide the verification problem.

A flow network (see for example [1]) is a directed graph with two special nodes. A source, the only node having no ingoing arcs, and a sink, the only node having no outgoing arcs. Each arc has a capacity. A flow is a function from the arcs to the non-negative integers assigning a value of flow to each arc. This flow function needs to respect the capacity of each arc and the so called flow conservation. The flow conservation says that the sum of flow reaching a node is equal to the sum of flow leaving a node for every inner node of the flow network. Thus, flow is only generated at the source and flows along different paths till it reaches the sink. The value of a flow function in a flow network is the sum of flow reaching the sink. The maximal flow problem is to find a flow function having a maximal value.

For a given place $p \in P$ we construct the associated flow network. Just like a tokenflow in an LPO a flow in the associated flow network describes the propagation of tokens between events. Each flow in the associated flow network coincides with a compact tokenflow. For each event in lpo we insert two nodes into the flow network. The first node is called the top-node, the second is called the bottom-node. A pair of such nodes represent an event $v \in V$. The flow at the top-node of v describes the value of tokenflow received by v and this value has to be greater than $W(p, l(v))$. Therefore, if flow arrives at the top-node of v a value of $W(p, l(v))$ is routed to the sink representing tokens consumed by the occurrence of $l(v)$. Additional flow can be distributed further flowing from the top-node of v to its bottom-node. The bottom-node of v distributes flow. The maximal number of flow this node can pass on is the number of flow received from its top-node plus $W(l(v), p)$. Therefore, flow in a value of $W(l(v), p)$ is routed from the source to the bottom-node of v representing tokens produced by the occurrence of $l(v)$. Additionally, nodes are connected according to the skeleton arcs of the lpo. For each skeleton arc (v, v') we add a corresponding arc in the associated flow network leading from the bottom-node of v to the top-node of v'. The value of a compact tokenflow on each event is considered by an additional node of the flow network. It can be seen as the bottom-node related to the initial marking. First, we add an arc from the source to this bottom-node having the capacity $m_0(p)$. Second, we add arcs from this bottom-node to all top-nodes associated to minimal events of lpo. Flow leaving this bottom-node is able to reach any top-node of the flow network via paths of skeleton arcs. This flow represents tokens consumed from the initial marking.

Figure 3 depicts the associated flow network of the LPO and the p/t-net show in Figure 1. Pairs of top- and bottom-nodes are drawn in rounded boxes labeled by the corresponding events. Inner arcs have no number attached, their capacity is equal to an upper bound for the maximal value of a possible flow functions. Such an upper bound will be defined in the following definition.

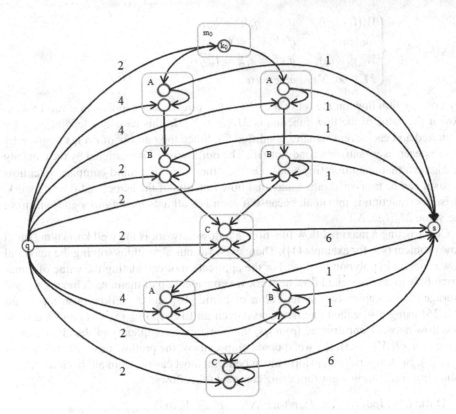

Fig. 3. The associated flow network of the p/t-net and the LPO shown in Figure 1

Definition 8. *Given a p/t-net* $N = (P, T, W, m_0)$, *an LPO* $lpo = (V, <, l)$ *with* $l(V) \subseteq T$, *the skeleton* $\lhd$ *of lpo and* $p \in P$ *a place. Let* $M_p(lpo, N) = \sum_{v \in V} W(p, l(v))$ *denote the sum of tokens consumed by all events of lpo. The associated flow network* $G = (K, F, c, q, s)$ *is defined by:*

$$K = \{k_0\} \cup \{t_v, b_v | v \in V\},$$

$$F = F_m \cup F_v \cup F_\lhd \cup F_q \cup F_0 \cup F_s \text{ with}$$

$$F_m = \{(k_0, t_v) | v \in min(V)\},$$

$$F_v = \{(t_v, b_v) | v \in V\},$$

$$F_\lhd = \{(b_v, t_{v'}) | (v, v') \in \lhd\},$$

$$F_q = \{(q, b_v) | v \in V\},$$

$$F_0 = \{(q, k_0)\},$$

$$F_s = \{(t_v, s) | v \in V\}.$$

$$c(k, k') = \begin{cases} W(l(v), p), & \text{if } k = q, k' = b_v, \\ m_0(p), & \text{if } (k, k') = (q, k_0), \\ W(p, l(v)), & \text{if } k' = s, k = t_v \\ M_p(lpo, N), & \text{otherwise.} \end{cases}$$

By construction flow on the set of arcs $F_{\vartriangleleft} \cup F_m$ directly corresponds to a valid token-flow if the value of the flow function is $M_p(lpo, N)$. In this case arcs leading to s are saturated and each event satisfies condition (i). Since the outflow of events is given by $F_v \cup F_q$ each event satisfied condition (ii). The outflow of k_0 is restricted by the capacity of F_0 such that condition (iii) holds. On the other hand, if a valid compact tokenflow exists it can be translated into a maximal flow function of the associated flow network. This flow function is maximal, because it saturates all arcs leading to s and will have the value $M_p(lpo, N)$.

Constructing a maximal flow function in a flow network is the well known maximal flow problem (see for example [1]). There exist various algorithms solving the maximal flow problem in polynomial time. For the application of calculating the value of a maximal flow in an associated flow network we consider two of them, each having a good average case complexity: the algorithm of Dinic [14] and a preflow push algorithm [22, 26] using a so called gap heuristic. Given an LPO $lpo = (V, <, l)$ and a associated flow network constructed from lpo the worst case complexity of the algorithm of Dinic is in $O(|V|^2 \, |<|)$, the worst case complexity of the preflow push algorithm is in $O(|V|^3)$. Both algorithms perform much better in most cases. All in all this leads to the following verification algorithm using compact tokenflows.

Data: LPO $lpo = (V, <, l)$, p/t-net $N = (P, T, W, m_0)$.
Result: Decides if $lpo \in L(N)$ holds.

for *each* $p \in P$ **do**
  $G \leftarrow$ associated compact flow network(p);
  $w \leftarrow$ value of maximal Flow(G);
  if $(w < M_p(lpo, N))$ **then**
    | RETURN *false*;
  end
end
RETURN *true*;

Algorithm 1. Verification algorithm using compact tokenflows

Remark, that using a preflow push algorithm the runtime of this verification algorithm is in $O(|P| \cdot |V|^3)$.

5 Experimental Results

In this section we will discuss some experimental results. To do so we implemented several existing and new verification algorithms into the tool VipTool [12]. We did extensive runtime tests and for this paper we show the most interesting results considering a selection of the implemented verification algorithms. For the new tokenflow algorithms we show results using both mentioned maximal flow algorithms, the algorithm of Dinic and the preflow push algorithm. We selected the algorithms as follows:

- Algorithm I checks if the LPO is an execution by considering all cuts,
- Algorithm II constructs a tokenflow as implemented in [4],
- Algorithm III constructs a tokenflow as described in [24] using the algorithm of Dinic,
- Algorithm IV constructs a tokenflow as described in [24] using the preflow push algorithm,
- Algorithm V constructs a compact tokenflow using the algorithm of Dinic,
- Algorithm VI constructs a compact tokenflow using the preflow push algorithm.

The following two experiments were performed using a Dual Core Prozessor, 1.7 GHz and 4GB RAM. In both experiments we consider the p/t-net shown in figure 4. It models a simple workflow performed by a group of students working within a collaborative learning environment. The p/t-nets structure is exemplary for simple workflows and we omit a description of this special workflow in this paper.

Figure 5 depicts an LPO of the partial language of the p/t-net shown in Figure 4. The LPO describes a cycle in the p/t-net by executing all the transitions once. In both experiments we will repeat and compose copies of this cycle of events to bigger and more complex LPOs. In the first experiment we will consider thin LPOs while in the second we will consider dense LPOs.

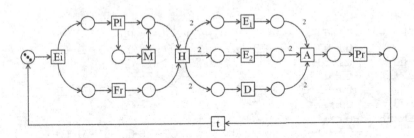

Fig. 4. A p/t-net describing a workflow

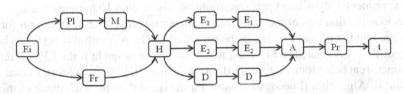

Fig. 5. Skeleton of an LPO of the partial language of the p/t-net shown in Figure 4

Experiment 1. *We consider 5 LPOs $lpo_1, \ldots, lpo_5$ of the partial language of the p/t-net of Figure 4. Each LPO is a composition of copies of the LPO shown in Figure 5. The number of copies varies from 6 to 120. The resulting LPOs are connected as shown in Figure 6, resulting in three parallel threads of events. We decide the verification problems using the algorithms I to VI.*

	lpo_1	lpo_2	lpo_3	lpo_4	lpo_5
number of copies	6	15	30	60	120
number of events	252	630	420	1260	5040
runtime in ms					
Alg I (cuts)	480	15091	-	-	-
Alg II (tokenflows, [4])	170	3002	25364	269862	-
Alg III (tokenflows, [24], Dinic)	3	15	55	250	1336
Alg IV (tokenflows, [24], preflow-push)	3	11	47	216	1130
Alg V (compact tokenflows, Dinic)	4	11	43	173	871
Alg VI (compact tokenflows, preflow-push)	3	12	36	148	699

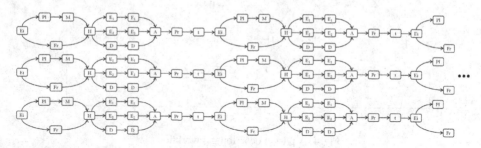

Fig. 6. A composition of copies of the LPO shown in Figure 5 leading to a thin LPO

In Experiment 1 algorithm I performs quite poorly. Within 10 minutes this algorithm only decides the first two of the given verification problems. The main reason for this is the huge number of cuts existing in the specified LPOs. Algorithm II performs better than algorithm I. As stated in [4] using tokenflows is reasonable if the LPO describes some concurrent behaviour. Algorithm III, as described in [24], is $|V|$-times faster then algorithm II. Algorithm II needs to construct a maximal flow for each place of the p/t-net and each event of the LPO. Algorithm III and IV only construct one maximal flow for each place. Algorithms V and VI use the new compact definition of tokenflows. As shown in the previous sections compact tokenflows regard only skeleton arcs instead of the transitive relation of the specified LPO. Experiment 1 contains a lot of concurrency, but even in this case the difference between the skeleton and the transitive relation matters. Both algorithms V and VI perform much better than the algorithms III and IV.

To further investigate the difference between all considered algorithms we provide a second experiment and increase the number of described dependencies between events of the tested LPOs.

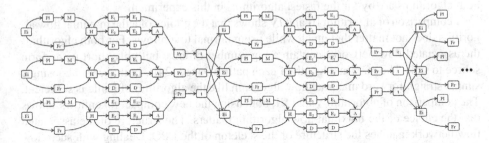

Fig. 7. A composition of copies of the LPO shown in Figure 5 leading to a dense LPO

Experiment 2. *We consider 5 LPOs $lpo_1, \ldots, lpo_5$ of the partial language of the p/t-net of Figure 4. Each LPO is a composition of copies of the LPO shown in Figure 5. The number of copies varies from 6 to 120. The resulting LPOs are connected as shown in Figure 7. We again decide the verification-problem using the algorithms I to VI.*

	lpo_1	lpo_2	lpo_3	lpo_4	lpo_5
number of copies	6	15	30	60	120
number of events	252	630	420	1260	5040
runtime in ms					
Alg I (cuts)	126	558	2424	12175	84605
Alg II (tokenflows, [4])	194	3602	32806	334647	-
Alg III (tokenflows, [24], Dinic)	3	16	70	334	1955
Alg IV (tokenflows, [24], preflow-push)	3	14	65	342	1936
Alg V (compact tokenflows, Dinic)	4	11	42	170	881
Alg VI (compact tokenflows, preflow-push)	4	9	35	146	648

In Experiment 2 algorithm I preforms better then in Experiment 1. The number of cuts is reduced and the algorithm is able to solve the first four tests within 10 minutes. Nevertheless, its runtime is poor. This time algorithm II performs worse then algorithm I. The number of transitive arcs to be considered is much higher than in Experiment 1. The notion of tokenflows highly depend on the number of arcs of a given

LPO. By the same reason Algorithm III and IV perform worse than in Experiment 1. Again, the runtime of algorithm II is $|V|$-times the runtime of algorithm III. Algorithm V and VI use the new compact definition of tokenflows. Their runtime almost equals the runtime given in Experiment 1. The LPOs shown in Figure 6 and 7 have almost an identical set of skeleton arcs. The runtime of algorithm V and VI is by this means independent of the number of described transitive dependencies between events. Again, both algorithms are by far the fastest algorithms in this experiment.

A comparison of algorithm III and IV shows that the choice of a the maximal flow algorithm does not matter that much while using normal tokenflows. For both algorithms the associated flow networks contain a big number of arcs, but only short paths form source to sink. The maximal length of each path is 3 (see [24]). In that case choosing a simple straightforward maximal flow algorithm like the algorithm of Dinic is sufficient. The comparison of algorithm V and VI, both using the new compact tokenflows, shows that the choice of the maximal flow algorithm matters. The structure of the associated flow network matches the structure of the skeleton of the LPO. Dealing with such flow networks using a preflow push algorithms with a gap-heuristic leads to the best runtime results. Notice, that the gap heuristic leads to an algorithm that only calculates the value of a maximal-flow and not the flow function itself. If a compact tokenflow should be constructed the algorithm of Dinic is the best choice.

6   Conclusion

We have shown a new compact definition of the partial language of a p/t-net. The new definition is based on the idea of tokenflows, since the number of arcs does not grow exponentially in the number of given events of an LPO. This new definition is only defined on the skeleton arcs of a given LPO such that it is not prone to the number of described dependencies. Compact tokenflows abstracts from the distribution of token as far as possible. In contrast to interlaced tokenflows only a single tokenflow is needed. All this leads to a definition which is optimized for deciding the verification problem. We have presented a corresponding verification algorithm and experimental results of its implementation in VipTool. We compared these results to all existing reasonable alternative algorithms.

An important topic of future research is to investigate if the new definition is applicable in the field of synthesis of p/t-net or unfolding a given p/t-net to its set of runs. In both fields algorithms using tokenflow lead to fast algorithms if the number of dependencies of the occurring LPOs are small. We hope that the new compact definition of tokenflows leads to similar results as for the verification problem, i.e. fast algorithms not prone to the structure of the given LPO or p/t-net.

References

[1]   Ahuja, R.K., Magnanti, T.L., Orlin, J.B.: Network flows: Theory, algorithms, and applications. Prentice-Hall, Englewood Cliffs (1993)

[2]   Baumgarten, B.: Petri-Netze: Grundlagen und Anwendungen. Spektrum, Heidelberg (1996)

[3] Bergenthum, R.: Algorithmen zur Verifikation von halbgeordneten Petrinetz-Abläufen: Implementierung und Anwendungen. Diplomarbeit, Katholische Universität Eichstätt-Ingolstadt (2006)

[4] Bergenthum, R., Desel, J., Juhás, G., Lorenz, R.: Can I Execute My Scenario in Your Net? VipTool Tells You! In: Donatelli, S., Thiagarajan, P.S. (eds.) ICATPN 2006. LNCS, vol. 4024, pp. 381–390. Springer, Heidelberg (2006)

[5] Bergenthum, R., Desel, J., Lorenz, R., Mauser, S.: Synthesis of Petri Nets from Finite Partial Languages. Fundamenta Informaticae 88(4), 437–468 (2008)

[6] Bergenthum, R., Desel, J., Mauser, S.: Comparison of Different Algorithms to Synthesize a Petri Net from a Partial Language. In: Jensen, K., Billington, J., Koutny, M. (eds.) ToPNoC III, LNCS, vol. 5800, pp. 216–243. Springer, Heidelberg (2009)

[7] Bergenthum, R., Desel, J., Mauser, S., Lorenz, R.: Synthesis of Petri Nets from Term Based Representations of Infinite Partial Languages. Fundamenta Informaticae 95(1), 187–217 (2009)

[8] Bergenthum, R., Juhás, G., Lorenz, R., Mauser, S.: Unfolding Semantics of Petri Nets Based on Token Flows. Fundamenta Informaticae 94(3-4), 331–360 (2009)

[9] Bergenthum, R., Lorenz, R., Mauser, S.: Faster Unfolding of General Petri Nets Based on Token Flows. In: van Hee, K.M., Valk, R. (eds.) PETRI NETS 2008. LNCS, vol. 5062, pp. 13–32. Springer, Heidelberg (2008)

[10] Best, E., Devillers, R.: Sequential and concurrent behaviour in Petri net theory. Theoretical Computer Science 55(1), 87–136

[11] Desel, J., Juhás, G.: "What Is a Petri Net?" Informal Answers for the Informed Reader. In: Ehrig, H., Juhás, G., Padberg, J., Rozenberg, G. (eds.) Unifying Petri Nets, LNCS, vol. 2128, pp. 1–25. Springer, Heidelberg (2001)

[12] Desel, J., Juhás, G., Lorenz, R., Neumair, C.: Modelling and Validation with VipTool. In: van der Aalst, W.M.P., ter Hofstede, A.H.M., Weske, M. (eds.) BPM 2003. LNCS, vol. 2678, pp. 380–389. Springer, Heidelberg (2003)

[13] Desel, J., Reisig, W.: Place/Transition Petri Nets. In: Reisig, W., Rozenberg, G. (eds.) APN 1998. LNCS, vol. 1491, pp. 122–173. Springer, Heidelberg (1998)

[14] Dinic, E.A.: Algorithm for Solution of a Problem of Maximum Flow in a Network with Power Estimation. Soviet Math Doklady 11, 1277–1280 (1970)

[15] Esparza, J., Römer, S., Vogler, W.: An Improvement of McMillan's Unfolding Algorithm. Formal Methods in System Design 20(3), 285–310 (2002)

[16] Ford, L.R., Fulkerson, D.R.: Maximal Flow Through A Network. Canadian Journal of Mathematics 8, 399–404 (1956)

[17] Goltz, U., Reisig, W.: Processes of Place/Transition-Nets. In: Díaz, J. (ed.) ICALP 1983. LNCS, vol. 154, pp. 264–277. Springer, Heidelberg (1983)

[18] Goltz, U., Reisig, W.: The Non-sequential Behavior of Petri Nets. Information and Control 57(2/3), 125–147 (1983)

[19] Grabowski, J.: On partial languages. Fundamenta Informaticae 4(2), 427–498 (1981)

[20] Janicki, R., Koutny, M.: Structure of concurrency. Theoretical Computer Science 112(1), 5–52 (1993)

[21] Juhás, G., Lorenz, R., Desel, J.: Can I Execute My Scenario in Your Net? In: Ciardo, G., Darondeau, P. (eds.) ICATPN 2005. LNCS, vol. 3536, pp. 289–308. Springer, Heidelberg (2005)

[22] Karzanov, A.V.: Determining the maximal flow in a network by the method of preflows. Soviet Mathematics Doklady 15, 434–437 (1974)

[23] Kiehn, A.: On the Interrelation Between Synchronized and Non-Synchronized Behaviour of Petri Nets. Elektronische Informationsverarbeitung und Kybernetik 24(1/2), 3–18 (1988)

[24] Lorenz, R.: Szenario-basierte Verifikation und Synthese von Petrinetzen: Theorie und Anwendungen, Katholische Universität Eichstätt-Ingolstadt. Habilitation (2006)

[25] Lorenz, R., Mauser, S., Bergenthum, R.: Testing the executability of scenarios in general inhibitor nets. In: Basten, T., Juhás, G., Shukla, S.K. (eds.) Proc. of Application of Concurrency to System Design 2007, pp. 167–176. IEEE Computer Society (2007)

[26] Malhotra, V.M., Kumar, M.P., Maheshwari, S.N.: An $O(|V|^3)$ Algorithm for Finding Maximum Flows in Networks. Information Processing Letters 7(6), 277–278 (1994)

[27] McMillan, K.L., Probst, D.K.: A Technique of State Space Search Based on Unfolding. Formal Methods in System Design, 45–65 (1992)

[28] de Oliveira Oliveira, M.: Hasse Diagram Generators and Petri Net. Fundamenta Informaticae 105(3), 263–289 (2012)

[29] Peterson, J.L.: Petri net theory and the modeling of systems. Prentice-Hall, Englewood Cliffs (1981)

[30] Petri, C.A.: Kommunikation mit Automaten, Dissertation, Technische Universität Darmstadt (1962)

[31] Pratt, V.: Modelling Concurrency with Partial Orders. International Journal of Parallel Programming 15 (1986)

[32] Reisig, W.: Petrinetze: Eine Einführung. Springer, Berlin (1986)

[33] Solé, M., Carmona, J.: Process Mining from a Basis of State Regions. In: Lilius, J., Penczek, W. (eds.) PETRI NETS 2010. LNCS, vol. 6128, pp. 226–245. Springer, Heidelberg (2010)

[34] Vogler, W. (ed.): Modular Construction and Partial Order Semantics of Petri Nets. LNCS, vol. 625. Springer, Heidelberg (1992)

[35] Winskel, G.: Event structures. In: Brauer, W., Reisig, W., Rozenberg, G. (eds.) APN 1986. LNCS, vol. 255, pp. 325–392. Springer, Heidelberg (1987)

Unifying the Semantics
of Modular Extensions of Petri Nets

Alexis Marechal and Didier Buchs

Centre Universitaire d'Informatique
Université de Genève
7 route de Drize, 1227 Carouge, Suisse

Abstract. Modularity is a mandatory principle to apply Petri nets to real world-sized systems. Modular extensions of Petri nets allow to create complex models by combining smaller entities. They facilitate the modeling and verification of large systems by applying a divide and conquer approach and promoting reuse. Modularity includes a wide range of notions such as encapsulation, hierarchy and instantiation. Over the years, Petri nets have been extended to include these mechanisms in many different ways. The heterogeneity of such extensions and their definitions makes it difficult to reason about their common features at a general level. We propose in this article an approach to standardize the semantics of modular Petri nets formalisms, with the objective of gathering even the most complex modular features from the literature. This is achieved with a new Petri nets formalism, called the LLAMAS Language for Advanced Modular Algebraic Nets (LLAMAS). We focus principally on the composition mechanism of LLAMAS, while introducing the rest of the language with an example. Our approach has two positive outcomes. First, the definition of new formalisms is facilitated, by providing common ground for the definition of their semantics. Second, it is possible to reason at a general level on the most advanced verification techniques, such as the recent advances in the domain of decision diagrams.

1   Introduction

Through the years the original version of Petri nets has been extended to integrate a wide range of notions such as token colors, time and probabilities. In the last decade an initiative has risen to tackle the diversity of the many Petri nets variants, to improve the common understanding of the paradigm and to facilitate the integration of Petri net tools. This initiative took the form of the international ISO-IEC 15909 standard [12], which includes Petri Net Markup Language (PNML) [8], a markup language meant to allow the communication between Petri net tools. Currently, the development of the ISO-IEC 15909 standard is focusing on extensions of the original Petri nets [7].

One of these extensions is modularity, that is, the ability to define complex systems by assembling smaller entities. There are many Petri nets formalisms that include the notion of modularity. This article proposes an approach to standardize the formal definitions of these formalisms. We describe this approach in

J.-M. Colom and J. Desel (Eds.): PETRI NETS 2013, LNCS 7927, pp. 349–368, 2013.

Section 2, where we propose to define a new modular formalism called LLAMAS. In Section 3 we study the state of the art of modular Petri nets formalisms. We then get to the core of the article by introducing the principles of the LLAMAS language. We describe first its composition mechanism in Section 4 and then the rest of the language by means of an example in Section 5. Section 6 shows an example of translation from Modular PNML to LLAMAS. We briefly mention in Section 7 how to perform verification in LLAMAS using Decision Diagrams. We conclude and mention ongoing and future work in Section 8.

2 Preliminaries

2.1 Motivation and Approach

There is a wide variety of mechanisms that have been defined to implement the notion of modularity in Petri nets. Two main approaches have been followed by their authors to define the semantics (i.e., the possible executions) of these mechanisms. Many authors define a translation from their modular formalisms to "flat" (non-modular) Petri nets. In some cases (e.g., [6]) this translation is quite complex and difficult to understand. In some other cases (e.g., [1,16]), such translation is simply not attempted as it would be too complex or yield Petri nets with infinitely many places or transitions. The semantics of these formalisms is usually defined as compositions of the transition relations of the leaf (non-hierarchical) modules. This low-level operation is usually very complex.

The solution we propose follows the path of the ISO-IEC 15909 standard and PNML. The approach of PNML consists in creating a language to serve as a syntactic platform (i.e., a metamodel) for the definition of Petri nets variants. To extend this syntactic standard to semantic considerations, we propose an approach comprised of two artifacts. First, a metamodel to grasp the syntactic concepts of modular formalisms (similarly to what is done in PNML). Second, a modular formalism to serve as a semantic platform for existing and new modular formalisms. We described a first proposition for the metamodel in [17]. Here, we focus on the most complex part of the approach: the semantic platform.

2.2 Formal View of the Approach

Modular formalisms allow to create complex models by composing smaller entities. Let F be a modular formalism, and let M_F be the set of modules that can be expressed with this formalism and, for each module $m \in M_F$, let $Sem_F(m)$ be its Labeled Transition System (LTS). We can define a composite module as an expression $m = \circ_i(m_1, \ldots, m_n)$ where $m_1, \ldots, m_n \in M_F$ are modules and $\circ_i$ is a composition function $\circ_i : \wp(M_F) \to M_F$. Thus, a module is the composition of modules, that in turn may be compositions of smaller modules, and so on, until reaching the level of leaf modules, i.e., modules that cannot be decomposed anymore. Let us note $LM_F \subseteq M_F$ the set of all the leaf modules in F and $Comp_F = \{\circ_1, \ldots, \circ_m\} \subseteq (\wp(M_F) \to M_F)$ its set of composition operators.

Let us go back to the two approaches we mentioned at the beginning of this section. A syntactic definition of the semantics of F consists in defining the semantics of the leaf modules and a function $Flatten : Comp_F \times \wp(M_F) \rightarrow LM_F$. This function takes a composite module and returns an equivalent leaf module. With it, we can define the semantics of composite modules by defining $Sem_F(\circ_i(m_1, \ldots, m_n)) = Sem_F(flatten(\circ_i, m_1, \ldots, m_n))$. While this approach is commonly applied to usual fusions of places and transitions, it is not always well suited to more complex cases. Indeed, as we mentioned previously, the $Flatten$ function can be extremely complex or sometimes give unwanted results (e.g., infinite nets). Moreover, because the result of this function is a non-modular model, the identity of the different modules is lost at runtime.

On the other hand, a semantic definition is based on an existing formalism SP (Semantic Platform), that has its own composition mechanisms $Comp_{SP}$ and semantics Sem_{SP}. To create a semantic definition of a formalism F, one must provide two functions, a function $Tr_{mod} : LM_F \rightarrow M_{SP}$ to translate the leaf modules and a function $Tr_{comp} : Comp_F \rightarrow Comp_{SP}$ for the composition operators. Then Tr_{mod} is extended to composite modules by defining inductively $Tr_{mod}(\circ_i(m_1, \ldots, m_n)) = Tr_{comp}(\circ_i)(Tr_{mod}(m_1), \ldots, Tr_{mod}(m_n))$. Finally, the semantics of F is defined as $\forall m \in M_F$, $Sem_F(m) = Sem_{SP}(Tr_{mod}(m))$. A semantic definition has the advantage of preserving the modules identity at runtime. Most formalisms that have semantic definitions use traditional LTS as a semantic platform SP, but they define their own LTS compositions $Comp_{SP}$. Moreover, working at the level of the LTS is a low-level operation, far from the modeling expressivity of Petri nets. This paradigm shift can have negative consequences. For instance, LTS are not suited for handling concurrency, and often the LTS resulting from the transformation of Petri nets are extremely large.

To avoid this paradigm shift while keeping the benefits of a semantic definition, we propose to use a single Petri nets formalism as a common semantic platform for Petri nets variants. This would facilitate the understanding of their definitions, and would allow reasoning at a general level on computational techniques. This approach is akin to virtual machines in the domain of programming languages. Of course, the semantic platform must be expressive enough to handle at least the existing formalisms. By this we mean that it should be possible to create a translation from each modular formalism to the semantic platform that would preserve its semantics. For any existing modular formalism F with a previously defined semantics $OldSem_F$, we should have:

$$\exists\, Tr : M_F \rightarrow M_{SP} \text{ s.t. } \forall m \in M_F,\ OldSem_F(m) \cong Sem_{SP}(Tr(m))$$

where $\cong$ is an isomorphism between LTS.

2.3 Boundaries of This Article

While an ideal solution would have been to use an existing formalism as a semantic platform, we did not find any expressive enough candidate in the literature. Because of this, we propose a new formalism specifically tailored to serve as a semantic platform for modular Petri nets formalisms. We named this formalism the LLAMAS Language for Advanced Modular Algebraic Nets (LLAMAS).

A formal definition of the language is out of the scope of this article, but it is given as a technical report in [19].

Our proposition is strictly limited to the definitions of modularity in the Petri nets formalisms. We rely on the notion of orthogonality as in [7] to limit ourselves to the notion of modularity. By this we mean that we expect other extensions of Petri nets (like time and probabilities) to be compatible with our approach. For instance, we consider that a modular temporal formalism can be defined by combining our modular mechanisms with temporal considerations. In particular, we do not consider the problem of the kind of data types used in the formalism. In the formal definition of LLAMAS we used Algebraic Abstract Data Types (AADTs) to define the data types, a general specification language that is already used in the ISO-IEC 15909 standard. To simplify, in the examples of this article we will only use trivial data types (black tokens, natural numbers).

2.4   State of the Art

Modular PNML [13] is an important candidate for a semantic platform. As a matter of fact, Modular PNML was the initial inspiration for the work presented in this article. Nevertheless, the composition mechanism used by Modular PNML is limited to the syntactical fusion of places and transitions. We consider that this is not enough to express the semantics of some important composition mechanisms defined in the literature, such as the non-deterministic compositions from [4], the parametric compositions from [6] or the complex compositions from [1]. It must be noted that Modular PNML defines a mechanism to compose the definitions of the data types used in the modules. This is an important feature that is not included in LLAMAS because, as mentioned previously, the definition of data types is out of our scope. We consider that our formalism could be easily extended to include Modular PNML's data types definitions, and thus both works can be seen as complementary.

By far, the most widely used modular formalism is the Hierarchical CP-nets [11]. It is a very general formalism, supported by a well established tool, and as such it is an interesting candidate to serve as a standard semantic platform. Nevertheless, similarly to the case of Modular PNML, some composition mechanisms from the literature cannot be represented by this formalism.

Our work can be compared to the process algebras such as CSP [9]. These languages aim to be universal languages that allow to define the behavior of complex formalisms. Similarly to the LTS, the modeling paradigm of process algebras is quite different from the one of Petri nets. By comparison, LLAMAS is itself a Petri nets formalism, and thus we avoid this paradigm shift.

3   Common Features of Modular Petri Nets Formalisms

To create a standard for modular Petri nets, it is obviously necessary to study the characteristics of the existing formalisms, which we did in [18]. Here we focus on some prominent ones about the nature of modules and the compositions.

3.1   The Nature of Modules

Encapsulation. Most modular formalisms include the notion of encapsulation. This means that each module defines a set of public elements that can be accessed by other modules, and a set of internal elements that are hidden and can only be accessed from within the module itself. The set of public elements of a module constitutes its *interface*. In some formalisms, the interface is a set of places and/or transitions of the Petri net itself (sometimes defined by labels on these places and transitions). This is the case of most formalisms that use the classical fusion of places and/or transitions, e.g., [13,11]. Formalisms that have more complex composition mechanisms usually define interfaces with additional elements that do not belong to the Petri net of the module, e.g., [1,16].

Hierarchy. The notion of encapsulation is usually linked to the notion of hierarchy. In hierarchical formalisms, a module (called *container*) can encapsulate other modules (called *submodules*). The submodules of a container can in turn be containers of their own submodules, and so on.

Instantiation. The instantiation of modules is a mechanism to define many identical or at least similar modules (instances) based on a blueprint (class). Many formalisms ([13,11,6,1,16]) define a static instantiation mechanism, which allows to instantiate modules during the definition of the model, prior to any computation. On the other hand, some formalisms define dynamic instantiation, that is, the ability to create instances during the computation of the model. Most of them (e.g., [1,16,15]) use Nets Within Nets (NWN) paradigm [21]. One exception is [11] which uses invocation transitions.

3.2   Compositions

The mechanisms that define the interactions between modules are called *composition mechanisms*. The Petri nets literature defines a wide variety of such mechanisms. To describe them, we could make very complex classifications but, for simplicity reasons, we will only mention one classification criterion: state-based vs. event-based mechanisms. Many formalisms define the compositions between modules by relating the places in one module with the places in another module. By far, the most common mechanism for this is the fusion of places. There are some mechanisms that define more complex compositions of places, e.g., the hierarchical transitions in the M-nets family [6]. Other formalisms compose the modules by connecting their transitions. In this case, the composition of a set of transitions produces a new event whose behavior depends on the transitions that were composed. Again, while fusion of transitions is the most common mechanism in this category, there exists more complex mechanisms, like the non-deterministic synchronizations from the CP-nets with channels [4]. Note that many formalisms define both kinds of compositions.

3.3   Summary

In short, the general structure of a hierarchical module may be represented by a tuple $HModule = \langle Spec, Net, I, Subs \rangle$, where $Spec$ is a definition of the data types used, Net is a Petri net that defines the internal behavior of the module, I is the interface of the module and $Subs$ is a set of submodules. A complete model may be represented as a tuple $HModel = \langle HModules, Comps, Inst \rangle$, where $HModules$ is a set of modules, $Comps$ is a set of compositions that define the interactions between these modules and $Inst$ is an instantiation mechanism.

The elements $Spec$ and Net are out of the scope of this article. A standard definition of the Petri nets (Net) is the main focus of the first part of the ISO-IEC 15909 standard, and [13] gives significant contributions to a standard for a modular data types specification $Spec$. Instead, LLAMAS focuses on the other elements, the interface I (see Section 5.3), the hierarchical mechanism $Subs$ (Section 5.4), the instantiation mechanism $Inst$ (Section 5.4 again) and the composition mechanism $Comps$, which is the subject of the next section.

4   The LLAMAS Composition Mechanism

In this section we will explore the composition mechanism of LLAMAS. A complete formal description of this mechanism is out of the scope of this article. Instead, we will explore its main features by means of simple examples. At the end of the section, we will give a partial semantics of our composition mechanism by defining a transformation to traditional Petri nets. The interested reader can find the complete formalization in [19].

The goal of this section is to define a single composition mechanism expressive enough to encompass most if not all the mechanisms that have been defined in the literature. We named it the LLAMAS Composition Mechanism (LCM). We mentionned in Section 3 that the composition mechanisms can be classified as state-based or event-based. One possibility was to define the LCM as a mixture of both categories, but we consider that this would be unnecessarily complex. Instead, we defined the LCM as an event-based mechanism, and we ensured that it is expressive enough to *simulate* state-based mechanisms (as explained in Section 2.2). A major advantage of defining the LCM solely as an event-based mechanism lies in the verification activity as it will be mentioned in Section 7.

In general terms, a composition in the LCM is a finite set of *events* that, combined, form another event. The result of a composition is a new event that may itself participate inside other compositions, forming a hierarchical structure. Let us start by the most simple example of composition, the fusion of two transitions t1 and t2. This fusion creates a new event c1 that behaves as both transitions fired simultaneously (we note this c1 = merge(t1,t2)). c1 may be in turn be fused with a third transition t3, forming a new event called c2 that behaves as the three transitions fired simultaneously (c2 = merge(c1,t2)= merge(merge(t1,t2),t3)). At this point, one may wonder if it is possible for a transition to participate in more than one composition. In our example, the question is if it would be possible to define two compositions c1 = merge(t1, t2) and c3 = merge(t1, t3), both using independently the transition t1. The answer to this question gives place to the first novel feature of the LCM: *bindings*.

4.1   Bindings

There are three possible approaches when considering if the events can participate independently in more than one composition:

- Each event can participate in at most one composition. This is the case of the usual fusion sets, and also of Modular PNML [13].
- Each event may participate in any number of compositions. This is the case of the CP-nets with channels [4] and the family of the M-nets [6].
- Finally, we have a hybrid approach, exemplified by CO-OPN [1] and the reference nets [15]. The events are synchronized by means of *calls*. If an event m1 calls another event m2, it means that m1 cannot be executed independently of this call, but m2 could be called by other events without involving m1.

In the latter case, we say that m1 was *bound* by the composition, and m2 was *called* by the composition (i.e., m1 cannot participate in other compositions but m2 is free to do so). In the LCM, we note this composition merge(bind(m1), call(m2)). Thus, as seen previously, the fusions of transitions in Modular PNML use only *bindings*, and the compositions in CP-nets with channels and the M-nets use only *calls*. Bindings are a purely syntactical feature that can be checked statically. We may refine now our previous definition of composition to include the notion of bindings. For now, we will say that a composition in the LCM is defined by two finite sets of events, a collection of *bound* events and a collection of *called* events. For instance, if we define a composition c4 = merge(bind(t1), bind(t2), call(t3)), t1 and t2 cannot participate in any other composition, while t3 can. Binding multiple times the same event (e.g., merge(bind(t1), bind(t1))) is also forbidden. Note that the notion of bindings is more general than the mechanisms found in the literature. Indeed, to the best of our knowledge, no composition mechanism allows to define a behavior like the one of c4, where two transitions are bound and one is called. Usually, all the transitions are bound, or all of them are called, or only one of

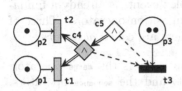

Fig. 1. Two compositions

them is bound and all the others are called. Graphically, we represent the compositions by diamonds, bindings by double arrows, and calls by dashed arrows. We gave a graphical representation of c4 in Fig. 1. This figure also shows a composition noted c5 = merge(bind(c4), call(t3)). Executing c5 is the same as executing simultaneously t1 and t2 once, and t3 twice.

The shades of the transitions and compositions in Fig. 1 represent their activity, which is the subject of the next paragraph. The symbol ∧ inside the compositions will be explained afterwards. In the following, to simplify the notation, we may omit the keyword call. For instance, c1 = merge(t1, t2) is the same as c1 = merge(call(t1), call(t2)).

4.2 Active vs Passive Events

We have seen that some events in the LCM are defined as aggregates of other events. In this context, it would make sense to define some basic events solely for the purpose of being part of a composition, and not with the intent of being executed independently. For instance, the port transitions in the Hierarchical CP-nets [11] are only executed if they are called by some external event. We call such transitions *passive*. While in most formalisms passive events are only allowed in the interface of the modules, we generalize this to allow any event (transition or composition) to be active or passive. An event is *active* if it can be executed by itself, and passive if it must be triggered by some composition to be activated. The LTS of a model is labelled only with active events. Graphically, active transitions are represented by black rectangles and active compositions by white diamonds, while their passive counterparts are grey. In Fig. 1, transition t3 and composition c5 are active, and all the other events are passive. The LTS of the net in Fig. 1 is shown in Fig. 2.

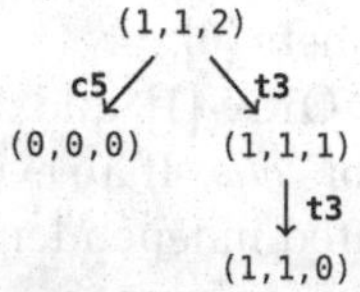

(1,1,2)

c5 / \ t3

(0,0,0)   (1,1,1)

↓ t3

(1,1,0)

Fig. 2. LTS of Fig. 1

4.3 Composition Operators

Up to now we have explored two features of the LCM: the bindings and the active/passive events. We will now explore a more complex feature, the *composition operators*. We said previously that a composition is a set of called events and a set of bound events, and we gave the example c1 = merge(t1, t2). Composition c1 behaves as a classical fusion of two transitions. This is indicated by the use of the composition operator merge. As we mentioned previously, the fusion is not the only way of combining the behavior of a set of events. There are other possible combinations, and each one will be indicated with a composition operator. Let us refine once again our definition of a composition in the LCM. A composition is now defined as a finite set of bound events and a finite set of called events, all linked by a composition operator. The composition operator determines how the behaviors of the individual events will be combined in order to define a new atomic event. Notice that this is consistent with the mindset of the usual Petri nets, where the atomicity of complex events is already a fundamental principle. In the LCM, we define five composition operators. Three of them are called *behavioral operators* and two of them are called *observers*.

Behavioral Operators. The first behavioral operator is the merge operator, mentioned previously. The other two are the any and the sequence operators. These three operators are taken from CO-OPN [1], a formalism with a powerful compositional mechanism. Let us start with the operator any. A composition c6 = any(t1, t2) corresponds to a non-deterministic choice between t1 and t2. This means that, whenever c6 is executed, one between t1 and t2 is executed, but not both. The merge and any operators are close to the operators from the ITS [20].

The last behavioral operator is the sequence. A composition c7 = sequence(t1, t2) is an event that first executes t1 and then t2. Thus, t2 can use the resources produced by t1. Let us stress out that c7 combines the behavior of t1 and t2 to form a single transactional event. If t2 cannot be fired the whole event is cancelled, and no other event can occur between the executions of t1 and t2. Note that, unlike the merge and any, the sequence operator is not commutative.

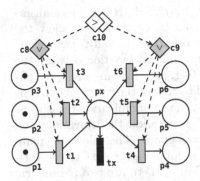

Fig. 3. any and sequence example

Consider for instance Fig. 3. It shows three compositions, c8 = any(t1,t2,t3), c9 = any(t4,t5,t6), and c10 = sequence(c8,c9). The any operator in c8 and c9 is represented with a logical disjunction ∨, and the sequence operator in c10 with the symbol >. Because the sequence operator is not commutative, we must show graphically the order of the elements that are composed. We do this by representing one diamond per composed element, to be read from left to right. There are only two active events in Fig. 3, the transition tx and the composition c10. Whenever c10 is executed, one of the three leftmost places loses its token, and one of the three rightmost places receives one token, all in one action. Both choices are non-deterministic, and thus the P/T Petri net equivalent of c10 would have 9 transitions (not counting tx). Because every execution of c10 is an atomic event, the transition tx will in fact never be executable.

The three behavioral operators are associative, the merge and any are also commutative and they are both distributive with respect to each other. A distinction between our work and the compositions in CSP [9] is our commitment to transactionality. In the LCM, atomic events are composed to build other atomic events. For instance, composition c10 in Figure 3 defined a single atomic event. While the parallel composition || from CSP has a similar behavior, it is not the case for the sequential composition ≫. Moreover, our compositions are always defined at the level of single events, whereas in CSP, as in most process algebras, they are defined at the level of the whole CSP processes.

Observers. The behavioral operators allow to create new events by combining the behaviors of other events. While this allows to represent a great number of composition mechanisms, it is not enough for the most complex ones. To cover the most extreme cases, we defined two operators called *observers*. They are the not and the read operators, which are respectively generalizations of the well known inhibitor arcs and read arcs. These operators allow to define special events that do not modify the state of the model, they only observe if some events *can* be fired or not. For instance, a composition not(t1) is enabled if and only if t1 is not. While the not operator checks that an event cannot be executed, the read checks if a given event can be executed. Please note that the observers are a central feature of the LCM. Without them, we could not handle many composition mechanisms such as the stabilization from CO-OPN [1], the complex hierarchy of the M-nets [6], or the special arcs from both the Object

Petri nets [16] and the reference nets [15]. Graphically, we represent the **not** operator with an exclamation mark ! and the **read** with the symbol ⊙.

4.4 Parametric Events

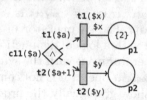

Fig. 4. Parametric events

Many High Level Petri Net (HLPN) formalisms that define event-based compositions allow the events to be parametric. Usually, parametric events allow to transmit information between the composed events (e.g. [4]). In some cases, the parameters are used to choose the recipient of a synchronization (e.g., the communication transitions in [6]). In the LCM, we allow the transitions *and the compositions* to define a list of parameters and a guard. This allows to create compositions like the the one in Fig. 4. We use a dollar sign $ next to the variables to distinguish them from values. In this figure, there are two parametric transitions, **t1($x)** which removes a token **$x** from place **p1** and **t2($y)** which adds a token **$y** to **p2**. **$x** and **$y** are the respective formal parameters of the transitions. There is also a composition **c11($a)= merge(t1($a), t2($a+1))**. In **c11**, **$a** and **$a+1** are the respective effective parameters of **t1** and **t2**. When **c11** is executed, the token with value 2 is removed from **p1** and a token with value 3 is added to **p2**.

4.5 Recursion

We saw in the example of Fig. 1 that an event can participate multiple times in the same composition (transition **t3** was called twice by composition **c5**). Similarly, a composition can participate inside itself, i.e., it can be recursive. As is usual in recursive frameworks, there is a danger of defining infinite computations that must be taken care of

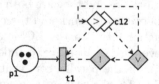

Fig. 5. Recursion example

by the modeler. Fig. 5 is an example of recursion. It shows a transition **t1** that removes a token from place **p1** and a composition **c12 = sequence(t1, any(not(t1), c12))**.

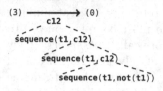

Fig. 6. Behavior of **c12**

This composition first calls **t1**, and then either **t1** cannot be executed again (i.e., **p1** is empty), or **c12** is called again. Thus, the place **p1** is emptied in a single action. This behavior is represented in Fig. 6. It shows the LTS of the example, labeled by **c12**, and the detail of the execution. With this pattern, we obtain the behavior of the well know reset arcs.

4.6 Graphical and Textual Notation - The Case against Labels

Through this section we have introduced the compositions with both a textual and a graphical notation. The graphical version of our compositions explicitly connects the elements that are composed by means of arrows. Some formalisms note their compositions by means of labels in the composed elements (e.g., [6,21,15]). Usually, this notation is difficult to read, as it defeats the intuition behind the graphical notation. On the other hand, an explicit representation of complex compositions may produce bloated graphics that can also be difficult to read. This is the reason for having both a graphical and a textual notation. Note that our current implementation (see section 7) only uses a textual notation, but we plan to implement a graphical version in the future.

4.7 Partial Semantics of the LCM

In Section 2.2 we mentioned that the modular mechanisms can have syntactic or semantic definitions. When they are applicable, syntactic definitions are usually more simple than semantic ones. The LCM is a complex mechanism that in practice cannot be fully defined with a syntactic definition. Nevertheless, the complete definition of the semantics of the LCM is too complex to be included in this article. Instead, we will give a syntactic definition of a reduced subset of the LCM, by translating some compositions to traditional Petri nets. For this translation to exist, the following conditions must be met:

a) Only the behavioral operators are allowed. The observers are generalizations of the inhibitor and read arcs. Thus, the compositions that use these operators cannot be represented with traditional Petri nets. Note that even if we used traditional inhibitor and read arcs it would be difficult (but not impossible) to represent the behavior of the observers in flat Petri nets.

b) No recursion is allowed, i.e., each composition must define a directed acyclic graph. Recursion leads to potentially infinite executions, which would be represented with infinitely many transitions.

The translation of the compositions that meet these conditions to flat Petri nets is done in two steps. First, the non-deterministic operator **any** is eliminated. This operation is similar to the search of the well known Disjunctive Normal Forms of propositional logics formulas. To illustrate this, we will for a moment represent our compositional operators with their graphical symbols. The **merge**, **any** and **sequence** are noted respectively as a conjunction $\wedge$, a disjunction $\vee$ and the symbol $>$. The operator $\vee$ follows the following distributivity rules for any events **ev1**, **ev2** and **ev3** and for any operator $op \in \{\wedge, >\}$:

- $(\text{ev1} \vee \text{ev2})\ op\ \text{ev3} = (\text{ev1}\ op\ \text{ev3}) \vee (\text{ev2}\ op\ \text{ev3})$
- $\text{ev1}\ op\ (\text{ev2} \vee \text{ev3}) = (\text{ev1}\ op\ \text{ev2}) \vee (\text{ev1}\ op\ \text{ev3})$

With these distributivity rules, it is easy to represent any composition as a composition **ev1** $\vee$ **ev2** $\vee$ ... $\vee$ **evn** with $n \in \mathbb{N}$ and no symbol $\vee$ inside any event **evi**. By using the inherent non-determinism of traditional Petri nets, we can replace that composition with the set of n events written separately. This set may be quite big, but it is always finite in the case of non-recursive compositions.

The second step is to translate the compositions with the operators $\wedge$ and $>$ to traditional transitions. For this, we extend the usual notions of Pre and Post conditions. A Petri net transition is defined by two sets Pre_t and $Post_t$, and the result of the firing of a transition t from a marking M is a marking M' defined by $M' = M - \sigma(Pre_t) + \sigma(Post_t)$, where σ is a substitution of the variables in t and $\sigma(Pre_t) \subseteq M$. To translate our compositions to flat Petri nets, we will inductively define pairs of Pre and Post conditions for each one of them. Let $+_m$ (resp. $-_m$) be the usual addition (resp. subtraction) of multisets.

- For any composition c = merge(ev1, ev2), we define
  $Pre_c = Pre_{ev1} +_m Pre_{ev2}$ and
  $Post_c = Post_{ev1} +_m Post_{ev2}$
- For any composition c = sequence(ev1, ev2), we define
  $Pre_c = Pre_{ev1} +_m (Pre_{ev2} -_m Post_{ev1})$ and
  $Post_c = Post_{ev2} +_m (Post_{ev1} -_m Pre_{ev2})$

If the events ev1 and ev2 are transitions, we use their usual Pre and Post conditions. If they are compositions, we apply our definition recursively. The process is guaranteed to be finite in the absence of recursive events. At the very end of the process the passive transitions are removed. Consider the composition c13 = merge(sequence(t1,t2),any(t3,t4)) in Fig. 7. The distributivity of any gives c13 = any(merge(sequence(t1,t2),t3),merge(sequence(t1,t2),t4)). Thus, the behavior of c13 will be represented by two transitions. With the equations given above, we obtain the two transitions in the right hand side of Fig. 7, where E1 is merge(sequence(t1,t2), t3) and E2 is merge(sequence(t1,t2), t4). To apply this technique to HLPN, the only restriction is that the variables must be renamed so that local variables in different events have different names. Note that this trivial example does not illustrate the methodological advantages of having a compositional mechanism.

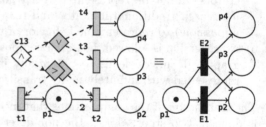

Fig. 7. From the LCM to flat Petri nets

5   The LLAMAS Language - An Example

In the previous section we explained in detail (even though informally) the composition mechanism of LLAMAS. In this section, we will introduce the rest of the language. For simplicity and space issues, we will only present the language by means of an example. A complete formalization is given in [19].

5.1   Internal Behavior of a Module - Petri Net

We will begin the description of our example with the Petri net from Fig. 8. Together with the compositions we will give in the next section, this Petri net will represent a bounded counter. It has two places named Bound and Counter. The initial value of the token in place Bound is a variable, and thus its actual value will be determined at instantiation (see Section 5.4). There are also three transitions

named GetBound, Reset, and Inc. All these transitions are passive (see Section 4.2). Notice
that, if it was an active transition, an execution of GetBound would remove the token from
place Bound. We will see in the next section that
this behavior will be restrained by the compositions. All three transitions have one parameter that unifies itself with the value taken
from the respective place.

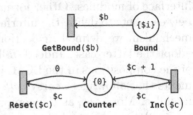

Fig. 8. Petri net example

5.2 Compositions

Fig. 9 shows four compositions that have been added to the example in Fig. 8.
The composition in the center is called ReadBound. It uses the observer read, whose
symbol is ⊙, and it binds the transition GetBound. The read operator indicates
that when the composition ReadBound is executed, it checks if transition GetBound
can be executed and it unifies the variable $b with the value taken from place
Bound, without any modification to the marking of the net (see Section 4.3).

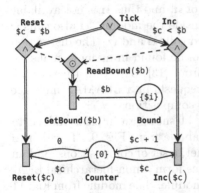

Fig. 9. Compositions example

Composition ReadBound binds GetBound (see Section 4.1), which means that GetBound cannot
participate in any other composition. Thus,
even though an execution of ReadBound removes
the token from place Bound, we ensured by
means of the composition ReadBound that this token will never actually be removed. The second composition, called Reset, *binds* the transition Reset, and *calls* the composition ReadBound,
and uses the merge operator. When it is executed, both composition ReadBound and transition Reset will be executed simultaneously, as
long as the guard of the composition ($c = $b)
is evaluated to *true*. Thus, the composition
Reset checks if the counter has reached its bound and resets it to 0. Composition Inc works in a similar way. It increases the value of the counter if it has
not reached the value of the bound. The final composition, Tick, binds both
compositions Reset and Inc with the operator any. Whenever Tick is executed, one
composition between Reset and Inc is also executed, but not both. If neither Reset
nor Inc are enabled, neither is Tick. All these compositions are passive and thus,
for now, we still did not define a single executable event in our example.

5.3 Interface - Basic LLAMAS Module

The notion of encapsulation is central in most modular formalisms. It implies
the existence of some elements in a module that are available to other modules (the interface), and some elements that are internal to the module. Some
modular Petri nets formalisms define some places and/or transitions as the

interface of modules. Other formalisms define entirely new elements to define the interface. The composition mechanism we defined in Section 4 requires that we adopt the latter case. Indeed, following the definition of the compositions in the LCM, we need interface elements that allow us to carry *calls* and *bindings* across modules. Moreover, as the compositions are directed, we need to distinguish between input elements and output elements (we respectively call them *services* and *requests*). Thus, we have four kinds of elements in our module interfaces: binding and non-binding services and binding and non-binding requests. Fig. 10

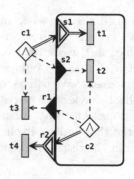

Fig. 10. Interface

illustrates the four possible combinations. It shows two transitions and one composition (resp. t1, t2 and c2) inside a module and two transitions and one composition (resp. t3, t4 and c1) outside the module. The interface of the module contains a binding and a non-binding service (resp. s1 and s2), represented by triangles directed towards the interior of the module. It also contains a binding and a non-binding request (resp. r2 and r1), represented by triangles directed towards the environment. c1 binds t1 by means of s1, and thus t1 is not available to other compositions, including compositions inside the module. c1 also calls t2 (by means of s2) and t3. Similarly, c2 binds t4 and calls t3 and t2. The distinction between input and output interface elements can be found in many formalisms, see for instance the import/export interfaces from [13], or the distinction between sockets and ports in [11]. Services and requests in a LLAMAS interface are parametric, similarly to the transitions and compositions.

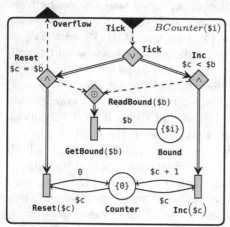

Fig. 11. A LLAMAS module

Fig. 11 shows an interface added to the example from Fig. 9. Together, a Petri net, a set of compositions, and an interface form a non-hierarchical LLA-MAS module. The module from Fig. 11 is called *BCounter*, and it takes one parameter $i to set the initial value of place Bound. The interface of *BCounter* is composed of a non-binding service called Tick and a non-binding request called Overflow. Service Tick calls composition Tick, thus making this composition available to the environment. The composition Reset now calls the request Overflow, apart from calling ReadBound and binding transition Reset. Thus, whenever there is an overflow, not only the counter will be reseted, but a notification will be sent to the container of *BCounter*. If the container is not able to respond to this request, the composition Reset will fail. The next section shows an example of such container.

5.4   Hierarchy and Static Instantiation

Another important feature of modular formalisms is hierarchy. In hierarchical Petri nets a module (called the *container*) may encapsulate other modules (called *submodules*), which in turn may encapsulate their own submodules. The hierarchy is a partial order relation between the modules in a system. Each container defines how its submodules communicate, both between themselves and with the container. A hierarchical LLAMAS module is thus composed of a Petri net, an interface, a set of submodules and a set of compositions (notice the similarity with our general definition in Section 3.3). The hierarchical mechanism we use in LLAMAS was inspired by Modular PNML [13]. In Modular PNML, each module defines a set of submodule sockets, and each submodule socket is denoted only by an interface, which defines the synchronization contract between the container and the submodule.

Before showing an example, let us mention one last important feature of LLAMAS: instantiation. An instantiation mechanism facilitates the definition of various identical (or similar) modules. It means that the modules in a system are copies of an initial definition, a blueprint. These copies may be fine tuned when they are created. Following the vocabulary of object-oriented languages, we call this blueprint a *class*, and its copies *instances* of the class. The instantiation may be *static*, i.e., performed once during the initial definition of the system, or *dynamic*, i.e., instances may be created during the execution of the system.

Fig. 12 shows a hierarchical module that uses three static instances of the module *BCounter* from Fig. 11 to model a clock. The module *Clock* contains three submodules sockets, all of them with the same interface as *BCounter*. The first submodule on the left (the hours) has a bound of 23, and the two others (minutes and second respectively) have a bound of 59. The module *Clock* defines a composition called **second**, which is active. It is the first and only active event we encounter in the whole example (remember that all the transitions and compositions in *BCounter* were passive). Each time **second** is executed, it calls the service **Tick** from the rightmost submodule (the seconds). Whenever this submodoule has an overflow (i.e., 60 seconds have passed), it calls the service **Tick** from the module in the center, by means of its **overflow** request. Similary, whenever there is an overflow on the center module (60 minutes), the service **Tick** from the leftmost submodule is called. Finally, whenever this submodule calls its own request **overflow** (24 hours), the transition **IncDays** is called, which is increases

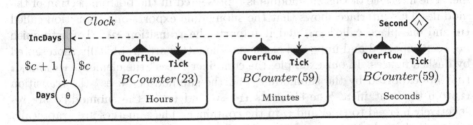

Fig. 12. A hierarchical LLAMAS module representing a clock

the counter in place **Days**. The module *Clock* has clearly an infinite number of reachable states. If we remove the place **Days** but keep the transition **IncDays**, the module will have $24 \times 60 \times 60 = 86400$ reachable states, without any deadlock (executing **Second** from the last state would reset the whole system). If we also remove the transition **IncDays**, the request **Overflow** from the leftmost submodule will not be executable, and this submodule would be blocked when its counter would reach the value 23. The whole system would then have the same 86400 states, but the last one, with marking $\langle 23, 59, 59 \rangle$, would be a deadlock.

5.5 Dynamic Instantiation

In this section we have seen a basic introduction to LLAMAS by means of an example. Some features of the language could not be described here for space reasons. The most important one is a reference-based implementation of the NWN paradigm [21], which adds dynamic instantiation to LLAMAS. This and other secondary features are included in the formal definition of the language.

6 Case Study: Modular PNML

LLAMAS was created with the intent of being able to handle most if not all the modular mechanisms from the Petri nets literature. We will show in this section one example of formalism whose composition mechanism can be translated to LLAMAS, Modular PNML [13]. We chose this example because Modular PNML, as mentioned previously, is an important candidate to serve as semantic platform in the approach we presented in Section 2. Moreover, Modular PNML handles modularity of the data types, which is out of the scope of LLAMAS. Thus, it is interesting to see the relation between the two formalisms.

Let us first briefly describe our exampe of Modular PNML. Details and a formal definition of the language are to be found in [13]. The composition mechanism of Modular PNML is a particular example of fusions of places and transitions. The module interfaces are composed of places and transitions of the Petri nets, which are either *imported* or *exported* by the modules. Modular PNML is a hierarchical formalism, and thus the modules can embed other modules. For instance consider the module on the left side of Fig. 13. The top layer of this module represents its interface. There is one exported transition called **t1**, and one exported place called **p2**. This Modular PNML module contains a submodule. The interface of this submodule is represented in the bottom section of the module. This interface shows that the submodule exports one transition called **t1s** and one place called **p1s**, and it imports the transition **t2s**. To distinguish between exported and imported elements, the latter are graphically represented by dashed figures. In our example, the container "uses" the elements **t1s** and **p1s** that are exported by the submodule, and the submodule "uses" the transition **t2** from the container. The elements **t1s**, **p1s** and **t2s** in the submodule are respectively merged to **t1**, **p1** and **t2** in the container. These merges are graphically represented by dashed lines. Note that the container re-exports the transition **t1** which was originally exported by the submodule. Please note also that, even

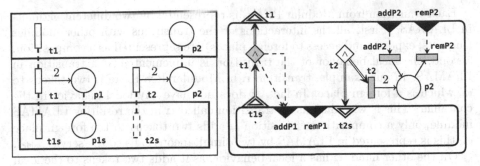

Fig. 13. A Modular PNML example and its LLAMAS translation

though our example does not show it, Modular PNML defines a static instantiation mechanism. This is not seen in the example because we only show the interface of the submodule but not its implementation.

We show the equivalent LLAMAS module in the right hand side of Fig. 13. There is a different representation for the sharing of transitions and the sharing of places. The LCM of LLAMAS is an event-based mechanism, which makes the representation of the sharing of transitions rather straightforward. Consider the case of the transition **t1s** which is exported by the submodule. Whenever the submodule executes this transition, it must inform the container of this execution. This is done by means of a binding *request* called **t1s**. Similarly, the imported transition **t2s** is represented by *service* **t2s**, and the export of **t1** in the interface of the container is represented by request **t1**.

Sharing places is a different problem. In Section 4, we claimed that even though the LCM is an event-based formalism, it is able to simulate state-based mechanisms. In our example, the submodule exports the place **p1s**. This means that the submodule is the owner of **p1s**, and it is offering its container the possibility to add and remove tokens from this place. We represent this with two services called **addP1** and **remP1**. Thus, the submodule is the only one who has a real access to his place **p1s**. When the container wants to interact with this place, it must request access to the submodule. Similarly, the container also exports one place called **p2**. This means that it is giving access to this place to its own container. This is represented again by two services, **addP2** and **remP2**, that call transitions to add and remove tokens from **p2**. Notice that our example is a simple P/T Petri net. The translation of HLPN places can be achieved in the LCM by means of parametric services and requests. Note that Modular PNML does not allow to share values with interface transitions, and the authors of [13] mentioned this as a possible improvement. This is already defined in LLAMAS, again by using parametric services and requests.

Up until now we described the translation of the module interfaces, let us now mention the translation of the internal Petri nets. First, only the local places are represented in the LLAMAS module, the foreign places will be accessed by means of the corresponding services and requests. In our example, **p1** belongs to the submodule, and thus it does not appear in the container. On the other hand, **p2** is a local place, and it remains in the result of the translation.

Each transition from Modular PNML is represented by two different artifacts in LLAMAS. First, all the interactions of the transitions with other modules (imports, exports, and access to foreign places) are represented as a composition. Second, the local behavior of the transition is implemented as a transition in LLAMAS. In our example, transition t1 in Modular PNML adds two tokens to p2, which is a foreign place. In fact, t1 does not have any local behavior in the container. This is why there is no transition called t1 in the resulting LLAMAS module, only a composition. Transition t1 adds two tokens to the foreign place p1, this is represented in LLAMAS by two simultaneous calls to the service addP1.

On the other hand, t2 has a local behavior, as it adds two tokens to the local place p2. In the LLAMAS module, this is represented by the transition called t2. Moreover, t2 is shared with the submodule. This is represented in LLAMAS by a binding from the composition t2 to the binding service t2s. Note that all the compositions in Modular PNML are fusions of places and transitions, which is translated by the use of the operator merge in every LLAMAS composition.

Because of readability and space reasons, we cannot include other use cases in this paper. Nevertheless, we considered complete translations of many formalisms, including the Hierarchical CP-nets [11], CO-OPN [1], the reference nets [15], the Object Petri nets [16], the Petri Box and M-nets families [6,14] and others. Stepping out of LLAMAS central objective, we also considered composition mechanisms from formalisms such as the ITS [20] and CSP [9].

7   Verification of LLAMAS Models

The objective of LLAMAS is to provide a common ground to define the semantics of modular Petri nets formalisms. This would improve the understanding and communication of formalisms and computational techniques in the scientific community. For this to be effective, LLAMAS must not only be an expressive enough semantic platform, it must be a complete modeling formalism on its own. To show that this is the case, we have developed a small prototype implementation of LLAMAS, available as an Eclipse plugin at http://goo.gl/BgONM. This prototype allows to create LLAMAS models and, for small models, to compute their state space. We plan to extend this tool to implement efficient verification of modular models by using the recent advances in model checking with hierarchical Decision Diagrams (DDs). DDs are data type structures specially designed to encode big sets of states. Recent years have seen the development of hierarchical variants of DDs such as the Σ Decision Diagrams (ΣDDs) [3], which were inspired by [5]. Our model checking tool AlPiNA [10,2] uses these structures to check properties on nets with sometimes over 10^{230} states.

In ΣDDs, the states of a Petri net are encoded as data structures designed to save memory. The behavior of the events is encoded by means of operations on these structures called *homomorphisms*. To encode a Petri net transition as a ΣDD homomorphism, we define a set of small operations that are combined with three operators that are built-in in the ΣDD framework: the union, the composition and the intersection (details are to be found in [3]). Let us consider an example. Let $t1$ be a transition that takes one token from a place $p1$

and adds two tokens to a place $p2$. A simplified version of the homomorphism that encodes the behavior of $t1$ can be noted $H_{t1} = H^+(p2,2) \circ H^-(p1,1)$, where the H^+ operation represents a post-condition, the H^- a pre-condition and the $\circ$ is the ΣDD composition operation. Consider a transition $t2$ encoded as $H_{t2} = H^+(p3,1) \circ H^-(p4,2)$. The LLAMAS merge of these transitions would be encoded as $H_{\mathtt{merge}(t1,t2)} = H^+(p2,2) \circ H^+(p3,1) \circ H^-(p1,1) \circ H^-(p4,2)$.

In short, in the ΣDD framework, the structure of the models and their behavior are defined by separate artifacts. Because of this, this framework is well adapted to represent event-based formalisms such as LLAMAS. Moreover, the execution of complex events is encoded as the combination of small operations by means of operators. This matches the mindset of the LCM, where complex events are defined as smaller events combined with five operators. Finally, the ΣDD support recursive and parametric executions. Because of these similarities, an implementation of LLAMAS in the ΣDD framework is a promising endeavor.

8   Conclusion and Perspectives

In this article we proposed an approach to create a unified definition process of modular extensions of Petri nets. With this approach new formalisms can have their syntax and semantics defined in the context of a common framework. For the definition of the semantics we propose a new formalism, called LLAMAS, specifically designed to include the characteristics of most if not all the existing modular Petri nets formalisms. To this day, we have not encountered a composition mechanism in the Petri nets literature that could not be translated to the LCM, even if sometimes the translation is not trivial. We believe that our approach would improve the understanding and communication of novel modeling and verification techniques in the scientific community.

This article only partially showed the work we developed in LLAMAS. Among the elements that had to be left out, let us cite the metamodel to standardize syntactic elements that we mentioned in Section 2, a complete formalization of the language that includes the NWN mechanism for dynamic instantiation, and other case studies than the one presented in Section 6.

We think that LLAMAS has promising perspectives. In Section 7 we briefly sketched some ideas about the verification of LLAMAS models using Decision Diagrams, an important research field. The language itself can be extended to include even more modular considerations, such as inheritance and subtyping relations between modules. Finally, while the current implementation of LLAMAS is a textual prototype, we plan to implement a graphical version of the language and integrate it with our tool AlPiNA.

References

1. Biberstein, O., Buchs, D., Guelfi, N.: Object-Oriented Nets with Algebraic Specifications: The CO-OPN/2 Formalism. In: Agha, G., De Cindio, F., Rozenberg, G. (eds.) APN 2001. LNCS, vol. 2001, pp. 73–127. Springer, Heidelberg (2001)

2. Buchs, D., Hostettler, S., Marechal, A., Risoldi, M.: AlPiNA: An Algebraic Petri Net Analyzer. In: Esparza, J., Majumdar, R. (eds.) TACAS 2010. LNCS, vol. 6015, pp. 349–352. Springer, Heidelberg (2010)
3. Buchs, D., Hostettler, S.: Sigma Decision Diagrams. In: Corradini, A. (ed.) TER-MGRAPH 2009: Preliminary Proceedings of the 5th International Workshop on Computing with Terms and Graphs, No. TR-09-05 in TERMGRAPH Workshops, pp. 18–32. Università di Pisa (2009)
4. Christensen, S., Hansen, D.: Coloured Petri Nets Extended with Channels for Synchronous Communication. In: Valette, R. (ed.) ICATPN 1994. LNCS, vol. 815, pp. 159–178. Springer, Heidelberg (1994)
5. Couvreur, J.-M., Thierry-Mieg, Y.: Hierarchical Decision Diagrams to Exploit Model Structure. In: Wang, F. (ed.) FORTE 2005. LNCS, vol. 3731, pp. 443–457. Springer, Heidelberg (2005)
6. Devillers, R., Klaudel, H., Riemann, R.C.: General parameterised refinement and recursion for the M-net calculus. Theoretical Computer Sc. 300, 259–300 (2003)
7. Hillah, L.-M., Kordon, F., Lakos, C., Petrucci, L.: Extending pnml Scope: A Framework to Combine Petri Nets Types. In: Jensen, K., van der Aalst, W.M., Ajmone Marsan, M., Franceschinis, G., Kleijn, J., Kristensen, L.M. (eds.) ToPNoC VI, LNCS, vol. 7400, pp. 46–70. Springer, Heidelberg (2012)
8. Hillah, L.M., Kordon, F., Petrucci, L., Trèves, N.: PNML Framework: An Extendable Reference Implementation of the Petri Net Markup Language. In: Lilius, J., Penczek, W. (eds.) PETRI NETS 2010. LNCS, vol. 6128, pp. 318–327. Springer, Heidelberg (2010)
9. Hoare, C.A.R.: Communicating sequential processes. CACM 21(8), 666–677 (1978)
10. Hostettler, S., Marechal, A., Linard, A., Risoldi, M., Buchs, D.: High-Level Petri Net Model Checking with AlPiNA. Fund. Informaticae 113(3-4), 229–264 (2011)
11. Huber, P., Jensen, K., Shapiro, R.M.: Hierarchies in coloured petri nets. In: Rozenberg, G. (ed.) APN 1990. LNCS, vol. 483, pp. 313–341. Springer, Heidelberg (1991)
12. ISO/IEC: Software and Systems Engineering – High-level Petri Nets. International Standard ISO/IEC 15909 (2004)
13. Kindler, E., Petrucci, L.: Towards a Standard for Modular Petri Nets: A Formalisation. In: Franceschinis, G., Wolf, K. (eds.) PETRI NETS 2009. LNCS, vol. 5606, pp. 43–62. Springer, Heidelberg (2009)
14. Klaudel, H., Pommereau, F.: M-nets: a survey. Acta Informatica 45(7-8), 537–564 (2009)
15. Kummer, O.: Referenznetze. Logos Verlag, Berlin (2002)
16. Lakos, C.: Object Oriented Modelling with Object Petri Nets. In: Agha, G., De Cindio, F., Rozenberg, G. (eds.) APN 2001. LNCS, vol. 2001, pp. 1–37. Springer, Heidelberg (2001)
17. Marechal, A., Buchs, D.: Modular extensions of Petri Nets: a generic template metamodel. Tech. Rep. 220, University of Geneva (2012), http://goo.gl/oZxYZ
18. Marechal, A., Buchs, D.: Modular extensions of Petri nets: a survey. Tech. Rep. 218, University of Geneva (2012), http://goo.gl/hnHhR
19. Marechal, A., Buchs, D.: The LLAMAS language, syntax and semantics. Tech. Rep. 221, University of Geneva (2013), http://goo.gl/PXCNf
20. Thierry-Mieg, Y., Poitrenaud, D., Hamez, A., Kordon, F.: Hierarchical set decision diagrams and regular models. In: Kowalewski, S., Philippou, A. (eds.) TACAS 2009. LNCS, vol. 5505, pp. 1–15. Springer, Heidelberg (2009)
21. Valk, R.: Object Petri Nets. In: Desel, J., Reisig, W., Rozenberg, G. (eds.) ACPN 2003, LNCS, vol. 3098, pp. 819–848. Springer, Heidelberg (2004)

Channel Properties
of Asynchronously Composed Petri Nets[*]

Serge Haddad[1], Rolf Hennicker[2], and Mikael H. Møller[3]

[1] LSV, ENS Cachan & CNRS & INRIA, France
[2] Ludwig-Maximilians-Universität München, Germany
[3] Aalborg University, Denmark

Abstract. We consider asynchronously composed I/O-Petri nets (AIOPNs) with built-in communication channels. They are equipped with a compositional semantics in terms of asynchronous I/O-transition systems (AIOTSs) admitting infinite state spaces. We study various channel properties that deal with the production and consumption of messages exchanged via the communication channels and establish useful relationships between them. In order to support incremental design we show that the channel properties considered in this work are preserved by asynchronous composition, i.e. they are compositional. As a crucial result we prove that the channel properties are decidable for AIOPNs.

1 Introduction

(A)synchronous composition. The design of hardware and software systems is often component-based which has well-known advantages: management of complexity, reusability, separation of concerns, collaborative design, etc. One critical feature of such systems is the protocol supporting the communication between components and, in particular, the way they synchronise. Synchronous composition ensures that both parts are aware that communication has taken place and then simplifies the validation of the system. However in a large scale distributed environment synchronous composition may lead to redhibitory inefficiency during execution and thus asynchronous composition should be adopted. The FIFO requirement of communication channels is often not appropriate in this context. This is illustrated by the concept of a software bus where applications push and pop messages in mailboxes. Also on the modelling level FIFO ordering is often not assumed, like for the composition of UML state machines which relies on event pools without specific requirements.

Compositions of Petri Nets. In the context of Petri nets, composition has been studied both from theoretical and practical points of view. The process algebra approach has been investigated by several works leading to the Petri net algebra [4]. Such an approach is closely related to synchronous composition. In [22] and [23] asynchronous composition of nets are performed via a set of

---

[*] This work has been partially sponsored by the EU project ASCENS, 257414.

J.-M. Colom and J. Desel (Eds.): PETRI NETS 2013, LNCS 7927, pp. 369–389, 2013.
© Springer-Verlag Berlin Heidelberg 2013

places or, more generally, via a subnet modelling some medium. Then structural restrictions on the subnets are proposed in order to preserve global properties like liveness or deadlock-freeness. In [21] a general composition operator is proposed and its associativity is established. A closely related concept to composition is the one of open Petri nets which has been used in different contexts like the analysis of web services [25]. Numerous compositional approaches have been proposed for the modelling of complex applications but most of them are based on high-level Petri nets; see [11] for a detailed survey.

Channel Properties. With the development of component-based applications, one is interested in verifying behavioural properties of the communication and, in the asynchronous case, in verifying the properties related to communication channels. Channel properties naturally occur when reasonning about distributed mechanisms, algorithms and applications (e.g. management of sockets in UNIX, maintaining unicity of a token in a ring based algorithm, recovery points with empty channels for fault management, guarantee of email reading, etc.).

Our Contributions. In this work we are interested in general channel properties and not in specific system properties related to particular applications. The FIFO requirement for channels potentially can decrease the performance of large scale distributed systems. Thus we restrict ourselves to *unordered* channels which can be naturally modelled by places of Petri nets. We propose asynchronously composed Petri nets (AIOPNs) by (1) explicitly representing channels for internal communication inside the net and (2) defining communication capabilities to the outside in terms of (open) input and output labels with appropriate transitions. Then we define an asynchronous composition operator which introduces new channels for the communication between the composed nets. AIOPNs are equipped with a semantics in terms of asynchronously composed I/O- transition systems (AIOTS). We show that this semantics is fully compositional, i.e. it commutes with asynchronous composition.

In our study two kinds of channel properties are considered which are related to consumption requirements and to the termination of communication. Consumption properties deal with requirements that messages sent to a communication channel should also be consumed. They can be classified w.r.t. two criteria. The first criterion is the nature of the requirement: consuming messages, decreasing the number of messages on a channel, and emptying channels. The second criterion expresses the way the requirement is achieved: possibly immediately, possibly after some delay, or necessarily in each weakly fair run. Communication termination deals with (immediate or delayed) closing of communication channels when the receiver is not ready to consume any more. We establish useful relations between the channel properties and prove that all channel properties considered here are compositional, i.e. preserved by asynchronous composition, which is an important prerequisite for incremental design.

From a verification point of view, we study the decidability of properties in the framework of AIOPN. Thanks to several complementary works on decidability for Petri net problems, we show that all channel properties considered in this work are decidable, though with a high computational complexity.

Related Work. To the best of our knowledge, no work has considered channels explicitly defined for communication inside composite components. However there have been several works where channels are associated with asynchronous composition. They can be roughly classified depending whether their main feature is an algorithmic or a semantic one.

From an algorithmic point of view, in the seminal work of [5], the authors discus several properties like *channel boundedness* and *specified receptions* and propose methods to analyse them. In [7], a two-component based system is studied using a particular (decidable) channel property, the *half-duplex property*: at any time at most one channel is not empty.

From a semantic point view, in [13] "connection-safe" component assemblies have been studied incorporating both synchronous and asynchronous communication. More recently in [2] *synchronisability*, a property of asynchronous systems, is introduced such that when it holds the system can be safely abstracted by a synchronous one. In the framework of Petri nets, E. Kindler has defined Petri net components where the interface is composed by places and composition consists in merging places with same identities [16]. He proposed a partial order semantics for such components and proved that the semantic is fully compositional. Furthermore, for a restricted linear temporal logic he established that properties of this logic are preserved by composition; however such a logic cannot express some of the channel properties we introduce here due to their branching kind. The Petri net based formalism of open nets is the closest formalism to ours. Several works [17],[25], and [24] address both the semantic point of view and the algorithmic one but only when the nets are assumed to be bounded which is not required here. We postpone to Section 2.2 a more detailed comparison with our formalism.

Organisation. In Section 2, we introduce AIOPNs and their asynchronous composition. Then, we provide a compositional semantic for AIOPNs in Section 3 in terms of AIOTSs. In Section 4, we define the channel properties and study their relationships and their preservation under asynchronous composition. In Section 5, we establish that all channel properties are decidable for AIOPNs. Finally, in Section 6, we conclude and give some perspectives for future work.

2   Asynchronous I/O-Petri Nets

2.1   Basic Notions

We recall some basic notions of labelled Petri nets and define their transition system semantics. A *labelled Petri net* is a tuple $\mathcal{N} = (P, T, \Sigma, W^-, W^+, \lambda, m^0)$, such that

- P is a finite set of places,
- T is a finite set of transitions with $P \cap T = \emptyset$,

- Σ is a finite alphabet,
- W^- (resp. W^+) is a matrix indexed by $P \times T$ with values in $\mathbb{N}$; it is called the *backward* (resp. *forward*) *incidence matrix*,
- $\lambda : T \to \Sigma$ is a transition labelling function, and
- m^0 is a vector indexed by P and called the initial marking.

The labelling function λ is extended as usual to sequences of transitions. The input (output resp.) vector $W^-(t)$ ($W^+(t)$ resp.) of a transition t is the column vector of matrix W^- (W^+ resp.) with index t. Given two vectors $\vec{v}$ and $\vec{v}'$, one writes $\vec{v} \geq \vec{v}'$ if $\vec{v}$ is componentwise greater or equal than $\vec{v}'$. A *marking* is a vector indexed by P. A transition $t \in T$ is *firable* from a marking m, denoted by $m \xrightarrow{t}$, if $m \geq W^-(t)$. The firing of t from m leads to the marking m', denoted by $m \xrightarrow{t} m'$, and defined by $m' = m - W^-(t) + W^+(t)$. If $\lambda(t) = a$ we write $m \xrightarrow{a} m'$. The firing of a transition is extended as usual to firing sequences $m \xrightarrow{\sigma} m'$ with $\sigma \in T^*$. A marking m is reachable if there exists a firing sequence $\sigma \in T^*$ such that $m^0 \xrightarrow{\sigma} m$.

Our approach is based on a state transition system semantics for Petri nets. A *labelled transition system* (LTS) is a tuple $S = (\Sigma, Q, q^0, \longrightarrow)$, such that

- Σ is a finite set of labels,
- Q is a (possibly infinite) set of states,
- $q^0 \in Q$ is the initial state, and
- $\longrightarrow \subseteq Q \times \Sigma \times Q$ is a labelled transition relation.

We will write $q \xrightarrow{a} q'$ for $(q, a, q') \in \longrightarrow$, and we write $q \xrightarrow{a}$ if there exists $q' \in Q$ such that $q \xrightarrow{a} q'$. Let $q_1 \in Q$. A *trace* of S starting in q_1 is a finite or infinite sequence $\rho = q_1 \xrightarrow{a_1} q_2 \xrightarrow{a_2} q_3 \xrightarrow{a_3} \cdots$. For $a \in \Sigma$ we write $a \in \rho$, if there exists a_i in the sequence ρ such that $a_i = a$, and $\sharp_\rho(a)$ denotes the (possibly infinite) number of occurrences of a in ρ. For $q \in Q$ we write $q \in \rho$, if there exists q_i in the sequence ρ such that $q_i = q$. For $\sigma = a_1 a_2 \cdots a_n \in \Sigma^*$ and $q, q' \in Q$ we write $q \xrightarrow{\sigma} q'$ if there exists a (finite) trace $q \xrightarrow{a_1} q_2 \xrightarrow{a_2} \cdots \xrightarrow{a_n} q'$. Often we need to reason about the successor states reachable from a given state $q \in Q$ with a subset of labels $\bar{\Sigma} \subseteq \Sigma$. We define $\mathrm{Post}(q, \bar{\Sigma}) = \{q' \in Q \mid \exists a \in \bar{\Sigma} . q \xrightarrow{a} q'\}$ and we write $\mathrm{Post}(q)$ for $\mathrm{Post}(q, \Sigma)$. Further we define $\mathrm{Post}^*(q, \bar{\Sigma}) = \{q' \in Q \mid \exists \sigma \in \bar{\Sigma}^* . q \xrightarrow{\sigma} q'\}$ and we write $\mathrm{Post}^*(q)$ for $\mathrm{Post}^*(q, \Sigma)$.

The semantics of a labelled Petri net $\mathcal{N} = (P, T, \Sigma, W^-, W^+, \lambda, m^0)$ is given by its *associated* labelled transition system $\mathrm{lts}(\mathcal{N}) = (\Sigma, Q, q^0, \longrightarrow)$ which represents the reachability graph of the net and is defined by

- $Q \subseteq \mathbb{N}^P$ is the set of reachable markings of $\mathcal{N}$,
- $\longrightarrow = \{(m, a, m') \mid a \in \Sigma \text{ and } m \xrightarrow{a} m'\}$, and
- $q^0 = m^0$.

2.2   Asynchronous I/O-Petri Nets and Their Composition

In this paper we consider systems which may be open for communication with other systems and may be composed to larger systems. Both the behaviour of primitive components *and* of larger systems obtained by composition can be described by asynchronous I/O-Petri nets introduced in the following. We assume that communication is asynchronous and takes place via unbounded and unordered channels such that for each message type to be exchanged within a system there is exactly one communication channel. The open actions are modelled by distinguished input and output labels while communication inside the system via the channels is modelled by communication labels. Given a finite set C of channels, an *I/O-alphabet over* C is the disjoint union $\Sigma = $ in $\uplus$ out $\uplus$ com of pairwise disjoint sets in of input labels, out of output labels and com of communication labels, such that $\Sigma \cap C = \emptyset$, com $= \{\triangleright a, a^\triangleright \mid a \in C\}$ and in and out do not contain labels of the form $\triangleright x$ or $x^\triangleright$. For each channel $a \in C$, the communication label $\triangleright a$ represents consumption of a message from the channel a and $a^\triangleright$ represents putting a message on a . Each channel is modelled as a place and the transitions for communication actions are modelled by putting or removing tokens from the channel places. Three examples of AIOPNs are shown in Fig. 1. The nets $\mathcal{N}_1$ and $\mathcal{N}_2$ model primitive components (without channels) which repeatedly input and output messages. The net $\mathcal{N}_3$ in Fig. 1c models a simple producer/consumer system with one channel *msg* obtained by composition of the two primitive components; see below. Here and in the following drawings input labels are indicated by "?" and output labels by "!".

Definition 1 (Asynchronous I/O-Petri net). *An* asynchronous I/O-Petri net *(AIOPN) is a tuple* $\mathcal{N} = (C, P, T, \Sigma, W^-, W^+, \lambda, m^0)$, *such that*

- $(P, T, \Sigma, W^-, W^+, \lambda, m^0)$ *is a labelled Petri net,*
- C *is a finite set of channels,*
- $C \subseteq P$, *i.e. each channel is a place,*
- $\Sigma = $ in $\uplus$ out $\uplus$ com *is an I/O-alphabet over* C,
- *for all* $a \in C$ *and* $t \in T$,

$$W^-(a,t) = \begin{cases} 1 & \text{if } \lambda(t) = \triangleright a, \\ 0 & \text{otherwise} \end{cases} \qquad W^+(a,t) = \begin{cases} 1 & \text{if } \lambda(t) = a^\triangleright, \\ 0 & \text{otherwise} \end{cases}$$

- *for all* $a \in C$, $m^0(a) = 0$.   ◇

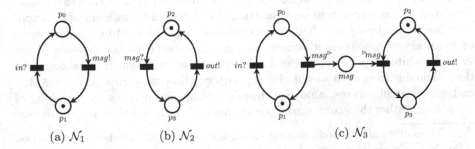

(a) $\mathcal{N}_1$ (b) $\mathcal{N}_2$ (c) $\mathcal{N}_3$

Fig. 1. Asynchronous I/O-Petri nets

Two I/O-alphabets are composable if there are no name conflicts between labels and channels and, following [1], if shared labels are either input labels of one alphabet and output labels of the other or conversely. For the composition each shared label a gives rise to a new communication channel, also called a, and hence to new communication labels $a^\triangleright$ for putting and $^\triangleright a$ for removing messages. The input and output labels of the alphabet composition are the non-shared input and output labels of the underlying alphabets.

Definition 2 (Alphabet composition). *Let $\Sigma_S = \text{in}_S \uplus \text{out}_S \uplus \text{com}_S$ and $\Sigma_T = \text{in}_T \uplus \text{out}_T \uplus \text{com}_T$ be two I/O-alphabets over channels C_S and C_T resp. Σ_S and Σ_T are composable if $(\Sigma_S \cup \Sigma_T) \cap (C_S \cup C_T) = \emptyset$ and $\Sigma_S \cap \Sigma_T = (\text{in}_S \cap \text{out}_T) \cup (\text{in}_T \cap \text{out}_S)$. The composition of Σ_S and Σ_T is the I/O-alphabet $\Sigma = \text{in} \uplus \text{out} \uplus \text{com}$ over the composed set of channels $C = C_S \uplus C_T \uplus C_{ST}$, with new channels $C_{ST} = \Sigma_S \cap \Sigma_T$, such that*

- *$\text{in} = (\text{in}_S \setminus \text{out}_T) \uplus (\text{in}_T \setminus \text{out}_S)$,*
- *$\text{out} = (\text{out}_S \setminus \text{in}_T) \uplus (\text{out}_T \setminus \text{in}_S)$, and*
- *$\text{com} = \{a^\triangleright, {}^\triangleright a \mid a \in C\}^1$*   ◇

Two AIOPNs can be (asynchronously) composed, if their underlying I/O-alphabets are composable. The composition is constructed by taking the disjoint union of the underlying nets and adding a new channel place for each shared label. Every transition with shared output label a becomes a transition with the communication label $a^\triangleright$ that produces a token on the (new) channel place a and, similarly, any transition with shared input label a becomes a transition with the communication label $^\triangleright a$ that consumes a token from the (new) channel place a. For instance, the AIOPN $\mathcal{N}_3$ in Fig. 1c is the result of the asynchronous composition of the two AIOPNs $\mathcal{N}_1$ and $\mathcal{N}_2$ in Fig. 1a and Fig. 1b resp. The newly introduced channel place is the place msg.

Our approach looks very similar to open Petri nets, see e.g. [17], which use interface places for communication. But there are two important differences: First, we explicitly distinguish channel places thus being able to reason on the communication behaviour between composed components; see Sect. 4. The second difference is quite important from the software engineer's point of view. We do not use interface places to indicate communication abilities of a component but we use distinguished input and output labels instead. We believe that this has an important advantage to achieve separation of concerns: The designer of a component has not to take care whether the component will be used in a synchronous or in an asynchronous environment later on; this should be the decision of the system architect. Indeed open Petri nets already rely on asynchronous composition while our formalism would also support synchronous composition, see [19], and mixed architectures. Since synchronous composition relies on matching of transitions rather than communication channels we have not elaborated this case

---

[1] $\Sigma = \text{in} \uplus \text{out} \uplus \text{com}$ is indeed a disjoint union, since for all $a \in C_{ST}$ the communication labels $a^\triangleright, {}^\triangleright a$ are new names due to the general assumption that input and output labels are not of the form $x^\triangleright, {}^\triangleright x$.

here. The difference between AIOPNs and modal I/O-Petri nets introduced in [8] is that AIOPNs comprise distinguished channel places but they do not support modalities for refinement (yet).

Definition 3 (Asynchronous composition of AIOPNs). *Let* $\mathcal{N} = (C_\mathcal{N},$ $P_\mathcal{N}, T_\mathcal{N}, \Sigma_\mathcal{N}, W_\mathcal{N}^-, W_\mathcal{N}^+, \lambda_\mathcal{N}, m_\mathcal{N}^0)$ *and* $\mathcal{M} = (C_\mathcal{M}, P_\mathcal{M}, T_\mathcal{M}, \Sigma_\mathcal{M}, W_\mathcal{M}^-, W_\mathcal{M}^+, \lambda_\mathcal{M},$ $m_\mathcal{M}^0)$ *be two AIOPNs.* $\mathcal{N}$ *and* $\mathcal{M}$ *are composable if* $\Sigma_\mathcal{N}$ *and* $\Sigma_\mathcal{M}$ *are composable and if* $P_\mathcal{N} \cap P_\mathcal{M} = \emptyset$, $(P_\mathcal{N} \cup P_\mathcal{M}) \cap (\Sigma_\mathcal{N} \cap \Sigma_\mathcal{M}) = \emptyset$, *and* $T_\mathcal{N} \cap T_\mathcal{M} = \emptyset$. *In this case their* asynchronous composition *is the AIOPN* $\mathcal{N} \otimes_{pn} \mathcal{M} = (C, P, T,$ $\Sigma, W^-, W^+, \lambda, m^0)$ *defined as follows:*

- $C = C_\mathcal{N} \uplus C_\mathcal{M} \uplus C_{\mathcal{N}\mathcal{M}}$, *with* $C_{\mathcal{N}\mathcal{M}} = \Sigma_\mathcal{N} \cap \Sigma_\mathcal{M}$,
- $P = P_\mathcal{N} \uplus P_\mathcal{M} \uplus C_{\mathcal{N}\mathcal{M}}$,
- $T = T_\mathcal{N} \uplus T_\mathcal{M}$,
- Σ *is the alphabet composition of* Σ_S *and* Σ_T,
- W^- *(resp.* W^+*) is the backward (forward) incidence matrix defined by:*

$$\text{for all } p \in P_\mathcal{N} \cup P_\mathcal{M} \text{ and } t \in T \qquad\qquad \text{for all } a \in C_{\mathcal{N}\mathcal{M}} \text{ and } t \in T$$

$$W^-(p,t) = \begin{cases} W_\mathcal{N}^-(p,t) & \text{if } p \in P_\mathcal{N}, t \in T_\mathcal{N} \\ W_\mathcal{M}^-(p,t) & \text{if } p \in P_\mathcal{M}, t \in T_\mathcal{M} \\ 0 & \text{otherwise} \end{cases} \qquad W^-(a,t) = \begin{cases} 1 & \text{if } \lambda(t) = {}^\triangleright a \\ 0 & \text{otherwise} \end{cases}$$

$$W^+(p,t) = \begin{cases} W_\mathcal{N}^+(p,t) & \text{if } p \in P_\mathcal{N}, t \in T_\mathcal{N} \\ W_\mathcal{M}^+(p,t) & \text{if } p \in P_\mathcal{M}, t \in T_\mathcal{M} \\ 0 & \text{otherwise} \end{cases} \qquad W^+(a,t) = \begin{cases} 1 & \text{if } \lambda(t) = a^\triangleright \\ 0 & \text{otherwise} \end{cases}$$

- $\lambda : T \to \Sigma$ *is defined, for all* $t \in T$, *by*

$$\lambda(t) = \begin{cases} \lambda_\mathcal{N}(t) & \text{if } t \in T_\mathcal{N}, \ \lambda_\mathcal{N}(t) \notin \Sigma_\mathcal{N} \cap \Sigma_\mathcal{M} \\ \lambda_\mathcal{M}(t) & \text{if } t \in T_\mathcal{M}, \ \lambda_\mathcal{M}(t) \notin \Sigma_\mathcal{N} \cap \Sigma_\mathcal{M} \\ {}^\triangleright\lambda_\mathcal{N}(t) & \text{if } t \in T_\mathcal{N}, \ \lambda_\mathcal{N}(t) \in in_\mathcal{N} \cap out_\mathcal{M} \\ {}^\triangleright\lambda_\mathcal{M}(t) & \text{if } t \in T_\mathcal{M}, \ \lambda_\mathcal{M}(t) \in in_\mathcal{M} \cap out_\mathcal{N} \\ \lambda_\mathcal{N}(t)^\triangleright & \text{if } t \in T_\mathcal{N}, \ \lambda_\mathcal{N}(t) \in in_\mathcal{M} \cap out_\mathcal{N} \\ \lambda_\mathcal{M}(t)^\triangleright & \text{if } t \in T_\mathcal{M}, \ \lambda_\mathcal{M}(t) \in in_\mathcal{N} \cap out_\mathcal{M} \end{cases}$$

- m^0 *is defined, for all* $p \in P$, *such that* $m^0(p) = m_\mathcal{N}^0(p)$ *if* $p \in P_\mathcal{N}$, $m^0(p) = m_\mathcal{M}^0(p)$ *if* $p \in P_\mathcal{M}$, *and* $m^0(p) = 0$ *otherwise.* $\diamond$

3   Compositional Semantics

We extend the transition system semantics of labelled Petri nets defined in Sect. 2.1 to AIOPNs. For this purpose we introduce asynchronous I/O-transition systems which are labelled transition systems extended by channels and a *channel valuation* function val : $Q \longrightarrow \mathbb{N}^C$. The channel valuation function determines for each state $q \in Q$ how many messages are actually pending on each channel $a \in C$. For $q \in Q$ and $a \in C$ we use the notation val$(q)[a{+}{+}]$ to denote the updated map

$$\text{val}(q)[a{+}{+}](x) = \begin{cases} \text{val}(q)(a) + 1 & \text{if } x = a, \\ \text{val}(q)(x) & \text{otherwise.} \end{cases}$$

The updated map $\text{val}(q)[a{-}{-}]$ is defined similarly. Instead of $\text{val}(q)(a)$ we will often write $\text{val}(q, a)$.

Definition 4 (Asynchronous I/O-transition system). *An* asynchronous I/O-transition system *(AIOTS) is a tuple* $\mathcal{S} = (C, \Sigma, Q, q^0, \longrightarrow, \text{val})$, *such that*

- $(\Sigma, Q, q^0, \longrightarrow)$ *is a labelled transition system,*
- C *is a finite set of channels,*
- $\Sigma = \text{in} \uplus \text{out} \uplus \text{com}$ *is an I/O-alphabet over* C,
- $\text{val} : Q \longrightarrow \mathbb{N}^C$ *is a function, such that for all* $a \in C, q, q' \in Q$:
  - $\text{val}(q^0, a) = 0$,
  - $q \xrightarrow{a^{\triangleright}} q' \implies \text{val}(q') = \text{val}(q)[a{+}{+}]$,
  - $q \xrightarrow{\triangleright a} q' \implies \text{val}(q, a) > 0$ *and* $\text{val}(q') = \text{val}(q)[a{-}{-}]$, *and*
  - *for all* $x \in (\text{in} \cup \text{out}), q \xrightarrow{x} q' \implies \text{val}(q') = \text{val}(q)$. $\quad\Diamond$

The first condition for val assumes that initially all communication channels are empty. The second condition states that transitions with labels $a^{\triangleright}$ and $^{\triangleright}a$ have the desired effect of putting one message on a channel (consuming one message from a channel resp.). The last condition requires that the input and output actions of an open system do not change the valuation of any channel. Sometimes we need to reason about the number of messages on a subset $B \subseteq C$ of the channels in a state $q \in Q$. We define $\text{val}(q, B) = \sum_{a \in B} \text{val}(q, a)$.

The semantics of an asynchronous I/O-Petri net $\mathcal{N}$ is given by its *associated* asynchronous I/O-transition system $\text{aiots}(\mathcal{N})$. It is based on the transition system semantics of a labelled Petri net (see Sect. 2.1) such that markings become states, but additionally we define the valuation of a channel in a current state m by the number of tokens on the channel under the marking m.

Definition 5 (Associated asynchronous I/O-transition system).
Let $\mathcal{N} = (C, P, T, \Sigma, W^-, W^+, \lambda, m^0)$ *be an AIOPN. The AIOTS associated with* $\mathcal{N}$ *is given by* $\text{aiots}(\mathcal{N}) = (C, \Sigma, Q, q^0, \longrightarrow, \text{val})$, *such that*

- $(\Sigma, Q, q^0, \longrightarrow) = \text{lts}(P, T, \Sigma, W^-, W^+, \lambda, m^0)$,
- *for all* $a \in C$ *and* $m \in Q, \text{val}(m, a) = m(a)$. $\quad\Diamond$

Example 6. The transition systems associated with the AIOPNs $\mathcal{N}_1$ and $\mathcal{N}_2$ in Fig. 1a and 1b have two reachable states and the transitions between them correspond directly to their Petri net representations. The situation is different for the AIOPN $\mathcal{N}_3$ in Fig. 1c. It has infinitely many reachable markings and hence its associated AIOTS has infinitely many reachable states. Fig. 2 shows an excerpt of it. The states indicate the number of tokens in each place in the order p_0, p_1, msg, p_2, p_3. The initial state is underlined.

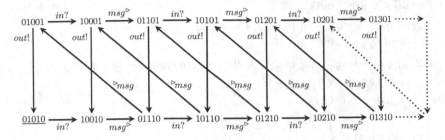

Fig. 2. Part of the associated AIOTS for $\mathcal{N}_3$ in Fig. 1c

Like AIOPNs also two AIOTSs can be asynchronously composed, if their underlying I/O-alphabets are composable. The composition is constructed by introducing a new communication channel for each shared input/output action and by appropriate transitions for the corresponding communication actions that modify the valuation of the new channels (see items 3 and 4 in Def. 7). Since the states of the composition must record the number of messages on the new channels $C_{\mathcal{S}\mathcal{T}}$, the state space of the composition adds to the Cartesian product of the underlying state spaces the set $\mathbb{N}^{C_{\mathcal{S}\mathcal{T}}}$ of valuations of the new channels. For a valuation $\boldsymbol{v} : C_{\mathcal{S}\mathcal{T}} \mapsto \mathbb{N}$ and channel $a \in C_{\mathcal{S}\mathcal{T}}$ we use the notation $\boldsymbol{v}[a{+}{+}]$ ($\boldsymbol{v}[a{-}{-}]$ resp.) to denote the updated map which increments (decrements) the value of a by 1 and leaves the values of all other channels unchanged.

Definition 7 (Asynchronous composition of AIOTS)

Let $\mathcal{S} = (C_{\mathcal{S}}, \Sigma_{\mathcal{S}}, Q_{\mathcal{S}}, q_{\mathcal{S}}^0, \longrightarrow_{\mathcal{S}}, \mathrm{val}_{\mathcal{S}})$ and $\mathcal{T} = (C_{\mathcal{T}}, \Sigma_{\mathcal{T}}, Q_{\mathcal{T}}, q_{\mathcal{T}}^0, \longrightarrow_{\mathcal{T}}, \mathrm{val}_{\mathcal{T}})$ be two AIOTSs. $\mathcal{S}$ and $\mathcal{T}$ are composable if $\Sigma_{\mathcal{S}}$ and $\Sigma_{\mathcal{T}}$ are composable. In this case their asynchronous composition is the AIOTS $\mathcal{S} \otimes \mathcal{T} = (C, \Sigma, Q, q^0, \longrightarrow, \mathrm{val})$ defined as follows:

- $C = C_{\mathcal{S}} \uplus C_{\mathcal{T}} \uplus C_{\mathcal{S}\mathcal{T}}$, with $C_{\mathcal{S}\mathcal{T}} = \Sigma_{\mathcal{S}} \cap \Sigma_{\mathcal{T}}$,
- Σ is the alphabet composition of $\Sigma_{\mathcal{S}}$ and $\Sigma_{\mathcal{T}}$,
- $Q \subseteq Q_{\mathcal{S}} \times Q_{\mathcal{T}} \times \mathbb{N}^{C_{\mathcal{S}\mathcal{T}}}$,
- $q^0 = (q_{\mathcal{S}}^0, q_{\mathcal{T}}^0, \boldsymbol{0}) \in Q$, with $\boldsymbol{0}$ being the zero-map,
- Q and $\longrightarrow$ are inductively defined as follows whenever $(q_{\mathcal{S}}, q_{\mathcal{T}}, \boldsymbol{v}) \in Q$:

  1: For all $a \in (\Sigma_{\mathcal{S}} \setminus C_{\mathcal{S}\mathcal{T}})$, if $q_{\mathcal{S}} \xrightarrow{a}_{\mathcal{S}} q_{\mathcal{S}}'$ then $(q_{\mathcal{S}}, q_{\mathcal{T}}, \boldsymbol{v}) \xrightarrow{a} (q_{\mathcal{S}}', q_{\mathcal{T}}, \boldsymbol{v})$ and $(q_{\mathcal{S}}', q_{\mathcal{T}}, \boldsymbol{v}) \in Q$.

  2: For all $a \in (\Sigma_{\mathcal{T}} \setminus C_{\mathcal{S}\mathcal{T}})$, if $q_{\mathcal{T}} \xrightarrow{a}_{\mathcal{T}} q_{\mathcal{T}}'$ then $(q_{\mathcal{S}}, q_{\mathcal{T}}, \boldsymbol{v}) \xrightarrow{a} (q_{\mathcal{S}}, q_{\mathcal{T}}', \boldsymbol{v})$ and $(q_{\mathcal{S}}, q_{\mathcal{T}}', \boldsymbol{v}) \in Q$.

  3: For all $a \in \mathrm{in}_{\mathcal{S}} \cap \mathrm{out}_{\mathcal{T}}$,

  3.1: if $q_{\mathcal{S}} \xrightarrow{a}_{\mathcal{S}} q_{\mathcal{S}}'$ and $\boldsymbol{v}(a) > 0$ then $(q_{\mathcal{S}}, q_{\mathcal{T}}, \boldsymbol{v}) \xrightarrow{\triangleright a} (q_{\mathcal{S}}', q_{\mathcal{T}}, \boldsymbol{v}[a{-}{-}])$ and $(q_{\mathcal{S}}', q_{\mathcal{T}}, \boldsymbol{v}[a{-}{-}]) \in Q$,

  3.2: if $q_{\mathcal{T}} \xrightarrow{a}_{\mathcal{T}} q_{\mathcal{T}}'$ then $(q_{\mathcal{S}}, q_{\mathcal{T}}, \boldsymbol{v}) \xrightarrow{a^{\triangleright}} (q_{\mathcal{S}}, q_{\mathcal{T}}', \boldsymbol{v}[a{+}{+}])$ and $(q_{\mathcal{S}}, q_{\mathcal{T}}', \boldsymbol{v}[a{+}{+}]) \in Q$.

4: For all $a \in \text{in}_T \cap \text{out}_S$,

4.1: if $q_S \xrightarrow{a}_S q'_S$ then $(q_S, q_T, v) \xrightarrow{a^\triangleright} (q'_S, q_T, v[a++])$
    and $(q'_S, q_T, v[a++]) \in Q$,

4.2: if $q_T \xrightarrow{a}_T q'_T$ and $v(a) > 0$ then $(q_S, q_T, v) \xrightarrow{\triangleright a} (q_S, q'_T, v[a--])$
    and $(q_S, q'_T, v[a--]) \in Q$.

− For all $(q_S, q_T, v) \in Q$ and $a \in C$,

$$\text{val}((q_S, q_T, v), a) = \begin{cases} \text{val}_S(q_S, a) & \text{if } a \in C_S \\ \text{val}_T(q_T, a) & \text{if } a \in C_T \\ v(a) & \text{if } a \in C_{ST} \end{cases}$$

For the rules (1),(3.1) and (4.1), we say that the resulting transition in the composition is triggered by S. Let ρ be a trace of $S \otimes T$ starting from a state $q = (q_S, q_T, v) \in Q$. The projection of ρ to S, denoted by $\rho|_S$, is the sequence of transitions of S, starting from q_S, which have triggered corresponding transitions in ρ.   ◊

The following theorem shows that the transition system semantics of asynchronous I/O-Petri nets is compositional. The proof is given in [12].

Theorem 8. *Let $\mathcal{N}$ and $\mathcal{M}$ be two composable AIOPNs. Then it holds that* $\text{aiots}(\mathcal{N} \otimes_{pn} \mathcal{M}) = \text{aiots}(\mathcal{N}) \otimes \text{aiots}(\mathcal{M})$ *(up to bijection between state spaces).*

4   Channel Properties and Their Compositionality

In this section we consider various properties concerning the asynchronous communication via channels. We give a classification of the properties, show their relationships and prove that they are compositional w.r.t. asynchronous composition, a prerequisite for incremental design.

4.1   Channel Properties

We consider two classes of channel properties. The first class deals with the requirements that messages sent to a communication channel should also be consumed; the second class concerns the termination of communication in the sense that if consumption from a channel has been stopped then also production on this channel will stop. The channel properties will be defined by considering the semantics of AIOPNs, i.e. they will be formulated for AIOTSs.

Some of the properties, precisely the "necessarily properties" of type (c) in Def. 10 below, rely on the consideration of system runs. In principle a system run is a maximal execution trace; it can be infinite but also finite if no further actions are enabled. It is important to remember, that we deal with open systems whose possible behaviours are also determined by the environment. Hence, the definition of a system run must take into account the possibility that the system may stop in a state where the environment does not serve any offered input of the system while at the same time the system has no enabled autonomous action, i.e. an action which is not an input from the environment. Such states will be called pure input states. They correspond to markings that "stop except for inputs" in [24]. Note that all possible communication actions inside the system can be autonomously executed. The same holds for output actions to the environment, since we are working with asynchronous communication such that messages can always be sent, even if they are never accepted by the environment. Formally, system runs are defined as follows.

Let $S = (C, \Sigma, Q, q^0, \longrightarrow, \text{val})$ be an AIOTS with $\Sigma = \text{in} \uplus \text{out} \uplus \text{com}$. A state $q \in Q$ is called a *pure input state* if $\text{Post}(q, \Sigma \setminus \text{in}) = \emptyset$, i.e. only inputs are enabled. A pure input state is a potential deadlock, as the environment of S might not serve any inputs for S. Let $q_1 \in Q$. A *run* of S starting in q_1 is a trace of S starting in q_1, that is either infinite or finite such that its last state is a pure input state. We denote the set of all runs of S starting from q_1 as $\text{run}_S(q_1)$.

In the following we also assume that system runs are only executed in a run-time infrastructure which follows a weakly fair scheduling policy. In our context this means that any autonomous action a, that is always enabled from a certain state on, will infinitely often be executed. Formally, a run $\rho \in \text{run}_S(q_1)$ with $q_1 \in Q, \rho = q_1 \xrightarrow{a_1} q_2 \xrightarrow{a_2} \cdots$, is called *weakly fair* if it is finite or if it is infinite and for all $a \in (\Sigma \setminus \text{in})$ the following holds:

$$(\exists k \geq 1 . \forall i \geq k . q_i \xrightarrow{a}) \implies (\forall k \geq 1 . \exists i \geq k . a_i = a).$$

We denote the set of all weakly fair runs of S starting from q_1 by $\text{wfrun}_S(q_1)$. It should be noted that for our results it is sufficient to use weak fairness instead of strong fairness.[2]

Example 9. Let $S = \text{aiots}(\mathcal{N}_3)$ be the associated AIOTS of the Petri net $\mathcal{N}_3$ in Fig. 1c. An excerpt of S has been shown in Fig. 2. The following are three traces of S starting in the initial state 01010:

$\rho_0 = 01010,$

$\rho_1 = 01010 \xrightarrow{in?} 10010 \xrightarrow{msg^\triangleright} 01110 \xrightarrow{{}^\triangleright msg} 01001,$

$\rho_2 = 01010 \xrightarrow{in?} 10010 \xrightarrow{msg^\triangleright} 01110 \xrightarrow{{}^\triangleright msg} 01001 \xrightarrow{out!} 01010.$

The traces ρ_0 and ρ_2 are runs of S while ρ_1 is not a run, since 01001 is not a pure input state. Now consider the infinite trace indicated at the bottom line in Fig. 2, that is an infinite alternation of $in?$ and $msg^\triangleright$. The trace is a run, since it is infinite. But the run is not weakly fair, since from the second state on ${}^\triangleright msg$ is always enabled but never taken.

Our first class of channel properties deals with the consumption of previously produced messages. We consider four groups of such properties (P1) - (P4) with different strength. In each case we consider three variants which all are parametrised w.r.t. a subset B of the communication channels.

---

[2] For a discussion of the different fairness properties see, e.g., [3].

Let us discuss the consuming properties (P1) of Def. 10 below for an AIOTS S and a subset B of its channels. Property (P1.a) requires, for each channel $a \in B$, that if in an arbitrary reachable state q of S there is a message available on a, then S can consume the message possibly after the execution of some autonomous actions. Let us comment on the role of the environment for the formulation of this property. First, we consider arbitrary reachable states $q \in \text{Post}^*(q^0)$ with q^0 being the initial state of S. This means that we take into account the worst environment which can let S go everywhere by providing (non-deterministically) all inputs that S can accept. Then, at some point at which a message is available on channel a, the environment can stop to provide further inputs and waits whether S can *autonomously* reach a state $q' \in \text{Post}^*(q, \Sigma \setminus \text{in})$ in which it can consume from a, i.e. execute $\triangleright a$. To allow autonomous actions before consumption is inspired by the property of "output compatibility" studied for synchronously composed transition systems in [14]. Property (P1.b) does not allow autonomous actions before consumption. It requires that S can immediately consume the message in state q, similar to the property of specified reception in [5]. Property (P1.c) requires that the message will definitely be consumed on each weakly fair run of S starting from q and, due to the definition of a system run, that this will happen in any environment whatever inputs are provided.

As an example consider the AIOTS $S = \text{aiots}(\mathcal{N}_3)$ associated with the AIOPN $\mathcal{N}_3$ in Fig. 1c and its reachable state 01101 such that one message is on channel *msg*. In this state S can autonomously perform the output *out*! reaching state 01110 and then it can consume the message by performing $\triangleright msg$. Since also in all other reachable states in which the channel is not empty S can autonomously reach a state in which it can consume from the channel, S satisfies property (P1.a) (for its only channel *msg*). However, S is not strongly consuming (P1.b). For instance in state 01101, S cannot immediately consume the message. On the other hand, S is necessarily consuming (P1.c). Whenever in a reachable state q the channel is not empty an autonomous action, either $\triangleright msg$ or *out*!, is enabled. Hence q is not a pure input state and, due to the weak fairness condition, eventually $\triangleright msg$ or *out*! must be performed in any weakly fair run starting from q. If $\triangleright msg$ is performed we are done. If *out*! is performed we reach a state where $\triangleright msg$ is enabled and with the same reasoning eventually $\triangleright msg$ will be performed. This can be easily detected by considering Fig. 2.

The other groups of properties (P2) - (P4) express successively stronger (or equivalent) requirements on the kind of consumption. For instance, (P3) requires that the consumption will lead to a state in which the channel is empty. Again we distinguish if this can be achieved after some autonomous actions (P3.a), can be achieved immediately (P3.b), or must be achieved in any weakly fair run (P3.c).

Definition 10 (Consumption properties). *Let $S = (C, \Sigma, Q, q^0, \longrightarrow, \text{val})$ be an AIOTS with I/O-alphabet $\Sigma = \text{in} \uplus \text{out} \uplus \text{com}$ and let $B \subseteq C$ be a subset of its channels.*

P1: (Consuming)

    a) S *is* B-*consuming, if for all* $a \in B$ *and all* $q \in \mathrm{Post}^*(q^0)$,

$$\mathrm{val}(q,a) > 0 \implies \exists q' \in \mathrm{Post}^*(q, \Sigma \setminus \mathrm{in}) \cdot q' \xrightarrow{\triangleright a} .$$

    b) S *is strongly* B-*consuming, if for all* $a \in B$ *and all* $q \in \mathrm{Post}^*(q^0)$,

$$\mathrm{val}(q,a) > 0 \implies q \xrightarrow{\triangleright a} .$$

    c) S *is necessarily* B-*consuming, if for all* $a \in B$ *and all* $q \in \mathrm{Post}^*(q^0)$,

$$\mathrm{val}(q,a) > 0 \implies \forall \rho \in \mathrm{wfruns}_S(q) \cdot {}^{\triangleright} a \in \rho .$$

P2: (Decreasing)

    a) S *is* B-*decreasing, if for all* $a \in B$ *and all* $q \in \mathrm{Post}^*(q^0)$,

$$\mathrm{val}(q,a) > 0 \implies \exists q' \in \mathrm{Post}^*(q, \Sigma \setminus \mathrm{in}) \cdot \mathrm{val}(q',a) < \mathrm{val}(q,a) .$$

    b) S *is strongly* B-*decreasing, if for all* $a \in B$ *and all* $q \in \mathrm{Post}^*(q^0)$,

$$\mathrm{val}(q,a) > 0 \implies \exists q' \in \mathrm{Post}(q, \Sigma \setminus \mathrm{in}) \cdot \mathrm{val}(q',a) < \mathrm{val}(q,a) .$$

    c) S *is necessarily* B-*decreasing, if for all* $a \in B$ *and all* $q \in \mathrm{Post}^*(q^0)$,

$$\mathrm{val}(q,a) > 0 \implies \forall \rho \in \mathrm{wfruns}_S(q), \exists q' \in \rho \cdot \mathrm{val}(q',a) < \mathrm{val}(q,a) .$$

P3: (Emptying)

    a) S *is* B-*emptying, if for all* $a \in B$ *and all* $q \in \mathrm{Post}^*(q^0)$,

$$\mathrm{val}(q,a) > 0 \implies \exists q' \in \mathrm{Post}^*(q, \Sigma \setminus \mathrm{in}) \cdot \mathrm{val}(q',a) = 0 .$$

    b) S *is strongly* B-*emptying, if for all* $a \in B$ *and all* $q \in \mathrm{Post}^*(q^0)$,

$$\mathrm{val}(q,a) > 0 \implies \exists q' \in \mathrm{Post}(q, \Sigma \setminus \mathrm{in}) \cdot \mathrm{val}(q',a) = 0 .$$

    c) S *is* B-*necessarily emptying, if for all* $a \in B$ *and all* $q \in \mathrm{Post}^*(q^0)$,

$$\mathrm{val}(q,a) > 0 \implies \forall \rho \in \mathrm{wfruns}_S(q), \exists q' \in \rho \cdot \mathrm{val}(q',a) = 0 .$$

P4: (Wholly emptying)

    a) S *is* B-*wholly emptying, if for all* $q \in \mathrm{Post}^*(q^0)$,

$$\mathrm{val}(q,B) > 0 \implies \exists q' \in \mathrm{Post}^*(q, \Sigma \setminus \mathrm{in}) \cdot \mathrm{val}(q',B) = 0.$$

    b) S *is strongly* B-*wholly emptying, if for all* $q \subset \mathrm{Post}^*(q^0)$,

$$\mathrm{val}(q,B) > 0 \implies \exists q' \in \mathrm{Post}(q, \Sigma \setminus \mathrm{in}) \cdot \mathrm{val}(q',B) = 0.$$

    c) S *is* B-*necessarily wholly emptying, if for all* $q \in \mathrm{Post}^*(q^0)$,

$$\mathrm{val}(q,B) > 0 \implies \forall \rho \in \mathrm{wfruns}_S(q), \exists q' \in \rho \cdot \mathrm{val}(q',B) = 0 . \qquad \Diamond$$

Note if the initial state of S is reachable from all other reachable states, i.e. the initial state is a *home state*, then S is B-wholly emptying.

The next class of channel properties concerns the termination of communication. We consider two variants: (P5.a) requires that in any weakly fair run, in which consumption from a channel a has stopped, only finitely many subsequent productions are possible, i.e. the channel is closed after a while. Property (P5.b) expresses that the channel is immediately closed.

Definition 11 (Communication stopping). *Let* S *be an AIOTS and* $B \subseteq C$ *be a subset of its channels.*

P5: (Communication stopping)

    a) S *is* B-*communication stopping, if for all* $q \in \mathrm{Post}^*(q^0)$, $\rho \in \mathrm{wfruns}_S(q)$
    and $a \in B$, $\sharp_\rho({}^{\triangleright} a) = 0 \implies \sharp_\rho(a^{\triangleright}) < \infty .$

    b) S *is strongly* B-*communication stopping, if for all* $q \in \mathrm{Post}^*(q^0)$, $\rho \in$
    $\mathrm{wfruns}_S(q)$ *and* $a \in B$, $\sharp_\rho({}^{\triangleright} a) = 0 \implies \sharp_\rho(a^{\triangleright}) = 0 .$ $\Diamond$

We say that an AIOTS S has a channel property P, if S has property P with respect to the set C of all channels of S.

Definition 12 (Channel properties of AIOPNs). *Let P be an arbitrary channel property as defined above. An AIOPN $\mathcal{N}$ has property P w.r.t. some subset B of its channels if its generated AIOTS* aiots($\mathcal{N}$) *has property P w.r.t. B; $\mathcal{N}$ has property P if* aiots($\mathcal{N}$) *has property P.*

Relevance of Channel Properties. The generic properties that we have defined fit well with properties related to specific distributed mechanisms, algorithms and applications. For instance:

- When sending an email the user must be confident that its mail will be eventually read. Such a property can be formalized as the necessarily consuming property.
- Most distributed applications can be designed with an underlying token circulation between the processes of the applications. This requires that at any time there is at most one token in all channels and that this token can be immediately handled. Such a property can be formalized as the strong wholly emptying property.
- Recovery points are useful for applications prone to faults. While algorithms for building recovery points can handle non-empty channels, the existence (and identification) of states with empty channels eases this task. Such a property can be formalized as the necessarily wholly emptying property.
- In UNIX, one often requires that a process should not write in a socket when no reader of the socket is still present (and this could raise a signal). Such a property can be formalized as the strong communication stopping property.

4.2   Relationships between Channel Properties

Table 1 shows relationships between the channel properties and pointers to examples of AIOPNs from Fig. 1 and Fig. 3 which have the indicated properties.

All the downward implications inside the boxes are direct consequences of the definitions. It is trivial to see that downward implication 3 is an equivalence, since *immediate* consumption leads to a decreasing valuation. Downward implications 9 and 16 are equivalences, since repeated decreasing of messages on a channel will eventually lead to an empty channel. The implications 4, 5, 7, 11-14 and 18 are proved in [12]. Additionally we have that all properties in box b) of Tab. 1 imply the strongest property in box a), since if S is strongly B-consuming we can by repeated consumption empty all channels in B.

Let us now discuss some counterexamples. As discussed in Sect. 4.1, a counterexample for the converse of implication 7 is the AIOPN $\mathcal{N}_3$ in Fig. 1c. An obvious counterexample for the converse of the implications 2, 10, 11, 12, 13 is given by the AIOPN $\mathcal{N}_4$ shown in Fig. 3a. $\mathcal{N}_4$ is also a counterexample for implication 6. The AIOPN $\mathcal{N}_5$ in Fig. 3b with channels a and b is a counterexample for the converse of implication 15. The net can empty each single channel a and

Table 1. Relationships between channel properties and examples

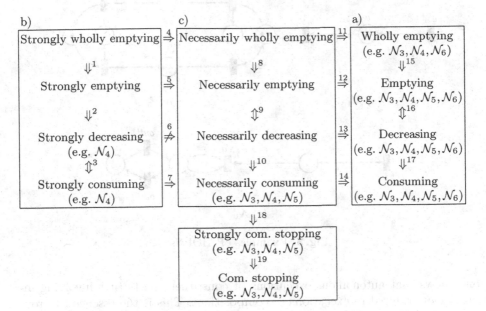

b but it can never have both channels empty at the same time (after the first message has been produced on a channel). A counterexample for the converse of implication 14 is shown by the net $\mathcal{N}_6$ in Fig. 3c. The net can put a token on the channel a, but afterwards the transition $\triangleright a$ is not necessarily always enabled which means there exists a weakly fair run such that there is always a token in a and $\triangleright a$ is never fired.

Counterexamples for the converse of implication 17 rely on the idea to produce twice while consuming once. A counterexample for the converse of implication 18 is provided by a net that first produces a finite number n of messages on a channel, then it consumes less than n of these messages and then it stops. Counterexamples for the remaining converse implications are straightforward to construct.

4.3 Compositionality of Channel Properties

Modular verification of systems is an important goal in any development method. In our context this concerns the question whether channel properties are preserved in arbitrary environments or, more precisely, whether they are preserved under asynchronous composition. In this section we show that indeed all channel properties defined above are compositional. This can be utilised to get a method for incremental design. The proofs of the results of this section are given in [12].

In order to relate channel properties of asynchronous compositions to channel properties of their constituent parts we need the next two lemmas. The first

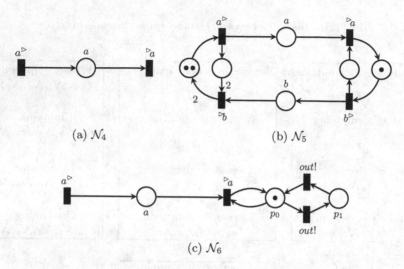

(a) $\mathcal{N}_4$ (b) $\mathcal{N}_5$

(c) $\mathcal{N}_6$

Fig. 3. Examples of AIOPNs

one shows that autonomous executions of constituent parts (not involving inputs) can be lifted to executions of compositions. This is the essence to prove compositionality of the properties of type (a) and type (b) in Def. 10.

Lemma 13. *Let $S = (C_S, \Sigma_S, Q_S, q_S^0, \longrightarrow_S, \mathrm{val}_S), \mathcal{T} = (C_T, \Sigma_T, Q_T, q_T^0, \longrightarrow_T, \mathrm{val}_T)$ be two composable AIOTSs, and let $S \otimes \mathcal{T} = (C, \Sigma, Q, q^0, \longrightarrow, \mathrm{val})$. For all $(q_S, q_T, v) \in \mathrm{Post}^*(q^0)$ and $\sigma \in (\Sigma_S \setminus \mathrm{in}_S)^*$ it holds that*
$$q_S \xrightarrow{\sigma}_S q_S' \implies \exists v' . (q_S, q_T, v) \xrightarrow{\bar\sigma} (q_S', q_T, v'),$$
with $\bar\sigma \in (\Sigma \setminus \mathrm{in})^*$ obtained from σ by replacing any occurrence of a shared label $a \in \mathrm{out}_S \cap \mathrm{in}_T$ by the communication label $a^\triangleright$.

The next lemma is crucial to prove compositionality of the "necessarily" properties of type (c) in Def. 10 and the communication stopping properties in Def. 11. It shows that projections of weakly fair runs are weakly fair runs again. This result can only be achieved in the asynchronous context.

Lemma 14. *Let $S, \mathcal{T}$ be two composable AIOTSs, and $S \otimes \mathcal{T} = (C, \Sigma, Q, q^0, \longrightarrow, \mathrm{val})$. Let $q = (q_S, q_T, v) \in Q$ and $\rho \in \mathrm{wfruns}_{S \otimes T}(q)$ be a weakly fair run. Then $\rho|_S \in \mathrm{wfruns}_S(q_S)$, is a weakly fair run.*

Proposition 15 (Compositionality of Channel Properties). *Let S and $\mathcal{T}$ be two composable AIOTSs such that C_S is the set of channels of S. Let $B \subseteq C_S$ and let P be an arbitrary channel property as defined in Sec. 4.1. If S has property P with respect to the channels B, then $S \otimes \mathcal{T}$ has property P with respect to the channels B. This holds analogously for asynchronous I/O-Petri nets (due to the compositional semantics of AIOPNs; see Thm. 8).*

Proposition 15 leads to the desired modular verification result for all properties except wholly emptying (P4): In order to check that a composition $\mathcal{N} \otimes_{pn} \mathcal{M}$

of two AIOPNs has a channel property P, i.e. P holds for all channels of the composition, it is sufficient to know that $\mathcal{N}$ and $\mathcal{M}$ have property P and to prove that $\mathcal{N} \otimes_{pn} \mathcal{M}$ has property P with respect to the new channels introduced by the asynchronous composition.

Theorem 16 (Incremental Design). *Let $\mathcal{N}$ and $\mathcal{M}$ be two composable AIOPNs with shared actions $\Sigma_{\mathcal{N}} \cap \Sigma_{\mathcal{M}}$ and let P be an arbitrary channel property but (P4). If both $\mathcal{N}$ and $\mathcal{M}$ have property P and if $\mathcal{N} \otimes_{pn} \mathcal{M}$ has property P with respect to the new channels $\Sigma_{\mathcal{N}} \cap \Sigma_{\mathcal{M}}$, then $\mathcal{N} \otimes_{pn} \mathcal{M}$ has property P.*

5 Decidability of Channel Properties

We begin this section by recalling some information related to semi-linear sets and decision procedures in Petri nets that we use in our proofs.

Let $E \subseteq \mathbb{N}^k$, E is a *linear set* if there exists a finite set of vectors of $\mathbb{N}^k$ $\{v_0, \ldots, v_n\}$ such that $E = \{v_0 + \sum_{1 \leq i \leq n} \lambda_i v_i \mid \forall i \; \lambda_i \in \mathbb{N}\}$. A *semi-linear set* [10] is a finite union of linear sets; a representation of it is given by the family of finite sets of vectors defining the corresponding linear sets. Semi-linear sets are *effectively* closed w.r.t. union, intersection and complementation. This means that one can compute a representation of the union, intersection and complementation starting from a representation of the original semi-linear sets. E is an *upward closed set* if $\forall v \in E. \; v' \geq v \Rightarrow v' \in E$. An upward closed set has a finite set of minimal vectors denoted $\min(E)$. An upward closed set is a semi-linear set which has a representation that can be derived from the equation $E = \min(E) + \mathbb{N}^k$ if $\min(E)$ is computable.

Given a Petri net $\mathcal{N}$ and a marking m, the *reachability* problem consists in deciding whether m is reachable from m_0 in $\mathcal{N}$. This problem is decidable [18] but none of the associated algorithms are primitive recursive. Furthermore this procedure can be adapted to semi-linear sets when markings are identified to vectors of $\mathbb{N}^{|P|}$. Based on reachability analysis, the authors of [9] design an algorithm that decides whether a marking m is a *home state*, i.e. m is reachable from any reachable marking. A more general problem is in fact decidable: given a subset of places P' and a (sub)marking m on this subset, is it possible from any reachable marking to reach a marking that coincides on P' with m?

In [20], the *coverability* is shown to be EXPSPACE-complete. The coverability problem consists in determining, given a net and a target marking, whether one can reach a marking greater or equal than the target. In [26] given a Petri net, several procedures have been designed to compute the minimal set of markings of several interesting upward closed sets. In particular, given an upward closed set *Target*, by a backward analysis one can compute the (representation of) upward closed set from which *Target* is reachable. Using the results of [20], this algorithm performs in EXPSPACE.

While in Petri nets, strong fairness is undecidable [6], weak fairness is decidable and more generally, the existence of an infinite sequence fulfilling a formula of the following fragment of LTL is decidable [15]. The literals are (1) comparisons between places markings and values, (2) transition firings and (3) their negations. Formulas are inductively defined as literals, conjunction or disjunction of formulas and $GF\varphi$ where GF is the infinitely often operator and φ is a formula.

The next theorem establishes the decidability of the strong properties of type (b) of Def. 10. Observe that their proofs given in [12] are closely related and that they rely on the decidability of reachability and coverability problems.

Theorem 17. *The following problems are decidable for AIOPNs: Is an AIOPN $\mathcal{N}$ strongly B-consuming, strongly B-decreasing, strongly B-emptying, strongly B-wholly emptying?*

The next theorem establishes the decidability of the properties of type (a) of Def. 10. Observe that their proofs rely on (1) the effectiveness of backward analysis for upward closed marking sets (2) the decidability of reachability and home space problems and (3) appropriate transformations of the net.

Theorem 18. *The following problems are decidable for AIOPNs: Is an AIOPN $\mathcal{N}$ B-consuming, B-decreasing, B-emptying, B-wholly emptying?*

Proof.

B-consuming. Given an AIOPN $\mathcal{N}$ and B a subset of its channels, one decides whether $\mathcal{N}$ is B-consuming as follows.

Let $a \in B$ and E_a be the upward closed set of markings defined by:
$$E_a = \{m \mid \exists t \in T \text{ with } \lambda(t) = {}^{\triangleright}a \text{ and } m \geq W^-(t)\}$$
E_a is the set of markings from which one can immediately consume some message a. Let F_a be the upward closed set of markings defined by:
$$F_a = \{m \mid \exists m' \in E_a \ \exists \sigma \in T^*. \ \lambda(\sigma) \in (\Sigma \setminus \text{in})^* \wedge m \xrightarrow{\sigma} m'\}$$
F_a is the set of markings from which one can later (without the help of the environment) consume some message a. One computes F_a by backward analysis.
Let G be defined by: $G = \{m \mid \exists a \in B.\ m(a) > 0 \wedge m \notin F_a\}$
G is a semi-linear set corresponding to the markings from which some message $a \in B$ will never be consumed. Then $\mathcal{N}$ is not B-consuming iff G is reachable.

B-emptying (and B-decreasing). Given an AIOPN $\mathcal{N}$ and B a subset of its channels, one decides whether $\mathcal{N}$ is B-emptying as follows. First one builds a net $\mathcal{N}'$:
- $P' = P \uplus \{run\}$
- $T' = T \uplus \{stop\}$
- $\forall p \in P \ \forall t \in T \ W'^-(p,t) = W^-(p,t), W'^+(p,t) = W^+(p,t), m'^0(p) = m^0(p)$
- $W'^-(run, stop) = 1,, \ W'^+(run, stop) = 0, \ m'^0(run) = 1$
- $\forall p \in P \ W'^-(p, stop) = W'^+(p, stop) = 0$
- $\forall t \in T$ such that $\lambda(t) \in \text{in } W'^-(run, t) = W'^+(run, t) = 1$
- $\forall t \in T$ such that $\lambda(t) \notin \text{in } W'^-(run, t) = W'^+(run, t) = 0$

$\mathcal{N}'$ behaves as $\mathcal{N}$ as long as $stop$ is not fired. When $stop$ is fired only transitions not labelled by inputs are fireable. Thus $\mathcal{N}$ is B-emptying iff for all $a \in B$ the set of markings $Z_a = \{m \mid m(a) = 0\}$ is a home space for $\mathcal{N}'$.

B-wholly emptying. Using the same construction $\mathcal{N}$ is B-weakly wholly emptying if $Z_B = \{m \mid m(B) = 0\}$ is a home space for $\mathcal{N}'$.

□

The next theorems, whose proofs are given in [12], establish the decidability of the necessarily properties of type (c) of Def. 10 and the communication stopping properties.

Theorem 19. *The following problems are decidable for AIOPNs: Is an AIOPN $\mathcal{N}$ necessarily B-consuming, necessarily B-decreasing, necessarily B-emptying, necessarily B-wholly emptying?*

Theorem 20. *The following problems are decidable for AIOPNs: Is an AIOPN $\mathcal{N}$ B-communication stopping, strongly B-communication stopping?*

6 Conclusion and Future Work

We have introduced asynchronously composed I/O-Petri nets and we have studied various properties of their communication channels based on a transition system semantics. Useful links between the channel properties are established. We have shown that the channel properties are compositional thus supporting incremental design. Moreover we have shown that the channel properties for AIOPNs are decidable. This work can be extended in at least three directions. The first direction would introduce new operations on AIOPNs, like hiding, to design component systems in a hierarchical way by encapsulating subsystems. The second direction concerns more general communication schemes like broadcasting. Finally, we want to establish conditions for the preservation of channel properties along the "vertical axis" namely by refinement, in particular within the framework of modal Petri nets as considered in [8].

Acknowledgement. We would like to thank the reviewers of the submitted version of this paper for many useful comments which led to a major restructuring of the paper.

References

1. Alfaro, L., Henzinger, T.: Interface-based design. In: Broy, M., Grünbauer, J., Harel, D., Hoare, T. (eds.) Engineering Theories of Software Intensive Systems. NATO Science Series, vol. 195, pp. 83–104. Springer, Netherlands (2005)
2. Basu, S., Bultan, T., Ouederni, M.: Synchronizability for verification of asynchronously communicating systems. In: Kuncak, V., Rybalchenko, A. (eds.) VMCAI 2012. LNCS, vol. 7148, pp. 56–71. Springer, Heidelberg (2012)
3. Bérard, B., Bidoit, M., Finkel, A., Laroussinie, F., Petit, A., Petrucci, L., Schnoebelen, P.: Systems and Software Verification: Model-Checking Techniques and Tools. Springer Publishing Company, Incorporated (2001)
4. Best, E., Devillers, R., Koutny, M.: Petri net algebra. Springer-Verlag New York, Inc., New York (2001)
5. Brand, D., Zafiropulo, P.: On communicating finite-state machines. J. ACM 30(2), 323–342 (1983)
6. Carstensen, H.: Decidability questions for fairness in petri nets. In: Brandenburg, F.J., Wirsing, M., Vidal-Naquet, G. (eds.) STACS 1987. LNCS, vol. 247, pp. 396–407. Springer, Heidelberg (1987)

7. Cécé, G., Finkel, A.: Verification of programs with half-duplex communication. Inf. Comput. 202(2), 166–190 (2005)
8. Elhog-Benzina, D., Haddad, S., Hennicker, R.: Refinement and asynchronous composition of modal petri nets. In: Jensen, K., Donatelli, S., Kleijn, J. (eds.) ToPNoC V, LNCS, vol. 6900, pp. 96–120. Springer, Heidelberg (2012)
9. de Frutos-Escrig, D., Johnen, C.: Decidability of Home Space Property. Rapports de recherche. Université Paris-Sud. Centre d'Orsay. Laboratoire de recherche en informatique (1989)
10. Ginsburg, S., Spanier, E.H.: Semigroups, presburger formulas and languages. Pacific Journal of Mathematics 16(2), 285–296 (1966)
11. Gomes, L., Barros, J.P.: Structuring and composability issues in petri nets modeling. IEEE Transactions on Industrial Informatics 1(2), 112–123 (2005)
12. Haddad, S., Hennicker, R., Møller, M.H.: Channel properties of asynchronously composed petri nets. Tech. Rep. LSV-13-05, Laboratoire Spécification et Vérification, ENS Cachan, France (2013)
13. Hennicker, R., Janisch, S., Knapp, A.: Refinement of components in connection-safe assemblies with synchronous and asynchronous communication. In: Choppy, C., Sokolsky, O. (eds.) Monterey Workshop 2008. LNCS, vol. 6028, pp. 154–180. Springer, Heidelberg (2010)
14. Hennicker, R., Knapp, A.: Modal interface theories for communication-safe component assemblies. In: Cerone, A., Pihlajasaari, P. (eds.) ICTAC 2011. LNCS, vol. 6916, pp. 135–153. Springer, Heidelberg (2011)
15. Jančar, P.: Decidability of a temporal logic problem for petri nets. Theor. Comput. Sci. 74(1), 71–93 (1990)
16. Kindler, E.: A compositional partial order semantics for petri net components. In: Azéma, P., Balbo, G. (eds.) ICATPN 1997. LNCS, vol. 1248, pp. 235–252. Springer, Heidelberg (1997)
17. Lohmann, N., Massuthe, P., Wolf, K.: Operating guidelines for finite-state services. In: Kleijn, J., Yakovlev, A. (eds.) ICATPN 2007. LNCS, vol. 4546, pp. 321–341. Springer, Heidelberg (2007)
18. Mayr, E.W.: An algorithm for the general petri net reachability problem. In: Proceedings of the Thirteenth Annual ACM Symposium on Theory of Computing, STOC 1981, pp. 238–246. ACM, New York (1981)
19. Peterson, J.L.: Petri Net Theory and the Modeling of Systems. Prentice Hall PTR, Upper Saddle River (1981)
20. Rackoff, C.: The covering and boundedness problems for vector addition systems. Theoretical Computer Science 6, 223–231 (1978)
21. Reisig, W.: Simple composition of nets. In: Franceschinis, G., Wolf, K. (eds.) PETRI NETS 2009. LNCS, vol. 5606, pp. 23–42. Springer, Heidelberg (2009)
22. Souissi, Y.: On liveness preservation by composition of nets via a set of places. In: Rozenberg, G. (ed.) APN 1991. LNCS, vol. 524, pp. 277–295. Springer, Heidelberg (1991)
23. Souissi, Y., Memmi, G.: Composition of nets via a communication medium. In: Rozenberg, G. (ed.) APN 1990. LNCS, vol. 483, pp. 457–470. Springer, Heidelberg (1991)
24. Stahl, C., Vogler, W.: A trace-based service semantics guaranteeing deadlock freedom. Acta Informatica 49(2), 69–103 (2012)
25. Stahl, C., Wolf, K.: Deciding service composition and substitutability using extended operating guidelines. Data Knowl. Eng. 68(9), 819–833 (2009)
26. Valk, R., Jantzen, M.: The residue of vector sets with applications to decidability problems in petri nets. Acta Informatica 21(6), 643–674 (1985)

MARCIE – Model Checking
and Reachability Analysis Done Efficiently

Monika Heiner, Christian Rohr, and Martin Schwarick

Computer Science Institute, Brandenburg University of Technology Cottbus
Postbox 10 13 44, 03013 Cottbus, Germany
marcie@informatik.tu-cottbus.de
http://www-dssz.informatik.tu-cottbus.de

Abstract. MARCIE is a tool for the analysis of generalized stochastic Petri nets which can be augmented by rewards. The supported analysis methods range from qualitative and quantitative standard properties to model checking of established temporal logics. MARCIE's analysis engines for bounded Petri net models are based on Interval Decision Diagrams. They are complemented by simulative and approximative engines to allow for quantitative reasoning on unbounded models. Most of the quantitative analyses benefit from a multi-threaded implementation. This paper gives an overview on MARCIE's functionality and architecture and reports on the recently added feature of CSRL and PLTLc model checking.

Keywords: generalized stochastic Petri nets, model checking, simulation.

1 Objectives

Generalized stochastic Petri nets ($\mathcal{GSPN}$) are a widely used formalism in application fields as performance evaluation of technical systems, or synthetic and systems biology. Augmented with rewards they permit intuitive modeling and powerful analyses of inherently concurrent stochastic systems. As their semantics can be mapped to Continuous-time Markov chains (CTMC), a wide range of quantitative analysis methods up to probabilistic model checking is available.

There are several tools supporting different kinds of efficient CTMC analysis, e.g., by applying symbolic techniques or discrete event simulation. However, their use is often restricted by specific constraints. There are tools which support only the analysis of bounded models, even if discrete event simulation is used. Some tools enable the augmentation of CTMC models by rewards, but do not provide model checking of related temporal logics as the Continuous Stochastic Reward Logic (CSRL). Most tools do not support multi-threading, although this could drastically decrease the runtime of the analyses. Often tools demand for skilled users with sophisticated insights how to specify the model best and how to set the most appropriate tool parameters to configure internal data structures and algorithms. Dedicated simulation tools generally support only the simple generation of traces, although more advanced evaluation would be desirable.

J.-M. Colom and J. Desel (Eds.): PETRI NETS 2013, LNCS 7927, pp. 389–399, 2013.
© Springer-Verlag Berlin Heidelberg 2013

MARCIE overcomes these problems and integrates all features into one tool dedicated to the analysis of $\mathcal{GSPN}$ extended by rewards.

2 Functionality

In this section we give an overview of MARCIE's functionality with special focus on its latest extensions. The numbers given in round brackets refer to Fig. 3.

2.1 Net Classes

Basically, MARCIE analyses $\mathcal{GSPN}$ augmented by rewards. However, MARCIE's internal net representation (1) distinguishes the following net classes as the range of supported analysis capabilities depends on them. The core build place/transition Petri nets extended by read and inhibitor arcs. As they do not contain any time information, we call them qualitative Petri nets ($\mathcal{QPN}$). Fig. 1 shows a very simple $\mathcal{QPN}$ for a producer and a consumer connected by an N-bounded buffer.

We speak of stochastic Petri nets ($\mathcal{SPN}$) if all transitions carry further information in terms of firing rates which govern exponentially distributed waiting times. We obtain generalized stochastic Petri nets ($\mathcal{GSPN}$) if additionally immediate transitions (no waiting time) are allowed. $\mathcal{SPN}$ and $\mathcal{GSPN}$ can be enriched by rewards which can be associated with states (rate rewards) or transitions (impulse rewards). They are specified by reward definitions in a style similar to [19]. A reward definition consists of a set of reward items – state reward items and transition reward items. A reward item specifies a set of states by means of a guard and a possibly state-dependent reward function defining the actual reward value. We call an $\mathcal{SPN}$ augmented with rate rewards a stochastic reward net ($\mathcal{SRN}$), and a $\mathcal{GSPN}$ augmented with arbitrary rewards a generalized stochastic reward net ($\mathcal{GSRN}$). Rewards do not change the state space, but prepare the ground for more convenient and powerful analyses.

2.2 Engines

IDD Engine. (3) A cornerstone of MARCIE is its efficient implementation of Interval Decision Diagrams (IDD) [29]. Three different state space generation algorithms (4) were implemented upon this. The first one is the common Breadth-First Search (BFS) algorithm. All transitions fire in one iteration once

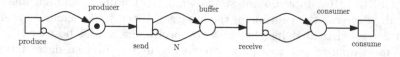

Fig. 1. A producer/consumer system with a buffer of size N

according to the transition order before adding the new states to the state space. The second algorithm is called Transition chaining. It works like BFS, but the state space is updated after the firing of each single transition. The Saturation algorithm is the last one. Transitions fire in conformance with the decision diagram. A transition is saturated if its firing does not add new states to the current state space. It should be noted that the efficiency of Chaining and Saturation depends on the transition order.

Having the state space, MARCIE permits to find dead states and to decide reversibility and liveness of transitions, which involves a symbolic decomposition of the state space into strongly connected components.

The implemented IDD engine enjoys several features to address efficiency issues, as for instance the concept of shared DDs, fast detection of isomorphic sub-diagrams by use of a unique table, and efficient operation caches; see [29] for a detailed discussion. Furthermore, the engine offers dedicated operations for forward and backward firing of Petri net transitions. It is well known that the variable order used for constructing a DD may have a strong influence on its size in terms of number of nodes, and thus on the performance of all related operations. To find an optimal variable order is an NP-hard problem. MARCIE's heuristic to pre-compute static variable orders has a simple underlying idea. It examines the structure of the given Petri net and arranges dependent places close to each other. Two places are dependent if there is a transition which affects both places [20]. MARCIE's order generator (2) offers seven options to control the generation of the place order, and six options to influence the transition order.

Symbolic CTMC Engine. (6) MARCIE provides exact quantitative analyses based on the computation of various probability distributions. Its symbolic engine is responsible for a compact representation of the real-valued state transition relation, a matrix, and some efficient numerical operations which are basically matrix-vector multiplications. The engine combines IDD-based state space encoding and "on-the-fly" generation of the state transitions which are labeled with the firing rates of stochastic transitions or the firing probabilities of immediate transitions. The computation vectors and the entries of the matrix diagonals are explicitly stored in arrays of double precision type and represent the actual limitation of applicability as their size equals the number of reachable states. On current workstations this still allows us to consider models with more than 10^9 states.

MARCIE computes the instantaneous and cumulative transient probability distribution and the steady state probability distribution for $\mathcal{GSPN}$, and the joint distribution of time and accumulated reward, a special case of Meyer's performability, for $\mathcal{SRN}$. For the latter, MARCIE makes use of Markovian approximation [5] and transforms the $\mathcal{SRN}$ description into an $\mathcal{SPN}$. For more details we refer to [26,28,25].

Approximative CTMC Engine. (7) To overcome the problem of an unmanageable state space size, MARCIE provides an approximative CTMC engine. This engine dynamically restricts the number of states. The basic idea is to combine a breadth-first variant of the state space construction with a transient analysis using uniformization [7]. During construction, all explored states having a probability below a specified threshold will be removed from the current state space. The default threshold is 10^{-11}, but can be changed by the user. Thus, only a finite subset of a possibly infinite state space will be considered. Contrary to the symbolic CTMC engine, it uses an explicit state space representation.

The approximative engine can handle $\mathcal{SPN}$ and $\mathcal{GSPN}$. The results can be exported in comma separated values (CSV) format for plotting or further analysis.

Stochastic Simulation Engine. (8) If the approximative numerical analysis exceeds the available memory, the method of choice has to be simulation. MARCIE provides two stochastic simulation algorithms – the direct method introduced by Gillespie [12], and the next reaction method introduced by Gibson & Bruck [11]. Stochastic simulation generates paths of finite length of a possibly infinite CTMC. In contrast to numerical analysis, simulation has a constant memory consumption, because only the current state is hold in memory.

Generally it is necessary to perform a sufficient number of simulation runs due to the variance of the stochastic behavior. We choose the confidence interval method as described in [22] to determine the required number of simulation runs. The user can specify the confidence interval by defining the confidence level, usually 95% or 99%, and the estimated accuracy, e.g., 10^{-3} or 10^{-4}. MARCIE calculates the required number of simulation runs to achieve this confidence interval. Alternatively, the user can set the number of simulation runs manually.

The individual simulation runs are done independently from each other. Thus, it is not challenging to parallelize stochastic simulations. MARCIE provides a multi-threaded simulation engine. Stochastic simulation results can be exported in CSV format for visualization, further analyses or documentation purposes.

This engine can not only treat $\mathcal{GSPN}$, but also $\mathcal{XSPN}$; see [14] for details.

2.3 Model Checkers

CTL (5) The Computation Tree Logic (CTL) [4] is a widely used branching time logic. It permits to specify properties over states and paths of a labeled transition system (LTS), the Kripke structure. Path quantifiers specify whether path formulas, which can be written by means of temporal operators, should be fulfilled on all paths or at least on one path starting in some state. One can interpret the reachability graph of a Petri net as a Kripke structure and thus apply CTL model checking algorithms. MARCIE supports symbolic CTL model checking for $\mathcal{QPN}$ based on its IDD engine, for details see [29].

CSL (9) The Continuous Stochastic Logic (CSL) [1] is the stochastic counterpart to CTL. The path quantifiers of CTL are replaced by the probability operator $\mathcal{P}$. The usual temporal operators are decorated with time intervals. In [2], CSL has been extended by the steady state operator $\mathcal{S}$ and by time-unbounded versions of the temporal operators. The basic CSL model checking algorithm is similar to the one for CTL, but additionally requires to compute steady state and transient probabilities. MARCIE supports CSL model checking of $\mathcal{GSPN}$ based on its exact symbolic engine. Unnested formulas can also be checked with the simulative engine. If CSL formulas are unnested and time bounded, it is also possible to use the approximative engine.

Reward Measures. In addition to CSL, special operators for the computation of expectations of instantaneous and cumulative state and transition rewards have been introduced [19]. MARCIE's symbolic CTMC and simulation engines support these measures, too. A genuine extension of CSL by rewards is presented in the next section.

2.4 New Functionalities

We now discuss in more details the latest features integrated into MARCIE.

Abstract Net Definition Language. MARCIE reads Petri net models defined in the Abstract Net Description Language (ANDL) [24]. ANDL is a lightweight and human readable description language with semantical and syntactical similarities to a guarded command language. However, contrary to, e.g., the PRISM language [16], ANDL enjoys an explicit Petri net semantics and defines additional transition types and rate function patterns.

ANDL complements model specification with bloated XML-based languages like PNML [15] (which MARCIE does support for $\mathcal{QPN}$) and serves as exchange format between MARCIE and its friend Snoopy [13], which can be used to construct $\mathcal{QPN}$, $\mathcal{SPN}$, $\mathcal{GSPN}$, and $\mathcal{XSPN}$.

The ANDL specification of the running $\mathcal{SPN}$ example in Fig. 1 and an additional reward definition are given in Fig. 2.

```
spn [procon] {                                              rewards [ r2r ] {
constants:                                                    /* states where the
  int      cap; // buffer capacity                            consumer is ready to
  double p_rate; // production rate                           receive, have
  double c_rate; // consumption rate                          a reward of 1
places:                                                       */
  producer = 1;                                               consumer > 0 : 1 ;
  consumer = 0;                                             }
  buffer   = 0;
transitions:
  receive : [consumer < 1] : [consumer + 1] & [buffer - 1]   : 1;
  send    : [buffer < cap] : [buffer + 1]   & [producer - 1] : 1;
  produce : [producer < 1] : [producer + 1]                  : p_rate;
  consume :                : [consumer - 1]                  : c_rate;
}
```

Fig. 2. The ANDL specification of a scalable $\mathcal{SPN}$ for the producer/consumer model. The buffer capacity and the rates of item production and consumption are defined by constants. The $\mathcal{SPN}$ is augmented by the separate reward definition $r2r$ which associates a reward of one to states where the consumer can receive an item.

Continuous Stochastic Reward Logic. (9) As a new feature MARCIE supports the Continuous Stochastic Reward Logic (CSRL) [5]. CSRL is a superset of CSL and augments the temporal operators with additional reward intervals. Re our example and the reward definition $r2r$, we can ask for instance for the probability to reach within t time units a state where the buffer is full. We may also expect that the consumer is ready to receive for at least half of the time period. Thus we specify the reward interval $[t/2, t]$, and obtain the CSRL formula

$$\mathcal{P}^{[r2r]}_{=?}[\mathbf{F}^{[0,t]}_{[t/2,t]}\, buffer = cap] \ .$$

The extent of MARCIE's support of CSRL model checking depends on the engines. The simulative engine supports unnested CSRL formulas for $\mathcal{GSRN}$. The symbolic engine currently supports full CSRL for $\mathcal{SRN}$. It uses Markovian approximation to transform the $\mathcal{SRN}$ into an $\mathcal{SPN}$ and maps the CSRL formula to a CSL formula with the reward bounds encoded as state properties. For more details we refer to [25].

Probabilistic Linear-Time Temporal Logic. (10) Besides the branching time temporal logics CTL and CSRL, MARCIE supports the Probabilistic Linear-time Temporal Logic with numerical constraints (PLTLc) [9]. In PLTLc, one can encode formulas on the future of paths through the state space of the model under study. So, it is quite obvious to deploy stochastic simulation to verify PLTLc formulas, because that is what stochastic simulation does – to compute paths through the model's state space. In contrast to branching time logics, PLTLc can not compute the probabilities of a given state, because it operates on paths, not on distributions. Therefore, it is impossible to nest the probability operator $\mathcal{P}$ in PLTLc. We recently extended PLTLc to check time-unbounded temporal operators, see [21].

Unlike symbolic model checking, simulative model checking computes a confidence interval of the expected probability rather than the concrete value, i.e., simulative model checking calculates probabilities up to a certain accuracy, which is the width of the confidence interval. Besides the standard return value of the $\mathcal{P}$ operator, the PLTLc model checker yields the expected probabilities of the domains of free variables (denoted with $) for which the formula holds. Back to our example, the maximum number of tokens on the place *buffer* up to time point t and their probabilities can be determined with the following formula

$$\mathcal{P}_{=?}[\mathbf{F}^{[0,t]}\, buffer > \$x] \ .$$

The PLTLc model checker works with any exact stochastic simulation algorithm, e.g., the direct method and the next reaction method, both are implemented in MARCIE. The model checking procedure is done on-the-fly, i.e., the formula is checked while the trace is generated. Furthermore, pre-computed integer traces given in CSV format can be verified.

3   Architecture

In the following we present the basic tool architecture, which is depicted in Fig. 3. We sketch the main ideas which we took into consideration during the development of MARCIE's components to achieve highly efficient analysis capabilities.

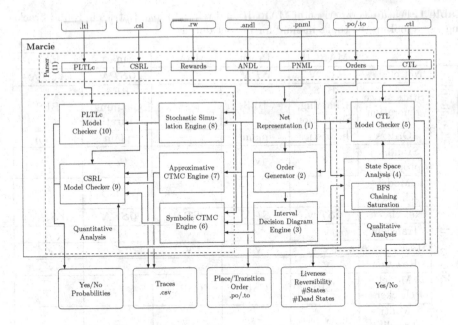

Fig. 3. MARCIE's architecture and its eleven components

MARCIE is entirely written in the programming language C++ with intensive use of template programming. It builds on the GNU multiple precision library and several parts of the boost library.

Currently, all parsers (11) for the actual Petri nets, CTL and CSRL formulas, reward definitions, as well as place and transition orders are built on the aging lexical analyzer Flex and parser generator Bison. We are about to move to the lightweight parser generator Spirit from the boost library, as it has already been done for the PLTLc parser. See [24] for detailed input syntax specifications.

4   Comparison with Other Tools

One could create a long list of tools, supporting the analysis of CTMCs and related formalisms and, thus, indirectly stochastic Petri nets as well. Due to the lack of space we confine ourselves to a very brief shortlist. Table 1 compares the main features. An elaborated comparison of CSL model checkers can be found in [17], comprising explicit, symbolic and simulative engines.

The probabilistic model checker PRISM [16] supports analysis of CTMCs, DTMCs and Markov Decision Processes by means of CSL, PCTL, and LTL, and exploits Multi-Terminal BDDs. It also permits the computation of expectations of reward measures and defines its own model description language in the style of guarded commands which can be easily used to specify bounded $\mathcal{SPN}$. An extensive performance comparison of MARCIE and PRISM was done in [27].

Table 1. Feature comparison of MARCIE and related tools. Entries in round brackets suggest a look in the tool's manual for further details.

<table>
<tr><th colspan="2"></th><th>MARCIE</th><th>Prism</th><th>MRMC</th><th>Smart</th><th>Möbius</th></tr>
<tr><td rowspan="5">Qualitative</td><td>Net classes</td><td>$\mathcal{XPN}$</td><td>$\mathcal{SPN}$</td><td>—</td><td>$\mathcal{GSPN}$</td><td>($\mathcal{XSPN}$)</td></tr>
<tr><td>State space generation</td><td>BFS, Chaining, Saturation</td><td>BFS</td><td>—</td><td>BFS, Chaining, Saturation</td><td>BFS</td></tr>
<tr><td>Orders</td><td>heuristics</td><td>plain</td><td>—</td><td>plain</td><td>plain</td></tr>
<tr><td>Standard properties</td><td>✓</td><td>(✓)</td><td>—</td><td>(✓)</td><td>—</td></tr>
<tr><td>Model checker</td><td>CTL</td><td>—</td><td>—</td><td>CTL</td><td>—</td></tr>
<tr><td rowspan="6">Numerical</td><td>Net classes</td><td>$\mathcal{GSPN}$</td><td>$\mathcal{SPN}$</td><td>($\mathcal{SPN}$)</td><td>$\mathcal{GSPN}$</td><td>($\mathcal{XSPN}$)</td></tr>
<tr><td>Transient</td><td>✓</td><td>✓</td><td>✓</td><td>✓</td><td>✓</td></tr>
<tr><td>Steady state</td><td>✓</td><td>✓</td><td>✓</td><td>✓</td><td>✓</td></tr>
<tr><td>Rewards</td><td>✓</td><td>✓</td><td>✓</td><td>—</td><td>✓</td></tr>
<tr><td>Model checker</td><td>CSRL</td><td>CSL</td><td>(CSRL)</td><td>—</td><td>—</td></tr>
<tr><td>Multi threading</td><td>(✓)</td><td>—</td><td>—</td><td>—</td><td>—</td></tr>
<tr><td rowspan="6">Simulative</td><td>Net classes</td><td>$\mathcal{XSPN}$</td><td>$\mathcal{SPN}$</td><td>($\mathcal{SPN}$)</td><td>—</td><td>$\mathcal{XSPN}$</td></tr>
<tr><td>Transient</td><td>✓</td><td>✓</td><td>✓</td><td>—</td><td>✓</td></tr>
<tr><td>Steady state</td><td>✓</td><td>—</td><td>✓</td><td>—</td><td>✓</td></tr>
<tr><td>Rewards</td><td>✓</td><td>✓</td><td>—</td><td>—</td><td>✓</td></tr>
<tr><td>Model checker</td><td>(CSRL), PLTLc</td><td>(CSL)</td><td>(CSL)</td><td>—</td><td>—</td></tr>
<tr><td>Multi threading</td><td>✓</td><td>—</td><td>—</td><td>—</td><td>✓</td></tr>
</table>

See also [26], where we compared PRISM and MARCIE's predecessor *IDD-MC* concerning transient analysis of biological models.

Another CSL model checker is the Markov Reward Model Checker (MRMC) [18]. It also offers analysis capabilities for CTMCs and related formalisms based on temporal logics. Besides MARCIE, it is the only tool supporting model checking of CSRL formulas. MRMC uses sparse representations to encode state space and matrices. Special features are bisimulation-based state space reduction and simulative steady state detection. MRMC provides simulative model checking of unnested CSL. It requires third party tools to generate the actual Markov model, which becomes prohibitive with increasing file size.

A further popular tool is SMART [3]. It offers qualitative and quantitative analysis of $\mathcal{GSPN}$ with marking-dependent arcs and defines its own model description language. SMART supports CTL, but not CSL model checking, in spite of its ability to compute transient and steady state probabilities. The user can choose between various explicit and symbolic storage strategies for the state space (e.g., AVL trees, Multi-valued Decision Diagrams (MDDs)) and for the rate matrix (e.g., Kronecker representations, Multi-Terminal MDDs, Edge-Valued MDDs, Matrix Diagrams (MxD)). However, some of these storage strategies force the user to obey some modeling restrictions. The use of MDDs, which, e.g., allow for saturation-based reachability analysis, requires to specify a suitable place partition.

A tool which offers explicit, symbolic (MDD, MxD, MTBDD, Lumping) and multi-threaded simulative analysis is Möbius [6].

None of these tools supports the numerical approximation algorithm for computing transient solutions of stochastic models as implemented in MARCIE. To the best of our knowledge, the tool Sabre [8] is besides MARCIE the only publicly available implementation. But in contrast to MARCIE, Sabre does not include any model checking capabilities.

The Monte Carlo Model Checker MC2 [9] validates PLTLc formulas, but does not include any simulation engine. MC2 works offline by reading a set of sampled trajectories, generated by any simulation or ODE solver software.

Furthermore, there exist a great variety of dedicated simulation tools, e.g., StochKit2 [23], but all lack advanced analysis methods.

5 Installation

MARCIE is available for non-commercial use. We provide statically linked, self-contained binaries for Mac OS X, and Linux. The tool, its manual and a benchmark suite can be found on our website http://www-dssz.informatik.tu-cottbus.de/marcie.html. Currently, MARCIE itself comes with a textual user interface. Tool options and input files can also be specified by a generic Graphical User Interface (GUI), written in Java, which can be easily configured by means of an XML description. The GUI is part of our Petri net analyzer Charlie [10].

References

1. Aziz, A., Sanwal, K., Singhal, V., Brayton, R.: Model checking continuous-time Markov chains. ACM Trans. on Computational Logic 1(1), 162–170 (2000)
2. Baier, C., Katoen, J.-P., Hermanns, H.: Approximate Symbolic Model Checking of Continuous-Time Markov Chains (Extended Abstract). In: Baeten, J.C.M., Mauw, S. (eds.) CONCUR 1999. LNCS, vol. 1664, pp. 146–161. Springer, Heidelberg (1999)
3. Ciardo, G., Jones, R.L., Miner, A.S., Siminiceanu, R.: Logical and stochastic modeling with SMART. Perform. Eval. 63(1), 578–608 (2006)
4. Clarke, E.M., Emerson, E.A., Sistla, A.P.: Automatic verification of finite state concurrent systems using temporal logic specifications. ACM TOPLAS 8(2), 244–263 (1986)
5. Cloth, L., Haverkort, B.R.: Five performability algorithms. A comparison. In: MAM 2006, pp. 39–54. Boson Books, Raleigh (2006)
6. Courtney, T., Gaonkar, S., Keefe, K., Rozier, E., Sanders, W.H.: Möbius 2.3: An extensible tool for dependability, security, and performance evaluation of large and complex system models. In: DSN, pp. 353–358 (2009)

7. Didier, F., Henzinger, T.A., Mateescu, M., Wolf, V.: Fast Adaptive Uniformization for the Chemical Master Equation. In: HIBI, pp. 118–127. IEEE Comp. Soc. (2009)
8. Didier, F., Henzinger, T.A., Mateescu, M., Wolf, V.: Sabre: A tool for stochastic analysis of biochemical reaction networks. In: Proc. QEST, pp. 217–218. IEEE Computer Society (2010)
9. Donaldson, R., Gilbert, D.: A Monte Carlo model checker for probabilistic LTL with numerical constraints. Tech. rep., University of Glasgow, Dep. of CS (2008)
10. Franzke, A.: Charlie 2.0 - a multi-threaded Petri net analyzer. Diploma Thesis (in German), BTU Cottbus, Dep. of CS (2009)
11. Gibson, M.A., Bruck, J.: Efficient exact stochastic simulation of chemical systems with many species and many channels. J. Phys. Chem. A 104, 1876–1889 (2000)
12. Gillespie, D.: Exact stochastic simulation of coupled chemical reactions. J. Phys. Chem. 81(25), 2340–2361 (1977)
13. Heiner, M., Herajy, M., Liu, F., Rohr, C., Schwarick, M.: Snoopy – a unifying Petri net tool. In: Haddad, S., Pomello, L. (eds.) PETRI NETS 2012. LNCS, vol. 7347, pp. 398–407. Springer, Heidelberg (2012)
14. Heiner, M., Rohr, C., Schwarick, M., Streif, S.: A comparative study of stochastic analysis techniques. In: Proc. CMSB 2010, pp. 96–106. ACM (2010)
15. Hillah, L., Kindler, E., Kordon, F., Petrucci, L., Trèves, N.: A primer on the Petri Net Markup Language and ISO/IEC 15909-2. PNNL 76, 9–28 (2009)
16. Hinton, A., Kwiatkowska, M., Norman, G., Parker, D.: PRISM: A tool for automatic verification of probabilistic systems. In: Hermanns, H., Palsberg, J. (eds.) TACAS 2006. LNCS, vol. 3920, pp. 441–444. Springer, Heidelberg (2006)
17. Jansen, D.N., Katoen, J.-P., Oldenkamp, M., Stoelinga, M., Zapreev, I.: How fast and fat is your probabilistic model checker? An experimental performance comparison. In: Yorav, K. (ed.) HVC 2007. LNCS, vol. 4899, pp. 69–85. Springer, Heidelberg (2008)
18. Katoen, J.P., Zapreev, I.S., Hahn, E.M., Hermanns, H., Jansen, D.N.: The ins and outs of the probabilistic model checker MRMC. Performance Evaluation 68(2), 90–104 (2011)
19. Kwiatkowska, M., Norman, G., Parker, D.: Stochastic model checking. In: Bernardo, M., Hillston, J. (eds.) SFM 2007. LNCS, vol. 4486, pp. 220–270. Springer, Heidelberg (2007)
20. Noack, A.: A ZBDD Package for Efficient Model Checking of Petri Nets (in German). Tech. rep., BTU Cottbus, Dep. of CS (1999)
21. Rohr, C.: Simulative Model Checking of Steady-State and Time-Unbounded Temporal Operators. In: Proc. BioPPN 2012, CEUR Workshop Proceedings, vol. 852, pp. 62–75. CEUR-WS.org (June 2012)
22. Sandmann, W., Maier, C.: On the statistical accuracy of stochastic simulation algorithms implemented in Dizzy. In: Proc. WCSB 2008, pp. 153–156 (2008)
23. Sanft, K.R., Wu, S., Roh, M., Fu, J., Lim, R.K., Petzold, L.R.: Stochkit2: software for discrete stochastic simulation of biochemical systems with events. Bioinformatics 27(17), 2457–2458 (2011)
24. Schwarick, M.: Manual: Marcie - An analysis tool for Generalized Stochastic Petri nets. BTU Cottbus, Dep. of CS (2010)
25. Schwarick, M.: Symbolic model checking of stochastic reward nets. In: Proc. CS&P 2012, CEUR Workshop Proceedings, vol. 928, pp. 343–357. CEUR-WS.org (2012)

26. Schwarick, M., Heiner, M.: CSL model checking of biochemical networks with interval decision diagrams. In: Degano, P., Gorrieri, R. (eds.) CMSB 2009. LNCS, vol. 5688, pp. 296–312. Springer, Heidelberg (2009)
27. Schwarick, M., Rohr, C., Heiner, M.: Marcie - model checking and reachability analysis done efficiently. In: Proc. QEST 2011, pp. 91–100. IEEE CS Press (2011)
28. Schwarick, M., Tovchigrechko, A.: IDD-based model validation of biochemical networks. TCS 412, 2884–2908 (2010)
29. Tovchigrechko, A.: Model Checking Using Interval Decision Diagrams. Ph.D. thesis, BTU Cottbus, Dep. of CS (2008)

CPN Tools 4: Multi-formalism and Extensibility

Michael Westergaard[1,2,*]

[1] Department of Mathematics and Computer Science,
Eindhoven University of Technology, The Netherlands
`m.westergaard@tue.nl`
[2] National Research University Higher School of Economics,
Moscow, 101000, Russia

Abstract. CPN Tools is an advanced tool for editing, simulating, and analyzing colored Petri nets. This paper discusses the fourth major release of the tool, which makes it simple to use the tool for ordinary Petri nets, including adding inhibitor and reset arcs, and PNML export. This version also supports declarative modeling using constraints, and adds an extension framework making it easy for third parties to extend CPN Tools using Java.

1 Introduction

CPN Tools [2] is a popular tool for modeling and analysis of colored Petri nets [6] (CP-nets). The large user base and several years of development has resulted in a stable and versatile tool for users and researchers working with CP-nets. CPN Tools incremental syntax check of models, making it very accessible for beginners, and provides tools for analysis both by means of simulation and state space generation, making it useful for research and industry alike. Unfortunately, some design choices made when starting CPN Tools development make it difficult to use for developers and researchers working with other formalisms similar to CP-nets. Furthermore, the modeling power of CP-nets imposes a certain mental overhead which is not desirable when performing simple modeling tasks. In this paper, we present the 4th major version of CPN Tools, which aims to make CPN Tools even more useful for regular users, and also provide developers and researchers with a solid base for simple extension development.

Colored Petri nets extend basic Petri nets (also known as place-transition Petri nets or PT-nets) [3] with distinguishable tokens and data types. This makes it possible to share net structure by relying on inscriptions. CP-nets constitute a pure extension and can hence simulate PT-net models. This often incurs a penalty in modeling complexity, however. CPN Tools 4 alleviates this by assuming sensible defaults for all inscriptions, which makes it possible to make a PT-net model in CPN Tools with the same syntax and amount of work required in a native PT-net editor.

---

* Support from the Basic Research Program of the National Research University Higher School of Economics is gratefully acknowledged.

J.-M. Colom and J. Desel (Eds.): PETRI NETS 2013, LNCS 7927, pp. 400–409, 2013.

CPN Tools 4 also allows users to use the full syntax of the Declare [17] work-flow language. This language does not explicitly focus on the control- or data-flow of models, but instead allows users to specify requirements on the order of execution of actions (transitions) such as "transition A cannot be executed before transition B" or "it is not possible to execute both transitions A and B". This is very useful for abstract specifications, especially in early phases, where the exact control-flow is of less significance. Some Declare constraints are very verbose to express explicitly, including the constraint stating that "it is not allowed to execute any transitions before A". Thus, by embedding the Declare language, CPN Tools 4 makes it possible to use the tool earlier in the development process, and later specify the declarative requirements more explicitly using classical net constructions or just retain the declarative specifications, as CPN Tools allows freely mixing of CP-nets, PT-nets, and Declare constraints.

Many low-level variants of Petri nets include extended syntax to extend the expressivity of the formalisms. CP-nets do not need these extensions as they can be expressed using common and documented patterns [1], but often these extensions constitute a shorthand which may be useful for recognition and ease of modeling. By embracing low-level nets, we think it is beneficial to support some of these extensions in CPN Tools as well. The focus has been on adding the most useful extensions in the most conservative way. We aim to make sure that a model working in a previous version of CPN Tools also works in future versions, and hence we prefer not to add extensions we are not sure will stand the test of time. For this reason, CPN Tools 4 adds support for inhibitor arcs and reset arcs in a limited form. We only allow an all-or-nothing semantics, which is contrary to the colored nature of CP-nets. The reason is that the semantics of colored inhibitor or reset arcs is not completely obvious, and we would prefer modelers to get experience with the limited versions before we extend support.

CPN Tools comprises two components, the user interface and the simulator, communicating using TCP. The simulator is developed is the functional language Standard ML, and the GUI is developed in the object oriented language BETA. While both languages are perfectly suitable for their uses in CPN Tools, they are not widely known. We have received a lot of requests to allow third party developers to contribute to CPN Tools. This includes researchers wanting to add their experimental extensions to CPN Tools instead of creating a tool from scratch, developers creating amazing extensions making the life easier for themselves and others, and our own students making projects of various complexity within a popular framework. While both components of CPN Tools are open source and the protocol between them public and documented, this is not an easy task due to the language barrier. CPN Tools 4 aims to make this easier by providing a back-end hook into the tool for Java developers.

CPN Tools has a history for allowing extensions. It uses a full programming language for inscriptions, which allows developers to develop libraries extending the annotation language of CPN Tools, including providing communication primitives using COMMS/CPN [4]. DESIGN/CPN, a predecessor of CPN Tools, even allowed such libraries to create elements elements in the GUI, leading to libraries such as a message-sequence chart library and MIMIC/CPN [10], which is a general-purpose tool for creating and animating graphical elements from CP-net models. CPN Tools has had external extensions, including the BRIT-NeY Suite [16], providing model visualizations in an external tool, and the AC-CESS/CPN library [13, 15] making it possible to interact with the simulator from external Java programs. While these tools have made it easier to *interact* with CPN Tools, they have not made it possible to *extend* CPN Tools aside from useful but simplistic annotation extensions. Simulator extensions in CPN Tools provide an architecture for directly extending CPN Tools without having to bother with the relatively unknown languages BETA and Standard ML.

An obvious way to allow extensions of CPN Tools is to provide a high-level macro language. This would be similar to providing a domain-specific inscription language, however. While suitable for some cases, it would not allow the truly creative uses of CPN Tools in the past. Instead, we have chosen to provide powerful low-level primitives in a regular and well-known language, The idea is to add hooks into the simulator allowing developers to modify the foundational behavior as necessary. Furthermore, we have added hooks making it possible to draw and control graphical elements directly in the CPN Tools GUI. Simulator extensions can also directly interfere with the syntax check and simulation of models, allowing them to extend the semantics of CP-nets. Extensions can seamlessly support new operations to the protocol between the CPN Tools GUI and simulator, making it easier to add new functionality to CPN Tools, as functionality can be prototyped in Java and subsequently be moved to Standard ML for improved performance if required. Finally, extensions can interact directly with the model, making it possible to create completely new CP-net-like formalisms inside CPN Tools without having to resort to esoteric languages.

Simulator extensions can work on many levels. They can provide extensions to the CPN inscription language written in Java, similar to what was previously possible using Standard ML libraries. Extensions can create and manipulate graphical elements in the CPN Tools GUI as was possible in DESIGN/CPN. Extensions can also filter the communication between the GUI and simulator, similarly to what the BRITNeY Suite previously offered [12], but much more tightly integrated with CPN Tools and easier to use. Extensions are also able to provide completely new functionality to the GUI, making it easier to implement certain features, including the Declare support, web-services integration, and PNML export (as specified in ISO/IEC 15909-1) of CPN Tools 4. Simulator extensions are designed to complement, not to replace, the ACCESS/CPN library. While extensions can present themselves as separate applications, the intention is that they present themselves inside the CPN Tools GUI and not as separate applications.

In the remainder of this paper, we go though the major new features of CPN Tools 4. We first summarize the existing and new architecture of CPN Tools in Sect. 2, and turn to the multi-formalism extensions of CPN Tools in Sect. 3. In Sect. 4, we present the simulator extensions framework and present several examples of its use. Finally, we conclude and provide directions for future work.

2   Architecture

CPN Tools and DESIGN/CPN before it have a bi-process architecture. This means they have a user-facing graphical user interface (GUI) and a lower level simulator component. The simulator component is responsible for the heavy algorithmic lifting whereas the GUI is responsible for allowing the user to indicate what the model should look like. While CPN Tools tries to hide this fact from the user, it can be useful to know. For example, this means that it is possible to design model on a relatively low-powered workstation and do heavy analysis on a more powerful grid, cloud or distributed architecture. This architecture is also important to understand simulator extensions in CPN Tools 4.

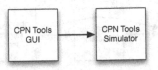

Fig. 1. Basic architecture of CPN Tools

The basic architecture of CPN Tools is shown in Fig. 1. Here, we see that the GUI is communicating directly with the simulator to provide editing and simulation of CP-net models. Here, the GUI initiates communication. Several tools have exploited this architecture to extend CPN Tools. Probably most prominent is the BRITNeY suite, which provided two modes of extensions; either it could be called as an external application to provide visualization, Fig. 2 (top), or it could mediate the communication between the GUI and the simulator (Fig. 2 (bottom)), allowing it to provide an even tighter integration such as performing inspection and on-the-fly modification of the constructed model and injection of appropriate inscription extensions.

Several extensions to CPN Tools have been proposed, both before and after the BRITNeY Suite, typically using the architecture in Fig. 2 (top) [4,8,9]. ACCESS/CPN instead replaces the user interface component in Fig. 1, and this architecture has also been used in ASAP [14]. MIMIC/CPN made it possible using DESIGN/CPN to provide an architecture similar to Fig. 2 (top) with bidirectional communication between the GUI and simulator.

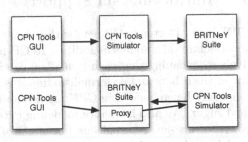

Fig. 2. CPN Tools and the BRITNeY Suite running in slave mode (top) and in filter mode (bottom)

With simulator extensions, we considered the least intrusive architecture change making it possible to reuse as much as possible of the existing code base, both ours and that developed by others. We wanted to make the extensions efficient and as far as possible transparent to end-users. We considered adding another process communicating with the user interface. The argument for this architecture is that it does not require (substantial) changes to the simulator component and it does not impose any overhead if unused. In the end, we decided to go with the architecture in Fig. 3. The main reason is that we do not wish to duplicate any implementation in ACCESS/CPN and we want to be able to load and simulate any model created in CPN Tools using ACCESS/CPN. We see that we add a new process for handling the extensions, much like the architecture in Fig. 2 (top). We allow bi-directional communication between the simulator and the GUI, as well as between the GUI and the extension manager. We also allow communication between the GUI and extension manager, but this is always mediated by the simulator. This architecture minimizes changes in the CPN Tools GUI, and the main challenge is a purely technical one, namely that the CPN simulator is inherently single-threaded. The architecture imposes minimal overhead as communication with the extension server only happens when communication with the simulator happens anyway.

CPN Tools uses an extensible protocol framework for communication between the GUI and the simulator, and we have extended this to also handle extensions and use the same framework for communication between the simulator

Fig. 3. Architecture of CPN Tools 4

and the extension manager. We extend the framework to handle call-back messages in the GUI (allowing the simulator or extensions to invoke procedures in the GUI) and to allow extensions to filter the communication between the user interface and the simulator.

3   Multi-formalism Support

CPN Tools 4 extends CP-nets with provisions for directly handling PT-nets [3] and Declare [17] models in addition to CP-net models. We do this in a conservative way, meaning that each formalism can be used completely independently of any of the others and no formalism imposes an overhead when not used. PT-nets can easily be embedded in CPN models and are handled by introducing syntactical sugar. We also introduce modeling extensions that traditionally extend the power of low-level formalisms but are just conveniences for CP-nets. Declare models deal only with the ordering of tasks, and have minimal or no handling of data flow. As such, we can see the join of a CP-net model and a Declare graph model as the synchronous product of the behavior of the CP-net model (projected onto just the transition instances) and the Declare model. Thus, Declare constraints are purely restrictions of the dynamic behavior of CP-nets, similar to the concept of time used in CPN Tools.

3.1   PT-Net Support

High-level net modeling formalisms, such as CP-nets, easily embed lower-level net formalisms, such as PT-nets. This is traditionally done by introducing a color set, or type, with just one element. In CPN Tools this type is called UNIT and it contains a single value, (). By making CPN Tools automatically recognize no explicit type as UNIT and no arc inscription as (), it is easy to emulate most common PT-net behavior as CPN models. By further allowing the shorthand n, where n is any integer, as a shorthand for $n'()$, or n tokens with the value (), we can also simulate weighted arcs and how initial markings are typically written in the setting of PT-nets. We also allow inhibitor arcs and reset arcs with the semantics known from PT-net literature [3], i.e., a transition connected to an arc with an inhibitor arc cannot be enabled if there are tokens on the place, and transitions connected to a place with a reset arc will not be inhibited from enabling and upon execution will consume all tokens from the connected place.PT-nets created using CPN Tools can be exported to PNML for analysis in other tools. In the model in Fig. 4, all places but one make use of this shorthand, and transition d has a reset arc and transition c an inhibitor arc. The save file dialog at the bottom-left exposes a save a PNML option.

3.2   Declarative Modeling

Declarative modeling has so far mostly focused on the control-flow perspective, i.e., the order of transitions. We can consider the embedding of declarative languages in CPN Tools as adding constraints to CP-nets or as adding a data perspective to declarative formalisms. From an implementation perspective, we prefer the former. Thus, simulation consists of considering whether a transition instance (or binding element) is enabled in the CP-net sense, and subsequently whether it is also allowed according to the declarative constraint. This prompts an easy means of simulation: we simply run the standard enabling check in CPN Tools (taking data into account) and subsequently (or in parallel) run a declarative check without considering data. In the example in Fig. 4, transitions a, b and d are enabled according to the Petri net semantics, but only a is enabled according to the Declare semantics (the init constraint indicates it must be the first transition to be executed).

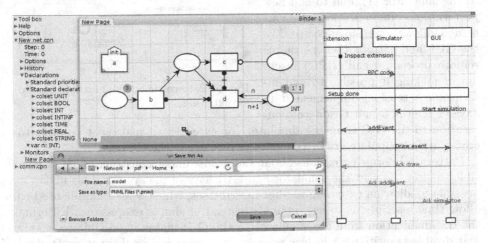

Fig. 4. CPN Tools 4 with a model mixing Declare, PT-nets, and CP-nets; to the right a model-generated message sequence chart

4 Simulator Extensions

Simulator extensions aim to make it possible to extend CPN Tools. The aim is not to allow external tools to interact with CP-net models, for that ACCESS/CPN is a much better tool, but rather aim to make it possible to make third party extensions of CPN Tools in Java with a look and feel as close as possible to the native feel of CPN Tools. Some features of CPN Tools 4 use extensions for simplified implementation; this includes support for Declare and PNML export.

We have already shown, in Fig. 3, the architecture of CPN Tools with extensions. The extensions can communicate directly with the simulator, and the simulator will take care of mediating communication between extensions and the GUI. To support extensions, we add 6 new kinds of communication, shown in Fig. 5. The first pattern (0) is the old kind of communication used in CPN Tools. The next two patterns augment pattern 0, and allow extensions to inspect and modify communication from the GUI to the simulator (1), and to add new patterns of communication (2). Pattern 3 allows extensions to act like the GUI, and pattern 4 adds a simple remote procedure call (RPC) mechanism making it possible to add inscription extensions implemented in Java. Patterns 5 and 6 add commu-

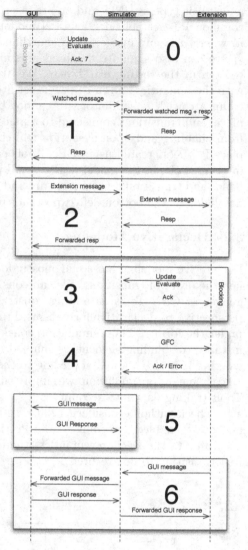

Fig. 5. Patterns of communication allowed for extensions

nication to the GUI from the simulator (or extensions), making it possible to create and manipulate graphical elements directly in the GUI.

Pattern 1 informs extensions about the model under construction. It is possible to alter the view the simulator and GUI have of the system. This makes it possible to allow an extension to alter inscriptions, which is for example used to allow time inscriptions to use intervals instead of simple expressions. Filtering the communication from the simulator to the GUI makes it possible to enable and disable transitions, which is used by the extension implementing Declare constraints to disable any transitions not allowed by the constraints.

Some functionality is easier written in Java than in BETA or Standard ML. This may be due to familiarity, a more suitable programing paradigm, or simply due to better library support. For example, PNML export is easily implemented using an XML transformation, but while common in Java, this feature is not available in BETA nor in SML. PNML export in CPN Tools 4 is implemented using a simulator extension written in Java. Similarly, Declare has already been implemented in Java, so using that implementation instead of writing one from scratch makes it much faster to make the implementation and the result is likely to be less error-prone as it has already received testing. This is done with communication pattern 2.

Pattern 3 allows extensions to completely control the simulator. This is for example useful for an extensions exposing a CP-net model as a web-service or by other means let external applications invoke a CP-net model. By allowing generic function calls to extensions using pattern 4, it is possible to expose code in extensions to models. This is used in CPN Tools to expose a Java library for drawing message sequence charts to CPN models.

The GUI callback mechanism (patterns 5 and 6) has existed for some time in CPN Tools, though it has only been used internally by, e.g., ACCESS/CPN 2 to implement cosimulation [13]. By extending this mechanism, it is now possible to directly invoke code in CPN Tools from extensions. Most importantly, it is possible to create pages and add graphical elements to them. This makes it possible to implement visualizations directly in the CPN Tools GUI. One such extension draws message sequence charts in the CPN Tools GU with the layout logic written in Java.

Often, extensions need several of the communication patterns; for example, the Declare extension adds new commands for syntax checking Declare constraints (pattern 2) and filters communication from the simulator to disable transitions (pattern 1). The message sequence chart extension asks the simulator to instantiate stub-functions for drawing message sequence charts (pattern 3). These functions call into the extension (pattern 4), which makes callbacks to

Listing 1. Parts of Declare extension

```
1   public class DeclareExtension
2       extends AbstractExtension {
3     static final int ID = 10001;
4     Option<Boolean> SMART =
5       Option.create("Smart", "smart",
6                 Boolean.class);

8     public DeclareExtension() {
9       addOption(SMART);
10      addSubscription(
11          new Command(400, 2)); }

13    public int getIdentifier() {
14      return ID; }
15    public Strig getName() {
16      return "Declare"; }

18    public Packet handle(Packet p) {
19      switch (p.getInteger)) {
20        case 400: ...
21        ...
22        case 10001: ... } }
23    ... }
```

Listing 2. Parts of MSC extension

```
31  public class MSCExtension
32      extends AbstractExtension {
33    public Object getRPCHandler() {
34      return new Dispatcher(channel); }
35    ... }

37  public class Dispatcher {
38    Channel c;
39    int serial = 0;
40    Map<Integer, Canvas> mscs = ...
41    public Dispatcher(Channel c) {
42      this.c = c; }

44    public Integer createMSC(String n) {
45      int id = serial++;
46      mscs.put(id, new Canvas(c, n));
47      return id; }

49    public void addEllipse(Integer id) {
50      Canvas cv = mscs.get(id);
51      cv.add(new Ellipse(10, 10, 60, 40)
52          .setBackground(Color.GREEN)); }
53    ... }
```

the GUI for performing the actual drawing inside the CPN Tools GUI (pattern 6). The chart at the right of Fig. 4 is created using this extension, and illustrates (simplified) the communication taking place to draw itself. In Listings 1 and 2, we see fragments of the implementations of extensions. Extensions must define a name (ll. 15–16) and a numerical identifier (ll. 3, 13–14) so the extension manager can tell all running extensions apart. Extensions can have options (ll. 4–6, 9) which are exposed in the GUI and automatically transmitted to the extension when changed. Handling packages received for communication patterns 1 and 2 is done by implementing a handle method (ll. 18–22). An extension says it wants to intercept packages using pattern 1 by subscribing to them (ll. 10–11). Handling pattern 4 is done by returning a RPC handler (ll. 33–34); all methods are automatically made available in the simulator. All communication (using pattern 6) for drawing is abstracted away by object oriented primitives (ll. 44–52).

5   Conclusion and Future Work

CPN Tools 4 improves on an already useful tool in two main areas: providing end-users with conveniences making them more efficient, and providing developers with an extension mechanism that can be used to make extensions of CPN Tools that feel close to native.

CPN Tools 4 provides syntactical sugar making it very easy to make Place/-Transition Petri net models. This is performed as a backward- and forward-compatible embedding of the formalism in CP-nets. In addition, CPN Tools adds support for common low-level special arcs, including inhibitor and reset arcs, and allows saving models using low-level constructs only in the PNML format. CPN Tools 4 also adds support for the Declare language, which allows modelers to focus less on the actual execution order but instead on constraints on the order of execution. Some of the user-facing features could not have been developed without the use of the other major new feature of CPN Tools, simulator extensions. This feature makes it easy to extend CPN Tools using Java code. Several extensions ship directly with CPN Tools; some features of CPN Tools appear as completely native features, but are realized using extensions, including Declare support, PNML export, and drawing message sequence charts from model executions. The default distribution includes a scene-graph-based library for maintaining visualizations in the CPN Tools GUI without worrying about the underlying protocol.

While we have extended CPN Tools to make it easy to make different kinds of models, we do not aim for a generic framework for Petri nets like the Petri Net Kernel [7]. We still deal primarily with CP-nets, and the embedding of PT-nets is just that, an embedding, which means we do not do any of the advanced analysis facilitated by looking at low-level nets, such as advanced state-space analysis as performed by LoLA [18], stochastic/timed analysis as performed by GreatSPN [5], or symmetry reduction as performed by CPN-AMI [11]. Adding Declare constraints is interesting as it provides Declare models with data (from the CP-nets) and provides CP-nets with a new description of control-flow at at higher level, making it easier to step-wise refinement by focusing on abstract control flow first, and add concrete control flow and data as necessary.

CPN Tools is available free of charge from cpntools.org; the GUI runs on Windows, and the simulator and extension manager on Windows, Mac OS X, and Linux.

References

1. van der Aalst, W., Stahl, C., Westergaard, M.: Strategies for Modeling Complex Processes using Colored Petri Nets. Transactions on Petri Nets and Other Models of Concurrency (to appear, 2013)
2. CPN Tools webpage, http://cpntools.org
3. Desel, J., Reisig, W.: Place/Transition Petri Nets. In: Reisig, W., Rozenberg, G. (eds.) APN 1998. LNCS, vol. 1491, pp. 122–173. Springer, Heidelberg (1998)
4. Gallasch, G., Kristensen, L.: A Communication Infrastructure for External Communication with Design/CPN. In: Proc. of Third CPN Workshop. DAIMI-PB, vol. 554, pp. 79–93 (2001)
5. GreatSPN webpage, http://gwww.di.unito.it/~greatspn/
6. Jensen, K., Kristensen, L.: Coloured Petri Nets – Modelling and Validation of Concurrent Systems. Springer (2009)
7. Kindler, E.: The ePNK: An Extensible Petri Net Tool for PNML. In: Kristensen, L.M., Petrucci, L. (eds.) PETRI NETS 2011. LNCS, vol. 6709, pp. 318–327. Springer, Heidelberg (2011), gdx.doi.org/10.1007/978-3-642-21834-7_18
8. Kristensen, L., Mechlenborg, P., Zhang, L., Mitchell, B., Gallasch, G.: Model-based Development of a Course of Action Scheduling Tool. STTT 10(1), 5–14 (2007)
9. Lindstrøm, B., Wagenhals, L.: Operational Planning using Web-Based Interfaces to a Coloured Petri Net Simulator of Influence Nets. In: Proc. of FMADS. CRPIT, vol. 12, pp. 115–124 (2002)
10. Rasmussen, J., Singh, M.: Mimic/CPN. A Graphical Simulation Utility for Design/CPN. User's Manual, http://www.daimi.au.dk/designCPN
11. The MARS Team: CPN-AMI webpage, http://www-src.lip6.fr/logiciels/mars/CPNAMI/
12. Westergaard, M.: The BRITNeY Suite: A Platform for Experiments. In: Proc. of 7th CPN Workshop. DAIMI-PB, vol. 579, pp. 97–116 (2006)
13. Westergaard, M.: Access/CPN 2.0: A High-level Interface to Coloured Petri Net Models. In: Kristensen, L.M., Petrucci, L. (eds.) PETRI NETS 2011. LNCS, vol. 6709, pp. 328–337. Springer, Heidelberg (2011)
14. Westergaard, M., Evangelista, S., Kristensen, L.M.: ASAP: An Extensible Platform for State Space Analysis. In: Franceschinis, G., Wolf, K. (eds.) PETRI NETS 2009. LNCS, vol. 5606, pp. 303–312. Springer, Heidelberg (2009)
15. Westergaard, M., Kristensen, L.: The Access/CPN Framework: A Tool for Interacting With the CPN Tools Simulator. In: Franceschinis, G., Wolf, K. (eds.) PETRI NETS 2009. LNCS, vol. 5606, pp. 313–322. Springer, Heidelberg (2009)
16. Westergaard, M., Lassen, K.: The BRITNeY Suite Animation Tool. In: Donatelli, S., Thiagarajan, P.S. (eds.) ICATPN 2006. LNCS, vol. 4024, pp. 431–440. Springer, Heidelberg (2006)
17. Westergaard, M., Maggi, F.: Declare: A Tool Suite for Declarative Workflow Modeling and Enactment. In: Proc. of BPMDemos. CEUR Workshop Proceedings, vol. 820. CEUR-WS.org (2011)
18. Wolf, K.: Generating Petri Net State Spaces. In: Kleijn, J., Yakovlev, A. (eds.) ICATPN 2007. LNCS, vol. 4546, pp. 29–42. Springer, Heidelberg (2007)

Author Index